Ethnic Boundaries in Neolithic Norway

Knut Andreas Bergsvik

BAR International Series 1554
2006

Published in 2016 by
BAR Publishing, Oxford

BAR International Series 1554

Ethnic Boundaries in Neolithic Norway

ISBN 978 1 84171 983 2

BAR Publishing is the trading name of British Archaeological Reports (Oxford) Ltd. British Archaeological Reports was first incorporated in 1974 to publish the BAR Series, International and British. In 1992 Hadrian Books Ltd became part of the BAR group. This volume was originally published by Archaeopress in conjunction with British Archaeological Reports (Oxford) Ltd / Hadrian Books Ltd, the Series principal publisher, in 2006. This present volume is published by BAR Publishing, 2016.

Printed in England

BAR
PUBLISHING

BAR titles are available from:

BAR Publishing
122 Banbury Rd, Oxford, OX2 7BP, UK
EMAIL info@barpublishing.com
PHONE +44 (0)1865 310431
FAX +44 (0)1865 316916
www.barpublishing.com

Preface

During the work with this thesis several people have been important. First of all, I would like to thank my supervisor, Lars Forsberg, for his enthusiasm for the project and for discussions and comments. I also express gratitude to Rolf Birger Pedersen for his great support. He performed the geological analyses of the raw materials, produced illustrations, gave me helpful insight in geology, and read and commented on chapter 8. Asle Bruen Olsen took the task of re-classifying the Vespestad adzes. I also thank him for his encouragement and for numerous discussions. Asle gave me access to data from the sites Neset II and Kleiva. I am also grateful to David N. Simpson for many discussions, for giving me access to the site 9 Bustadvika, and for presenting me with databases from Kotedalen.

Randi Barndon, Jostein Bergstøl, Håkon Glørstad, and Marianne Skandfer read drafts of chapter 2 and 3. Randi also read chapter 9 and 10. I am very grateful for their comments and for many illuminating conversations on ethnicity and archaeology.

Together with Einar Østmo, the entire Stone Age collection at Bergen University Museum was surveyed in order to identify TRB axes and pottery. I thank him for sharing his knowledge. Over the last few years he has also willingly discussed several aspects of the Neolithic in eastern Norway. I also thank Leif Inge Åstveit for his help with the Neolithic pottery at Bergen University Museum.

I profited greatly from discussions with Sigmund Alsaker, Hein Bjartmann Bjerck, Anders Hagen, Fredrik Hallgren, Trond Klungseth Lødøen, Gro Mandt, Arne Johan Nærøy, Morten Ramstad, and Guro Skjelstad.

I thank Kari K. Kristoffersen, who provided me with data from her excavations at Vatlestraumen and Stord-Bømlo. Similarly, Svein Indrelid helped me during the analysis of the sites at Hardangervidda, and Live Johannessen helped with the site Synnaland. The majority of the work with the data was carried out in the Stone Age collection at Bergen University Museum. During this process, Else Johansen Kleppe was extremely helpful by giving me full confidence and by organising a favourable place to work. Bergen University Museum also granted the applications for taking geological samples of artefacts. I thank Arthur Fasteland for valuable help in the topographical archive. At the Museum af Cultural History, Oslo, Einar Østmo was very helpful. I also received great help from Anne Haug at the Museum of Natural History and archaeology in Trondheim.

During the entire process of writing, the staff at the University Library in Bergen was always very accommodating and effective, for which I am very grateful. I thank Ellinor Hoff, who drew several artefacts and helped me organising the figures. Per Viggo Bergsvik also helped with the illustrations. Svein Skare photographed the raw materials. The manuscript was read and the English corrected by Melanie Wrigglesworth.

My place of work has been at the Dept. of Archaeology, University of Bergen. I thank Lollol Hordnes and Arne Mykkeltveit for their help on several matters.

I commenced writing this thesis in January 1999. The majority of the work has been financed by the Research Council of Norway (NFR). I was also granted 3 months from the Faculty of Arts at the University of Bergen. The geological investigations were financed partly by the Research Council of Norway, partly by the University of Bergen (Dept. of Earth Science, Dept. of Archaeology, and Bergen University Museum). The thesis was defended February 24th 2004. The opponents were Richard Bradley, University of Reading and Charlotte Damm, University of Tromsø. The chair of the committee was Gro Mandt, University of Bergen. I thank them for inspiring discussions at the defence and for comments on the thesis. I thank David Davidson of BAR for the pleasant cooporation and Christian Bakke, Media & Communication Centre, University of Bergen, for performing the lay-out. I also thank the Faculty of Arts, University of Bergen, and the Norwegian Archaeological Society for supporting the lay-out.

I express a warm gratitude to my mother and father, Marit and Per Bergsvik, for their help during the final phases of the work. I am happy that my children Per Viggo, Sigurd, and Frida have been there to make me think about other things. Last, but not least, I thank Gitte for being my best colleague and my best friend.

Contents

APPENDIX 1

APPENDIX 2

APPENDIX 3

APPENDIX 4

APPENDIX 5

APPENDIX 6

APPENDIX 7

APPENDIX 8

APPENDIX 9

APPENDIX 10

List of figures

List of tables

"A few days after hearing of the capture of the ship, there arrived at Nootka a great number of canoes filled with savages from no less than twenty tribes to the north and south. Among these from the North were the Ai-tiz-zarts, Shoo-mad-its, Neu-wit-ties, Savin-nars, Ah-owz-arts, Mo-watch-its, Suth-sets, Neu-chad-lits, Mitch-la-its and Quay-u-quets; the most of whom were considered tributary to the Nootka. From the South, the Aytch-arts and Esquitates also tributary with the Kla-oo-quates, and the Wickanninnish, a large and powerful tribe about two hundred miles distant. The last were better clad than the others and their canoes wrought with much greater skill...

...As soon as the dance was finished, Maquina began to give presents to the strangers in the name of his son Sat-sat-sok-sis. There were pieces of European cloth generally of a fathom in length, muskets, powder, shot &c. Whenever he gave them any thing, they had a peculiar manner of snatching it from him with a very stern and surly look, repeating each time the words, *Wocash Tyee.* This I understood to be their custom, and was considered a compliment which if omitted would be supposed as a mark of disregard for the present. On this occasion, Maquina gave away no less than one hundred muskets, the same number of looking glasses, four hundred yards of cloth, and twenty casks of powder, besides other things."

John R. Jewitt (Captive of Nuu-cha-nulth Chief Maquina 1803–1805)

Chapter 1

Introduction

During the transition to the early Neolithic, a number of changes took place among the hunter-fishers of southern Norway. New resources were exploited, and some groups took up agriculture. Several new artefacts were invented and more lithic raw materials were quarried. People became increasingly sedentary and distinctive regional traditions developed. At the same time, long distance trade was initiated between these regions. The fact that the changes happened more or less simultaneously in many regions was probably not coincidental. They were, most likely, intimately linked to changes that also took place in the social lives of these people.

One of the most important social changes may have been the development of more marked ethnic boundaries, which were related to increasing social inequality among the local groups. Such boundaries are important, because they enable control of own populations, and because they increase the social status for the ones who are able to cross them. In this thesis, the main problem is to investigate whether such ethnic boundaries can be delineated. I will try to identify them archaeologically, and I will discuss how and why they were established and maintained.

Cultural differences are important resources for the establishment of ethnic differences. Consequently, it is of vital importance to investigate whether such differences can be traced in the archaeological data. This is done by recording as many early Neolithic cultural practices as possible and by subjecting them to a multivariate analysis. I will document and quantify site locations, raw materials, tool types and lithic reduction techniques in order to decide whether the distributions on these practices co-vary across space. Another means of distinguishing boundaries is investigating whether there have been ruptures or major changes in communication. This will be done by recording the distribution of the lithic raw materials with known sources.

Although the development of more marked social boundaries probably took place throughout southern Norway, my main area of analysis only includes western Norway between Sunnmøre and Hordaland, and the mountain range between eastern and western Norway. Regions adjacent to this area will, however, be drawn into the analysis and discussions.

The most important data is from 37 excavated early Neolithic sites, which have been selected and re-analysed by means of a reference system for raw materials for the purpose of this thesis. Some of these sites have been excavated by me, but the large majority have been investigated and published by others. In addition to the site-data, I will apply stray finds (artefacts mainly delivered to the museums by non-archaeologists). Geological isotope analysis and mineralogical studies are important elements of the investigation of the lithic data. Furthermore, data from six site location analyses will be applied.

The thesis concentrates on the early Neolithic, which is approximately dated to between 5200 and 4700 BP (4000-3500 calibrated BC) in western Norway and 5100 and 4500 BP (3950-3300 BC) in eastern Norway. The late Mesolithic will, however, be referred to both as a contrast to and as a historical background for the boundaries that emerged during the Neolithic.

Chapter 2

History of research

Introduction

The problem of cultural variation has received considerable interest among archaeologists dealing with the Norwegian Neolithic. During the last 130 years, regionally distinct distributions of artefact styles, technologies, raw materials, and subsistence-settlement patterns have been interpreted as indicating differences in races, ethnic groups, cultures and social territories. In this chapter, I will describe how the cultural variation has been approached by different scholars from the 1870's until today.

With a few early exceptions, this research history concentrates on southern Norway (the area to the south of Trøndelag – see Fig. 1). Such delimitation is made, first of all because this area is covered by this thesis. Furthermore, I regard the Neolithic research in southern Norway as a relatively autonomous field (e.g. Glørstad 2002a:168 ff). This means that, although broad international chronological insights and theoretical trends have been influential, significant developments within Neolithic Norwegian archaeology have also taken place independently of these insights and trends. I will therefore try to describe the research history as a reflexive process between the international and the local. I find it necessary to present the long term developments of the field in some detail, because these constitute important conditions for my own research, and also because the unresolved contradictions and problems in the field roused my interest in the topic of ethnicity in the first place.

My main focus is on the early Neolithic. However, due to a lack of chronological resolution before the 1970's, it is often not possible, for that period, to distinguish between research on early Neolithic and middle Neolithic.

The late 1800's

For the earliest scholars working on the Stone Age, ethnicity was an important theme. The archaeological material was explicitly related to present day ethnic groups. The earliest authors that commented on the first occupants of Norway argued that they were predecessors of today's "Lapps" or "Finns" (Sámi), and that they lived in the entire Scandinavian Peninsula throughout the Stone Age. The Swedish[1] zoologist Sven Nilsson argued, for example, that the Sámi were agriculturists and that they had constructed the megaliths in southern Sweden. However, being physically and culturally inferior to the "Celtic" and "Germanic" peoples that later migrated into this area during the Bronze Age and Iron Age, the Sámi were forced to withdraw to the northern outskirts and to live by hunting, fishing and reindeer herding (e.g. Munch 1852, Keyser 1868, Nilsson 1938, see Baglo 2001 for a more detailed review of the literature before 1860).

In 1866, Oluf Rygh and Hans Hildebrand both suggested that the Stone Age of Norway and Sweden had a dualistic nature (Hildebrand 1866, Rygh 1866). Rygh's point of view was elaborated in an article from 1872 on the excavated shell midden at Steinkjer in Nord-Trøndelag, where he argued that the midden had been left by people belonging to an "Arctic Stone Age", which were predecessors of the present day Lapps (Sámi). This interpretation was supported by the slate knives, spears and projectile points that had been found at the site. Slate tools had, until then, mainly been found in northern Scandinavia, which was the main territory of the Sámi population (O. Rygh 1872). In southern Norway, however, fewer slate tools had been found. The Neolithic material mainly consisted of different types of axes, chisels, sickles and projectile points made of flint. Similar types were frequently found in southern Sweden and Denmark (southern Scandinavia). Rygh argued that these implements had been left by a "Scandinavian Stone Age" people that practised agriculture (O. Rygh 1876a). It was generally held that these people were the predecessors of the present day Swedish and Norwegian populations (e.g. Montelius 1874:152). Contrary to Nilsson's view, it was argued that they could not have been Lapps, because it is impossible first to have been agriculturalists and the builders of megaliths, and thereafter hunter and fishers (Winther 1878:108). The dualistic theory of Rygh and others was most likely inspired by social Darwinism, which had been introduced to archaeology by John Lubbock in England some years before (e.g. Trigger 1989:110-118). An implication of this view was that one did not ascribe the Sámi population as much potential for development as the Norwegians, because they were seen as culturally inferior and also biologically inferior with respect to the capability to utilise culture (see Storli 1993 and Baglo 2001 for discussions of this problem).

1 Sweden and Norway were politically united between 1814 and 1905. Sven Nilsson, Hans Hildebrand, and Oscar Montelius, who can be considered as Swedish-Norwegian, are therefore included in this part of the research history.

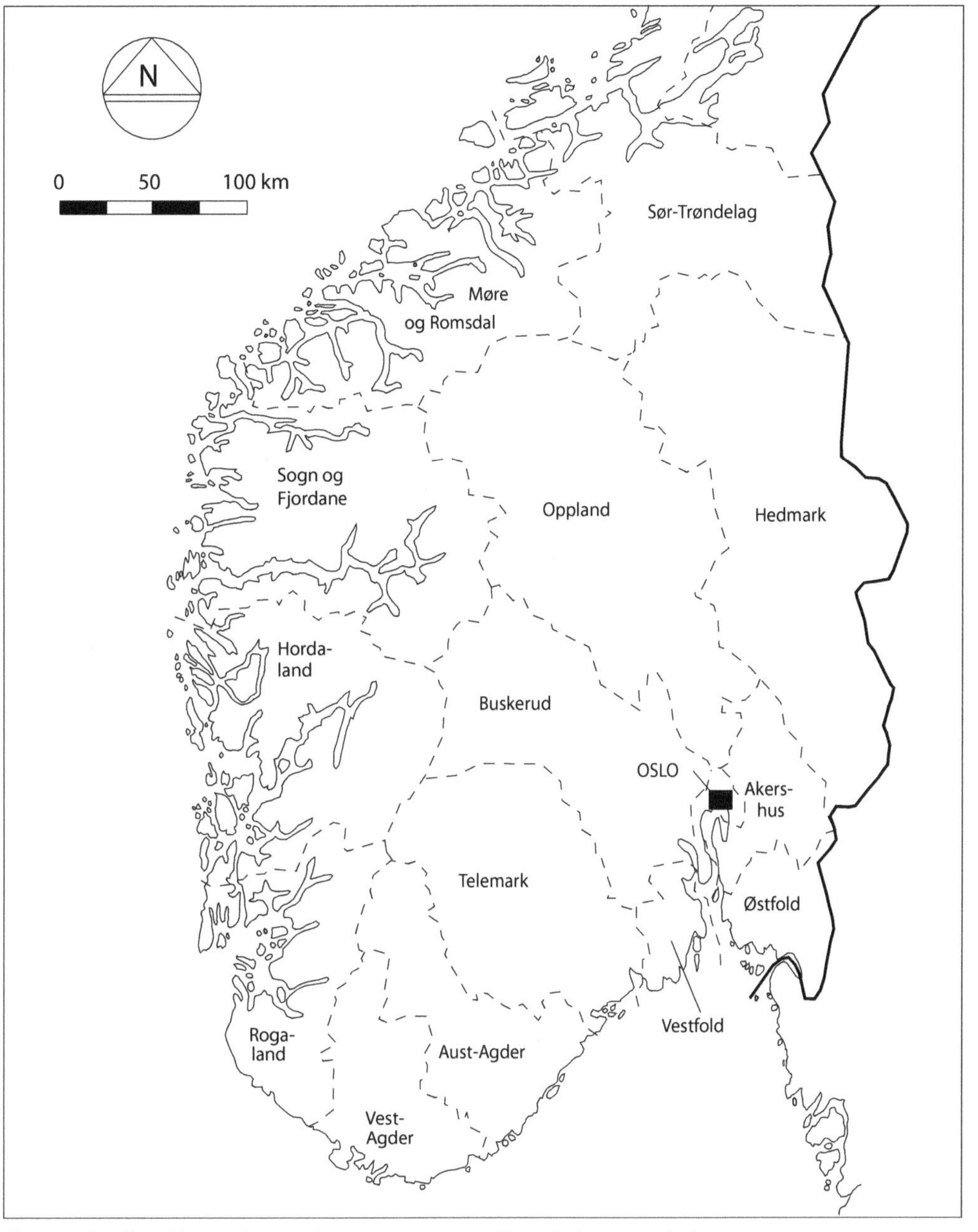

Figure 1. *Southern Norway. Present day county names and boundaries are marked.*

Important for Oluf Rygh's division into two different cultures was the fact that flint and slate implements never occurred in the same contexts (O. Rygh 1876b). However, as the museum collections grew during the late 1800's, several flint axes turned up in northern Norway, and slate knives and projectile points were found in the south. In a number of cases, these types were found together. This situation created a problem for Rygh, and in a lecture in 1895, which was referred and supported by Haakon Schetelig, he changed his previous view on the matter, by arguing that the slate forms were originally "Scandinavian" and that they should be seen as a "cultural loan" to the Lapps (Schetelig 1901:12). Two years later this seemingly ad hoc position of Oluf Rygh was countered by his brother, Karl Rygh, who concluded, on the basis of several arguments, that slate should clearly be seen as an original "Arctic" implement, and that flint tools in the north and slate tools in the south should instead be interpreted as results of travelling, trade and social interaction between these two populations (K. Rygh 1903:60-63). The "Arctic" and "Scandinavian" cultures were, according to Karl Rygh "…not sharply defined, but were partly involved in each other" (K. Rygh 1897:21 my translation).

Nevertheless, Oluf Rygh's change in the question was followed up by several influential scholars. In a book which was published 1904, Andr. M. Hansen vigorously argued against the Sámi being descendants of the "Arctic" Stone Age populations. In line with the contemporary view among ethnographers and linguists (e.g. von Düben 1873, Y. Nielsen 1891), Hansen held that the Sámi had immigrated into (southern) Norway at a much later stage (Hansen 1904). The "Arctic group" of artefacts was instead interpreted as results of an "Anarian" hunter-fisher people, which occupied all of Norway until the end of the Neolithic period, and which developed out of the Mesolithic "Nøstvet culture". This extinct people were gradually replaced by a new and superior race, the agricultural "Arians", which he saw as the ancestors of present day Norwegians. From the Bronze Age and onwards "Arians" migrated into Norway from the south. Oluf Rygh's "Scandinavian" Neolithic flint axes found in Norway were therefore, according to Hansen, traded or lost by "Arians" on forays to the north, and could not be seen as results of permanent occupations.

After the Second World War, Hansen has largely been neglected by the archaeological research community, mainly because of his explicit racist viewpoints. Still, as Anders Hesjedal points out in his doctoral thesis on the Sámi in Norwegian archaeology (Hesjedal 2000), he was a respected figure among his contemporaries, and he was more influential than most archaeologists today tend to think.

The political union between Norway and Sweden was dissolved in 1905 and the first archaeological synthesis on Norwegian prehistory, "Norges oldtid" (The prehistory of Norway), was published already the following year (Gustafson 1906). Although Gabriel Gustafson supported a general distinction between the "Arctic" and the "Scandinavian", he argued that the relation between these two cultures was unclear. As an example, he referred to Holeheia, Rogaland, a site that he had excavated some years before (Gustafson 1899). Here, artefacts of both slate and flint had been found together. Furthermore, contrary to Hansen, he argued that no physical anthropological data could support large-scale immigrations to the area from the south. He therefore held that the Stone Age inhabitants of Norway were probably the ancestors of the present day inhabitants of Norway.

The early 1900's

Until now, the research on Neolithic Norway had not had achieved full attention from any scholar, but this changed with A. W. Brøgger. In a period of four years, Brøgger published three seminal works on the Neolithic (Brøgger 1906; 1907; 1909). In the introduction of the book from 1906 he strongly argued against Hansen's viewpoints on the Neolithic flint implements. He suggested that these 5500 artefacts were not lost or traded. They should instead be seen as the results of a Neolithic sedentary farming population that lived in eastern Norway on an annual basis. With regard to the "Arctic" Stone Age, he argued that this was a result of a Mesolithic eastern migration into Norway, and that the slate forms originally had been manufactured in bone among these groups. Thus, in line with K. Rygh, he rejected Hansen's and O. Rygh's late idea of slate points being "Scandinavian". The racial question was not commented on by Brøgger, however, he stated that "...the arctic forms (are) *not* Danish, nor Swedish or from southern Norway, in other words *not Scandinavian*... (Brøgger 1906:85 my translation, italics original). Consequently, in 1906, he supported Oluf Rygh's original thesis of the "Arctic" being early Sámi and the "Scandinavian" being Norwegian predecessors.

According to Brøgger, the Neolithic in western Norway most likely belonged to the "Arctic Stone Age", mainly because of the slate points that had been found at the two excavated sites Holeheia, Rogaland and Vespestad, Hordaland (Gustafson 1899, Schetelig 1901), and also because of the particular ground adzes of the "Vestland type" that he thought were inspired by the "Arctic" slate technology. Important in this context is a letter from Haakon Schetelig, which was published by Brøgger in his book from 1906. Here the term "Vespestad culture" is introduced, and Schetelig argued that this is a late Nøstvet (late Mesolithic) local development in western Norway.[2] According to Brøgger, the "isolation" of western Norway continued into the Neolithic period. The contact between east and west was minimal because of the natural topographical obstacles, and took primarily place along the coastline (Brøgger 1907:57). The imported flint axes showed, on the other hand, that the contacts between different "Scandinavian" populations had been extensive (Brøgger 1906:110).

Up to this point, the terms cultures and ethnic groups had been vaguely defined, and were seen as coincidental with races. However, in his doctoral thesis from 1909 "Den arktiske stenalderen i Norge" (The Arctic Stone Age in Norway), Brøgger applied a more precise conceptual framework, which was taken from the Austro-German "Kulturkreis Lehre" (e.g. Kossina 1902). He followed Gustav Kossina by saying that archaeological cultures may be seen as representative of distinct ethnic groups in prehistory. In contrast to Kossina's explicit connection between race and culture, however, he stated that it is not possible to relate any of the Stone Age cultures in Norway to races (Brøgger 1909:28,170). On the other hand, for Brøgger, this was a problem of lack of good physical anthropological data, not of a principle that archaeologists should avoid racial questions altogether. For example, he presented several arguments that the "Arctic" was not the Sámi, but an unknown "x-people"

2 Contrary to the "Nøstvet culture", the "Vespestad culture" was never accepted in the research community. As I will point out in Chapter 5, the Vespestad adze has proved to be a Neolithic type, but this does not affect Brøgger's argument here.

(Brøgger 1909:171). This indicates that he nonetheless saw the different peoples as different races. Brøgger summed up the "Arctic Stone Age" as a:

> *"...separate Stone Age group that showed a people of hunters and fishers, with a distinguished ceramic style, particular tool types in slate and greenstone, a naturalistic art for sculptures and drawings of humans and animals, and finally, developed within a relatively limited area (Baltic-Arctic Sweden, Arctic-Atlantic Norway)" (Brøgger 1909:132, my translation).*

Brøgger argued that, during the early Neolithic, this culture had spread through immigration from Russia through Finland and into Sweden and Norway. Regional traditions developed within the culture. For example, slate knives were particularly common along the Norwegian coastline. Mainly on the basis of the distribution of these knives, Brøgger argued that the proper province of the "Arctic Stone Age" of Norway began in Romsdal and continued northwards (Brøgger 1909:70). Western Norway, where only slate projectile points (no knives) had been found, and also where a particular type of Vespestad adzes and Vestland adzes could be distinguished, was seen as a regional development within the "Arctic". In contrast to Hansen, he believed that the "Arctic" did not develop out of the late Mesolithic in this region, but that these two cultures met in western Norway. Later, when the "Arctic" and the "southern Scandinavians" met, the "Arctic" experienced a pressure. Having a "higher" culture than the "Arctic", the immigrating "Scandinavians" took over the footholds of the "Arctic", which withdrew or melted into the latter and disappeared (Brøgger 1909:127).

More recently, Brøgger's thesis from 1909 has been critically commented, particularly with regard to his discussion of the Sámi. Anders Hesjedal has pointed out that while Brøgger on the one hand held that it is not possible to relate today's ethnic groups to "Arctic" or "Scandinavian", he *only* put forward arguments against the Saami being descendants of the "Arctic", *not* against the Norwegian – "Scandinavian" connection. In this subtle way, he effectively removed the Saami from Norwegian prehistory without doing the same to the Norwegians. According to Hesjedal, his position with regard to the Saami was readily accepted by colleagues as well as in the general public. This should be related to the important role that the archaeologist Brøgger played for the creation of a national identity during the early years of the independent Norwegian state (Hesjedal 2000:90,109). A similar interpretation has been forwarded by Narve Fulsås (1999) on the historian Ernst Sars. The engagement in the "who-came-here-first" questions, which was a marked feature of the Norwegian Stone Age research at the time, and which endured for quite a while, may therefore be seen as partly politically motivated; it was important to clear away the "dangerous" Saami history, not only from northern Norway, but from the entire Scandinavian peninsula in order to construct a pure and safe past with which the liberated Norwegians could identify.[3]

The inter-war period

With great success, V. Gordon Childe introduced his version of the "Kulturkreis Lehre", the "culture-historical" archaeology in England in 1925 (Childe 1925). In 1929, he defined culture as follows:

> *"We find certain types of remains – pots, implements, ornaments, burial sites, house forms – constantly recurring together. Such a complex of regularly associated traits we shall term a "cultural group" or just a "culture". We assume that such a complex is the material expression of what today would be termed a "people" (Childe 1929: v-vi).*

Childe made much the same reservations with regard to Kossina's link between race and culture/ethnicity as Brøgger had done in 1909 (Helliksen 1996:53). However, in contrast to Childe's approach, Brøgger's variant had not been unreservedly adopted by the research community in Norway. Helge Gjessing and Anathon Bjørn, although applying the term culture, never defined it and used it in the encompassing way as Brøgger had done in 1909. As Helliksen (1996:49) has pointed out, these archaeologists looked at individual traits and were more preoccupied by typological and chronological questions than by delimiting cultures and discussing migrations. H. Gjessing and Bjørn were generally sceptical to the theory of an "Arctic" immigration. In their works from Sunnmøre and Rogaland they both argued that the earlier Neolithic Stone Age culture in western Norway was a continuation of the Mesolithic. This theory, which had earlier been put forward by Hansen, was partly supported by Brøgger's own idea from 1906, namely that the slate

3 See also L. N. Myhre (1994:80), Ramstad (2000a), Baglo (2001:54), and Skandfer (2002) for a critical review of Brøgger on the question of the Sámi, and Helliksen (1996) for his evolutionary viewpoints. Even if Montelius maintained it as late as 1919, the "theory of the Lapps" was more or less disposed of from now on, but it was brought up again by Gutorm Gjessing (1935) and later in a new context by Povl Simonsen (1959) for northern Norway. Today there is a relatively broad agreement (in the research community) that the Sámi ethnicity developed during the early Iron Age in the northern half of the Scandinavian peninsula.

industry was a typological development of the Mesolithic bone points (both had been produced by sawing). They also suggested, in line with O. Rygh's changed theory (slate as a cultural loan to the Arctic), that some of the Mesolithic flint implements were probably prototypes for the slate points. Furthermore, H. Gjessing held that the Neolithic adzes typologically could be traced back to the Mesolithic adzes in this region. In H. Gjessing's view, Brøgger's theory was also weakened by the new habitation sites and rock-art sites with "Arctic" tools and motives that had recently been uncovered well within the domains of the eastern Norwegian territories of the "Scandinavian" older Stone Age (H.Gjessing 1920:168 ff, Bjørn 1921b:44,45).

The theory of an "Arctic" immigration was also rejected by Haakon Shetelig[4] in his Stone Age synthesis "Primitive tider i Norge" (Primitive times in Norway) (Shetelig 1922:262 ff). Nevertheless, it appears that Shetelig was more sympathetic than H.Gjessing to Brøgger's new concept of culture as it was applied to the "Arctic Stone Age" (Shetelig 1922:288). However, instead of "Arctic", he preferred to use the term "northern Scandinavian Later Stone Age", which:

> *"...gathers the entire circle of this primitive civilisation in Norway and Sweden, as a whole a parallel to the Finnish Stone Age, but within its area rich in nuances; Gotland with its distinctive stamp, the Swedish east coast with a different stamp, in Norway the western dwelling sites with greenstone and small flint artefacts, further north the great slate finds. Future investigations will delineate these boundaries clearer. What is clear is the common origin, a common level of life form within this northern area of the European Neolithic, outside the Nordic-megalithic culture" (Shetelig 1922:293, my translation).*

His rejection of the immigration theory, however, made Shetelig more flexible than Brøgger about the "northern Scandinavian" as a "package", and allowed for much more regional variation with regard to economic basis (which included agricultural practice in eastern Norway) as well as differences in typology. Still, it is apparent that he drew a firm line between the "northern Scandinavian" and the "southern Scandinavian Stone Age" as regards the socio-political level and ability to make progress (Shetelig 1922:289). Although he allowed for "an ability to independent innovations" within the "northern" culture,[5] these abilities could not match the developments that simultaneously took place in southern Scandinavia. In cases of interaction between the two, he therefore often described the "northern" as recipients, but seldom the other way around (e.g. Shetelig 1922:312). One important result of these southern influences was the establishment of "sub-megalithic" agricultural groups in eastern Norway. Shetelig did not believe in a large scale movement of people (Shetelig 1922:246). In a later work he stated that the immigrants in this and later phases of the Neolithic "...were relatively few, and they assimilated with the older population, with the hunter-gatherers... which were the basic component of the Stone Age people in Norway" (Shetelig 1930:40 my translation).

In 1925 Brøgger launched his book "Det norske folk i oldtiden" (The Norwegian people in prehistory), in which he suggested that, from the Neolithic onwards, a combination of agriculture and hunting/fishing characterised the entire prehistory of Norway. He found that there was no contradiction between these two subsistence patterns, because this combination was the most convenient way of life in Norway (Brøgger 1925:21). The shaping of the specifically Norwegian mentality was therefore a result of an interaction between the land and its people. It could not be explained by immigrations (Brøgger 1925:16).

Brøgger's new ideas and Shetelig's thesis of a common Later Stone Age culture in northern Scandinavia throughout the Neolithic was soon to be challenged. Already in 1924, Anathon Bjørn re-introduced the dualistic view for eastern Norway.[6] He argued that the relatively large amount of thin-butted flint axes from the south were at least partly brought to this district by immigrating agricultural groups, who established themselves as an upper class together with locally distinguished groups from the aboriginal population, which he termed the "dwelling site culture" (Bjørn 1924:37 ff). In later contributions on the subject, he focused more strongly on the dualistic view by linking it explicitly to different races. According to Bjørn, the

4 Haakon Shetelig had changed the spelling of his name from Schetelig to Shetelig.

6 In his doctoral thesis, Håkon Glørstad constructs the "Brøgger school" and "Shetelig school" partly on the basis of contrasting viewpoints on the problems of cultural dualism and early Neolithic immigration (Glørstad 2002a:210). In my view, however, (the late) Brøgger and Shetelig were largely in agreement on these issues, while Bjørn represented the opposing view. Alternatively, one might therefore construct a "Bjørn school" and a "Brøgger/Shetelig school".

5 Elsewhere in the text, however, Shetelig is open to Hansen's idea that the "Arctic" was a very primitive and basically non-inventive race and that unchanged pockets of this Stone Age culture was present as late as during the Iron Age at the mountain plateaus and in caves and rock-shelters (Shetelig 1922:312:191, 202). With regard to the mountain sites, these speculations were firmly put to an end by Johs. Bøe (1942) in his book about the excavations at Sumtangen, where he established that the "surviving" stone implements had been deposited at the same time as other Neolithic artefacts in southern Norway. For the caves and rock-shelters, the dualistic theory has been debated until recently (e.g. Hagen 1967:180 ff, Bakka 1973, Bergsvik 2001a).

dualism, which lasted throughout the Neolithic, and which was a result of immigrations by both "megalithic" and "battle axe peoples":

> *"...clearly originates in a racial difference. The evidence from the period treated above should in reality reflect the difference between a superior Germanic group of peoples and a culturally inferior non-Germanic people, which must be the direct descendants of the aboriginal inhabitants of the country, and which have left the oldest known dwelling sites along the coast" (Bjørn 1931:239, my translation).*

The early post-war period

Bjørn's theory of two immigrations and the resulting "cultural dualism" in the south was supported by Gutorm Gjessing in his comprehensive work "Norges steinalder" (The Stone Age in Norway) (G. Gjessing 1945:353). G. Gjessing also assumed that the immigrating populations were of different races than the existing hunter-gatherers (G. Gjessing 1945:336), but he clearly downplayed the racial question and did not necessarily see a connection between a new race and change in ethnicity or culture (G. Gjessing 1945:419). He held that although the immigrating farmers may have had an "aristocratic" attitude, this was most likely a major drawback in their relations with the aboriginal population (G. Gjessing 1945:444). In terms of the long lasting dogma of the cultural influences only moving from south to north, Gjessing advocated a much more balanced view (on this point see particularly G. Gjessing 1942:14 ff). For example, he argued against the theory that slate was a substitute for flint. Slate was instead seen as superior to flint in a "Circumpolar" marine adaptation. As Bjørnar Olsen (1991:76 ff) and Morten Ramstad (2000a:61) have pointed out, however, it appears that he also supported the traditional view of the "Circumpolar" peoples being conservative and non-inventive compared to the southern Scandinavians. Regarding the dualism, G. Gjessing held, in line with Brøgger (1925) that the "Circumpolar" hunter/fisher culture and the farmer's culture fused and became one (G. Gjessing 1945:444 ff.). This fusion, however, appears to have happened much later, and not something that was particularly common during the Neolithic period. G. Gjessing clearly maintained that there also existed an ethnic dualism between these two opposite ways of living into the Bronze Age and Iron Age.

G. Gjessing pointed out that Neolithic Norway was characterised by a number of cultural areas, each of which were intimately connected by a complex network of cultural relations (G. Gjessing 1945:333). Concerning the specific regional differences in Neolithic southern Norway, G. Gjessing pointed, as Brøgger before him (1909), to the marked differences in material culture in the coastal regions to the north and south of Romsdal (G. Gjessing 1935: 44-56; 1945). He held that these differences were a result of ethnic differences between the "sub-Arctic" culture and a hunter-fisher (dwelling site) culture in coastal western Norway. The latter culture, he argued, "... was more strongly bound up by local traditions than in other parts of the country" (G. Gjessing 1945:321). The southern boundary of these conservative coastal hunter-fishers was not discussed specifically by G. Gjessing, but he held that Rogaland was different from the area immediately to the north in that it had been strongly influenced by northern, southern and eastern impulses (G. Gjessing 1945:330).

The stray finds of flint axes that had been retrieved from the early Neolithic of lowland eastern Norway were either interpreted as indicating small farming settlements or as results of hunting expeditions into the interior. Slate implements and some habitation sites had also been found, particularly in eastern Hedmark, but these were generally not dated as far back as the early Neolithic (Hagen 1946, Hougen 1947). This position was supported by Erik Hinsch in his work about the "funnel beaker culture" (TRB)[7] in Norway (Hinsch 1955). Hinsch produced several arguments that TRB groups had emigrated from Denmark, and that they occupied sandy soils in several districts around the Oslo fjord. In the course of time, the relations towards the southern "core area" of the TRB culture weakened, while the relations between the farmers and aboriginal coastal hunter-fishers grew stronger. This process resulted in a "de-neolithification" of the farmers and in a cultural mixture between these two groups during the middle Neolithic period (Hinsch 1955:104).

It appears from the above that three main cultural areas were delineated: The "slate culture" mainly occupied the areas on the coast to the north of Sunnmøre, and also possibly the northern parts of the inland in eastern Norway. The "dwelling site culture" inhabited the west coast to the north of Sunnmøre, and the coast of southern Norway. The "TRB culture" occupied the districts around the Oslo fjord. Rather than a dualism, there was, in other words, a situation of three main cultures in early Neolithic southern Norway

7 Hinsch introduced the term "TRB" as a replacement of the "megalithic", mainly because almost no megaliths had been found in Norway. Ironically, TRB shards are also rare in this area.

After 1955

Among the Norwegian archaeologists who applied the culture concept before the 1960's, Hinsch was closest to Childe's understanding of it (Håland 1977:3). However, he was also the last one to apply it on a broad interregional scale. During the 1960s and 1970s, specialists on the Stone Age focused more on chronology and economy, and here the concept of culture became less relevant.[8] One important reason for this change in perspective was that the number of excavated sites increased immensely during this period, particularly in the mountain zone and at the west coast. It was therefore more convenient to concentrate the research to regions or districts. Secondly, the method of radiocarbon dating gave new possibilities for the establishment of regional chronologies independently of the Danish and Swedish system, which until now had been applied as a lead for the chronological framework in Norway. A third reason was that the general international intellectual climate at the time (see next chapter for details), exposed an increased scepticism to the concept of culture, and it was partly replaced by ecological-functional and neo-evolutionary interpretations and a new terminology. One of the early critical Norwegian archaeologists was Anders Hagen, who held that the Neolithic cultures were purely archaeological constructions which may have had little or nothing to do with ethnic groups (Hagen 1970:114 ff). Nevertheless, I find that it is fair to say that, despite of these criticisms and changes of focus, the concept of culture was also applied in the succeeding years for the early Neolithic assemblages in southern Norway. The general idea of three main cultures was also maintained by several scholars. Below, I will briefly describe the relevant research in the different regions. For the area of analysis, a more detailed presentation of previous research is given in Chapter 9.

For eastern Norway, several problems of cultural affinities and change were brought up in regional Neolithic studies. Egil Mikkelsen did not exclude the possibility that agriculture was introduced partly by an immigration of farming groups, but he held that it was rather a matter of adaptation: local hunter-gatherer groups had integrated agriculture in their traditional subsistence patterns mainly because of population pressure (Mikkelsen 1982; 1984). Based on pollen data from interior valleys in Telemark collected and analysed by Helge Høeg, he suggested that groups with a mixed economy utilised both coastal and interior areas (Mikkelsen 1989). See also Bostwick Bjerck (1988) and Indrelid (1994). Mikkelsen's theory of adoption of agriculture by local hunter-gatherers was partly supported by Einar Østmo (1988:224). However, on the basis of early Neolithic data from Østfold, Østmo placed much more emphasis on the theory of an immigration of TRB groups into this area from the south, and that these had a "TRB identity" at least in the early Neolithic. The immigration hypothesis was also supported by Bergljot Solberg (1989b).

During the late 1980s and early 1990s, a number of hunter's sites from the interior lowland valleys were surveyed and excavated (Boaz 1997; 1998). Pollen data was also collected and analysed (Høeg 1996). It soon became clear that these areas had been used continuously from the late Mesolithic and into the early and middle Neolithic, and that agriculture was not introduced before the middle Neolithic. Although thin-butted axes had been retrieved at the sites, these were not related to the hunting expeditions of farming groups from the coast. Joel Boaz argued that the sites had been occupied by traditional inland hunter-gatherer populations which were significantly different from the coastal TRB groups.

For the "dwelling-site culture"[9] in coastal western Norway, Egil Bakka found that the cylindrical core technique, the production of pottery and the slate industry - which were found together at several excavated coastal sites - should be seen as the results of strong external cultural impulses from the middle Neolithic eastern "pitted ware culture" and the northern "slate culture" (Bakka 1964). Later chronological work has shown that the cylindrical core technique and the slate industry appeared as early as about 5200 BP (4000 BC) in this area, and this made it possible to independently distinguish an early Neolithic in western Norway (Indrelid 1973, Nærøy 1987; 1993a, A. B. Olsen 1992).

The above results from western Norway manifested the old hypothesis of long-term indigenous cultural development among the western groups. Today, it is generally held that there was a demographic continuity in this region during the late Mesolithic - early Neolithic transition, that no significant immigrations took place at this time, and that the populations were mainly hunter-fishers (e.g. Nygaard 1987, Bostwick Bjerck 1988, A. B. Olsen 1992).

During the last few years, several scholars have suggested that the coastal region was territorially subdivided. Asle Bruen Olsen and Sigmund Alsaker argued that the barely overlapping distribution patterns of greenstone adzes

8 It should be noted that several scholars during the 1960's and 1970's applied the culture-concept for early Neolithic site data (e.g. Odner 1965, Ingstad 1970, Skjølsvold 1977). They are not referred here, because these scholars explicitly related the material to the middle Neolithic "battle-axe culture" and "pitted-ware culture". A discussion of the affiliations to these cultures lies beyond the scope of the current work.

9 Bakka applied Meinander's (1961) term "sub-Neolithic" which designated the entire northern hunter-gatherer culture groups to the north of the TRB culture. The term has sporadically been used later for the Neolithic in southern Norway (e.g. Prescott 1996), but has generally not gained acceptance.

from Hespriholmen, Bømlo, and diabase adzes from Stakaneset, Flora, were coincidental with the location of two distinct "social territories" (A. B. Olsen & S. Alsaker 1984). This concept had initially been defined by Grahame Clark in his work "The Earlier Stone Age Settlement of Scandinavia" :

> *"... the total territory drawn upon for supplies, including raw materials, and finished products as well as food stuffs, by a given community by virtue of belonging to a larger social grouping... " (Clark 1975:14).*

and

> *"... the most extensive territory of which the ordinary individual would probably at this time be aware. The households united in social territories would have been knit together in two distinct ways, by sharing the redistribution of raw materials and products and by displaying certain idiosyncratic styles" (Clark 1975:22).*[10]

A. B. Olsen and S. Alsaker pointed out that there was a marked fall-off in the southern distribution of diabase adzes and in the northern distribution of greenstone adzes in the district of Nordhordland, which A. B. Olsen and S. Alsaker defined as a "boundary area" between the territories. For the northern territory, they argued that the quarry at Stakaneset and the contemporary rock art in Vingen and Ausevik together formed a social and religious centre for the entire social territory (A. B. Olsen & S. Alsaker 1984:100).

The boundary between the "slate culture" and the "dwelling site culture", which Brøgger and G. Gjessing located between Romsdal and Sunnmøre, has been discussed by Hans Christian Søborg on the basis of the distribution of slate knives (Søborg 1986; 1988). In his analysis, however, Søborg moved the proposed boundary to Stad, about 60 km to the south. Søborg stated that the boundary first of all was socially determined. He noted that there was in fact a quite large "boundary area" between the southern distribution of slate knives and the northern distribution of Neolithic diabase adzes, and that a third social territory and exchange network may have been situated in the coastal area to the north of Stad (Søborg 1988:237). A fourth social territory was also suggested for the southern coast of southern Norway (S. Alsaker 1987, Ballin & Jensen 1995:236, Landmark & Stylegar 1998).

Several scholars have discussed the social interaction between the different cultures. Svein Indrelid argued that a certain degree of communication took place between eastern and western populations at Hardangervidda (Indrelid 1994:302). This was supported by Asle Bruen Olsen, who found it likely that TRB pottery as well as agricultural products were distributed to western Norway through these networks (A. B. Olsen 1992:155). In line with Bakka and Søborg, A. B. Olsen held that there were significant social networks from western Norway towards the "slate culture" in the north (A. B. Olsen 1992:254). On the basis of cross-cutting distributions of different traits, Christopher Prescott and Lisa Bostwick Bjerck have also pointed out that there were "social channels" and "social networks" between east and west (Bostwick Bjerck 1988, Prescott 1996).

Concluding remarks

The founding father of Neolithic research in Norway, Oluf Rygh, has had a considerable impact on the field. Rygh was mainly preoccupied with the delineation of cultures and in cultural origins, and he established a basic dualistic framework of an aboriginal hunter-gatherer culture and a farming culture of southern origin.

Later research has provided large amounts of new data, and these theories have been modified and adjusted. It soon became clear that instead of two, there were three cultures in southern Norway. In recent years, the concept of culture – which may be seen as the backbone of the entire approach – has been under pressure from Norwegian archaeologists influenced by the "New Archaeology". First of all, it has been suggested that there were significant differences in the economic basis, and that the spatial variations in the artefact material may largely have been a result of functional, not cultural, differences. Secondly, the spatial distribution of types that previously were assumed to have been distinctive of different cultures, overlapped with each other. This has made it difficult to delineate the precise location of some of the cultural boundaries. Thirdly, it has been suggested that, instead of immigration, the changes during the transition to the Neolithic were largely the results of a mixture between old and new forms, and this has made it difficult to operate with clear-cut cultural distinctions. Finally, the large number of new data has indicated that, within the broad "slate culture", "dwelling site culture" and "TRB culture" culture, there were spatial subdivisions. On the basis of these new insights and results, one might therefore think that the culture concept has had its day.

10 The term "social territory" was developed on the basis of ethnographic studies of hunter-gatherer "tribal" social organisation (e.g. Birdsell 1953, Tindale 1974). See Price (1980) and Gendel (1984) for a further development of the concept for use in Mesolithic research, and Bergsvik (2003) for critical comments.

Nevertheless, despite of these problems, no scholar has been willing to abolish the culture concept altogether. Today, more than 130 years after Rygh's first article on the subject, one largely accepts his divisions, and the concept is still in use in Childe's sense as equivalent to prehistoric ethnic groups. The main reason is, as Arne Skjølsvold (1977:222) and Einar Østmo (1989:12) have pointed out, there are often marked patterns of discontinuity and fall-off in the distribution of material culture traits. As long as such patterns can be observed, the concept may be relevant as indicating socially constructed differences.

These socially constructed differences are, however, until now, poorly understood. It is not clear why the "slate culture" would be markedly different from the "dwelling site culture". It is also unclear why there were internal subdivisions within these larger cultures. How did these differences develop, and why were they maintained for such a long time? Theories of immigration are seldom quite satisfactory. It is also often hard to explain these discontinuities by factors such as regional differences in adaptation and function. Furthermore, the character of the relations is not developed very well. Interpretations like "cultural influences", "contact networks", and "social channels" are not very specific to what happened "on the ground".

I suggest that one solution to the problem may lie in a proper understanding of the ethnic relations among these populations. In the next chapter, I will explore this theme by distinguishing conceptually between ethnicity and culture. My main theory is that the differences emerged in contexts of ethnicity in the boundary areas and that different cultural tradition constituted important resources for such processes to take place.

Chapter 3
Ethnicity and archaeology

Introduction

In the previous chapter, I presented the research history on cultural variation in the early Neolithic in southern Norway. It was concluded that, although several scholars have uncovered problems with the concept of culture, it is still in frequent use as an equivalent to ethnicity. In this chapter, I try to develop a theory of ethnicity that may be applicable for interpreting my own data from southern Norway. I argue that, rather than being a static, homogeneous phenomenon which coincided nicely with cultural values and which persisted across time and space, it was more likely situational and local, and a function of boundary processes between different groups. Taking this position, ethnic groups and cultures cannot be documented in a straightforward fashion by mapping the distribution of randomly chosen traits, as is too often the case with many studies. In the approach attempted here, a conceptual distinction is made between ethnicity and culture. Furthermore, it is critically considered whether ethnicity was a relevant identity among Neolithic populations. This is done by investigating a variety of ethnic relations among hunter-gatherers of the recent past. On the epistemological side, the approach acknowledges the active role that material culture plays in human discourse, and it advocates an archaeological methodology that draws attention away from circumscription of ethnic groups to the boundaries between them.

Problems with cultures, ethnic groups and social territories

The definitions of culture, ethnic groups and social territories that were introduced in Chapter 2 are hampered by two fundamental problems. The first problem is ontological and concerns the basic understanding of these concepts. The second problem is epistemological and concerns the relationship between cultures and archaeological data.

Ontological problems

The term social territory was introduced by Grahame Clark because it was closer to ethnographically known demographic units and because it was meant to describe social groups that were too small to be classified as a culture. One might therefore argue that it is not affected by the critiques which have been launched towards the concept of culture. But when one examines Clark's definition of a social territory from 1975 (see Chapter 2), it largely corresponds to the way culture was defined before the 1960s. I therefore regard it as basically being of the same order and should therefore be subjected to the same critiques as that of culture.

Prior to the 1960's, archaeologists and anthropologists shared similar viewpoints regarding this field of research. It was generally held that a culture, tribe and ethnic group had more or less corresponding significance and was composed of implicit and explicit patterns of behaviour. This behaviour constituted the distinctive achievement of human groups, which consisted of patterns of material culture, beliefs, myths, ideas and values, all of which could be delineated, compared and classified (Singer 1968:530). Ethnic identity was in this context seen as a passive reflection of cultural similarity and was therefore assumed to coincide with discrete, homogenous, integrated cultures (Jones 1997:48). During the late 1950s and 1960s, however, several important critiques were raised against this simple classification.

The critiques, which were most explicitly formulated by Fredrik Barth (1969), shifted the focus from passive representations such as language and material culture, to self-definitions of particular ethnic groups in opposition to other groups. According to Barth, ethnicity is a form of social organisation and a characteristic of self-ascription and ascription by others:

> *"A categorical ascription is an ethnic ascription when it classifies a person in terms of his basic, most general identity, presumptively determined by his origin and background. To the extent that actors use ethnic identities to categorise themselves and others for the purposes of interaction they form ethnic groups in the organisational sense" (Barth 1969:13ff).*

Instead of being passive products of cultural differences, ethnic groups are instead seen as collective organisational strategies, formed as a result of competition over socio-economic resources. An important element of his model was that the boundaries usually do not stop people from interacting with one another. On the contrary, people frequently cross the boundaries, which in Barth's approach are metaphorical rather than physical. This leads, according to Barth, to a complex situation with regard to cultural content:

> *"...although ethnic categories take cultural differences into account, we can assume no simple one-to-one relationship between ethnic units and cultural similarities and differences. The features that are taken into account are not the sum of "objective" differences, but only those which the actors themselves regard as significant" (Barth 1969:14).*

In accordance with this new understanding of ethnicity, several specialists on hunter-gatherers questioned the existence of tribes as meaningful units as equivalent to ethnic groups (this critique was not directed towards "tribe" in the sense of an evolutionary stage in social organisation e.g. Service 1971). First of all, it was argued that tribes are ethnographical constructions, mainly based on the distributions of linguistic differences. They held that even linguistic boundaries can seldom be delineated on a map. There is usually a large degree of overlap around the border areas where multilingualism and dialect fuzziness dominate. Another argument is that there is a constant flow of cultural traits such as artefacts, art styles, and techniques, across the tribal borderlines. Except for very broad trends, few correlations have been documented between cultural traits and linguistic boundaries. Furthermore, tribes are almost never politically and territorially unified. Even their names are imposed on them by traders or early colonists. Finally, few ethnographers are able to document in-group notions referring to a social unit at this level. It was therefore concluded that tribes are not homogenous demographic units of which the local populations are particularly conscious or with which they systematically identify (e.g. Berndt 1959, Fried 1968, McKennan 1969, Riches 1982, Suttles 1987).

These critiques of the concepts of culture, tribe, social territory and ethnic group affect many of the sited studies of the early Neolithic southern Norway (Chapter 2), because in most of these studies the notions of cultures and social territories are accepted as basic, monolithic units of social organisation among hunter-gatherers and primitive farming communities in general. The recent discussions and critiques within the social sciences demonstrate clearly that such an understanding of ethnicity is no longer acceptable. Therefore, if the concept of ethnicity is to be applied to interpretations of Neolithic data, it needs an ontological re-orientation.

Barth's critiques and insights have certainly influenced later anthropological studies of ethnicity. But the critiques have deep consequences also for archaeological epistemology, which brings us to the second problem: How do we *know* that we are dealing with Neolithic ethnic groups and boundaries in the archaeological data?

Epistemological problems: archaeology and ethno-archaeology

The "normative" perspective of the culture-historical approach was already the subject of massive attacks during the 1960s from archaeologists such as Lewis (Binford 1965) and David Clarke (1968). Clarke questioned the very existence of archaeological cultures as "monothetic" units and pointed out the arbitrary selection and crude classification of artefacts as representative of whole cultural entities (Clarke 1968:35 ff). Instead he presented the "polythetic" model, which embraced a much larger variety of artefacts and attributes. These often presented themselves as untidy nesting hierarchies of artefact groupings. Nevertheless, Clarke, like Childe before him, still saw these groupings as representative of cultural traditions of human groups (Clarke 1968:13). Binford, on the other hand, pointed out that there is no predictive relationship between archaeological cultures and ethnic groups. Binford's key argument was that variations in archaeological assemblages are more likely to be explained as individuals or groups of people performing different tasks at different locations, rather than as ethnic diversity (Binford 1965:205). Binford and his followers did not altogether abandon the idea that such distributions could correlate with ethnic groups, but this could only be the case within the *stylistic* domain (Conkey 1990:10). James Sackett (1985) however, later questioned this dichotomization between function and style. While adopting a similar basic premise concerning style and ethnicity as other processualists, he introduced the term "isocrestic variation" ("isocrestic" meaning "equivalent in use"). He argued that there are several equivalent ways of doing things, but people choose those which are in accordance with their own cultural tradition. These "isocrestic choices" result in isocrestic variation, of which style is only a subset; "a butchering technique may potentially convey as much ethnically stylistic information as a pottery decoration" (Sackett 1986:630). He therefore held that functional and nonfunctional aspects of human behaviour should not be analytically separated because they may have the same relevance on interpretation of social boundaries. A similar argument was forwarded by Heather Lechtman (1977), who in contrast to Sackett's more passive isocrestic variation, considered Andean textile artisans as instrumental in expressing ethnicity. Martin Wobst (1977) also stressed the active role of the material culture in the context of ethnic signalling. While supporting the distinction between style and function, he predicted that some highly visible artefacts that entered social contexts probably carried stylistic messages about social status. In contrast, small artefacts were not actively used as media for this type of information exchange.

Inspired by Wobst, Polly Wiessner (1983) undertook an ethno-archaeological study of projectile point styles among different Kalahari San language groups. A correlation between these groups and different styles lead her to conclude that these different projectile points actively communicated ethnicity ("emblemic style") and that the individual ("assertory") styles were subordinate to this in order to ease mobility between the ethnic groups. These interpretations were questioned by Sackett (1985), who held that Wiessner's emblemic styles probably did not actively communicate ethnicity, and that they were rather examples of his "isocrestic variation".

Barth's new perspective on ethnicity as a process has resulted in two main directions in archaeology and ethno-archaeology (Jones 1997:28).

One direction focused strongly on the relationship between material culture and ethnic symbols. In an influential ethno-archaeological study from Baringo, Kenya, Ian Hodder (1982) further complicated archaeological study of material culture. Here, he discovered that some items were reserved for members of particular ethnic groups, while other items crossed the ethnic boundaries. Based on this data, Hodder held that although material culture is often loaded with symbolic connotations, these are not necessarily related to *ethnic* identity. A similar case had already been made some years before by Randi Håland (1977:12 ff) on the basis of data from the Sudanese Nubia. Another point connected to this was made by Richard Gould (1980). Gould's ethno-archaeological studies from Australia demonstrated that his archaeological styles made little sense to the aborigines. For them, the origin, colour and sizes of the tools were often more important for their symbolic properties than the artefact classifications that had been performed by Gould. These arguments and cautions were further elaborated by Pierre Lemonnier (1986) on data from the Anga groups in New Guinea. Lemonnier held, in line with Sackett, that styles of technology and production processes are just as informative about ethnicity as the end products themselves. He acknowledged that some of the technological choices and items in his case were related to the distribution of ethnic groupings. However, similarly to Hodder, he said that their particular significance is dependent on the context, and that it is not possible to predict beforehand which of these choices or products has ethnic significance.

The other direction was more preoccupied with ethnicity as a structuring principle for economy and politics. Early examples of this direction can be taken from Northern Norway. Else Johansen Kleppe pointed out that the populations that inhabited house sites in the Varanger fjord, Finnmark between 1 and 1700 AD probably had a Saami ethnic identity (Kleppe 1977). Later, Knut Odner and Bjørnar Olsen focused on ethnicity as a process; they held that the Saami ethnicity in northern Fenno-Scandinavia emerged as a result of a process of economic specialisation and differentiation (Odner 1983; 1985, B. Olsen 1985b).

These approaches, which expose a far from simple relationship between material culture and ethnicity, have considerable consequences in terms of how research on prehistoric ethnicity can be carried out. As Hodder has effectively shown, archaeologists cannot *a priori* assume what kinds of objects or styles were used to signal ethnic identity by a prehistoric group. If ornaments, tools and raw materials were not related to such identities, they might easily have been distributed across ethnic boundaries. According to Sackett's discussion of style and function, neither the degree of energy input nor the lack of utilitarian value are sufficient arguments for *these* artefacts to be associated with ethnic identity. In fact, *any* object or technological choice is likely to symbolise ethnicity. In my view, recent distribution studies of selected types and raw materials in Neolithic western Norway (e.g. A. B. Olsen & S. Alsaker 1984, Søborg 1988) have not seriously considered these epistemological problems; no convincing arguments have been presented that these greenstone/diabase adzes and slate knives should be interpreted as reflecting the distributions of prehistoric social territories and ethic groups.

As a result of these problems, many regard ethnicity as an impossible field of research for prehistoric archaeologists. Social anthropologists remain largely negative (e.g. Lemonnier 1986:181), pointing to the interpretative problems that face them in *living* societies. A number of archaeologists also express scepticism. Relating to the ethno-archaeological discussions in particular, many take a restrictive stance (e.g. G. A. Clark 1994, Bågenholm 1996, Anfinset 1998, Stark 1998). The most explicit example of such a restrictive attitude is Bruce Trigger, who states that although ethnicity *was* of importance for prehistoric peoples,

> *... it was a subjective concept that archaeologists cannot hope to study to any significant degree in the absence of relevant historical or ethnographic data. Fortunately, there are many more appropriate problems that archaeologists who lack access to other sorts of data are equipped to investigate (Trigger 1995:277).*

Possible solutions

So where does this leave the study of Neolithic ethnicity? Is it too difficult for prehistoric archaeologists? Should we follow Trigger's recommendation and find easier and more adequate problems to discuss? In my view, Trigger's attitude to the problem is too restrictive. I agree in principle that some prehistoric social relations might be difficult to grasp, simply because they cannot be related to the archaeological data in any convincing manner. But I am not sure if ethnicity should be included among these at the outset. In my view, ethnicity is a relevant field, including periods and regions that are not covered by historical or ethnographic sources. But if the problem is discussed, the approach must be altered considerably in comparison to the studies from early Neolithic southern Norway that I have referred to above. Below, I will distinguish conceptually between culture and ethnicity on the basis of recent research in the social sciences. Furthermore, I will critically discuss the relevancy of ethnicity in a Neolithic (hunter-gatherer) context. I will also attempt to develop an archaeological methodology, which focuses on the boundaries instead of on the circumscription of prehistoric ethnic groups.

The concepts of ethnicity and culture

I find it necessary to accept the consequences of Fredrik Barth's main thesis (1969), and look upon ethnicity as a boundary phenomenon, primarily concerning how people think about themselves as groups and how they set themselves apart from other groups. Nonetheless, Barth's seminal work does not cover all aspects of ethnicity, and during the last few years, sociologists, anthropologists and archaeologists have extended his approach.

One central aspect of Barth's approach is the active role of the individual as entrepreneur (cf. Barth 1966). In the context of interaction between ethnic groups who occupy different ecological or social niches, entrepreneurs may apply their ethnicity or even change ethnicity as a strategy in order to reach their social or economic goals. Ethnicity is therefore viewed as an "organisational vessel which may be given varying amounts and forms of content in different socio-cultural systems" (Barth 1969:14). Barth's approach proved to be very influential for the "instrumental" approaches, where ethnicity is regarded as "constituting the shared beliefs and practices that provide the groups with the boundary maintenance and organisational dimensions necessary to maintain, and compete for, socio-economic resources" (Jones 1997:74) Others have also pointed out the fluid and situational aspects of ethnicity. A person's ethnic identification, they argue, can vary in different situations depending on the context and scale of identification, resulting in a series of nesting dichotomizations of inclusiveness and exclusiveness (R. Cohen 1978:387).

During the last few years, many have become increasingly aware that common interests and self-maximising economic and political aspects have been over-emphasised by the instrumentalists. Psychological factors and senses of self, myths of origin, common institutions, and cultural background, although acknowledged by the instrumentalists, are taken for granted and play secondary roles compared to economic and political relations in the formation and transformation of ethnicity. The ultimate implication of this perspective may therefore be that ethnicity is constantly changing and fluid; that it only comes into existence in order to serve the purpose of interest groups (B. Olsen & Kobylínski 1991: 20 ff, Jenkins 1997, Jones 1997:77 ff). Such an understanding of ethnicity may not seem very workable for prehistoric archaeology, and may partly explain some of the scepticism of the concept. B. Olsen & Kobylínski (1991:22) therefore call for an adequate theory of ethnicity applicable to archaeology; one which not only considers the function of ethnicity but also cultural factors of descent. Recently, the anthropologist Thomas Hylland Eriksen (1992), the sociologist Richard Jenkins (1997) and the archaeologist Sîan Jones (1997) have published such approaches, drawing mainly on Pierre Bourdieu's practice-theory (Bourdieu 1977) and Anthony Giddens's theory of structuration (Giddens 1984).

Eriksen, Jones, and Jenkins all maintain, in line with the instrumentalists, that ethnicity is socially constructed as a result of boundary processes. They therefore reject that it is "primordial" in the sense of a fundamental, essentially unchanging and an unchangeable aspect of human existence. On the other hand, as Jenkins points out, ethnicity may under a number of circumstances take on a "primary identity", along with gender, kinship and selfhood (Jenkins 1996:65). It may therefore matter a great deal for the local populations and it may be more resistant to change than many of the instrumentalists are willing to admit. It is also made clear; however, that ethnicity is first of all a local phenomenon: a product of local circumstances and situations. It is not likely to have the same force everywhere.

Drawing mainly on the thinking of Handelman (1977), A. P. Cohen (1985), and Cornell (1996), Jenkins argues that, in addition to the transactional, one must also pay attention to the symbolic construction of similarity that goes on *within* the boundary: to the conjunction between the institutional and the cultural.

The "cultural" as a concept needs some clarification here. Jenkins defines "culture" as a domain of symbols and meanings which is participated in unevenly across a population. Instead of a singular form, it is a model of different *cultures* of social differentiation based on language, religion, cosmology, symbolism, morality, and ideology (Jenkins 1997:14). This definition, which is applied here, differs significantly not only from that of "culture" as an all-embracing characteristic of humanity, but also from culture in the traditional sense of a "people" as discussed in the above paragraph. Culture in Jenkins's sense is obviously an important part of the package of socialisation during the early stages of a child's life. It constitutes a basic and embodied "sense of self" which often goes unquestioned, but which is essential for understanding the world and for being understood by the world. Depending on the circumstances, and the salience of ethnicity, early socialisation is likely to include an ethnic component. Such a component will refer explicitly to common cultural aspects that characterise the ethnic group in contrast to other groups (the vast amount of cultural aspects that are *shared* with the opposing group in question will not be relevant for this purpose, but may be important in defining a third group). These common cultural aspects will be applied as symbols for group similarity in the context of early socialisation:

> *"Cognitively, if nothing else, the child will develop a point of view on a social world which is axiomatically organised in terms of ethnic classifications. She will learn not only that she is an 'X', but also what this means: in terms of her esteem and worth in her own eyes and in the eyes of others; in terms of appropriate and inappropriate behaviour and in terms of what it means not to be an 'X', what it means to be an 'Y' or 'Z' perhaps (Epstein cited in Jenkins 1997:59).*

The social construction of cultural difference is, therefore, dependent upon the social construction of cultural similarity. In fact, these two aspects are likely to strengthen one another; a cultural awareness or categorisation of the Other – relating to what others believe, how others do things and how others behave – affect the internal ethnic identifications, by creating stronger consciousness about own language, non-verbals, dress, food or structure of space. At the boundary, culture is made explicit and transformed from something that is *known* to something that is known *about* (Jenkins 1997:76). This means that culture is dual: on the one hand, it provides a frame within which action can be meaningful. On the other, it is dependent upon intentional action in order to be reproduced (Eriksen 1992:47). This interplay between culture and intentional agency is a necessary condition for change to take place. Based on these considerations, ethnicity can be defined as:

> *"...the systematic and enduring social reproduction of basic classificatory differences between categories of people who perceive each other as being culturally discrete" (Eriksen 1992:3).*

This definition is adopted here because it embraces the changeable and situational nature of ethnicity. At the same time it opens up for an analysis of the cultural content, without specifying the *nature* of this content at the outset. Ethnicity is, in other words, "empty" of substantial content. It is therefore applicable to different kinds of ethnicities. This brings us to the next point, which considers whether ethnicity is an adequate concept for the study of small scale prehistoric societies.

Is ethnicity a modern phenomenon?

Several writers hold that ethnicity is a relatively modern phenomenon and a product of colonialism and capitalist world systems (e.g. Fried 1968, Wolf 1982, Shennan 1989, Comaroff & Comaroff 1992). The authors of a recent collection of articles on the subject Miriam Stark (1998), for example, argue that social groups in the past may have been different from and more flexible than ethnic groups today. Rather than being a member of a bounded group, people may also have had networks of overlapping identities (Hegmon 1998:273). In line with this, MacEachern (1998:130) says that it would be a mistake to focus only on one level - the "ethnic"- at the expense of others. More neutral concepts are therefore presented, such as "social fields" (Welsch & Terrell 1998) or "social boundaries" (Goodby 1998).

There is no doubt that colonialism, trade and other western influences have contributed to a disintegration of pre-existing forms of ethnic identity. These factors have resulted in the creation of a number of new, strongly self-conscious groups who use their ethnicity as an argument in economic and political issues. But does this imply that ethnicity as a collective social identity was *not* at work before the colonists and traders arrived and that ethnicity was not embedded in the political relations among prehistoric populations? Today most anthropologists realise that hunter-gatherers cannot be seen as a "cultural type", and that basic social and psychological processes within such societies are not different from those found in humanity in general (e.g. Kuper 1988). Furthermore, modern as well as prehistoric hunter-gatherers lived under a variety of environmental, demographic, socio-economic and cultural conditions, creating diversity in the past as well in the present (Kelly 1995:340). This diversity is likely to have constituted an important background for ethnicity. I therefore agree with Jones when she maintains that:

> *"...it is therefore no reason why ethnicity should be restricted to the context of European colonialism, or to any other macro socio-historical developments, if it is seen as the kind of group consciousness that is based on the dialectical opposition of different cultural traditions in the process of social interaction" (Jones 1997:102).*

Acknowledging that ethnicity is an effect of both isolation *and* interaction, I therefore find it a key concept for the analysis of prehistoric social life. Depending on the context, it can function as a "primary identity" (Jenkins 1997:65) and may thus have been crucial in determining the organisation of many other types of inter-group phenomena. Instead of escaping from the problem, by applying more general or neutral terms, I therefore consider ethnicity to be an adequate and important field of research for prehistoric archaeology. Still, one can not deny that some of the most influential models for how ethnicity works are taken from complex modern multi-ethnic societies such as the City of London or the island of Mauritius (A. Cohen 1974, Eriksen 1992). The conditions constraining and enabling ethnic groups in such places (urban centres, extensive travelling, and visual media) are obviously very different from those affecting the Neolithic hunter-gatherers, even if basic social and psychological behaviours may have been the same. In order to modify and supplement these theories, but also to avoid a determinist view, it is therefore necessary to investigate the nature of socially constructed boundaries among hunter-gatherer groups in the recent past.

I only refer to hunter-gatherer studies, partly in order to delimit the wide range of ethnographic studies that may be relevant for this problem, partly because the Neolithic populations that are investigated in this thesis lived mainly by hunting and fishing.

Ethnic relations among recent hunter-gatherers

One problem regarding many ethnographic accounts on ethnicity is that the fieldwork, the majority of which was conducted during the first half of the 20th century, was done within the "etic" or "scientific" paradigm, i.e. one was less occupied with agent's notions than with one's own notions and classifications (Riches 1982:3). They are therefore subject to the critiques of Barth and others referred to above. On the other hand, more recent and "emic" studies (those more concerned with the agent's notions) are usually society-centric and focused on locally distinct behaviour rather than on boundary processes and inter-regional relations (Wobst 1978:303). Considering that most, if not all, of the hunter-gatherers today are deeply embedded in the modern world, emic approaches on inter-ethnic relations cannot entirely replace the etic data today. On the other hand, it is argued that it is important to bridge the gap between etic and emic approaches anyway (Riches 1982:3). In the context of ethnicity, this means that data concerning how groups of people identify themselves and categorise others should be regarded alongside data on "scientifically" documented boundaries and interaction across boundaries. During the past few years, such issues have been addressed for hunter-gatherers of the north-west coast of America, in northern Canada/Alaska, and Australia. Below, I will evaluate how in-group identities and out-group categories are constructed among these peoples.

Regarding the first, it has recently been argued that topological and territorial overtones of the term "ethnic boundary" should be downplayed; that "boundary" instead should be seen as a metaphor for a process that takes place between members of different identities, and as something that can occur anywhere or in any context (Jenkins 1996:98). The critique of the "tribe" as a general, inclusive and territorial unit among hunter-gatherers can probably be seen as a consequence of this line of thinking. However, the abolishment of the "tribe" as a general, inclusive unit, does not mean that in-group identity related to spatial boundaries is absent among hunter-gatherers. On the contrary, there seems to be an agreement that such notions are important, but they are primarily found at lower levels, which also constitute the largest and most inclusive units in terms of kinship-ancestor relations, territorial rights, and political organisation. For North America it is argued that collective in-group notions are strongest at the level of the "locational band" or "regional band" (Helm 1968:118, McKennan 1969:105, Riches 1982:110). These political and territorial units are more or less equivalent to the "local groups" of the north-west coast of America, which serve as comparable reference points for in-group identities in this area (e.g. Drucker 1983:88). Peterson & Long (1986:26 ff) point out that the "band" generally has a similar significance in Australia. But although in-group references to the "local group" or "band" may prove to be the most pronounced, it does not follow that in-group attitudes are exclusively found here. The cases of the Australian clan-estate relationship (Stanner 1965) and the sib-system among the Tlingit at the north-west coast of America (de Laguna 1972) demonstrate that strong in-group attitudes may cut across boundaries of hunting territories. In-group notions may also refer to larger alliances of several bands/local groups against common enemies (e.g. Berndt & Berndt 1964, Burch & Correll 1972, Drucker 1983, Lourandos 1998:61). Furthermore, in-group identification sometimes refers to encompassing units such as "Eskimo" as opposed to "Indian" in northern Canada and Alaska (e.g. Graburn & Strong 1973, Townshend 1979). Berndt & Berndt (1964:42) are therefore probably generally correct when they characterise the Australian Aborigines as having nesting ethnic identities depending on the context of the situation.

Data on out-group categories indicates that such attitudes flourish among hunter-gatherer groups. Often they relate to specific, named, often adjacent groups. In other instances, general, categorical terms are applied, like "mountain people" or "inland people" (Lourandos 1998:45). They may also refer to "strangers" in general, which is a much-used term for people who are not band, kin, or allies. These are usually treated with hostility (Burch & Correll 1972). Although relations may be friendly in practice, negative attitudes are typical features of the out-group attitudes and categories. "Less-than-human", "half-human" or other derogatory remarks are popular, often used in opposition to the term "human" by which the local populations designate themselves (Helms 1988:22 ff). These prejudices are instilled in the members of the in-group at an early age. Ernest Burch illustrates this point very well: "Countless evenings of hearing about treachery and warfare, or hilarious renderings of other speech style could only reinforce such attitudes, serving to differentiate further the members of one's own society from all other human beings" (Burch 1980:278). Hostile relations notwithstanding, the situation is not generally characterised by isolation and lack of contact. Although mountain ranges or dry basins separate people, resulting in fairly endogamous populations (e.g. David & Cole 1990), such natural hindrances usually do not prevent contact (e.g. McCarthy 1939; 1940, Burch 1988). The degree and nature of the inter-group relations, however, vary considerably. In the case of Australia, many groups participate in long-distance social networks. Some of these networks are "open" and cut across linguistic groups and major drainage basins, involving exchanges of both spouses and artefacts (Tindale 1974:75 ff, Hamilton 1980:8 ff, Lourandos 1998:56, 66). Such systems are, however, most common in the arid zone. In the south-west, but also in the tropical north, more "closed" social networks operate, which usually involve only the neighbouring territorial bands (Lourandos 1998). In North America, these more closed networks are most common in some parts of Alaska and on the north-west Coast, where they include trading partnerships, exchanges of marriage partners, as well as aggregations, feasting, competitions, and war (e.g. Burch & Correll 1972, Townshend 1979, Drucker 1983, Burch 1988). In arctic boreal Canada, on the other hand, the networks are more open (e.g. Damas 1963).

This brief account indicates that, similar to modern ethnic groups, hunter-gatherers have pronounced in-group identities as well as out-group categories. Hunter-gatherers also frequently interact with out-groups. In addition, it appears that, although most prominent at local levels, ethnic identity can be found at several levels of the social organisation at the same time. I will therefore argue that ethnic relations among recent hunter-gatherers are essentially the same as in "modern" societies. Still, there is one important aspect that separates hunter-gatherer ethnicity from that of populations in the industrialised world. Not only among modern ethnic groups, but also among groups related to early state formation, origin myths often refer to places far away from where people actually live (e.g. Eriksen 1992, Hedeager 1997). In reality, however, several of these ethnic groups are co-residential. Although the clan-estate relationship in Australia and the sib-system of the north-west coast of America may indicate a complexity of man-land relationships, such a "melting-pot" situation generally does not fit the hunter-gatherers I referred to above. In these cases, ethnic boundaries are "real" in the sense that they relate to spatially confined boundaries; the in-groups are usually co-residential. The out-groups live somewhere else. Here one might argue that I embrace the "primordial" approach, and see particular hunter-gatherer groups always as descendants of the pristine occupants of a given territory, but this is not my position. I acknowledge that origin myths are quite as "invented" among hunter-gatherers as they are in the modern world, and that migrations often take place. On the other hand, one should not always take for granted, as Sîan Jones (1997:104) appears to do, that ethnic identities constantly are constructed, re-invented, and contested, or that the group in question just recently formed such an identity and that the historical development in the region has been one of constant migration and flux. As Jenkins points out, "To say that ethnic identity is transactional and changeable, is really to say that it *may* be; it doesn't mean that it *always* is or *has* to be" (Jenkins 1997:51 italics original). In this context, it should be considered that the Neolithic populations that are studied in this thesis lived in small-scale, local arenas, detached from the global world as we know it today. They were certainly not confronted with powerful and aggressive colonist states. If they had ethnic identities, these were products of relations with other, neighbouring groups on more or less similar socio-economic levels. Such relations may of course have been rapidly changing and fluctuating, but they may also have been stable over long periods of time.

Ethnic boundaries and material culture

Based on the ethnographic data, it is possible to assume that ethnic boundaries among hunter-gatherers often coincide with spatially defined or territorial boundaries. If this is the case, I expect that such spatially defined ethnic boundaries could also have been relevant in the Neolithic period (see Hallgren 1998 for a similar position with regard to the Swedish Neolithic). However, bearing in mind the epistemological discussion on material culture and ethnicity, how can we possibly come to grips with such boundaries? A combination of broad geographical distribution studies and detailed studies of local contexts and circumstances is recommended here.

At the broad level, I find that the old method of documenting the regional variation in artefacts, styles, and technologies still serves this purpose. What the early archaeologists saw was a tendency towards an increased regional diversity in the archaeological data: functionally similar tools were made differently in different regions. Raw materials and ornaments tended to vary in similar fashions and over time, the regions were subdivided. Of course this is a perfect example of the "culture-historical" methodology that was criticised above. Stephen Shennan, for example, argues that an archaeological culture defined in this way is a pure analytical construct, and the probability that it coincides with any socio-cultural unit of the past is very small (Shennan 1989). But I cannot agree with this. I agree with Shennan that it would be wrong, as Childe did, to interpret these regions automatically as reflecting ethnic groups, but I think it would be a mistake to abandon the methodological approach altogether.

It is absolutely certain that prehistoric archaeologists cannot base themselves on emic data; they do not have direct access to what people felt and meant. On the other hand, it is the privilege, and the obligation of archaeology, on the basis of etic data, to construct these feelings and attitudes. Instead of neglecting and abolishing these different regional distributions of material culture, I would therefore like to take two steps back and ask what they mean, once again.

In my opinion, the cultural variation that Brøgger (1906; 1909), Hinsch (1955), Søborg (1988), A. B. Olsen & S. Alsaker (1984) and others were able to delineate in Neolithic southern Norway are extremely interesting, first of all because they are very good examples of Sackett's isocrestic variation (Sackett 1985). If we follow Sackett's reasoning, this does not mean that we are dealing with different ethnic signalling. Instead it probably tells us something about decreased mobility; that people move less than before and that they stay more permanently within smaller regions than before. Due to decreased inter-group contact and communication about cultural values, different styles and traditions develop independently of each other. Such traditions would not be restricted to particular raw materials or particular axe shapes. As Sackett points out, they would also include subsistence practices, hunting techniques, cooking techniques, architectural solutions, location of sites, rock-art styles etc. Along with a large number of aspects lost for us today, these traits would constitute the culture (local meanings) of the populations. As many such cultural traits as possible should therefore be recorded and subjected to multivariate analysis.

Although we are able to map the distribution of all of these traits, we would still be far from our purpose, because neither isocresticism nor culture is equivalent to ethnicity (Shennan 1989:20). As pointed out above, culture may be shared unevenly, because different segments of the population participate in multiple, and not always shared, spheres of social interaction. As a result, archaeological entities or isocrestic traits will not constantly be found together, and the pattern is likely to be an untidy one (Clarke 1968:299). Clarke developed classificatory expedients in order to remove the untidiness in the cross-cutting distributions and to reach some kind of essence in the material culture, which he thought, would be representative of a prehistoric ethnic group. But such a procedure is criticised by Shennan (1989:20:13), who argues that one should instead recognise untidiness *itself* as the essence of the situation. I believe Shennan is correct in that one should consider all archaeological traits, not only the ones that fit. On the other hand, I find his critique unsatisfactory if it involves a rejection of every type of concurring spatial distribution of different archaeological traits as meaningful in terms of ethnicity. Instead one should focus more on another point made by Shennan, namely that culture or isocrestic variation provides the raw materials for the forming of ethnic identities by a local population and that it has the potential to be received as messages of ethnicity by another population (Shennan 1989:20, Conkey 1990:11). This implies that artefacts should be seen as more than passive reflections of ethnicity. They may rather – as Wiessner points out in her discussion with Sackett – be activated into meaningful style in certain contexts and situations (Wiessner 1985:162). Still, an even more radical step should be taken here. It is argued that the very distinction between active and passive in this matter puts too much weight on humans as controllers of their life worlds and that it underestimates the role that material culture plays for human beings. Although we are conscious and purposeful agents, we are also continuously restricted and structured by the material world. We get new things and we change habits, but things and habits also work back on us and take part in constructing our selves. According to John Barrett (1994:36), such "structures" are in a sense pre-understandings; they are not questioned in day-to-day activities, but they orientate the subjects and enable them to act effectively. Tradition and memory thus become, according to Barrett, necessary conditions of agency.

In a context of ethnicity, this probably means that material culture plays a significant role in the relations, but as Sîan Jones and Morten Ramstad point out, the choices of values and styles are not coincidental within the socio-historical context (Jones 1997:125, Ramstad 1998:357). Therefore, when a group of people with common values, technology and subsistence practices enter into contact situations (conflict, competition over scarce resources, barter, or gift exchanges) with groups that have different habits and different material culture, they interpret their own traditions and habits anew; depending on the situation, they strengthen them or they discard them. If ethnicity turns out to be a relevant organisational feature in such situations, it is likely that people draw on their traditional cultural values; origin myths, attachment to places and learned technological styles that separate them from other groups, and that they use these values, styles and attachments to signal their ethnic belonging.

On the basis of these assumptions, I would maintain that regional distribution studies of similarities and differences in the archaeological record are of some help, because they would delineate areas with greater and lesser habitual or cultural overlap. In so doing, they would indicate where possible ethnic boundaries might have been situated in the terrain.

At even more detailed levels of the archaeological analysis, such possible boundaries should be investigated further. It is important, however, to stress that identification of boundaries between ethnic groups is not the same as *circumscription* of such groups. Although particular ethnic groups may have lived within single territories, it is not likely that boundary processes between neighbouring groups were of the same kind, either in terms of in-group/out-group relations or the symbolic meaning of the material culture. *The boundary processes can therefore best be studied at a local level*, where the symbolic meaning of particular artefacts and practices are likely to be less equivocal than on a regional scale. Rather than trying to circumscribe groups by means of archaeological or other criteria, concentrating on local behaviour on both sides of possible boundaries of a more limited length is more fruitful.

It should be relatively clear from the above that one cannot automatically assume an idiomatic status for any cultural trait, and it is not possible to uncover ethnicity by means of a "checklist approach". Nevertheless, as Bjørnar Olsen points out (1985a:27), there is a need for some archaeological criteria which should be satisfied if ethnicity is to be seen as a relevant interpretation. In the following, I will discuss a set of archaeological criteria that may help distinguish between arbitrary boundaries and socially constructed boundaries during the early Neolithic period in southern Norway. The criteria that will be discussed here are co-variation, sudden fall-off, boundary crossing and historical continuity.

(1) Co-variation of cultural traits. Hunting techniques, architectural solutions, lithic industries, cuisine and other traditions are normally shared unevenly across space. Thus, one may not expect to find

a clear-cut correlation of such traits. On the other hand, common culture may also be a powerful resource for ethnicity. If the distribution boundaries of many independent archaeological variables are coincidental, this might indicate that one is dealing with the delineation of a group that has strong, indigenous traditions.

(2) Sudden fall-off in raw material distributions. In the case of the distribution of particular raw materials, gradual fading is the normal or expected, because increasing distance from the source will lead to increasing transportation costs and more limited requests. Gradual fading towards a distribution boundary therefore says little about the social significance of such a boundary, because it may have been the same people who moved back and forth. On the other hand, if the disruptions of the raw material gradients are marked, and if this boundary is not coincidental with natural boundaries such as mountain plateaus or long stretches of water that would prevent further spread, one might be dealing with a situation where the raw material was not permitted to cross, or was not wanted across the boundary. In such cases, we are dealing with a major rupture in communication and a relative isolation (some crossing items should be present – see next point) that may have been caused by ethnic boundary maintenance.

(3) Crossing. Ethnicity is first of all a result of boundary processes. A discussion of the topic should therefore consider not only the archaeological traits that constitute the boundary, but also the ones that go across. If there are no crossing items, this might be due to lack of contact (for example due to insurmountable natural hindrances), and the evidence for an ethnic relation would be weak. Distribution studies of lithic materials provide the best data on crossing. Considering that the raw materials have to be moved physically, one would presume that a wide distribution of a particular stone would indicate the maintenance of contact between groups.

(4) Historical continuity. One might expect that boundaries with deep traditions, going back to times immemorial, would be less changeable and more important than other more recently formed boundaries. Such "old" boundaries would provide spatial references for in-group/out-group attitudes. They would also – if the mobility across them traditionally were lesser than within – constitute a boundary between landscapes that people knew well (in which there would be numerous attachments), and areas with which they were less familiar. A boundary indicated by distributions of styles, types, or raw materials in one particular period should therefore coincide with a similar type of boundary in the preceding and/or succeeding period(s).

In addition to moving back and forth between these different levels of archaeological analysis, it is also necessary to investigate the socio-historical circumstances that would make ethnicity relevant in the first place. In my view, two types of circumstances are particularly likely to generate strong ethnic relations in hunter-gatherer societies: sedentism and contrasting subsistence patterns.

(1) Sedentism. Although residentially mobile peoples may have ethnic relations with others, one may perhaps expect that sedentary hunter-gatherers and early farming communities would be more predisposed for developing this type of inter-group relations. Neighbouring sedentary hunter-gatherers usually exploit similar ecological niches. In a context of competition over resources (particularly in the boundary areas) between such groups, ethnicity is likely to appear as an organisational feature, because it provides the ideological basis for boundary maintenance between the groups. In such situations, local leaders are likely to make ethnicity an important issue, often in order to "mask" internal conflicts that are the results of increased circumscription (Ramstad 1998). The sense of being different may also have stronger cultural foundations among sedentary than among mobile peoples, because sedentism implies that people stay permanently around certain sites and territories. Groups therefore develop close attachments to the area itself and to locales within the area. Such attachments are likely to provide strong resources for in-group identity (but see Ingold 1986:131 ff) for examples of mobile peoples with strong attachments to particular places). Although sedentism may allow for task groups or individual messengers to be mobile over long distances, it generally implies restrictions on mobility for the groups as a whole. Most people who live outside the territory would thus be regarded as unfamiliar strangers, toward which negative out-group attitudes are likely to be most pronounced.

(2) Contrasting subsistence patterns. In areas with major differences in subsistence practices, such as between neighbouring inland residential hunter-

gatherers and coastal hunter-fishers, or between coastal hunter-fishers and inland farmers, ethnicity is likely to be an important factor. It is to be expected that the type of subsistence would have determined a major part of the cultural repertoire. In a case of coastal hunter-fishers vs. farmers, these different populations would eat different food and use different tools, but they would also most likely have quite different attitudes towards animals, landscapes and seasons. In the zones of interaction in between (e.g. B. Olsen 1988, Zvelebil 1998), these attitudes are likely to have been perceived as very different. Between coastal hunter-fishers and inland hunter-gatherers, the differences would not be that dramatic, but it is easy to imagine that inland elk-hunters would have different tool kits as well as different preferences for food compared to salmon-fishing groups on the nearby coast. From what is said about the "cultural stuff" being ethnically relevant, it is to be expected that such groups with contrasting subsistence patterns would easily mobilise the cultural differences as symbols of ethnic belonging in contexts of interaction between the two.

Obviously, these lists of criteria and circumstances are not exhaustive and are certainly not absolute in the sense that ethnicity is excluded if one or more of the criteria cannot be satisfied, or that ethnicity is proved if they all are satisfied. They should primarily be seen as working tools in the process of exploring the problem in an early Neolithic context. Hopefully, such a multi-level approach will produce plausible arguments for the spatial location of prehistoric ethnic boundaries. But it will never produce one single valid interpretation on the subject. Undoubtedly, an ethnic group only exists in the consciousness of those who include themselves within and exclude themselves from this group. It can therefore not be "found" in the archaeological data (B. Olsen & Kobylínski 1991:12). When investigating the problem, one must be open to a situation where several incompatible, but potentially valid (multivalent) interpretations of the same data can exist side by side. Some archaeologists (e.g. Kohl 1993:15 ff, Trigger 1995) see a number of dangers with such an open epistemological position, arguing that if we accept a plurality of interpretations particularly with regard to ethnicity, we may also be forced to accept a number of racist, chauvinist and nationalist readings. An alternative is that archaeologists, while pleading objectivity, present their interpretation as the only valid solution. However, this may lead to an even more dangerous situation for existing minorities (Härke 1995:56, Jones 1997:11). A third alternative is to avoid the problem altogether (Trigger 1995), but such a strategy leaves the field open to non-professionals, entirely beyond the influences of archaeological research. The open position, which is advocated here, should not, however, be seen as an excuse for launching badly-supported hypotheses or for performing sloppy methodology. On the contrary, only careful quantitative and descriptive analyses will make it possible to delimit the number of potentially valid interpretations. The approach recommended here is to apply a multivariate approach, which includes both regional and local levels of study. This implies that investigations of different contexts for production, use and discard of the lithic tools should accompany distribution-studies of stylistic elements, raw materials and other cultural traits. Such a shift of emphasis between different perspectives on the archaeological data is likely to improve the understanding of both ethnicity and contact between different ethnic groups in Neolithic Norway.

Conclusion

To conclude, I think it is important to engage in Neolithic ethnicity. But when dealing with the concept, it is necessary to heed the consequences of the insights from the social sciences, and look upon ethnicity as a situational, socially constructed phenomenon, primarily concerned with how people think about themselves as groups and how they set themselves apart from other groups. It is a social construction of "us" and "them" which is marked in cultural terms.

In order to operate this approach, the concepts "ethnicity" and "culture" were held to be of different orders. Ethnicity was, in line with recent anthropological thinking, seen primarily as an organisational feature, devoid of cultural content at the outset. Culture was, differently from the old culture-concept and the above concept of ethnicity, seen as learnt practices and traditions of for example language, technique, landscape perceptions, or symbolism which may be shared across ethnic boundaries, but which also, when they are not shared with opposing groups, provide the resources for the emergence, maintenance and renewal of ethnicity.

Four criteria should be satisfied if a boundary was to be interpreted as ethnic. There should be (1) co-variation of several cultural traits, (2) a sudden fall-off in the distribution of lithic raw materials (3) evidence for crossing, and (4) deep history of the boundary. Furthermore, it was suggested that such an analysis would also have to consider the socio-historical circumstances that would make ethnicity relevant in the first place. In conclusion, when the concept is applied, it should therefore come as a result of thorough and broadly informed empirical studies, not from spatial distributions of randomly selected tool types or raw materials alone.

In the next chapters, I will investigate possible ethnic boundaries by comparing as many early Neolithic cultural traits as possible. First, I will perform a site location analysis in order to find out if there were different cultural traditions for the location of sites in different local areas in the coastal region. Second, I will examine the distribution of all lithic raw materials, types, and technologies at 37 different excavated Early Neolithic residential sites on the west coast and at the mountain plateau between east and western Norway.

Chapter 4
Site location analysis

Introduction

It is assumed that the way the early Neolithic populations located their sites in the terrain was influenced by their cultural traditions. Detailed information about site locations in different parts of the region might therefore reveal differences or similarities in these traditions. In this chapter, I will perform a site location analysis at both macro-level and micro-level. The macro-level analysis will compare distribution of long-term sites and short-term sites in four different types of environment on the west coast (Fig. 2): tidal current channels (Skatestraumen, Fosnstraumen, and Bjorøy), exposed outer coast island (Kollsnes), protected inner coast island (Stord-Bømlo-Sveio connection) and inland bay (Soltveit). The micro-level analysis will compare seven site location factors in these same areas: harbour conditions, slope, view over the sea, size of available site area, shelter from dominant winds, distance to the shoreline, distance to fresh water. I will describe the method of site location analysis. The results of the analysis will be analysed together, and they will be discussed in a comparative perspective. First, however, I will briefly present the natural environment in western Norway.

Climate, topography, vegetation, and resources of western Norway

In spite of the northerly location of the west coast of Norway, the Gulf Stream, and westerly winds strongly influence the climate, which is characterised by mild winters, cool summers, and heavy precipitation rates. Towards the mountain plateau between east and west, the climate is dryer and colder (Wallén 1968).

Topographically, the outer coast is relatively different between Sunnmøre and Sunnhordland (see Fig. 58 for location of districts). The coast of Sunnmøre is characterised by large islands with relatively high altitudes (up to 600 m above sea level). These high islands often have broad rims of habitable, easily cultivated, sandy soils. The outer coast of Sogn og Fjordane lack the broad land-rims. Instead, habitable land is concentrated to sandy bays and lower islands. The coast of Hordaland (Nordhordland, Midthordland, and Sunnhordland), is different from Fjordane in that it consists of a large number of small and large low-lying islands with channels of water and small fjords in between. Further east, the entire region differs in that it consists of a large number of long fjords that are mainly oriented in the east-west direction. Although there are several low-lying areas in which agriculture is practised today, the mainland in the fjords is generally steep and has relatively high altitudes (up to 800 m above sea level). Further inland, the fjords are deeper with steep slopes up to the mountain plateau. The ridge between eastern and western Norway is characterised by high and steep mountains with altitudes up to 2800 m above sea level in the northern part of the analysed area (Breheimen-Jotunheimen). Further south (Lærdal-Hemsedal) the mountains are lower and several areas between the peaks are relatively level at about 12-1300 m above sea level. The plateau Hardangervidda is relatively level at about 1200 m above sea level. The majority of Haukelifjell has more or less the same altitudes as Hardangervidda, but the terrain is generally steeper.

Pollen analyses show that the entire coastal and low-lying inland was covered with dense mixed leaf-forests during the Atlantic and the Subboreal periods. Significant areas of the mountains were also, in contrast to the present situation, covered with trees (Gjærevoll 1992).

It is apparent that, during these periods, western Norway was rich in terrestrial and marine resources, especially along the coast and the lower part of the fjord valleys where a variety of resources was available throughout the year (see detailed discussion in Warren 1994:39 ff). Fish, sea birds, and marine mammals tend to concentrate in tidal current channels at the coast. As well as being strategic for communications, these channels are known as excellent fishing sites by the local populations today (Bergsvik 2001c). The seasonal and topographic distribution of available food resources structures or constrains, but does not determine, past settlement patterns. The seasonal and spatial distribution of resources allows for much choice. In consequence, both year-round sedentary settlements and more transient settlements were possible. Several different areas (coast, fjord, mountains) were accessible by boat over relatively short distances. These areas would be easy to use by groups involved in permanent residential moves and task groups involved in shorter trips launched from year-round base camps. From the perspective of a

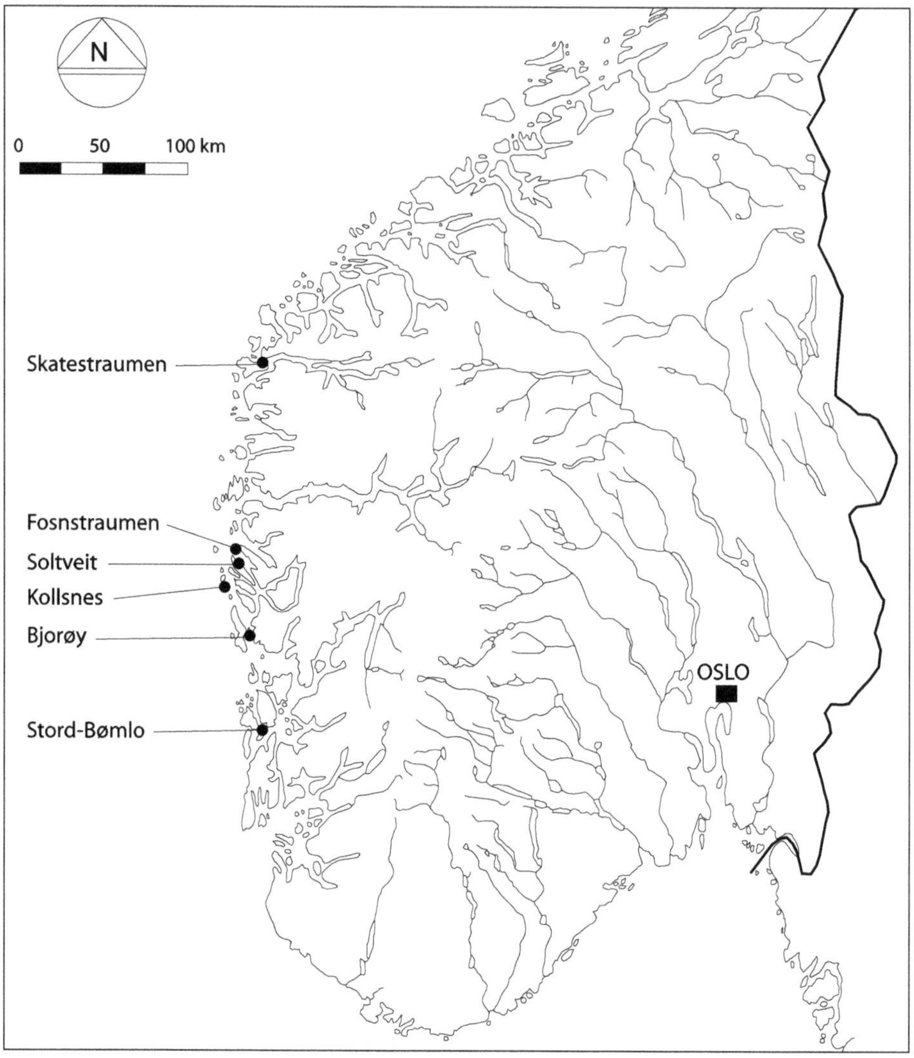

Figure 2. *Southern Norway. The areas with location analyses are marked.*

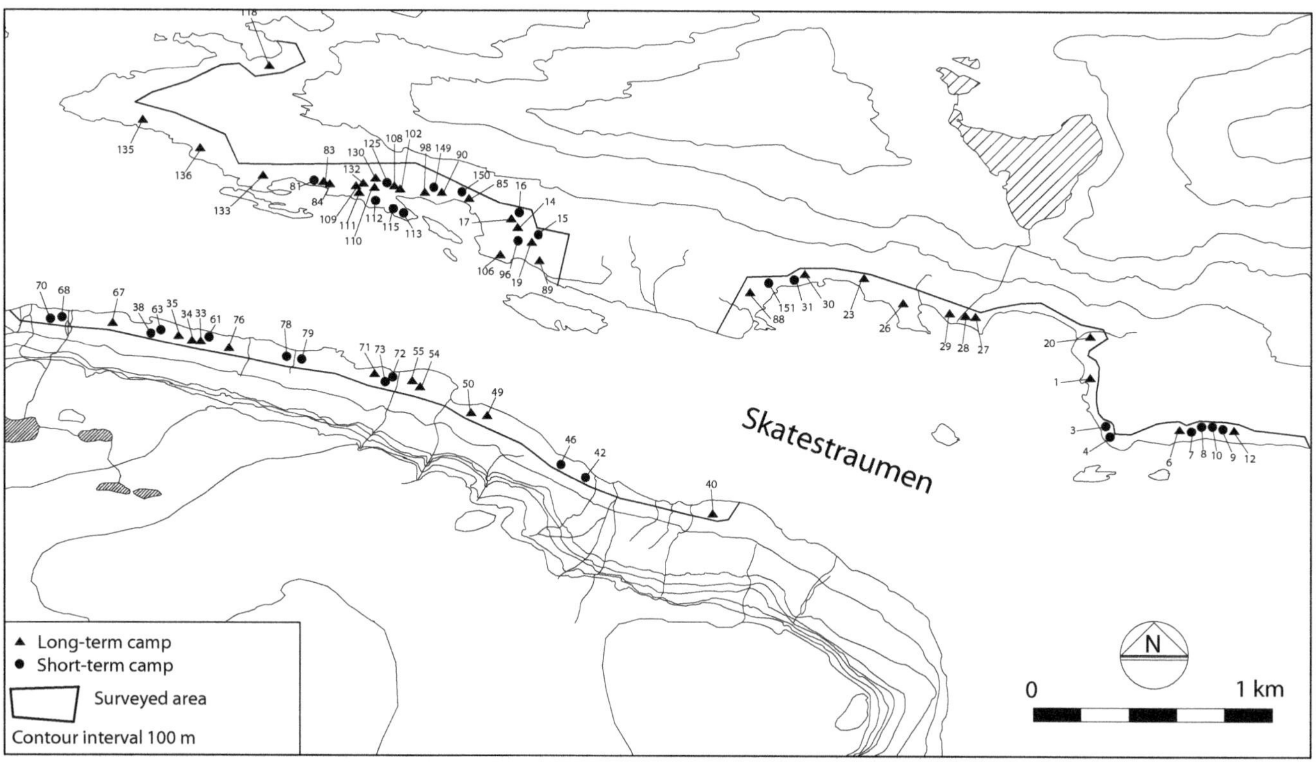

Figure 3. *Skatestraumen, Nordfjord. Early and middle Neolithic sites.*

coastal group, moving along the coast into districts north or south of Nordhordland would not provide access to any new food resources, whereas moving in th east-west direction would provide access to avariety of new resource (Warren 194:86 f).

Figure 4. *Skatestraumen, Nordfjord. The site 1 Haukedal during excavation. In the background is the mountain Hornelen. The photo is taken towards SSW.*

Method

Macro-level analyses are very common in archaeology. In these types of analyses, one usually investigates how the distribution of archaeological remains correlates with different macro-environmental factors. One example of such an analysis from Hordaland, western Norway, is Egil Bakka and Peter Emil Kaland's diachronic study of the distribution of stray finds in areas that were favourable for either hunting/fishing or agriculture (Bakka & Kaland 1971).

Micro-level analyses are also effective means for recording variations in land-use. They are developed in order to have precise descriptions of the factors that may have led to occupation of a particular place. The methods have been has been applied for late Palaeolithic and Mesolithic settlement sites in Germany and France (Hahn 1979, White 1985, Kvamme & Jochim 1989, Eriksen 1991), Northern Sweden (Meschke 1977, Forsberg 1985), Northern Norway (Bjerck 1989, Barlindhaug 1996), and on the southern coast of Norway (Berg-Hansen 2001).

During the last few years, micro-level site location analyses have been performed within the current area of analysis in coastal western Norway at Fosnstraumen, Nordhordland (Bergsvik 1995), Kollsnes, Nordhordland (Bergsvik 1995), Bjorøy, Midthordland (K. Kristoffersen 1995) and at the Stord-Bømlo-Sveio highway connection (hereafter called Stord-Bømlo) (K. Kristoffersen 2001a). The Neolithic[1] data from these four areas cover the southern part of the region, and will be compared below. I will also use Neolithic data from a site location analysis that has been performed for the present work at Soltveit, Nordhordland, and Skatestraumen, Nordfjord. Ideally, the analysis should include Sunnmøre and the mountain zone as well. Site location analyses are, however, only carried out in the six areas cited above (Fig. 3-10).

The aim of the present chapter is to discuss whether different site locations may be a result of different cultural practices. However, even if differences between the areas may be interpreted as different culturally influenced preferences, there are surely other factors that may result in inter-regional variability on a macro-level as well as a micro-level. The most important of these variables are (1) site types, (2) representativity, (3) background environment, and (4) physiographic conditions. A possible theory of culturally influenced differences depends upon a sufficient control of these variables.

Site types

There is usually a large variation in prehistoric site types, and it is to be expected that different site types have different locations, depending on the individual functions. For example, slaughter sites or transitory sites would probably be placed differently in the terrain than residential sites. Test pits do not provide substantial information about the character of occupation, such as types of activities and group composition. Based on surveys, it is often not possible to give more than a general characteristic of the site, where length of the occupation is the clearest separating variable. Here, large sites with thick black deposits, great artefact density, and great raw material variation are classified as long-term camps. It is assumed that most of these sites were residential sites.[2] Small sites with thin deposits, low artefact density, and low raw material variation are classified as short-term camps. This category probably covers a large variety of

1 The above studies from western Norway compare early Mesolithic, late Mesolithic, early Neolithic, middle Neolithic, and late Neolithic site locations. Here, I combine the early and middle Neolithic data.

2 An alternative interpretation of the long-term sites might be that they were accumulations of many short-term occupations (e.g. Binford 1982). See Bergsvik (2001b) for a discussion of this problem with regard to the site Kotedalen in Nordhordland.

site types, from short-term field camps to slaughter sites, but most of them were probably field camps (see Chapter 6 for a more detailed discussion of site types). Short-term sites and long-term sites will be distinguished in all of the analysed areas.

Representativity

When arguing that some topographical factors were more important than others for the location of prehistoric sites, this argument should be based upon a sufficient test of areas that are not characterised by these factors. In other words, the analysis should be representative of the real distribution of prehistoric sites in the particular area. In western Norway, where prehistoric sites generally are covered with layers of turf or soil, such a control can only be attained as a result of thorough test-pit surveys.

The survey methods in the five areas were somewhat different, partly because the surveys were lead by different persons, partly because the methods had to be adjusted as the size of the different areas was different. Still, all of the areas have been thoroughly surveyed, and the results are fairly comparable. All of the surveys and subsequent excavations were performed prior to development of the areas (mainly road construction). The following methods were applied in the different areas: At Fosnstraumen and Soltveit, the survey covered 1,5 and 1 km² at levels between 0 and 30 m above sea level. A "systematic sampling" was applied, where as a general rule, test pits were dug at 20 m intervals. The surveyed area at Kollsnes covered 2,5 km². The shoreline between 5 and 15 m above sea level was thoroughly surveyed by different independent crews, and smaller, selected areas higher than these levels were also test-pitted. At the Stord-Bømlo connection, the survey followed the projected road alignment, which mainly followed the present shoreline, and in some cases went at 40-50 m above sea level. At Bjorøy, the projected road alignment, which went entirely through prehistoric shorelines, was thoroughly surveyed by several different independent crews. The work at Kollsnes, Stord-Bømlo and Bjorøy did not follow a systematic pattern, but I find that the qualities of the surveys are satisfactory for a comparison to the systematically surveyed Fosnstraumen. The results from Skatestraumen are more problematic. A total of 4 m² was test-pit surveyed at Skatestraumen. Levels between 3-4 and 15 m above sea level were covered. Earth-borers were, to a large extent, used in order to locate sites. This has probably resulted in representative data with regard to long-term large sites with thick black deposits (which is detectable with the earth-borer). However, compared to the other areas, where spades and water-meshes were used in order to locate sites, relatively few of the short term sites were probably found at Skatestraumen.

With regard to the macro-scale analysis, it should be apparent from the above that all of the surveyed areas are situated at a relatively close distance from the

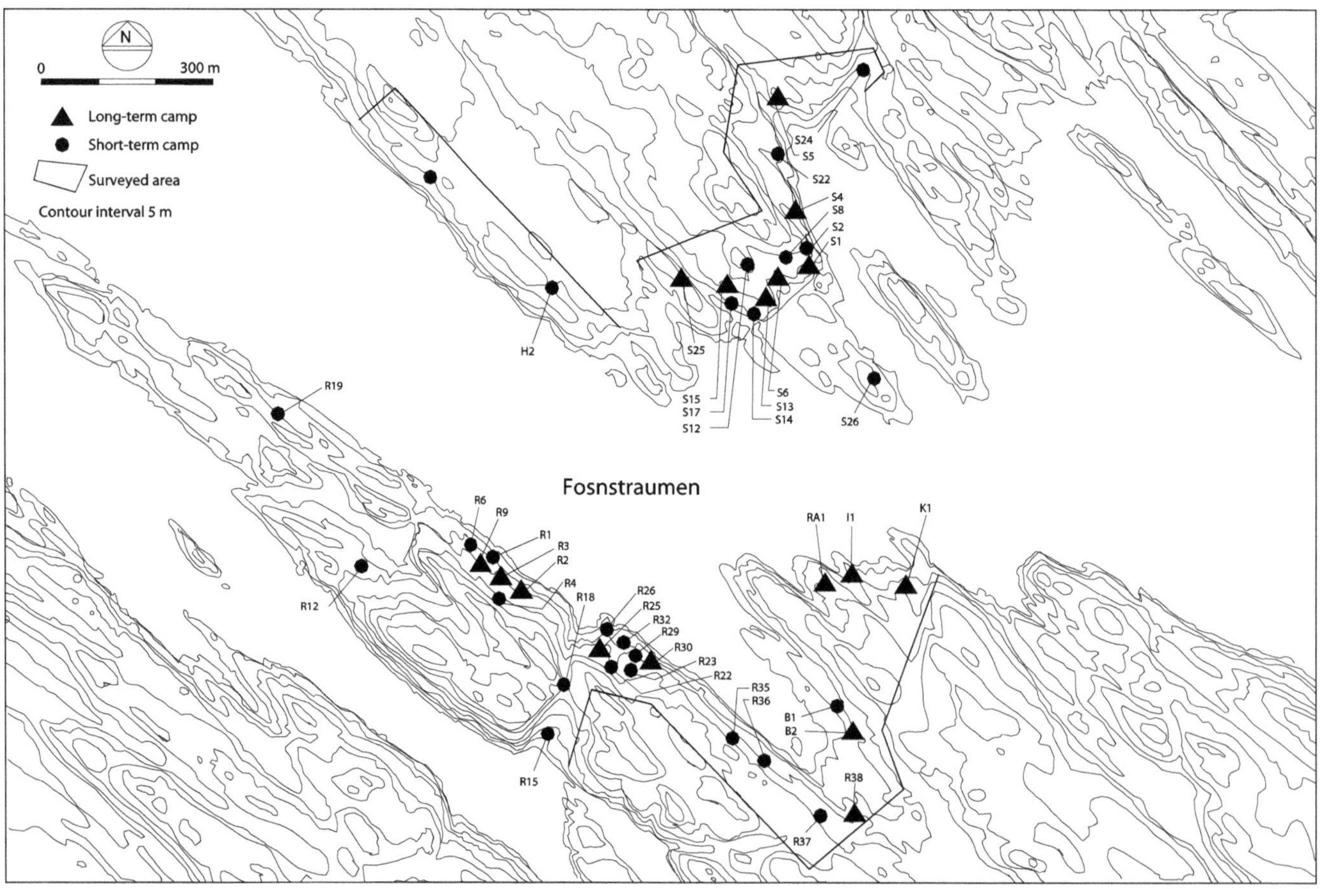

Figure 5. *Fosnstraumen, Nordhordland. Early and middle Neolithic sites.*

Figure 6. *Fosnstraumen, Nordhordland. The site Kotedalen is situated at the southern side of the channel, just underneath the bridge construction. Photo taken towards the south.*

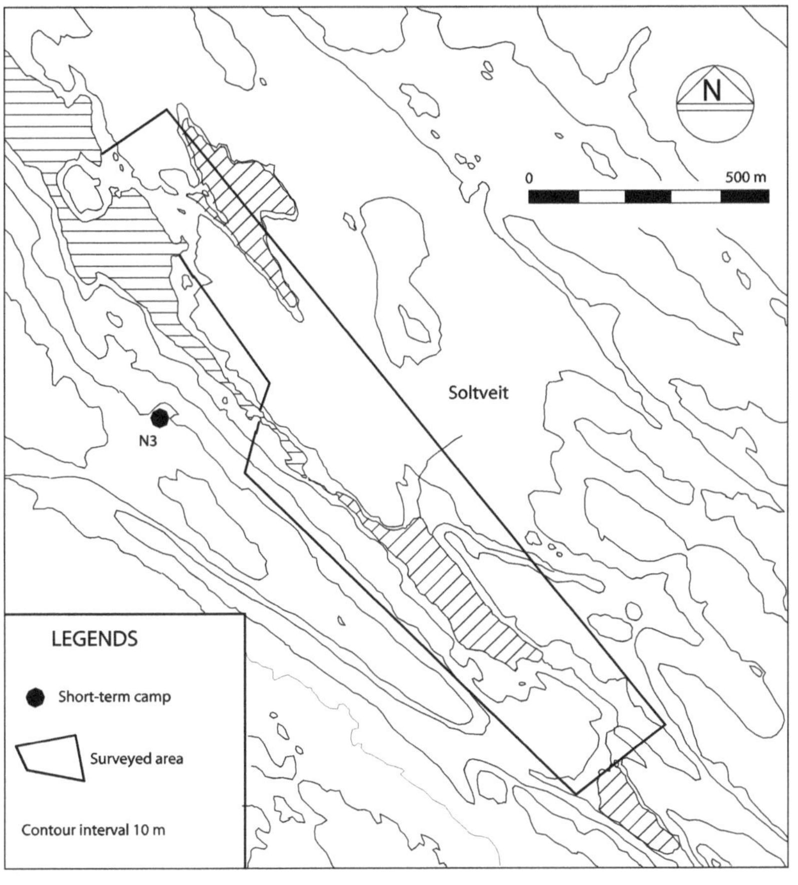

Figure 7. *Soltveit, Nordhordland. Early and middle Neolithic sites.*

contemporary shoreline at the coast. None of the areas can be regarded as particularly attractive for prehistoric cultivation or inland hunting or fishing. Bakka & Kaland (1971) argue that many areas in western Norway are favourable as regards primitive agriculture or lowland foraging. Until now, however, none of these areas have been subjected to thorough test-pit surveys. One might argue that these types of environments should have been considered, because they would have added important dimensions to the analysis. With regard to farming, the problem may not be particularly pressing, considering the faunal and botanical data indicating that agriculture was not introduced in this region before the middle Neolithic (Hjelle et al. 1992). For inland hunting and fishing however, the problem is greater. Recently, Inger Berg-Hansen (2001) has surveyed a number of sites of Lista, Vest-Agder, which clearly show that inland (lowland) hunting sites were quite frequent there. Until

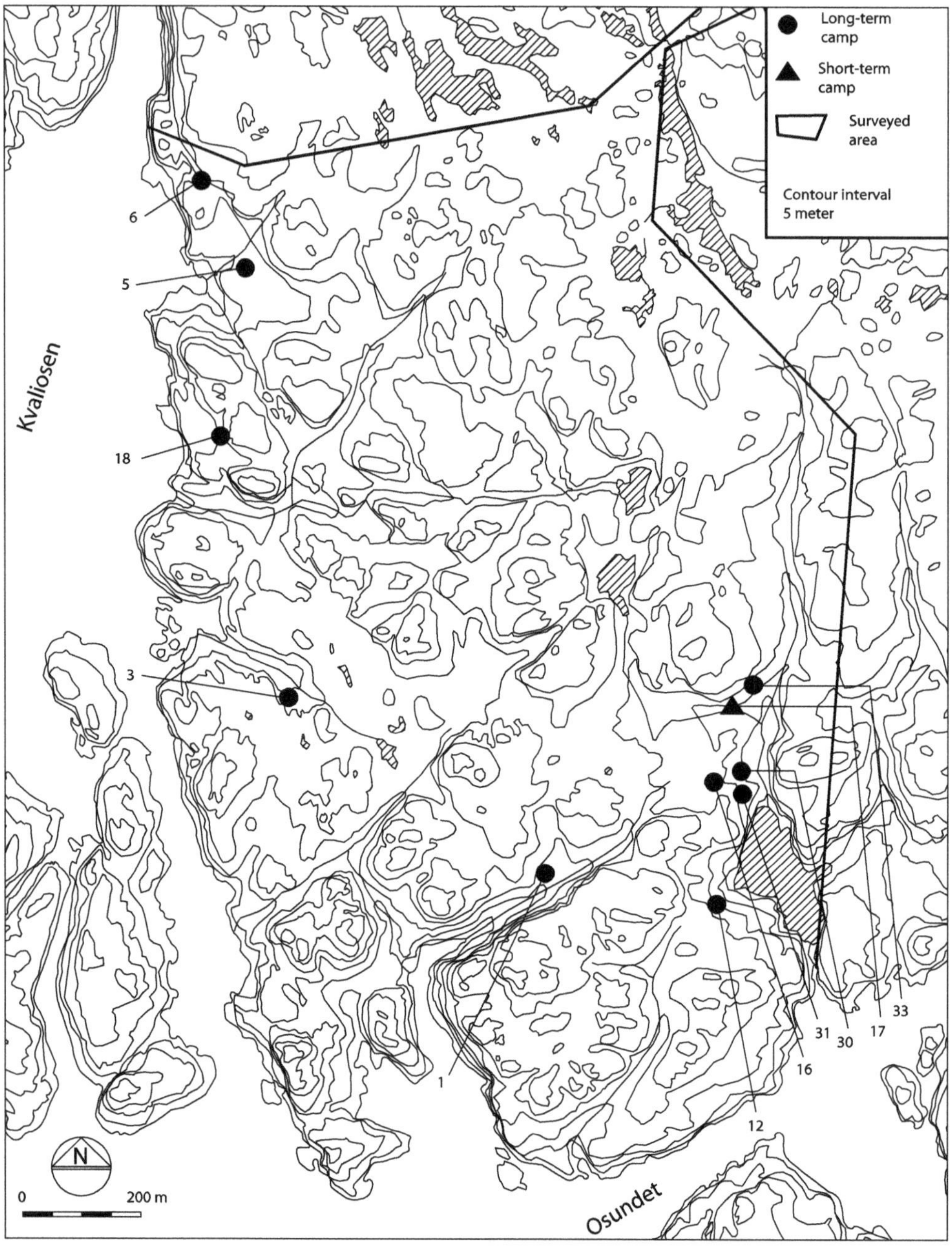

Figure 8. *Kollsnes, Nordhordland. Early and middle Neolithic sites.*

such a survey has been performed in my area of analysis, however, a comparison of site locations at shore-bound sites will have to suffice.

The physiographical conditions

Site location analyses are performed in the terrain. Several factors or variables, which are assumed to have been important for the location of prehistoric sites, are evaluated and classified individually at the sites themselves based on a number of clearly defined criteria.

As Kvamme & Jochim (1989:2) point out, a site location analysis results in geographical site data which is comparable to data from other sites, in contrast to earlier more personal and imprecise impressions of Stone Age site locations. Still, it is not evident that data from this type of analysis in one area is directly comparable to data from other areas. Universal criteria are difficult to set up, mainly because different combinations of similar environmental variables create different physiographic conditions. Two sites with similar topography and different exposure to sea-waves will, for example, have completely different harbour conditions. In a comparative perspective, these differences have to be considered (see below: Micro-analysis: criteria for evaluation).

Background environment

A significant problem is that the background environment may be different. One area may, for example, have plenty of rivers and lakes, while other areas have only a few fresh water sources. In a similar fashion can local topography in one area create numerous viewpoints, while few viewpoints are found in other areas. Kvamme & Jochim (1989:2) point out that such factors have to be described and considered, because it would be meaningless to argue that proximity to fresh water was important for site locations if fresh water is found everywhere.

This means that that every analysis should consider the local environment, based on personal experience at the sites themselves, preferably in different weather conditions. If data is to be compared to other areas, the environmental differences between the local areas have to be adjusted for. Below, the background environment in the different local areas will be compared.

Background environment in the analysed areas

Skatestraumen is situated in the district Nordfjord just south of the outlet of the fjord Nordfjord, between the islands Bremangerlandet and Rugsundøy. It is a strong tidal current channel. The currents are created by the meeting of the fjords Frøysjøen and Nordfjord. Rugsundøy as well as Bremangerlandet drop from altitudes of several hundred meters. Only narrow strips of land along the shores of Skatestraumen are habitable. The habitable area, which is partly forested, partly pasture-land today, is characterised by cliffs and low ridges. Bare rock-outcrops and bays with small, stony beaches dominate the shoreline.

Fosnstraumen, also a strong tidal current channel, lies between the relatively low islands Fosnøy and Radøy just at the outlet of Lygrefjorden in Nordhordland. The topography around Fosnstraumen, which is mostly pasture land today, is characterised by low ridges, deep bays and marshland. The shoreline mostly consists of slopes of naked rock, sometimes broken by small gravel beaches.

Soltveit lies at the bottom of a deep bay, three km to the south of Fosnstraumen. It has a similar topography as Fosnstraumen.

Kollsnes, which is directly exposed towards the North-Sea, is situated at the southern tip of the relatively low

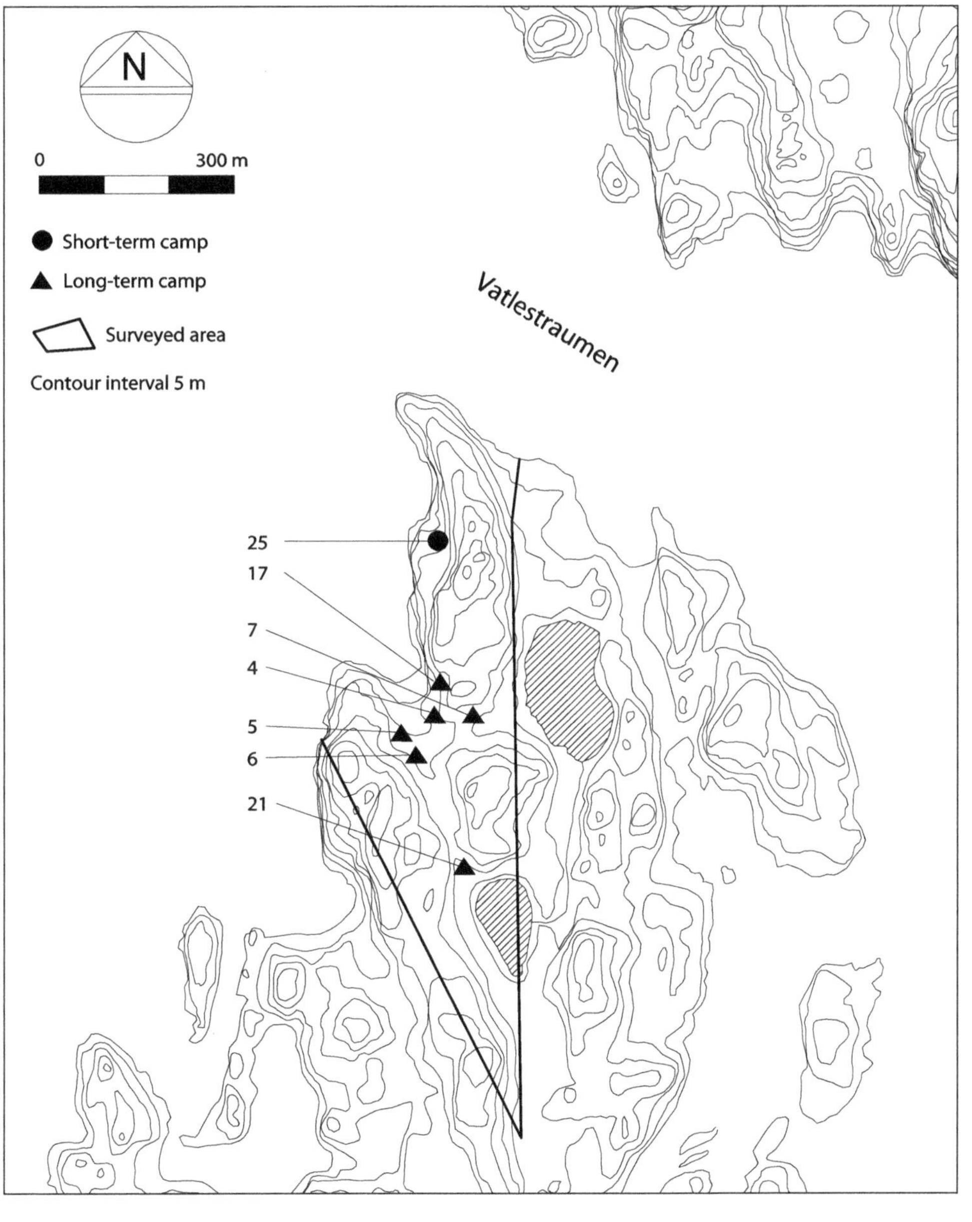

Figure 9. *Bjorøy, Midthordland. Early and middle Neolithic sites.*

island Oen in Øygarden. Kollsnes, which until recently was used for pastures, is almost treeless. It is characterised by marshes/small lakes and low naked ridges. The shoreline mostly consists of slopes of naked rock broken by deep bays.

Bjorøy is a small, relatively low island in Midthordland. It lies in the sound between the large island Sotra and the mainland. Its northern tip and the mainland form the tidal current channel Vatlestraumen. The topography, which is mostly forested today, is characterised by low naked ridges and marshland. The shoreline is dominated by one broad gravel beach, which is flanked by slopes of naked rock.

The Stord-Bømlo connection consists of the small islands Nautøy, Spissøy and Føyno and parts of the large islands Stord and Bømlo. The area is partly pastured, partly forested. The topography is generally low and is characterised by naked ridges and marshland. The shoreline is dominated by small beaches, flanked by slopes of naked rock.

One may say that the background environment is relatively different in the five areas with regard to several factors.

Harbour conditions are generally good at Stord-Bømlo. There are numerous small, protecting islets and islands, and the number of bays is generally very large. At Bjorøy, the area of analysis mainly consists of one large, protected bay. This bay offers a very good harbour for almost all of the surveyed sites. At Kollsnes, Fosnstraumen, and Skatestraumen, on the other hand, the Neolithic shoreline mostly offered unprotected slopes of rock and few good harbours. The possibilities for view are relatively restricted in all of the areas except at Stord-Bømlo. Here, small islands and small viewpoints often enable view of more than 180 degrees from the vicinity of the site.

In all of the areas, large portions are heavily exposed to the dominant wind directions. Other locations are more sheltered.

The fresh water sources are relatively few in all areas except at Skatestraumen, where numerous creeks and rivers come down from the high mountains on both sides of the channel.

There are generally good conditions for level ground and availability of small and large site areas.

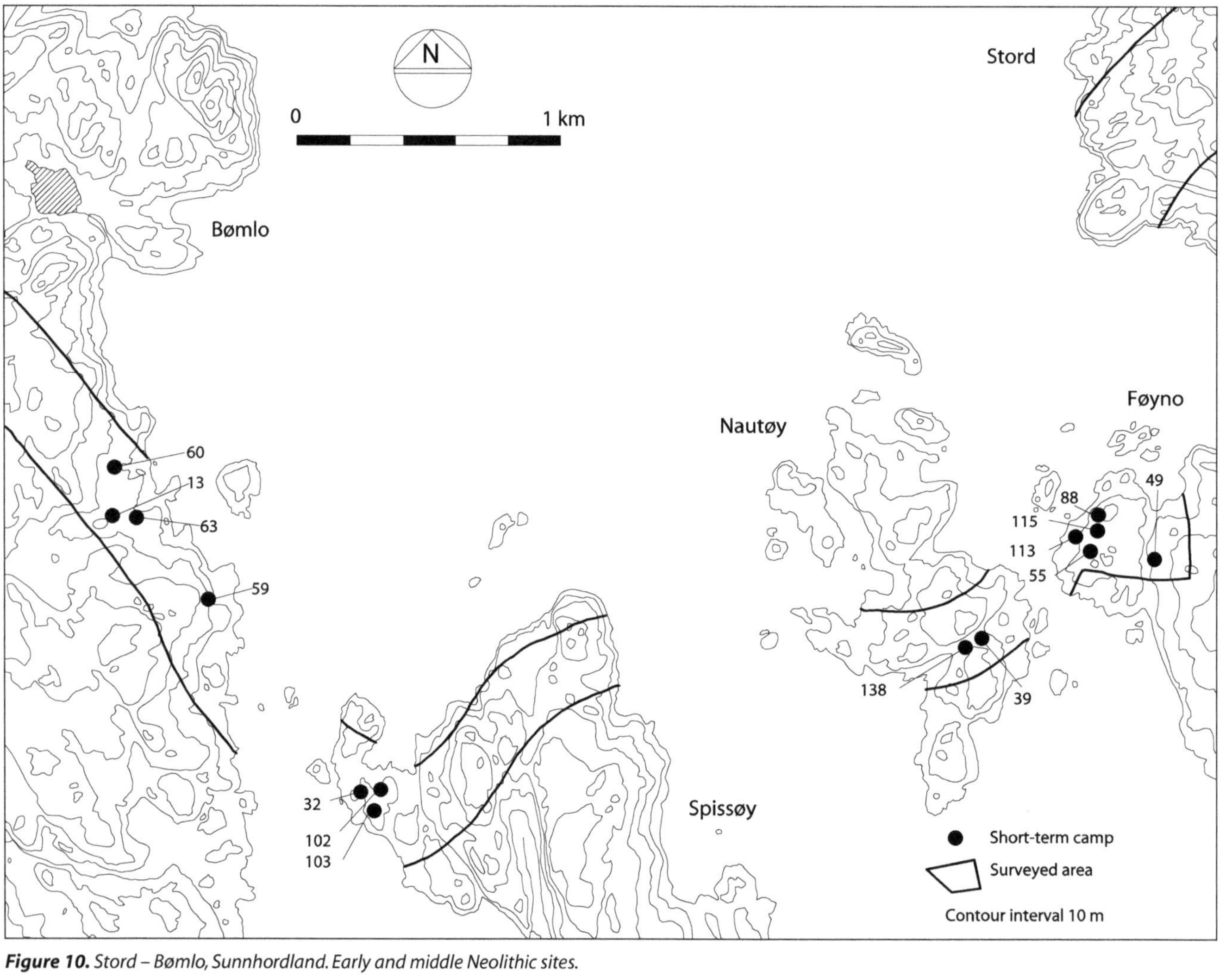

Figure 10. *Stord – Bømlo, Sunnhordland. Early and middle Neolithic sites.*

There are several habitable areas in all levels above the shoreline. At Bjorøy, however, the area is dominated by the large bay Nilsvik, which first of all offers habitable levels below 4 m above the contemporary shoreline. The maximum high tides and minimum low tides are different. When the harbour of the city of Bergen is used as a measure, the (astronomically determined) differences at the harbour at Måløy (just north of Skatestraumen) is 24% larger than that of Bergen, while the harbour at Leirvik (just north of Stord-Bømlo) is only 75% of that of Bergen (Statens Kartverk 1997).

With regard to the macro-analysis, it is apparent that, during the Neolithic, the entire region was rich in resources, especially along the coast and the lower part of the fjord valleys where a variety of resources were available at all seasons of the year. Fish, sea birds, and marine mammals tend to concentrate in tidal current channels like Skatestraumen, Fosnstraumen, and Bjorøy. As well as being strategic for communications, these channels are known as excellent fishing sites by the local populations today (see Bergsvik 2001c for a more detailed discussion of the channels).The resource-situation at Soltveit, Kollsnes and Stord-Bømlo may have been more seasonal or unpredictable.

Macro-analysis

The goal of the macro-level analysis was to investigate whether the short-term camps and long-term camps were situated in different types of environments in different parts of the analysed region.

The result of the analysis (Fig. 11) shows that the sites classified as long-term camps clearly concentrate around the tidal current channels. This type of sites is almost absent in the other environments. Short-term camps, on the other hand, are present both at the channels and at inland bays, exposed island and protected island.

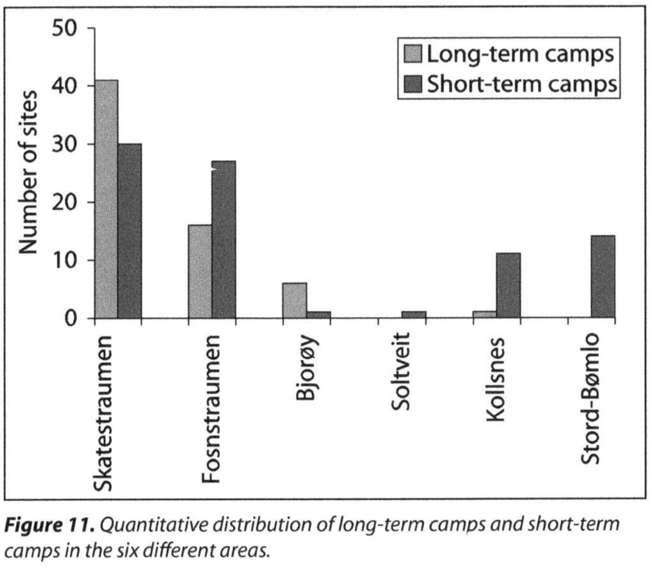

Figure 11. *Quantitative distribution of long-term camps and short-term camps in the six different areas.*

There are marked differences between the different channels with regard to the relative proportion of short-term camps and long-term camps. It is likely that this largely is a result of different survey-methods (see above). The systematic test-pit survey at Fosnstraumen produced many short-term sites. If this method had been applied at the other channels, the relative quantitative distribution of sites would probably have been much more in line with that at Fosnstraumen.

In conclusion, the macro-analysis show that the Neolithic populations preferred the tidal current channels for location of long-term sites in the entire coastal region. These areas were also used on a short-term basis. The other types of environment were used almost only on a short-term basis.

Micro-analysis: criteria for evaluation

Some of the variables were measured fairly accurately. Others were difficult to determine precisely, either because of changed circumstances since the Stone Age or because of their composite nature. In order to make the data quantitatively comparable, a score-system was established. The criteria for determining the scores were as follows.[3]

Harbour conditions

The harbour conditions may have changed as a result of sea-level changes. Fortunately, shoreline displacement curves have been constructed for all of the relevant areas (Kaland 194, Krzywinski & Stabell 1984, Svendsen & Warren 2001, Bergsvik 2002a). It is therefore reatively easy to reconstruct the sea level in the terrain. On the other hand, even if the shoreline is known, the quality of the harbour is difficult to evaluate. The protected and the and the unprotected harbours are often easy to determine. The middle group is more troublesome. It is, for example, easy to go ashore and take the boats in on a broad sandy beach, but it may be difficult to enter a boat, particularly if the waves are big. A quiet bay with steep sides, while perfectly safe, may be difficult for going ashore.

The following topographical situations (Fig. 12) result in three categories of harbour conditions:

> Protected harbour: 3 points. A protected harbour can be characterised as "... a site where the local topography stabilises the waves at the shore independently of the direction of winds and waves" (Bjerck 1989:92). It is also a place where it is fast and easy to go ashore, to pull a small boat ashore, and to enter boats.

3 Proximity to fresh water and size of available site area was not evaluated in my original work from Fosnstraumen. These evaluations have been performed later in connection with the present work. Kari Kristoffersen's criteria for evaluations on view over the sea at Stord-Sveio and for proximity to fresh water at Bjorøy have been altered to make them comparable to my own data.

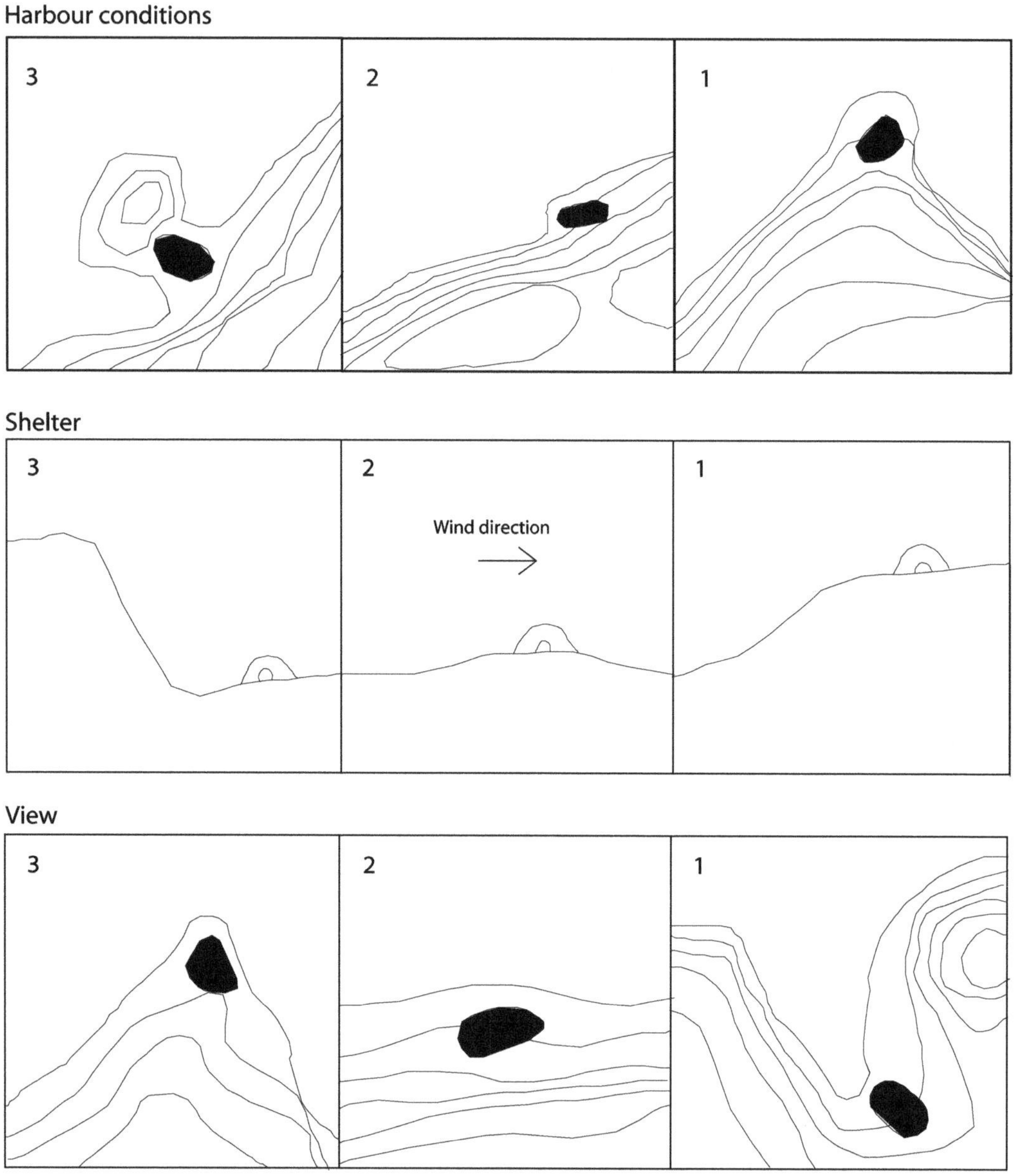

Figure 12. *Examples of different conditions of harbours, view, and shelter. The numbers 3, 2 and 1 indicate scores.*

Medium harbour: 2 points. Sheltered against currents, wind and waves from only one side.

Unprotected harbour: 1 point. Unsheltered from currents, wind and waves, or the conditions for going ashore, pulling ashore, and to enter small boats are bad.

Local topography, wind and waves create different harbour conditions in the different areas. Kollsnes is generally very exposed towards the North Sea. Sea waves and south-western winds therefore create difficult harbour conditions at sites that elsewhere would be quite sheltered. Strong winds which alternate between western and eastern directions sometimes also create difficult harbour conditions at Skatestraumen. The areas Fosnstraumen, Bjorøy and Stord-Bømlo are, on the other hand, sheltered against wind and sea-waves.

Slope

As a result of later natural processes like the growth of turf or soil slides, it is often difficult to determine the original slope of the terrain. The slope may also vary within the site itself. The evaluation of the angle of inclination must therefore be based on a general impression of the site:

Flat (0-5°): 3 points
Sloping (5-10°): 2 points
Steep (10-20°): 1 point

Shelter for dominating wind directions

During the Preboreal and Atlantic periods (ca 9000-6000 BP, 8200-4900 BC), global reconstruction indicate that the wind in western Norway mainly had a western direction, as today, even if it was somewhat stronger (Cohmap Members 1988:1046). Recent local measurements of winds were undertaken at 850 mb.

at the oil-refinery Mongstad, Nordhordland between 1973 and 1976. These measurements show that the dominating direction on a year-round basis was south-western (Førland 1984:35), which is in accordance with the global reconstruction. This indicates that the main wind direction at 850 mb. has been the same for about 9000 years.

The measurements at Mongstad were taken at a great altitude. As a result of less friction, the wind turns clockwise from the ground and upwards in the atmosphere (Førland 1984:33). The effective wind direction at ground level is therefore different. Daily measurements are taken at the lighthouse at Hellesøy, Nordhordland (Andresen 1979:65 ff). The tendencies here are held to be representative of the entire coastline. The measurements taken between 1961 and 1975 are used as a reference by selection of the most dominating wind direction during the winter (I assume that shelter is more needed during winter than in other seasons). During the period between November and February, the dominating wind-direction at ground level is between south-southeast and south.

On the basis of this information, one may expect that the SSE-S winds (see below) were dominant at Fosnstraumen, Kollsnes, Bjorøy and Stord-Bømlo. Skatestraumen, on the other hand, is situated between two high mountain massifs which channel and strengthen the south-eastern and southern winds in eastern or western directions respectively.

Several factors may have influenced the effect of the wind during the Neolithic. The average temperature was about 1° C warmer than today (Nesje & Kvamme 1991), and this may have had a moderating effect on the wind, even if it was stronger. Another factor is the vegetation. Today there are mostly bushes and small trees in the area, but during the Subboreal, western Norway was probably covered with a dense, primeval forest. This forest would lessen the effect of the wind. Still, even if these factors should be considered in the discussion, the evaluation at the sites only accounts for the exposure or shelter which is created by the local topography (Fig. 12):

Sheltered: 3 points. A natural ridge, cliff or hill, at least 2 m high and 20 m long immediately sheltering the site in the dominating wind direction(s).

Medium shelter: 2 points. Level terrain in the dominating wind direction(s).

Unsheltered: 1 point. Sloping terrain downwards from the site or water surface (at least 100 m broad and 200 m long) in the dominating wind direction(s).

Size of accessible site area

The size of accessible site area is defined as the area which is continuously habitable or possible to use for other domestic purposes. The area may be delimited by shorelines, steep cliffs, or hills.

Large area (>200m²): 3 points
Medium area (50-200m²): 2 points
Small area (0-50m²): 1 point

View of the sea

The objects that one wanted to see from a site may have been as different as marine mammals, swimming terrestrial mammals, fish, birds of prey, enemies, own people or simply the sea itself.

The evaluation of view considers the view from the site itself, and also the view from points that are clearly associated with the site. These points should not be situated more than 10 m from the site-boundary (Fig. 12).

Wide view (>180): 3 points
Medium view (90-180°): 2 points
Narrow view (<90°): 1 point

Distance to fresh water

The distance to fresh water is the linear distance between the site and the closest brook/river or lake. The estimation of this distance is difficult, because the situation today is very much the result of a significant growth of turf during the last 2000 years. Still, the lakes in the relevant areas are all natural basins, and these basins were probably also lakes during the Neolithic. The number of natural basins and lakes was probably much larger than the ones that can be observed today. On the other hand, the growth of turf has created a much more stable situation, because they function as "stores" for water in periods of little rainfall today. In these periods of drought, the lakes and brooks in Neolithic western Norway may have supplied very little water.

On the basis of these considerations, it seems most secure to measure the distance to the closest fresh-water source today:

Short distance (<25m to fresh water): 3 points
Medium distance (26-75m to fresh water): 2 points
Long distance (>75m to fresh water): 1 point

Distance to contemporary sea level

Reliable and detailed shoreline displacement curves have been constructed for all of the relevant areas (see under harbour conditions). It is therefore easy to estimate the location of the contemporary Neolithic shorelines. Here, the lowest levels during the early/middle Neolithic are used as reference points:

Skatestraumen: 3 m above sea level
Fosnstraumen: 7 m above sea level
Bjorøy: 8 m above sea level
Kollsnes: 7 m above sea level
The Stord-Bømlo connection: 7 m above sea level

The most accurate measurement would be the real (horizontal) distance from the site to the shore. However, considering that the Neolithic shorelines in all of the areas were below 10 m above sea level and that the relief of the terrain is generally sharp in all of the areas, a measurement of the vertical distance to the sea is sufficient for this purpose.

It should be kept in mind that only levels up to about 30 m above sea level have been test-pit surveyed. All of the relevant areas are therefore situated close to the shoreline. The possible variability between the areas therefore only reflects a variation within sites that already should be interpreted as shore-bound. The criteria are as follows:

Close (<4 m): 3 points
Medium (4-8): 2 points
Distant (> 8): 1 point

Results and discussion

The analysis shows that there are significant differences in site locations between the areas (see appendix 1 for individual scores at sites). The question is how these differences should be interpreted. As noted above, the four factors representativity, site functions, physiographic conditions and background environment should be considered along possible intra-regional cultural differences. In the following, I will describe and discuss the results for each factor.

Harbour conditions

The harbours at sites at Bjorøy, Kollsnes, and Stord-Bømlo are generally well protected, while several sites at Skatestraumen and Fosnstraumen have medium scores (Fig. 13). In the cases of Bjorøy and Stord-Bømlo, the background environments offer plenty of good harbours. It is therefore not surprising that the sites are protected as well. The results at Kollsnes, however, need other explanations, because the number of protected harbours offered by the Neolithic local environment was more limited. In this area, it is evident that the protected harbours were deliberately chosen for location of sites. Interestingly, the backgrond environments at Fosnstraumen and Skatestraumen are more or less the same as at Kollsnes: there are generally few naturally protected harbours in the area. The large number of sites with medium harbours at these two areas may be interpreted as a result of culturally determined practices in Kollsnes vs. Skatestramen/Fosnstraumen. Nevertheless,

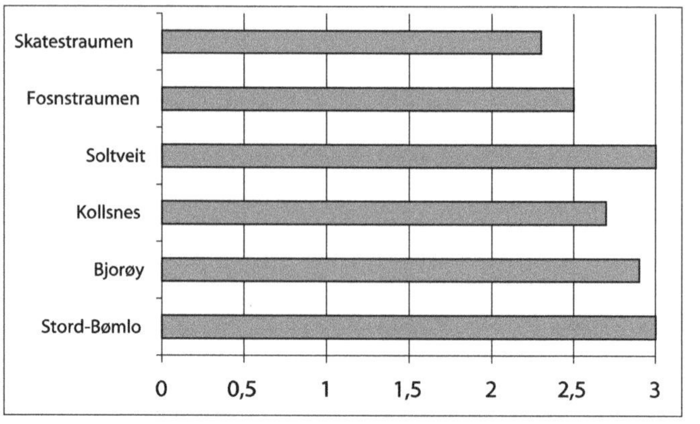

Figure 13. *Average scores for harbour conditions.*

although this option cannot be excluded, I do not find it very likely. It should be remembered that Kollsnes is extremely exposed to sea-waves, which created great risk of lives and boats when entering an landing. Entering and landing was probably not so risky at the more proteced Fosnstraumen and Skatestraumen. The difference between medium and protected harbours may thus have been less distinctive and important in the latter areas. Another point is that Kollsnes was occupied on a short-term basis only. When people arrived, they may have been free to choose the sites in the most protected harbours. Skatestraumen and Fosnstraumen, on the other hand, seem to have been much more crowded: numerous long-term camps have been found. Although people ideally would have preferred the protected harbours, these may not have been accessible, and newcomers/visitors may have been forced to accept less attractive locales. In sum, the different scores of harbour conditions in the different areas are likely to have been caused by other factors than different cultural preferences.

Slope

In contrast to the background environment, which everywhere is dominated by sloping and steep sides, sites in all areas are evenly distributed between slightly sloping and level ground (Fig. 14). This means that level ground was a significant site location factor in all areas.

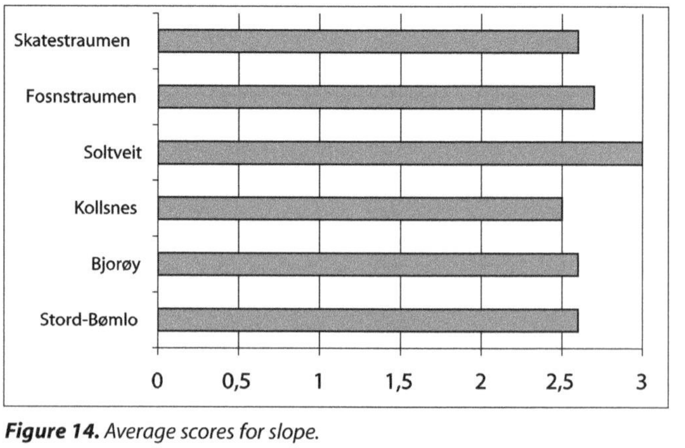

Figure 14. *Average scores for slope.*

Shelter

The Kollsnes sites are generally sheltered, while sites in the other areas, similar to the different background environments, have medium or little shelter (Fig. 15). This indicates that shelter for dominant winds was not significant in other areas than Kollsnes.

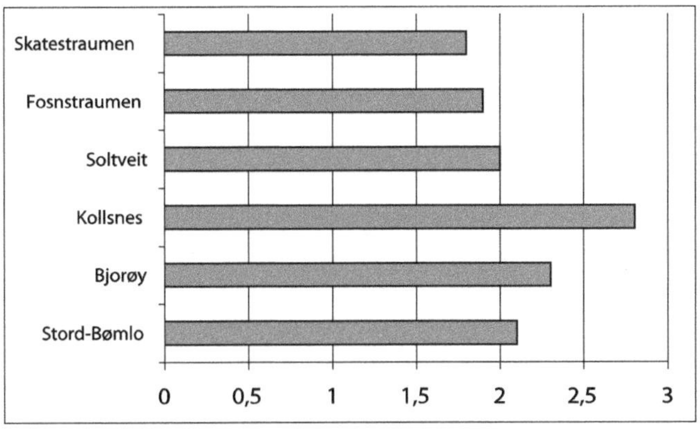

Figure 15. *Average scores for shelter from dominant winds.*

Although shelter might be culturally meaningful, it seems more likely – considering the lack of sheltered sites in the other areas – that good shelter was not significant for the local populations. The good shelter at Kollsnes may instead be seen as a bonus or an unfortunate side effect (mosquitoes) of the choice of protected harbours.

Available site area

While the background environment in all areas is characterised by an almost even distribution of small, medium and large spaces, there is generally much accessible space around the sites in all of the areas (Fig. 16). Availability of a large site area was therefore a significant site location factor in all of the areas.

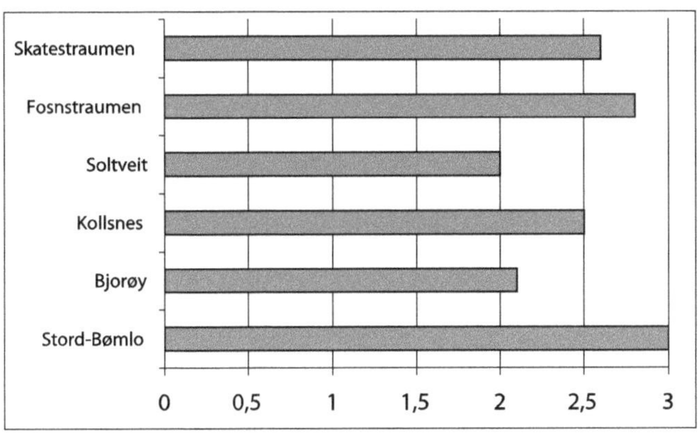

Figure 16. *Average scores for size of available site area.*

View of the sea

View from most of the Stord-Bømlo sites is wide, while most of the sites at Fosnstraumen, Skatestraumen and Bjorøy have medium view of the sea. Sites at Kollsnes generally have little view (Fig. 17). The background environment generally offers good conditions for view at Stord-Bømlo. The high scores for view may therefore not be significant. At Fosnstraumen, Skatestraumen, and Bjorøy, on the other hand, view should be interpreted as a significant site location factor. The difference between Kollsnes and the rest of the local areas is interesting, because the background environment offers similar conditions. Again, I believe that this difference should be related to other factors than cultural ones. At Kollsnes, the protected harbours are all situated at the bottom of deep bays, which are flanked by relatively high, rocky hills. The view is therefore restricted. In other words, if they wanted good harbours, they could not have good view at the same time. This conflict between interests is not as apparent at Bjorøy, Skatestraumen and Fosnstraumen, where the view is wide from the protected sites.

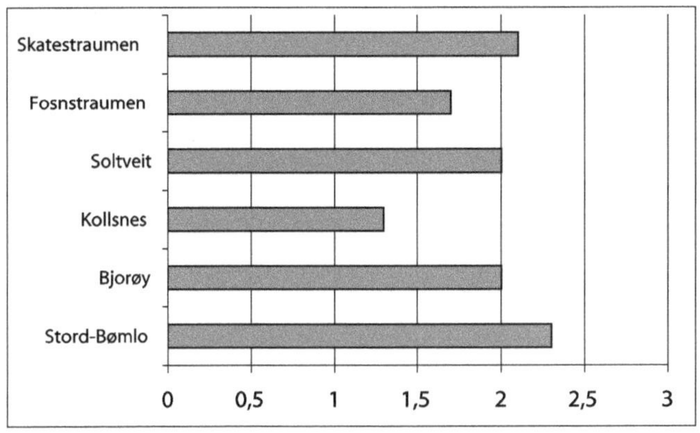

Figure 17. *Average scores for view of the sea.*

Distance to fresh water

The background environments show that Skatestraumen has more fresh-water sources than the other areas (Fig. 18). The results from Skatestraumen, which show that most sites have medium distances, may therefore not be significant. The same may also be the case at Bjorøy and Fosnstraumen, where there are fewer sources, and where distances are generally longer. On the basis of this data, one might suggest that proximity to fresh water sources was not a significant site location factor. A radically different conclusion should then perhaps be drawn from the results at Kollsnes and Stord-Bømlo. Here, most of the sites have short distances to the few fresh water sources that are found. Still, I find it difficult

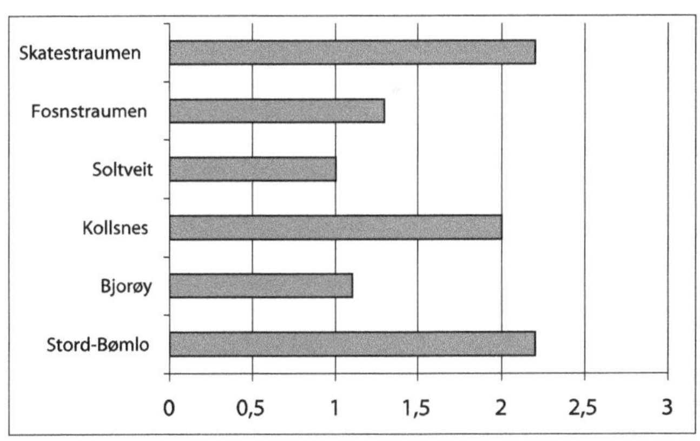

Figure 18. *Average scores for distance to fresh water.*

to accept such a conclusion, considering that the sites in these two areas are solely short-term camps, where people were less dependent on stable fresh water sources than populations at long-term sites. It is perhaps more likely that proximity to fresh-water in these areas was not a particularly important factor; it should rather be seen as a bonus for establishing camps in protected harbours. The best harbours at Kollsnes are deep bays, and this is where most of the creeks run into the sea.

Distance to the shoreline

The background environment in all of the areas except at Bjorøy offers similar comparable conditions for occupation in all levels above the Neolithic shorelines. Despite thorough surveys of levels up to 15-20 m above sea level, very few sites were found at levels above 8 m. This indicates strongly that the shoreline was an important site location factor in all of the areas (Fig. 19).

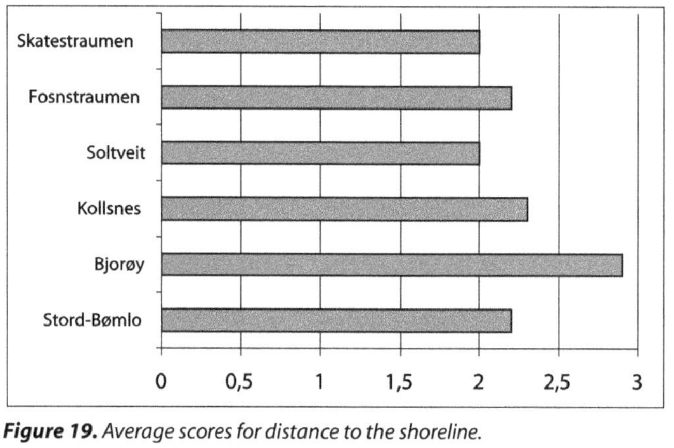

Figure 19. *Average scores for distance to the shoreline.*

There are some differences between the local areas which might be relevant for the problem at hand. The sites at Bjorøy are almost only situated close to the contemporary shoreline, while at Fosnstraumen, Stord-Bømlo, and Kollsnes, the elevations vary between medium and close. At Skatestraumen, a large majority of the sites have medium distances to the shoreline. These differences are interesting, but should probably be explained by differences in the background environment. The levels at the Bjorøy sites may be a result of an over-representation of low-lying possible site-areas (the bay of Nilsvik). The high levels at Skatestraumen may also be a result of environmental differences: The local populations have located their sites higher because of the larger differences between high and low tides.

Conclusion

On the basis of the above discussions, the site locational data indicates that at the macro-level, the Neolithic populations preferred the tidal current channels for location of long-term sites in the entire coastal region. The other types of environment were used primarily on a short-term basis. In my view, this is indicative of marked similarities in subsistence-settlement patterns along the coast (see Chapter 9 for a discussion of these patterns).

The data on the micro-level also indicates great similarities: harbour conditions, view over the sea, level ground, accessibility of large site areas and proximity to the shoreline were important site location factors in the entire region. Shelter for dominating winds and proximity to fresh water seem to have been less important. I find that, although there are differences between the five areas, these differences may be explained by variations in site functions, background environment or physiographic factors. There are no differences that cannot – in some way or another – be explained by such factors. On the basis of the above, it is therefore safe to conclude that the traditions for site locations and the perceptions of the landscape were relatively similar, and that this should be related to concurring subsistence patterns in the five local areas. However, this lack of divergence does not mean that there were no ethnic boundaries within the area. It is quite possible to have similar subsistence practices and still perceive each other as different. In the following chapters, I will examine the lithic data in order to detect such possible differences.

Chapter 5

Early Neolithic raw materials, types, and techniques

Introduction

The data that is analysed in the remainder of this thesis consists of raw materials, types and techniques. These categories are from selected and re-classified units at 37 early Neolithic excavated sites in the area of analysis. In addition, distribution studies of Vespestad adzes, thin-butted axes, and slate knives are presented. The classification of raw materials at the sites followed a reference system. This system will be described, and the raw materials that have been distinguished by the reference system in the present work are presented. Secondly, I will discuss the chronological significance of the types and techniques that are categorised and compared at residential sites and distribution studies.

A reference system for raw materials

All raw materials that are applied in this analysis are classified for the present project according to a reference system for raw materials. The system was developed during the excavations at Skatestraumen, Nordfjord (Bergsvik 1999, Bergsvik 2002a). Below, I will briefly present the background for the system, the development of the method, and how it was applied in the current project.

Previous methods and problems in western Norway

During the early and middle Mesolithic in western Norway, flint deposited by icebergs was the dominant raw material used in the production of stone tools. During the Late Mesolithic, (to the south of Stad) flint was gradually replaced by a variety of alternative raw materials. By the transition to the Neolithic only about 10% of the lithics were made of flint (e.g. Bjørgo 1981, Nærøy 1988, A. B. Olsen 1992). Some of the alternative raw materials were found as water rolled cobbles along river banks and sea shores, but most of them were probably quarried from bedrock outcrops. A few of these quarried outcrops have been identified, and prehistoric utilisation has been documented at four geological sources: greenstone at Hespriholmen and Stegahaugen, Bømlo (S. Alsaker 1987), diabase at Stakaneset, Flora (A. B. Olsen 1981), rhyolite at Siggjo, Bømlo (S. Alsaker 1987) and quartzite at Kjølskarvet II , Lærdal (A. B. Johansen 1978). The outcrops were defined geologically, and these definitions made it possible to (mainly visually) distinguish artefacts made of these raw materials in the collections at the museums (see Chapter 8). Regional studies of rhyolite, greenstone and diabase artefacts demonstrate that they have been spread strikingly far from their sources (A. B. Olsen & S. Alsaker 1984, S. Alsaker 1987).

When the lithics from Stone Age excavations in western Norway are catalogued, these easily recognisable raw materials are distinguished as separate groups. As a result, they make up a data set which is accessible both for intra and inter site comparisons. These four categories represent, however, only a smaller portion of the total amount of raw materials in use. The remainder consists of different qualities and colour variations of jaspers, chalcedonies, quartzes, quartzites, mylonites, cherts, slates, sandstones, basalts, pumices and soapstones. Some of these seem to have been spread over long distances; others appear to have had a more limited range. Their sources, however, remain unknown. This is probably the reason why so little effort has been directed towards distinguishing between them geologically for the purpose of classification. Although there are exceptions (e.g. A. B. Olsen 1992:77), a general problem is that raw materials are classified only very superficially. The term "quartz" is often used for quartz, quartzite and mylonite. Fine blue and coarse green mylonite are also often categorised as one and the same type. Another problem, mainly caused by different levels of geological skills, is that identical raw materials from different sites are categorised as different types. The result of this superficial and inconsistent cataloguing is that important research potential is masked; the data is hard to access and difficult to compare.

The reference system

When the large excavation project at Skatestraumen in Bremanger was initiated in 1992, one of the goals was to create a better system for raw material classification. During the project, more than 100,000 lithic artefacts were collected from 11 excavated and 138 test-pit surveyed sites (Bergsvik 2002a). As a test, a new method, which is termed a *reference system*, was applied on all of the excavated sites and 43 of the test-pit surveyed sites.

There are two basic differences between the reference system and traditional methods. First, it aspires to a more detailed level of classifiation by adding the quality and colour to the geological types. Second it attempts to increase consisteny through identifcation based on hand-samples, by visually comparing the new artefacts with specimens in a reference collection. Below, I will describe in more detail how the system was organised.

Greenstone*	Quartz crystal
Diabase	Fine, white quartz
Basaltic rock	Other fine quartzes*
Flint*	Medium quartz
Fine flint	Coarse quartz
Medium flint	Medium grey mylonite 1
Coarse flint	Medium, grey mylonite 3*
Fine, green quartzite 1	Coarse, grey mylonite*
Fine, brown quartzite 1	Blue mylonite*
Fine, black quartzite 1	Green mylonite*
Other fine black quartzites*	Other mylonites*
Fine, blue quartzite 1	Rhyolite
Fine, blue quartzite 7	Chert
Fine, blue quartzite 8	Jasper
Fine, grey quartzite 2	Red slate
Fine, grey quartzite 3	Green slate
Fine grey quartzite 11	Grey slate
Other fine grey quartzites*	Blue slate 1
Fine, white quartzite 16	Blue slate 2
Medium green quartzite	Slate other
Medium, grey quartzite 1	Anorthosite
Medium, grey quartzite 8	Soapstone
Medium, grey quartzite 9	Soapstone 8
Medium, grey quartzite 10	Sandstone
Medium, grey quartzite 11	Pumice
Other medium quartzites*	Amber
Coarse quartzites	Undetermined rock

Table 1. *Raw materials distinguished by the reference system.*
** Refer to raw material category consisting of several individual raw materials (grouped). See appendix 2 for a complete list of all raw materials distinguished, and appendix 10, Colour Figure 1–3 for colour photographs.*

Three aspects of the raw material were characterised: geological type, quality and colour. In addition, a number was added for individual types. The *geological type* was first determined by the archaeologists. For some of the raw materials, such as slate, pumice, soapstone and sandstone, this was rather easy. Mylonite, quartz and quartzite were more problematic to characterise. As the cataloguing proceeded, I therefore consulted geologists about individual types. The majority of the raw material categories used have also been determined geologically at the Dept. of Earth Science in Bergen (E. Alsaker 2002).

The *quality* of the raw material was often difficult to determine, and was therefore done somewhat intuitively. Most of the time, differences in grain size provide a useful key. However, there are good raw materials at both ends of the spectrum. Fine grained raw materials are usually best, but quartz crystal and hydrothermal quartz, while extremely large grained, are also excellent raw materials for some tool types. The problem with using grain size as an indicator of quality is primarily restricted to portions of the middle group. Quartzites, which are sedimentary sandstone in origin and thus should have identifiable grain/crystals, have sometimes been metamorphosed such that grains/crystals are not always visible in hand samples and therefore cannot be placed precisely on a grain size scale. Some hydro-thermal quartzes have grain structure due to other formation processes. The determination of quality, therefore, was primarily an evaluation of *how effective the raw material is for its supposed range of use.* Three categories were applied: fine, medium, and coarse.

The original *colour* was also difficult to determine, due to often unpredictable post-depositional chemical processes affecting the raw material. David Simpson's documentation of radically different colours of refitted pieces of rhyolite clearly demonstrates this problem (Simpson 1996). Often only the surface of the flakes and tools has been affected by these processes. By cutting some problematic pieces, we were to a large extent able to document the original colour. Seven categories of colour were applied: red, blue, yellow, grey, black, green, and white. In addition to geological type, quality and colour, *a number* was added to characterise sub-categories.

Some of the raw materials in the collection were very difficult to distinguish from each other. The raw materials that caused most trouble, the coarse and middle grained white and grey quartzes and quartzites were therefore represented in the collection as unspecified groups, not as individual types.

If the new reference system is evaluated against the traditional methods for cataloguing in western Norway, two aspects are important to compare: level of detail and level of consistency. As for level of detail, despite the revisions discussed above, the result is a much more detailed catalogue, which probably is much closer to the raw material categories of the prehistoric people, than is the more superficial impression of the lithic raw material use provided by traditional methods. Despite an increase in the number of types, the results are also more consistent.

Using this system, mapping both locally distinct and widely distributed raw materials will be possible. As will be apparent in Chapter 8, this kind of survey can be done whether the exact provenance of the specific raw materials are known or not.

The application of the reference system in this thesis

The reference system was used for raw material classification at all of the 37 sites in the analysis. This implied that when new raw materials were identified at sites outside of Skatestraumen, these were added to the reference collection as new categories. In Colour Figures 1-3, Appendix 10, and Table 1, I present the raw material categories that are applied in this project. Altogether 68.626 artefacts were classified at the residential sites during the present work.

Compared to the original system which was developed during the excavation project at Skatestraumen, significant modifications have been made regarding the level of detail in quality as well as colour and individual numbers. In addition to the groupings of coarse types mentioned above, several other individual types have been grouped under general terms like "other fine grey quartzites" or "grey slate". There are basically two reasons for these modifications of the method. In some cases (mainly fine and medium quartzites and quartz) the grouping was done because the specific raw materials made up fairly small shares. The categories may therefore consist of a number of different raw materials from a variety of sources. In other cases (mainly mylonites, slates and soapstone) it was done because there were difficulties in distinguishing between sub-categories visually. Still, this type of grouping refers to raw materials that probably originate at the same source (the grouped raw material categories are presented in appendix 2).

It may be argued that these groupings reduce the potential of the method, which is to present the regional distribution of only individual raw materials. One is in danger of grouping raw materials that, for example, originate at different quarries in the region, and this will seriously affect the analysis. However, when evaluating the reference system that was developed at Skatestraumen, it must be admitted today that too much "splitting" of visually similar types was done in that project, and this made it difficult to get an overview of the situation (but see Bergsvik 2002a:283). A degree of "lumping" of similar types makes the method more operational, particularly when applied on a regional scale.

In Chapter 7, the regional distribution of the raw materials is analysed, and their provenance is discussed in Chapter 8.

Types and techniques

The distributions of types and techniques make up a significant part of the analysis. It is therefore necessary to describe and define these elements in some detail. Another objective of the presentation is to discuss the chronological relevance of the different traits. Such a discussion is necessary, because several of them will be used in order to date the excavated sites archaeologically.

The analysis includes both eastern and western Norway. At the outset it is therefore important to note that the recent chronological work in these two regions have been performed independently of each other. This has resulted in somewhat different chronological frames. In eastern Norway, the framework is closely related to that of southern Sweden and Denmark, in which the early Neolithic starts about 5100 BP/3800 (calibrated) BC and ends about 4500 BP/3300 BC (Larsson & Olsson 1997). In western Norway, the period starts 100 radiocarbon years earlier, about 5200 BP/4000 BC, and ends about 4700 BP/3500 BC (A. B. Olsen 1992, Nærøy 1993a). The main emphasis of this work is on western Norway. This means that some of the elements that constitute the latter part of the late Mesolithic in *eastern* Norway will be discussed below. The chronology of the types and techniques are presented in Fig. 23.

Vespestad adzes

Vespestad adzes (Fig. 20) are made of different basaltic rocks. According to Nærøy, they "... have arched rectangular, round or oval sections, are shorter than 12 cm, the edge angle is less than 26°, the edge width is larger than 40% of the length; and the BT (width/thickness) proportion is less than 20 (Nærøy 1993a:82-83). Asle Bruen Olsen has also pointed out that some of the adzes have rounded or faceted sides, others have plan sides (A. B. Olsen 1981:142-143). Several of the Vespestad adzes are only superficially ground. Some of them may be difficult to distinguish from the late Mesolithic adzes with blunt necks and round sections (Nærøy 1988:96). Vespestad adzes are separated from the Vestland adzes/ chisels, which have " ... rectangular or slightly arched rectangular sections, are shorter than 12 cm, the edge angle is less than 26°, the edge width is greater than 40% of its length and the BT proportion is greater than 20" (Nærøy 1993a:83).

Nærøy did not find any changes in the frequency of use of this type across these two periods and argued that the Vespestad adzes may have been used in both early and middle Neolithic periods. A. B. Olsen, on the other hand, has argued that while the Vespestad adzes dominated during the early Neolithic, they were less frequent during the middle Neolithic (A. B. Olsen 1981:148; 1992:101). During my own investigations at Skatestraumen, several Vespestad adzes were associated with early Neolithic layers, while none were identified in middle Neolithic deposits (Bergsvik 2002a:289). In this thesis, Vespestad adzes will primarily be considered early Neolithic.

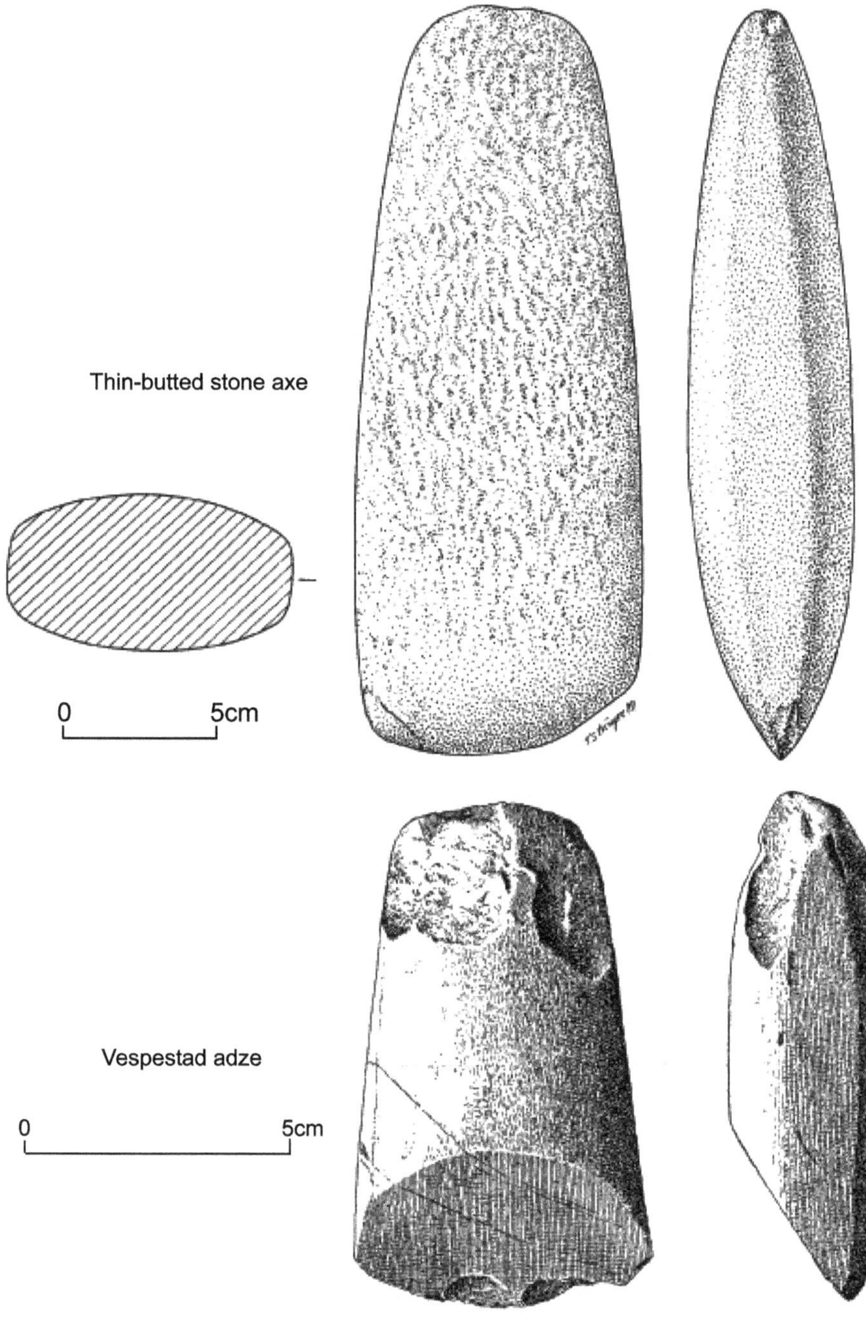

Figure 20. *Early Neolithic axes and adzes. Reproduced from Østmo (1988) and Brøgger (1907).*

There is general agreement that the Vestland adzes and chisels were introduced during the early/middle Neolithic transition about 4700 BP (3500 BC) and that they were used throughout the middle Neolithic A.

Flint axes and stone axes

The early Neolithic flint axes include point-butted and thin-butted types. These two main types are separated into 3 and 7 subtypes respectively. Both main types have arched surfaces straight and usually rounded edges. The point-butted type I has a bi-convex section, all other types have four-sided sections (P. O. Nielsen 1978:65-66, 72-82). Point-butted and thin-butted axes are distinguished from the thick-butted axes in that they lack a regularly formed, plane, neck-surface which is delimited by square angles.

Thin-butted axes (Fig. 20) are distinguished from point-butted axes in that their neck-width is more than 4 cm (P. O. Nielsen 1978:63).

According to Poul Otto Nielsen, the point-butted types can be restricted to the Danish early Neolithic A, dated to between 5100 and 4900 BP (3800-3700 BC). The thin-butted types were used from early Neolithic A, ca 5100 BP until the end of middle Neolithic A II, ca 4400 BP (3000 BC) (P. O. Nielsen 1978:109). Point-butted and thin-butted types were also made of stone. Chronologically, the stone variants are mainly related in the same periods as the corresponding axe-types of flint (Ebbesen 1984).

Tanged blade points and tanged flake points

Tanged points (A-points) are points at which the distal end forms a natural point and where a tang is formed by retouching both sides of the proximal end (Fig. 21).

Tanged points have traditionally been related to early Mesolithic and middle Neolithic assemblages, and were generally held to be absent from the intermediate periods. In 1972, however, Svein Indrelid argued that the tanged points "survived" in the southern Norwegian highlands during the middle- and late Mesolithic. This theory was questioned by others (e.g. Bakka 1976, A. B. Johansen 1976, Mikkelsen 1976), and was rejected by Indrelid himself in a later contribution (Indrelid 1994:187), in which he instead suggested that they were reintroduced during the late Mesolithic-early Neolithic transition. The theory of continuity could not be upheld because of the complete lack of tanged points at the many late Mesolithic coastal sites in western Norway that had been excavated in the meantime. Although tanged points in western Norway may have occurred sporadically during the late parts of the late Mesolithic (Nærøy 1993:91), they were not (re) introduced to any significant degree before about 5200 BP (Olsen 1992, Nærøy 1993b). Another reason for Indrelid's change of opinion was that he first of all related tanged points to the techniques for producing ordinary blades and small blades. These techniques were also absent in southern Norway between ca 9000 and 5200 BP (8200-4000 BC).

Still, the large number of tanged points at sites radiocarbon dated to the late Mesolithic in the highlands (e.g. Indrelid 1973:36 ff) is hard to explain. They should probably not all be related to secondary Neolithic occupations. Recent chronological work from eastern Norway may be relevant for this problem. In 1975, Egil Mikkelsen distinguished a late Mesolithic phase 4 (Kjeøy Phase), dated to between ca. 6300/5600 (5300/4400 BC) and 5000 BP (3750 BC), in which he argued that tanged points were present (Mikkelsen 1975, Lindblom 1984). Later work has confirmed the existence of phase 4, and also of the presence tanged points (Glørstad 1998). According to Håkon Glørstad, the transition between phase 3 (Nøstvet) and phase 4 should be fixed at about 5800 BP (4650 BC). Interestingly, the tanged points in phase 4 and early Neolithic eastern Norway are small and broad and seem to be made from flakes, *not* blades (see examples of illustrations in Østmo 1976, Berg 1995:43).

In my view, a *distinction between blade technique and flake technique* may solve Indrelid's chronological tanged point problem in the highlands. The illustrations of points from excavated sites in the highlands indicate that significant amounts were made from flakes e.g. (Martens & Hagen 1961:29, Odner 1965, Indrelid 1994:189, 191). Earlier authors were aware that tanged points were made from both blades and flakes (Odner 1965, A. B. Johansen 1969, Indrelid 1973:36), but this distinction was not made explicit and was not applied in their chronological analyses, which instead were based on differences in retouch and breadth/length-index. (Ballin & Jensen (1995:189) distinguish between these points in their report, but do not discuss the implications). In his comparative analysis from 1973, Indrelid found that the retouch type changed through time, but the main difference between his Mesolithic and Neolithic assemblages was that the Neolithic tanged points generally were longer and thinner than the Mesolithic points (Indrelid 1973:65, see also Odner 1968). I find it likely that this difference is largely representative of the difference between points made from blades and points made from flakes.

A consequence of the above is that blade technique should not be seen as a premise for the production of tanged points; to say that cylindrical core technique appear ca 5200 BP (4000 BC), is not the same as saying that tanged points are (re) introduced this late. Tanged blade points and tanged flake points had different chronological significance as well as different regional distribution patterns (see Chapter 7). Consequently, it is necessary to distinguish between these different types. In the following, such a distinction is made on the basis of an evaluation of whether the points are produced from blades or flakes.

In sum, it appears that tanged flake points were (re) introduced between 5800 and 5200 BP (4650-4000 BC), while tanged blade points were introduced about 5200 BP (4000 BC). Both types were used throughout the middle Neolithic period, and possibly also during the late Neolithic (Indrelid 1994:188). It should be noted that blade points constitute a significant element in the middle Neolithic "pitted-ware culture" (e.g. Becker 1951).

Single-edged points

Single-edged points (Fig. 21) have two retouched sides. The shortest side retouch is shorter than 60% of the longest retouch (which separate them from transverse points). They were produced from both blades and flakes (Helskog et al. 1976:25).

Single-edged points were (re) introduced in the late Mesolithic phase 4 (Mikkelsen 1975, Lindblom 1984). Glørstad (1998) has documented them from about 5500 BP (4300 BC). They were frequently used in the early Neolithic, and have also been found at middle Neolithic sites (e.g. Skjølsvold 1977, Østmo 1985).

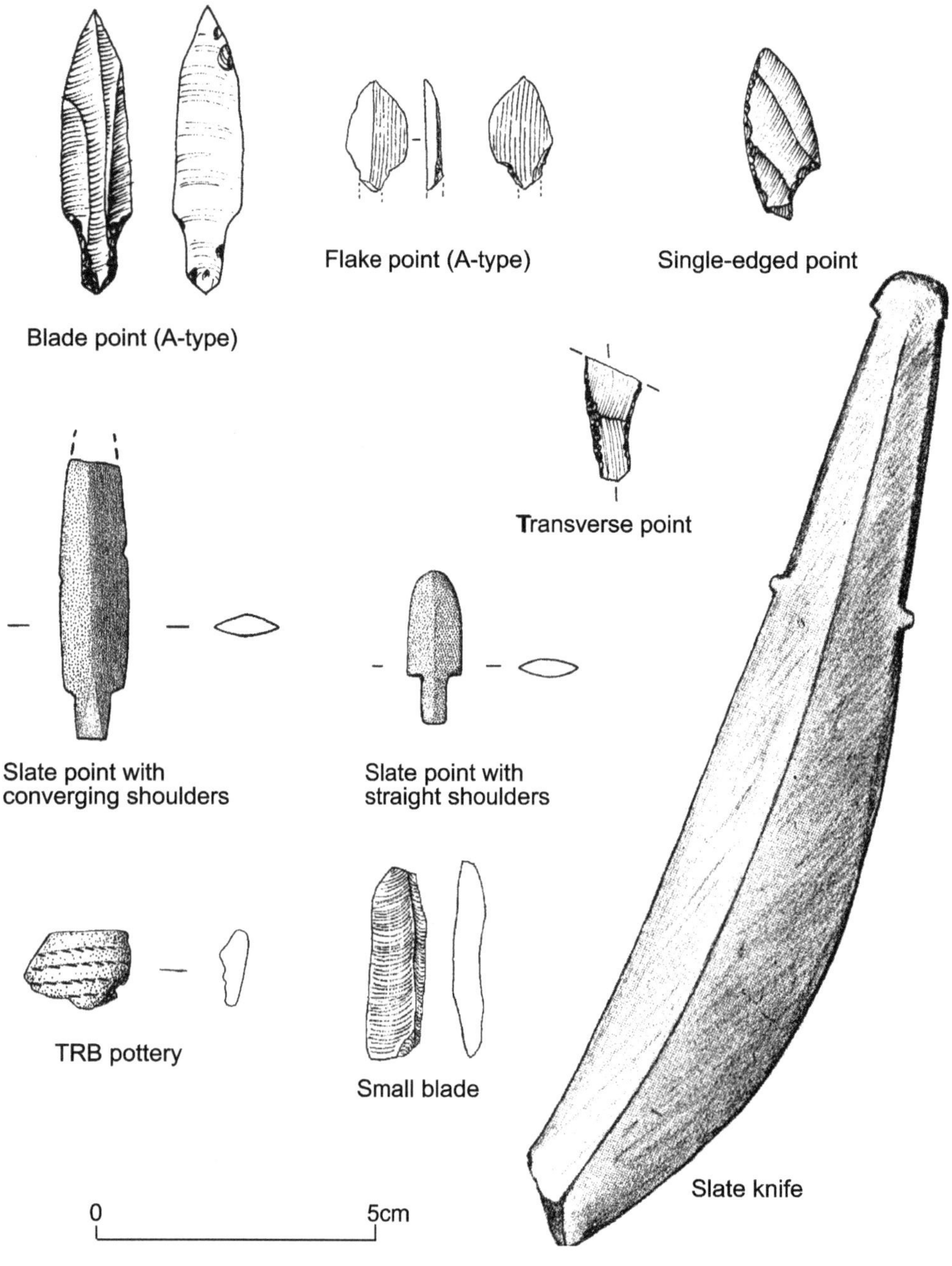

Figure 21. *Early Neolithic projectile points, slate knife and TRB shard. Reproduced from Brøgger (1909), Martens & Hagen (1961), Helskog et al. (1976), Olsen (1992), Bergsvik (2002).*

Transverse points

Transverse points (Fig. 21) have two retouched sides from the base to a straight or oblique edge. The shortest side retouch is longer than 60% of the longest retouch (which separate them from single-edged points) (Helskog et al. 1976:25). In Norway, they are mainly made from flakes.

Transverse points were introduced in the late Mesolithic phase 4 (Mikkelsen 1975, Lindblom 1984). Glørstad (1998) has documented them from about 5800 BP (4650 BC). They were used throughout the early Neolithic, and they occur sporadically in middle Neolithic contexts (Indrelid 1994:184, Persson 1998).

Slate projectile points

Different types of slate points are distinguished based on attributes such as the shape of the blade-section, outline of the blade, and presence/absence of tang (e.g. Nærøy 1993b). These attributes have chronological significance.

In Finland, Sweden and northern Norway, slate points were used during the late Mesolithic. In southern Norway, however, they were introduced ca 5200 BP (4000 BC). They appeared throughout the middle Neolithic, and may have been used also in the late Neolithic. The main chronological work in southern Norway has been performed in the Nordhordland district, western Norway (A. B. Olsen 1992; Nærøy 1993). Several typological attributes are characteristic of early Neolithic and middle Neolithic points respectively. The only qualitative criterion is the shape of the shoulders; the middle Neolithic points have hanging shoulders, while this element is not found in the early Neolithic. The other differences between early and middle Neolithic points are only based on quantitative tendencies: The majority of the early Neolithic points are comparatively small, they have bi-convex blade sections, converging outlines and square or contracting shoulders (Fig. 21). The middle Neolithic points are generally longer, most of them have diamond shaped cross-sections and parallel sides. Several have square or contracting shoulders. Notches along the ridges are restricted to the middle Neolithic B slate points (Bergsvik 2002a:288).

The Pyhensilta/Nyelv point was not treated by Nærøy and A. B. Olsen. This type has a rhombic or lens-shaped cross-section, has a flattened and tapering base and a ground basal facet, triangular in shape (Meinander 1939). In Finnmark, northern Norway it was mainly early Neolithic (Helskog 1980), in Finland it appeared much later, during the late Neolithic/early Bronze age (Edgren & Törnblom 1992), while in southern Norway it is found at both early, middle and late Neolithic sites (e.g. Bøe 1942, Østmo 1984, Indrelid 1994). It is therefore not possible to date the Pyhensilta/Nyelv point more precisely than generally to the Neolithic period.

Slate knives

Slate knives are both single-edged and double-edged. Six main types are distinguished by Hans Christian Søborg (1986; 1988); three single edged (narrow bladed, broad bladed and short-bladed) and three double edged (asymmetrical, near-asymmetrical and completely symmetrical) (Fig. 21).

The typological-chronological tendencies are discussed by both Søborg and Ramstad (1999). The knives were introduced during the late part of the late Mesolithic. Søborg holds that the knives were used between 5000 and 3500 BP (3750-1800 BC) south of the Polar Circle. Ramstad argues that, except for the late Neolithic-Bronze Age short-bladed type, the other types were used during the early and middle Neolithic periods, between ca 5300 and 4200 BP (4100-2800 BC).

Pottery

The early Neolithic (TRB) pottery in southern Scandinavia have S-shaped profiles (funnel-shaped), thin walls, and fine grained temper, and are burned in relatively high temperatures. They are first of all decorated by impressions of real cords, but impressions of cord stamps, mussels, fingernails, and sticks are also found. The decoration appears mainly around the rims on the early vessels, while later decoration covers more of the surfaces (e.g. Madsen & Petersen 1984, Larsson 1985, Koch 1998). Shards of early Neolithic TRB-type vessels (Fig. 21) have been found in southern Norway at the sites Ramsvikneset, and Kotedalen, Nordhordland (A .B. Olsen 1992, Bakka 1993), Sumtangen, Hardangervidda (Bøe 1942), and Gjellestadvige, Rogaland (Bang-Andersen 1981). Middle Neolithic TRB types have been found at Narestø II, Telemark (Nummedal & Bjørn 1930, Mikkelsen 1984, Amundsen 2000), and Rognlien, Telemark (Ingstad 1970, Mikkelsen 1989, Amundsen 2000).

Another type of pottery found at Neolithic sites in southern Norway appears to have been locally produced in the different districts. The date of this pottery is still not settled. Considering that this artefact type might represent a very useful variable for evaluating the problem of this thesis, an extensive discussion on the dating of the locally produced pottery will appear below.

These local vessels have different types of profiles, and thick walls. They are tempered with different types of sharp-edged small rocks, and they are burned in relatively low temperatures. The vessels are mainly decorated by impressions of cord stamps and sticks. The decoration is found on the inside as well as on the outside of the vessels (e.g. Skjølsvold 1977, Nærøy 1987, A. B. Olsen 1992, Åstveit 1999, Amundsen 2000). Recently, direct radiocarbon AMS-dating of the organic coating (food crust) from a number of the locally produced middle Neolithic vessels from the sites Auve (Vestfold), Slettabø (Rogaland), Kotedalen and Ramsvikneset, Hordaland, Narestø I and II (Telemark), and Eg (Vest-Agder) have produced unexpectedly early results. A significant number of the dates are early Neolithic (Østmo 1993:46, Glørstad 1996:45, Åstveit 1999:58, Amundsen 2000:61, 68). These results may indicate that production of local pottery started in southern Norway already during the late Mesolithic-early Neolithic transition, i.e. that the local populations imported pottery and produced their own, regionally distinct types at the same time. Obviously, these results are intriguing for the problems that are explored in this thesis. Nevertheless, contextual as well as methodological problems have to be considered when evaluating these early radiocarbon dates.

The contextual problem relates to the fact that none of the directly dated shards of the local types, *nor* any other shards of local types have been found in contexts that are dated to the early Neolithic. They are all found at sites or in layers that are typologically dated or (conventionally) radiocarbon dated to the transition to the middle Neolithic or to the middle Neolithic period. If local production for pottery really took place earlier than 4700-4800 BP (3500-3400 BC), one would expect to find at least some shards at the many early Neolithic sites that have been excavated in southern Norway. This indicates that local production of pottery did not start before 4700 BP (3500 BC).

One methodological problem is that the samples may have been contaminated by older carbon. This is most likely the case for one of the food crust samples from Narestø II, which produced a date to 6185±90 BP (Amundsen 2000:61, 62). This source of error might also have affected other samples that have less deviation than this particular result.

Another methodological problem relates to possible marine elements in the sample material. Due to the reservoir-effect, the dates may be significantly older than the real age of the sample. A possible marine element is indicated by a high C13-content in the sample-material (e.g. Persson 1999:31-39). The C13-content was low for all of the dated sherds from Norway, which may indicate an exclusively terrestrial content in all of the food crusts. This again might indicate that the old results are correct. However, new chemical analyses show that marine remains are in fact present in many of the samples. At Auve, only terrestrial animal fat was found (Isaksson 1996), however, at Ramsvikneset and Kotedalen, remains of marine fat were traced in three, possibly four from a total of six samples (Isaksson 1999, Åstveit 1999:56-57). On the basis of these results, Åstveit adjusted for the possible reservoir-effect by subtracting 440 radiocarbon years from all of the dated samples with marine content. When this was done, the oldest date of locally produced pottery overlapped with the conventional dating from middle Neolithic phase 14 at Kotedalen. Åstveit carried out the same procedure for the Slettabø-dates, and the result was then in accordance with Skjølsvold's conventional dating of layer III to the early-middle Neolithic transition. If this procedure is performed at the dates from Auve, Narestø, and Eg, the results would correspond with the conventional/archaeological dates at these sites as well.

Nevertheless, both Amundsen and Åstveit have argued that, even if the possible reservoir-effect and other sources of error are accounted for, some of the results are still older than expected. In their opinion, a reasonable conclusion is that production of local pottery started during the early Neolithic period, about 5000-4900 BP (3750-3600 BC) (Åstveit 1999:59, Amundsen 2000:48).

In my view, however, the above discussion demonstrates that there are still many unsolved methodological problems related to direct dating of food crusts. Consequently, I therefore hesitate to accept their conclusion until these problems are controlled and the method is refined, *or* locally produced pottery is documented in conventionally dated early Neolithic contexts. Until new methods or better data is available, the only convincing early Neolithic pottery is the TRB-type of sherds referred to above. In line with A. B. Olsen (1992:142), I argue that the first local production of pottery in southern Norway started during the early-middle Neolithic transition, about 4800-4700 BP (3550-3500 BC). The "locally produced" pottery will, therefore, not be analysed below.

Blade technique

A blade should, according to Helskog et al. (1976) have approximately parallel sides over at least 2/3 of the length and/or the sides should be approximately straight. The length/breadth relation should be at least 2:1 or more, and there should be one or more ridges parallel to the sides. Blades are produced by different techniques, and the blade types vary accordingly.

Blades were applied directly as scraping tools and knives or were transformed into projectile points.

The cylindrical core technique is restricted to the Neolithic period. It was most likely performed from opposite platforms on cores that were prepared beforehand to have approximately perpendicular angles between the platforms and the sides. Most of the cores/blades are between 4 and 6 cm long. If two or more blades were produced from each side, the core received a cylindrical form. Blades produced by this technique are usually between 8 and 14 cm wide, have thick proximal ends and have characteristic triangular cross-sections (3-5 millimetres thick) (Fig. 22). The blades were used directly as knives or scraping tools, or they were converted into blade points (see above). The technique appeared ca 5200 BP (4000 BC). It was practised through the middle Neolithic period (A. B. Olsen 1992, Bakka 1993, Nærøy 1993a), possibly also during the late Neolithic (Indrelid 1994:188).

During the Neolithic, blades were also produced from single faced cores with two opposing platforms which have acute edge angles. The blades produced by this technique have the same length/breadth variation as the blades produced from cylindrical cores, but are generally thinner (ca 2 millimetres) (Simpson 1996). This technique has, until recently, only been related to the early Mesolithic (e.g. Bjerck 1986), but Simpson's investigations at Krossnes, Hordaland, indicate that it was re-introduced in the early Neolithic. The Krossnes example may, of course, be considered an anomaly. However, judging from the large portion of thin Neolithic blades, one may perhaps expect that a number of them have been produced from single faced cores.

In this analysis, blades will be used as a Neolithic chronological marker. It is therefore important to distinguish blades produced from cylindrical core technique from blades produced by Mesolithic unifacial technique, bipolar technique and micro blade technique

Neolithic single faced cores may be difficult to separate from those produced from early Mesolithic cores of the same type. However, the latter cores and the

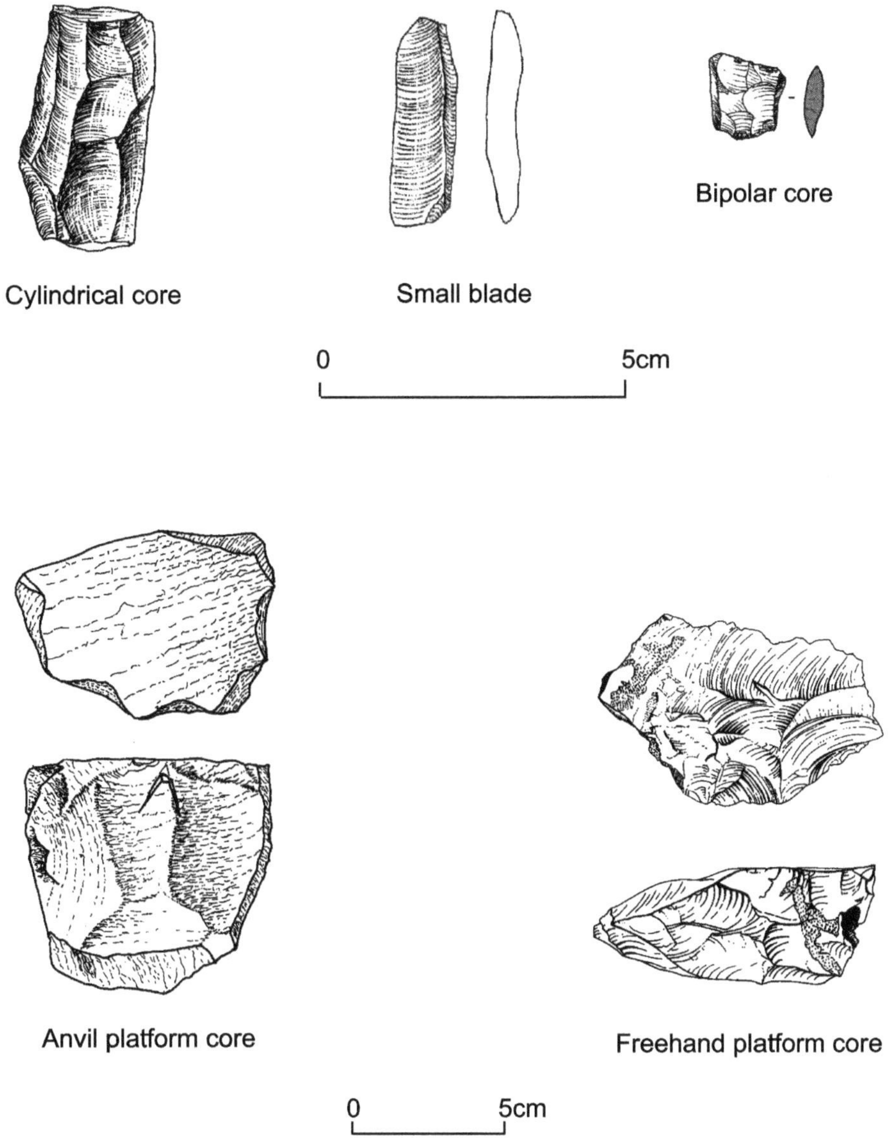

Figure 22. *Blade and cores (techniques). Reproduced from Bjerck (1986), Callahan (1987, Olsen (1992).*

corresponding blades tend to be large (Bjerck 1986:110). In the current analysis, the blades are divided into three groups according to size: ordinary blades >12 millimetres, small blades 8 >< 12 millimetres and micro blades: <8 millimetres. Considering that early Mesolithic blades are predominantly ordinary blades, ordinary blades will not be used as a chronological marker for the Neolithic.

The micro blade technique, which probably was performed by pressure technique on conical cores or handle cores, results in narrow and very even blades (Bjerck 1986:110). The blades also appear to be quite thin, presumably because of the pressure technique. The blades produced by this technique are easy to distinguish from the other types of blades.

Although first of all regarded as a flake technique (see below), bipolar technique may also result in blades or blade-like flakes. It was performed by holding one end of the core against an anvil stone while striking the other end with a hammer stone. It produced irregular blades and flakes. The cores received an elliptic form with traces of crushing at each end (Fig. 22). The technique was practised throughout the lithic period, but was particularly common during the late Mesolithic. Bjerck (1986:116) has argued that a fairly large portion of the late Mesolithic blades were produced by this technique. There might be a problem in distinguishing between Mesolithic "bipolar" blades and blades produced by Neolithic techniques. However, according to Bjerck (1986:115), the majority of late Mesolithic blades, disregarding whether they are the results of the micro blade technique or the bipolar technique, are less than 8 millimetres wide. Thus, blades measuring less than 8 millimetres can not be used as a chronological marker for the Neolithic.

As a result of this discussion, it may be concluded that the Mesolithic blade industries mainly produced ordinary blades and micro blades, while the Neolithic industries produced mainly small blades. Consequently, a dominance of small blades at a site may only be used as a chronological marker for the Neolithic.

Chronological framework for Southern Norway

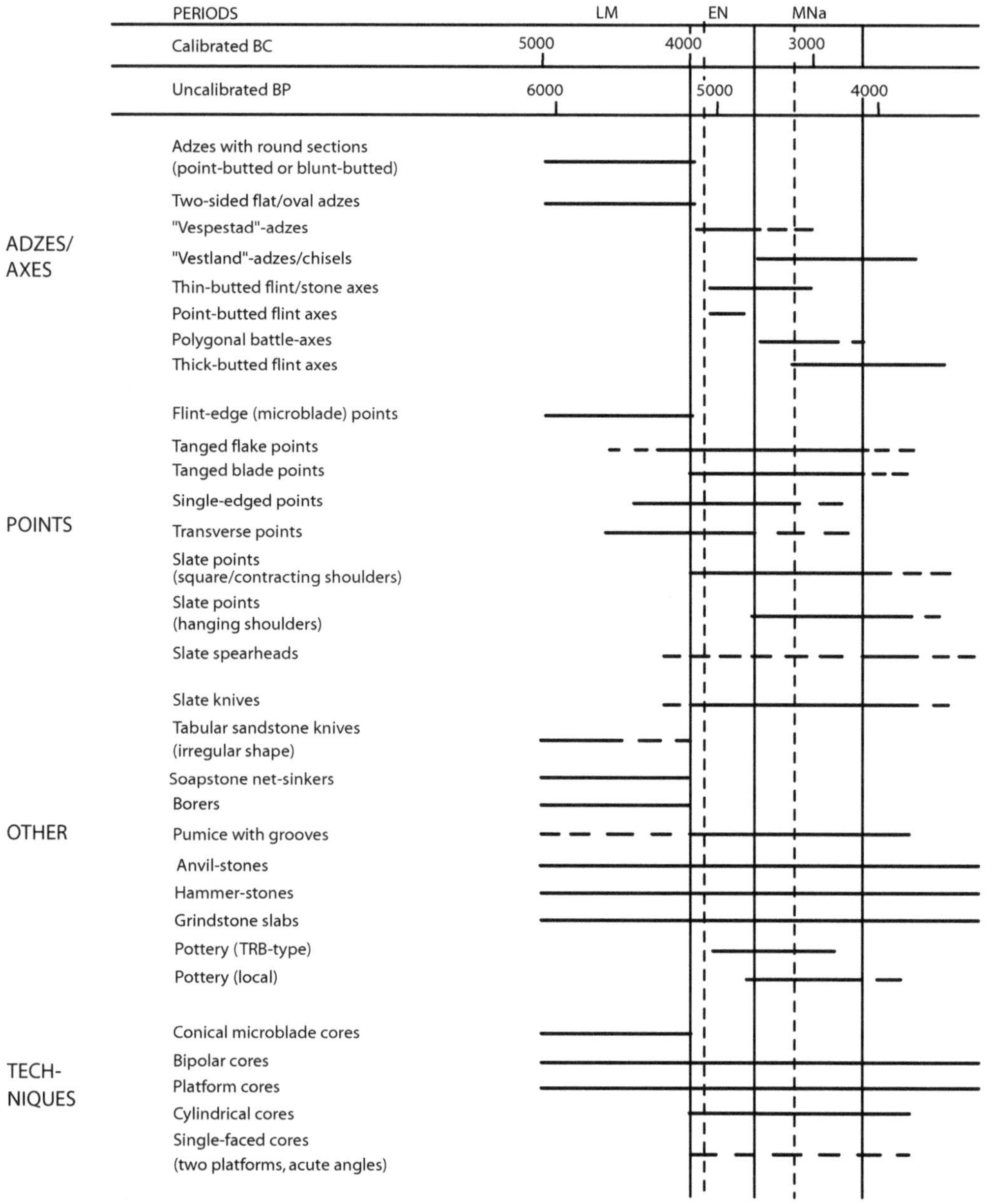

Figure 23. *Chronological typological and technological framework for southern Norway from the latter part of the late Mesolithic throughout the middle Neolithic A. The diagram only shows secure (solid line) and insecure (broken line) presence, not relative frequencies. The chronological divisions in western Norway is drawn with a solid line, while those in eastern Norway is drawn with a broken line. The Mesolithic framework is based on several authors (Mikkelsen 1975, Bjørgo 1981, A. B. Olsen 1981, Bjerck 1983, Gjerland 1984, A. B. Olsen & Alsaker 1984, Bjerck 1986, Alsaker 1987, Nærøy 1987, A. B. Olsen 1992, Nærøy 1994, Glørstad 1998, Bergsvik 2002). For references on the Neolithic chronology, see text.*

Flake technique

The flake technique is mainly practised in three ways: by freehand percussion, anvil percussion, and bipolar percussion (Callahan 1987:15, 16). These different techniques resulted in cores and flakes of different shapes and sizes (Fig. 22). The flakes were used directly as knives or scrapers or they were converted into flake-points, single-edged points, and transverse points (see above). The technique was practised throughout the lithic period.

For the purpose of this thesis, I operate with the easily recognisable bipolar core as a particular type. However, distinguishing between cores that have been made from anvil percussion and freehand percussion is more difficult. The terms for different platform cores and the category "other cores" that are applied in this thesis may thus be the results of freehand percussion as well as anvil percussion (some of the platform cores may, in fact, also be related to early stages of the cylindrical core technique).

I distinguish between large (>4 cm), medium (><4-1 cm) and small (<1 cm) flakes. The flakes are not related to any of the core techniques.

Slate production

Production of slate knives or projectile points is done (1) by preparation of quarried blocks either by striking or sawing the blocks into the desired shapes and (2) by grinding the blanks into the finished products.

Adze production

Production of axes/adzes took place by (1) preliminary coarse flaking (with large hammer stones) of quarried blocks in order to produce rough-outs, (2) secondary reduction of rough-outs into blanks by smaller hammer stones, and (3) grinding on grinding slabs into desired axe/adze shapes (A. B. Olsen & S. Alsaker 1984:81-82).

Conclusion

The goal of this chapter was to present the raw materials, types, and techniques that are compared in the analysis. In the first section, I presented the reference system that is used for classification of raw materials. Compared to earlier methods in western Norway, this system produces a more detailed, consistent, and reliable result. The raw materials distinguished by this system in the current work are presented in Appendix 10.

In the second section, the early Neolithic types and techniques were described and frames of dating were discussed. For the majority of the elements, I supported a previously established framework, but a relatively thorough discussion on classification and dating was needed for some types and techniques. For the purpose of this thesis, I distinguished between blade points and flake points of the A-type. It was also held that dominance of small blades could be applied as a dating element for the early Neolithic period. Relating to the recent discussion of the early dating of locally produced pottery, I generally rejected these early dates and suggested that the production did not start until the transition to the early Neolithic. The frames of dating for types and techniques applied here are presented in Fig. 23.

Chapter 6

Excavated Sites

Introduction

In this chapter the excavated sites are described. A short presentation of general site data is found here, and there is a discussion of dating and site function. When comparing sites on a regional scale, there is always a danger of comparing elements that are of different orders, not representative or which are not adequate for the issue at hand. The sites were selected on the basis of five criteria. They should: (1) be archaeologically dated to the early Neolithic, (2) be functionally classified as residential sites or field camps, (3) have representative sizes of analysed units, (4) be excavated by using similar techniques, and (5) be distributed relatively evenly throughout the area of analysis.

Before the site-presentation, I will describe the methods that were applied in order to date the sites and to classify them functionally. After the presentation, I will evaluate how the sites correspond to the above criteria.

Dating methods

The dating of the excavated sites is partly done on the basis of previously existing chronological work on types and techniques (see Chapter 5), partly on the basis of radiocarbon determinations.

For several of the sites, archaeological dating is the only method applied. This is problematic for two main reasons: first of all, sites in southern Norway were usually frequently re-occupied; few of the selected units can be regarded as having been deposited exclusively during the early Neolithic. Secondly, as one may see from the chronological work that has been undertaken in southern Norway (Fig. 23), none of the elements (except possibly for the Vespestad adze) are restricted to the early Neolithic alone. Some types overlap with the late Mesolithic period, others with the middle Neolithic. The challenge is thus to evaluate to which degree older or younger elements – that do not overlap with the early Neolithic – are present. If

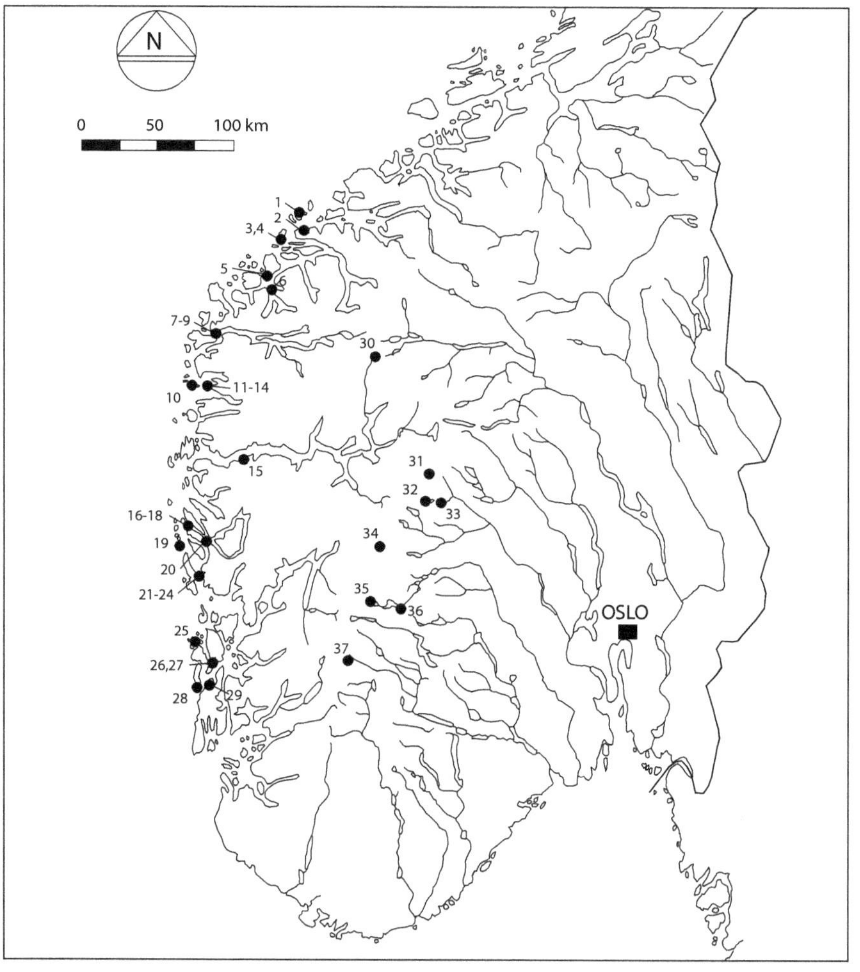

Figure 24. *Southern Norway. Distribution of excavated sites used in the analysis of raw materials, types, and techniques. The numbers refer to list in Table 2.*
1: *Nordøy grendahus*
2: *Synnaland*
3: *Valderøya vest 13*
4: *Valderøya vest 26*
5: *9 Bustadvika*
6: *Holvikshaugen*
7: *17 Havnen*
8, 9: *1 Haukedal*
10: *Kleiva*
11: *Botnaneset III*
12, 13: *Vikja I*
14: *Neset II*
15: *Masnesberget*
16: *Snekkevik 1*
17, 18: *Kotedalen*
19: *Torsteinsvik 11*
20: *Flatøy VIII*
21: *17 Nilsvik*
22-24: *4 Nilsvik*
25: *Austvik III*
26: *88 Føyno*
27: *115 Føyno*
28: *Sokkamyro*
29: *3 Tjernagel*
30: *Styggvasshelleren*
31: *Mørkedøla I*
32: *Gyrinos IV*
33: *Blånut IV*
34: *760 Finnsbergvatn*
35: *526 Nordmannslågen*
36: *1020 Bjornesfjorden*
37: *Vivik*

late Mesolithic or middle Neolithic elements are frequent, the site cannot be dated primarily to the early Neolithic: a broader frame of dating has to be applied. If such elements are few, a mainly early Neolithic date may be suggested.

Sites that are radiocarbon dated might be easier to determine. The C14-method is, however, pregnant with problems of its own. A radiocarbon dating cannot always be trusted to be a real age determination of a site or a layer. The C14-results are particularly problematic when the samples are taken from sites that have a complex stratigraphy. Many of the sites in this analysis have several occupational layers. This problem might have been solved by dating several samples from the same layer, but such a procedure has only been carried out at a few sites. Below, radiocarbon determinations will, therefore, only be considered in cases where there are few elements that are not typologically and technologically dated to the early Neolithic. When such elements are few, however, the result of the radiocarbon dating will be used as an indicator of a more specific date; whether the unit was deposited during the early, middle, or late part of the early Neolithic period. In the site presentation, only uncalibrated dates BP are referred to. Calibrations of the individual dates are presented in appendix 4. All calibrations have been performed by using the programme OxCal version 3.8. (Bronk Ramsey 2000). Atmospheric data is from Stuiver et al. (1998).

Site functions

There is a relatively large variety in site types and site functions among hunter-gatherers and hunter-fisher populations in general. One usually distinguishes three main categories of sites: "residential sites" (sites used by households for a relatively long time), "field camps" (temporary operational centres for small, mobile households or crews), and "special-purpose sites" (hunting stands, observation points, slaughter sites, adze-workshops and caches) (e.g. Binford 1980). During the last few years, most of these site-types have been identified in southern Norway (e.g. Bjerck 1990, A. B. Olsen 1992, Indrelid 1994, Boaz 1998, Nærøy 2000). Consequently, it is expected that there is a great variety of site-types to choose from in the area of analysis.

For the purpose of the analysis, which is to compare the regional distribution of raw materials, types, and techniques, residential sites and to a minor degree field camps probably provide the best contexts. The reason is that at these types of sites, many different activities took place, and this resulted in a generalised or varied set of raw materials and tool types.

Such a generalised pattern would represent the broadest section of the material culture at a given time. Although a hunting-stand, a adze-workshop or a slaughter-site in some respects might be informative for the problem at hand, the raw material use and typological/technological

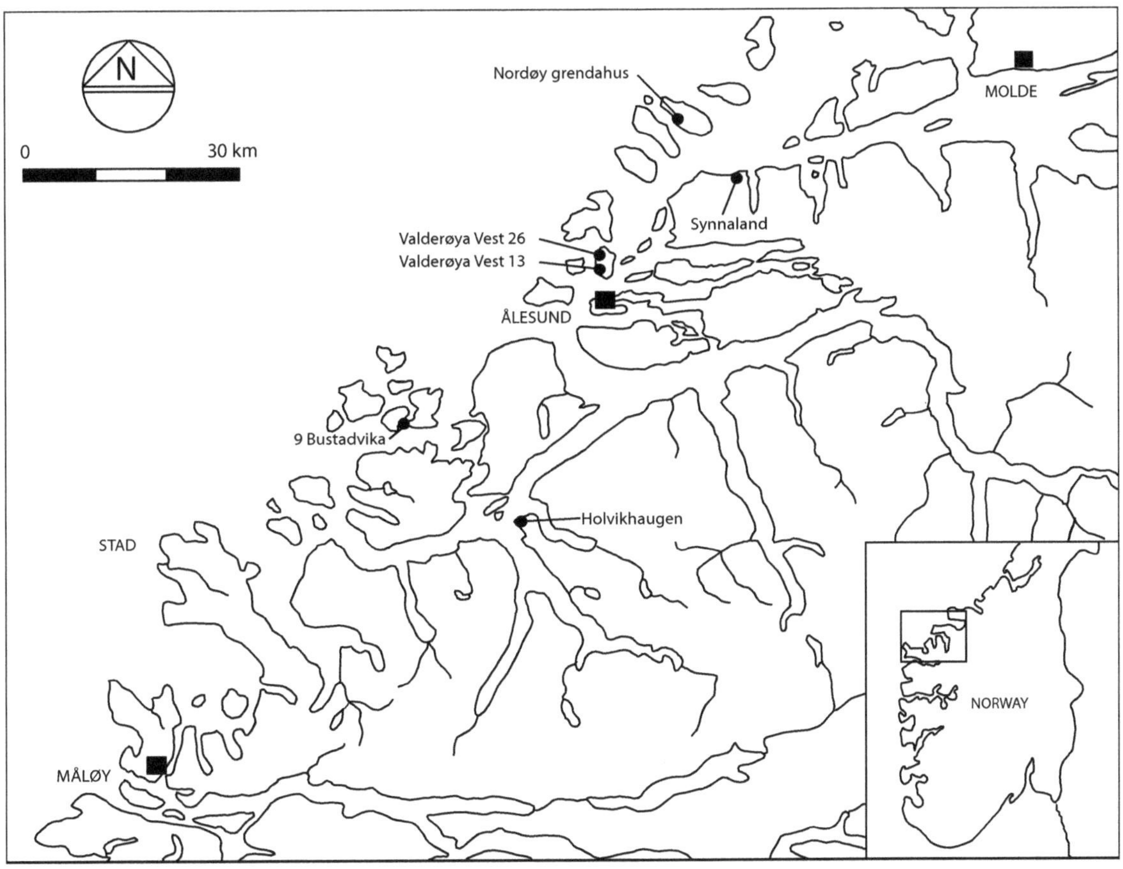

Figure 25. *Sunnmøre. Distribution of excavated sites used in the analysis.*

variation at special purpose sites would be skewed considerably compared to the general patterns at the residential sites and field camps. When selecting sites for the analysis, I have primarily looked for residential sites and secondarily field camps. Special purpose sites are avoided.

A problem is that distinguishing between these different types of sites is not always easy. Residential sites are usually quite distinctive because of thick layers and large sizes. On the other hand, as Lewis Binford (1982:16 ff) has pointed out, such layers may sometimes be the results of redundancy of a large number of short-term occupations. I find it likely that many of the sites with thick cultural layers on the coast of western Norway were used a number of times on a short-term basis as well. However, several of them clearly also localised sedentary residential camps for long periods. A case in point is Kotedalen (see below), where archaeological, faunal, and botanical data indicates that the site was used on a sedentary basis in the late Mesolithic and early/middle Neolithic periods (A. B. Olsen 1992, Bergsvik 2001b).

The differences between field camps and special purpose sites are less obvious. Ideally, an interpretation of site-type should be based on detailed analysis of the individual sites. Such detailed studies have been performed for some of the sites in this study (A. B. Olsen 1992, S. Kristoffersen 1995b, Trones 1998, Bergsvik 2002a), but was impossible to carry out for all of the excavated sites. Still, there are other means by which one may – at least at a general level – approach the problem of site-types. One way is to investigate the site locations, another is to quantify the archaeological features.

Site locations

As shown in Chapter 4, quite a few sites interpreted as "long-term camps" and "short-term camps" have been surveyed and excavated in western Norway. At the macro-topographical level, only long-term sites were found at tidal current channels, while short-term sites were found in all of the test areas. Although long-term sites scored highest on the micro-level, short-term camps also had high scores on factors such as harbour conditions, view

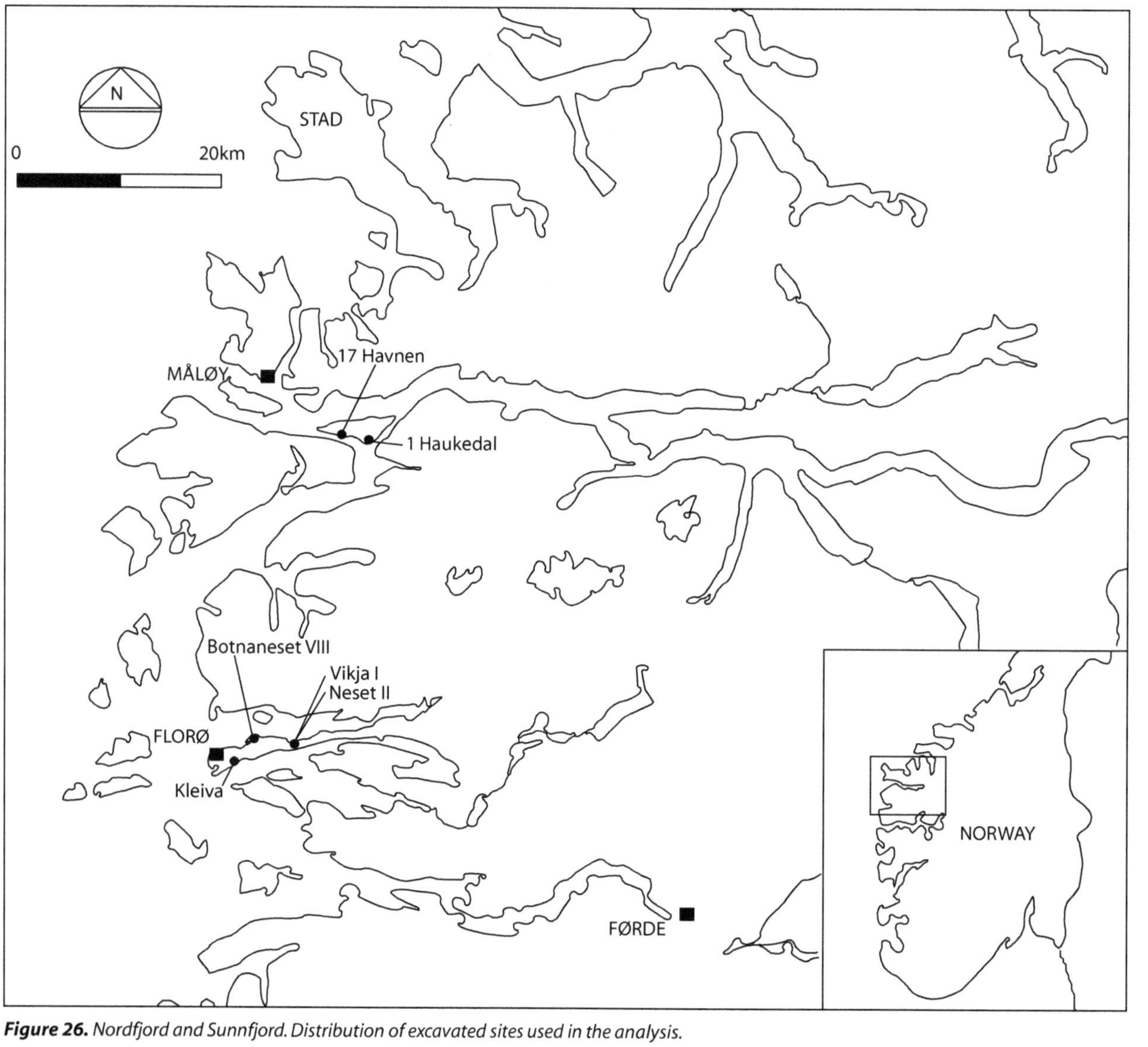

Figure 26. *Nordfjord and Sunnfjord. Distribution of excavated sites used in the analysis.*

of the sea, proximity to the shoreline, proximity to fresh water, and level and drained ground.

It was assumed that the long-term camps could probably be generally interpreted as residential sites, while the short-term camps may have been both field camps and special purpose sites. With regard to the short-term camps, however, I find it likely that the majoriy of these were probably field camps rather than special purpose sites. One would expect that if most of them were of the latter category, their micro-locations would be characterised by their particular functions; slaughter sites would be situated where the game had been taken down, which, in principle, could be at any location, and observation points would be expected to primarily have a good view. The fact that the large majority of the short-term camps in Chapter 4 have locations that are generally good indicates, in my view, that they were field camps left by households or crews who occupied them on a temporary basis. I therefore suggest that favourable site locations on a macro-level as well as a micro-level may indicate that the site belongs to the residential site or field camp categories.

Ideally, a site location analysis should have been performed on all of the excavated sites that are presented below. However, such an investigation must be performed in the terrain, and it was not possible to visit all of the sites in this analysis. In the following presentation of sites, these factors will partly be evaluated from maps, partly from observations made by the excavators.

Admittedly, favourable site locations may also characterise a number of different special-purpose camps. Thus, it is not sufficient to distinguish between residential sites, field camps and special purpose sites on the basis of topographical variables only. Archaeological variables also have to be considered.

Archaeological features

On the basis of excavated and test-pit surveyed sites at Fosnstraumen, Nordhordland, I have previously recorded a set of archaeological characteristic features at 45 Neolithic surveyed sites (Bergsvik 1991). This information was used in order to distinguish between long-term camps and short-term camps (see Chapter 4).

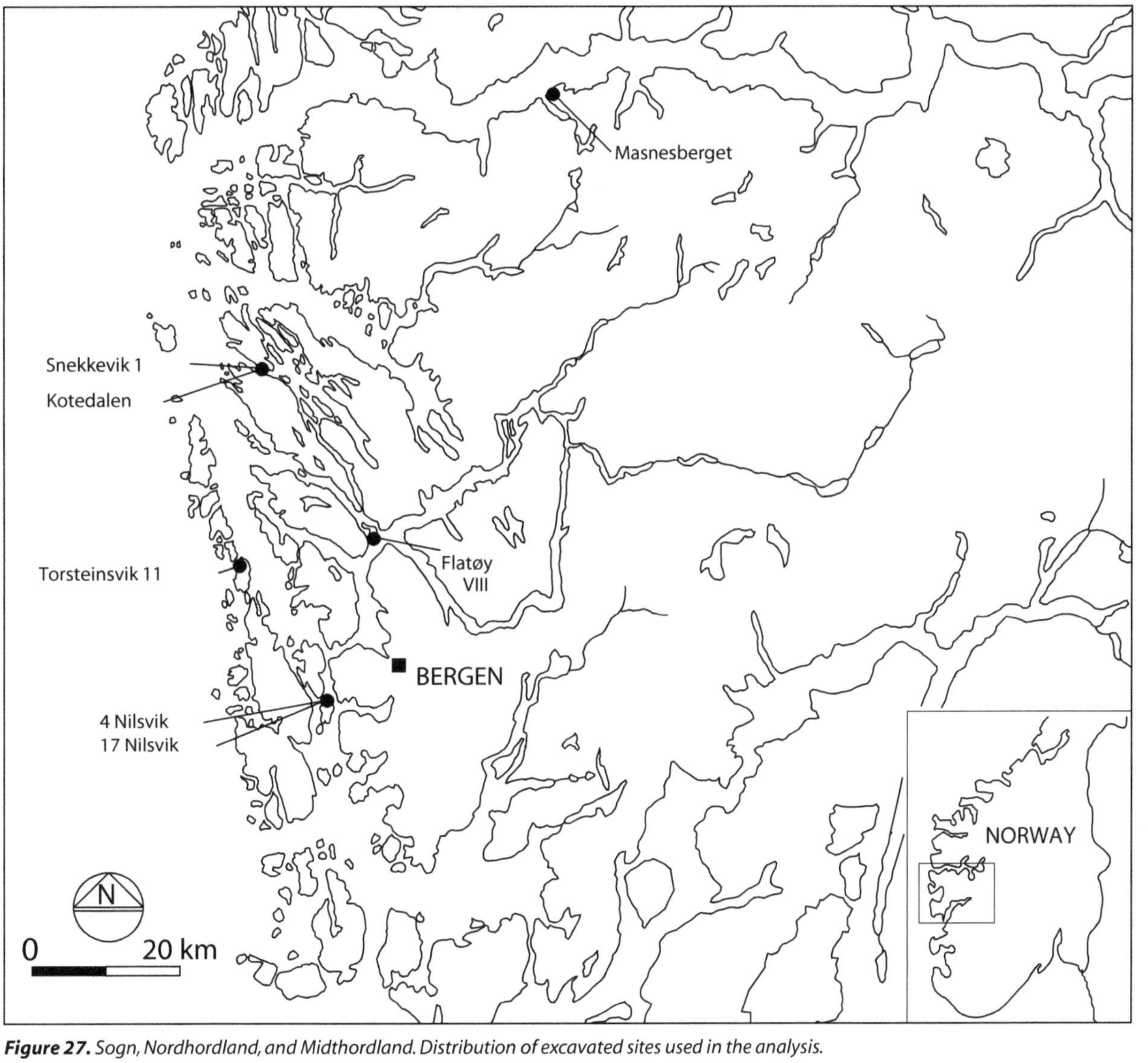

Figure 27. *Sogn, Nordhordland, and Midthordland. Distribution of excavated sites used in the analysis.*

Below, the archaeological data from Fosnstraumen will be applied in order to distinguish between residential sites, field camps and special purpose sites among the 37 sites in this analysis. At the outset, it is expected that sites that are large, have thick layers, high artefact density and cultural layers were likely to have been used as residential sites. Sites that generally score low on these attributes will generally be considered as special purpose sites. The attributes from Fosnstraumen: site size, layer thickness, and artefact density are therefore applied in the following. Layer type is also applied. On the basis of the distribution of measurements within the 45 Neolithic sites at Fosnstraumen, I will distinguish three size groups for each variabe.

The sites had *sizes* ranging between ca 12 m^2 and 2200 m^2, but the main body of sites ranged between ca 50 m^2 and 300 m^2. On the basis of these numbers, the following size-groups could be established:

Small: <50 m^2
Medium: 50 >< 300 m^2
Large: >300 m^2

The *layer thickness* at sites at Fosnstraumen varied between 7 cm and 200 cm. Most of the layers ranged between 15 and 50 cm. On the basis of these numbers, the following groups could be established:

Thin layer: <15 cm
Medium thick layer: 15 >< 50 cm
Thick layer: >50 cm

The maximum/minimum values of *artefact density* were 0,05 and 13,8 artefacts per litre. Most of the numbers ranged between 0,20 and 1,50 artefacts per litre. The following groups could be established:

Low artefact density: <0, 20 artefacts per litre
Medium artefact density 0,20 >< 1,50 artefacts per litre
High artefact density: >1,50 artefacts per litre

The sites that score highest on these variables at Fosnstraumen generally also have *cultural layers*. Such layers are deposited artificially, presumably as results of intensive and long-term activities at the sites. These

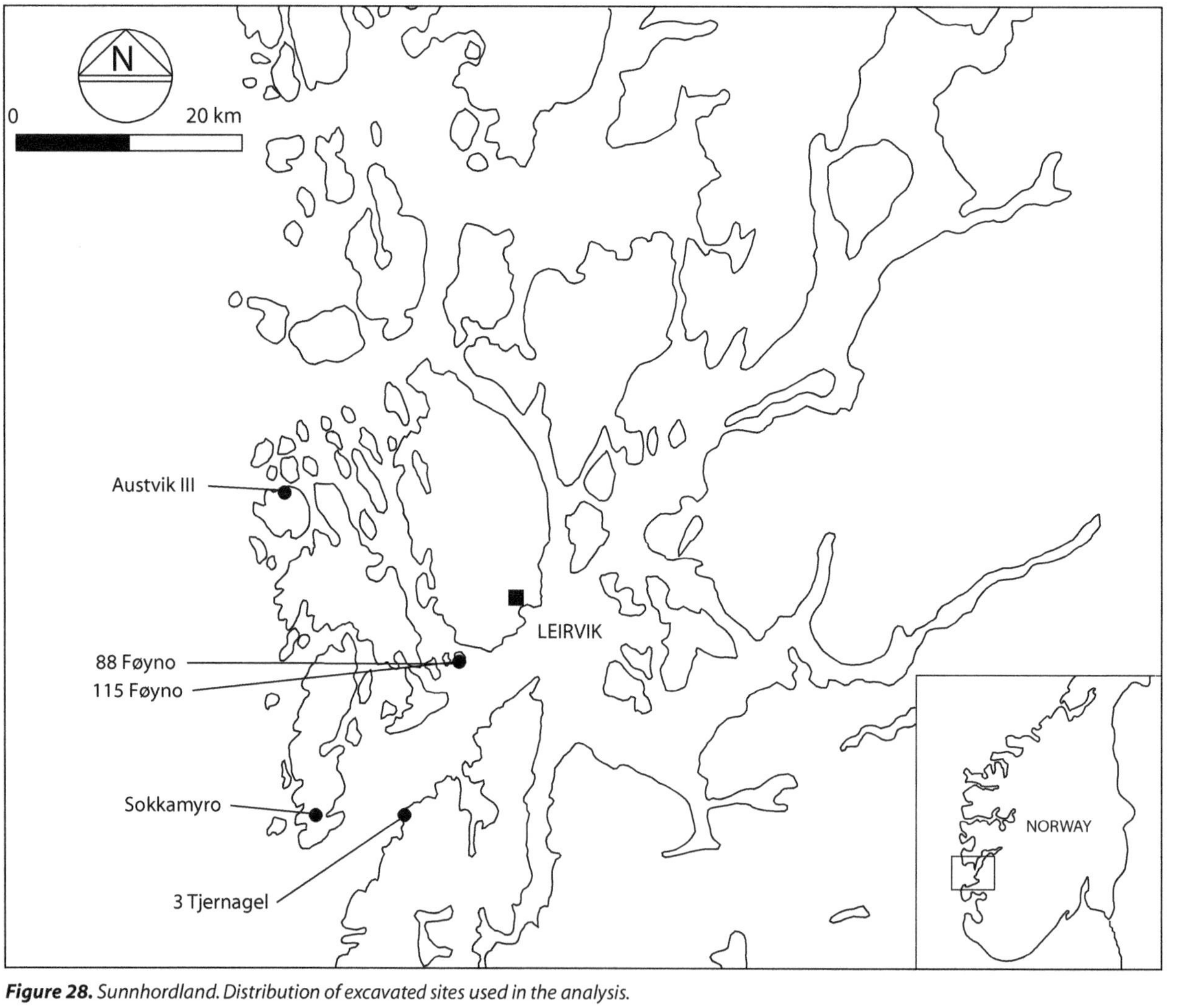

Figure 28. *Sunnhordland. Distribution of excavated sites used in the analysis.*

layers appear as "fat" and dark, and they contain large amounts of charcoal, decomposed organic particles (often burned bone fragments), mineral particles, and often large amounts of fire-cracked stones.

These results from Fosnstraumen will be used in order to distinguish between different site types in the following presentation. In addition to these factors, I will also add a fifth variable: *number of artefact types*, which is assumed to be a good indicator on how few or how many types of activities that took place at a site. This variable was not measured at Fosnstraumen, because the samples (mainly from test-pits) were too small. The data in the current analysis, however, only includes excavated sites. They are therefore probably representative of the artefact variation at the site. If the data from the 37 excavated sites in the current analysis is compared, the maximum and minimum number of artefact forms is 15 and 45. The majority of the numbers range between 18 and 27. The following size-groups may be established on the basis of these numbers:

Low number of artefact forms: <18
Medium number of artefact forms 18 >< 27
High number of artefact forms: >27

In order to distinguish between residential sites, field camps and special purpose sites in the current thesis, the above size groups will serve as a base-line. If the excavated sites are to be characterised as residential sites or field camps, they cannot score low on several of these features.

The excavated sites

Below, 29 coastal sites and 8 mountain sites are presented, starting with the coastal sites in the north. Geographical location, excavator, date of the excavation and number of analysed artefacts are shortly presented. I will also present the site plan and the size/location of units selected for re-classification for the present work. There are also short discussions on dating and site function. The main data and interpretations from each site are presented together in Table 2. Cross-tables on raw materials and lithic categories are found in appendix 3. Calibrated C14-dates are pesented in appendix 4. The locations of the sites are presented on a regional map and district maps, Fig. 24-28.

Nordøy grendahus phase 3

The site is situated on the SW side of Flemsøy, Haram municipality (Fig. 25). It was excavated by Hans Christian Søborg in 1984 (Søborg 1994). Three occupational phases could be distinguished at Nordøy grendahus from the late Mesolithic (phase 1 and 2), early/middle Neolithic (phase 3), and pre-roman Iron Age (phase 4). The extension of the site is about 1500 m^2, of which 27 m^2 (2 %) was excavated. 7 m^2 were selected for analysis from layers 1 – 3 in squares 100x/99-105y (Fig. 29). Altogether 2374 lithic tools and debris were analysed.

Dating: Typological/technological elements: late Mesolithic: 3 micro blades produced by micro blade technique, 1 net-sinker. Late Mesolithic/Neolithic: 1 transverse point. Early/middle Neolithic: 1 blade tanged projectile point (A-type), 1 slate point. Middle Neolithic: Pottery, 1 blade tanged projectile point (B-type), 1 fragment of a tabular sandstone knife. These elements indicate a broad late Mesolithic - early/middle Neolithic dating. 5 radiocarbon samples were dated, but only one of these were collected from layers deposited during phase 3:

T-5967, 6080±80 BP, 101 x/100y, layer 3 (for calibration, see appendix 4).

The result corresponds well with three other radiocarbon determinations which date phase 2 to between 6000 and 6300 BP at the site. Considering that T-5967 was collected from the stratigraphical transition (layer 3) between phase 2 and 3 at the site, I find the result likely to be representative of late Mesolithic phase 2 rather than phase 3.

Site function: The macro-environment is favourable for communications and marine hunting and fishing. There are good harbour conditions, good view of the sea. The site is close to the sea on level and drained ground. The site is very large (1500 m^2), the number of artefact forms is medium (18), the artefact density is high (2,26 per litre), the layer thickness is medium (30 cm), and the site has a cultural layer. Seen together, these attributes are interpreted to indicate that the site was a residential site.

Synnaland

The site is situated at Brattvåg, Haram municipality (Fig. 25). It was excavated by Anne Merete Knudsen in 1980 (Knudsen 1981) and by Live Johannessen in 2000 (Johannessen 2002). Both late Mesolithic and Neolithic occupational phases can be distinguished at Synnaland. Altogether 53, 5 m^2 were excavated. 20 m^2 were selected for analysis from layers 4 – 6 in squares 6 - 11x/8 – 13y (only from the 1980 excavation) (Fig. 30). 1182 lithic tools and debris were analysed.

Dating: Typological/technological elements: late Mesolithic: 3 blades produced by the micro blade technique. Early/middle Neolithic: 1 ground adze with rectangular cross-section, 1 cylindrical core, 1 slate projectile point. These elements indicate a mainly early/middle Neolithic date. No radiocarbon samples were dated.

District		Site	Dating	Analysed area (m²)	Analysed number of artifacts	Site size (m²)	Number of artifact forms	Artifact density (per litre)	Layer thickness (cm)	Layer type	Site location (Macro/Micro)	Site function
Coastal area												
N. Sunnmøre	1	Nordøy grendahus, ph.2	LM/EN/MN	7	2374	Large (1500)	Medium (18)	High (2,26)	Medium (30)	Cultural layer	Good/Good	Residential site
	2	Synnaland	LM/EN/MN	20	1182	Medium (50)	Low (15)	Medium (0,59)	Medium (20)	Sand layer	Good/Medium	Field camp
	3	Valderøya vest 13, ph.2	EN/MN	17	656	Large (300)	Medium (19)	Medium (0,26)	Medium (15)	Gravel layer	Good/Medium	Residential site
	4	Valderøya vest 26, ph 1	EN/MN	9	2400	Large (375)	Medium (24)	High (1,77)	Medium (35)	Cultural layer	Good/Good	Residential site
S. Sunnmøre	5	9 Bustadvika	LM/EN	17,5	1847	Medium (200)	High (32)	Medium (0,58)	Medium (20)	Sand layer	Good/Good	Field camp
	6	Holvikshaugen	EN/MN	17	507	Large (700)	Medium (20)	Medium (0,2)	Medium (20)	Cultural layer	Good/Good	Field camp
Nordfjord	7	17 Havnen, phase 3	EN (late)	7	1635	Large (300)	High (27)	High (1,55)	Medium (20)	Cultural layer	Good/Good	Residential site
	8	1 Haukedal, phase 2	EN (early)	2,5	490	Medium (150)	Medium (22)	Medium (1,3)	Medium (15)	Cultural layer	Good/Good	Residential site
	9	1 Haukedal, phase 3	EN (middle)	2	629	Medium (150)	Low (17)	High (2)	Medium (15)	Cultural layer	Good/Good	Residential site
Sunnfjord	10	Kleiva	EN (middle)	7,5	803	Medium (150)	Medium (22)	Medium (0,71)	Medium (25)	Cultural layer	Good/Good	Field camp
	11	Botnaneset VIII, ph. 3	EN (middle)	2	2068	Medium (225)	Low (17)	High (10,34)	Thin (10)	Sand layer	Good/Good	Field camp
	12	Vikja I, phase 2	EN/MN	4	1767	Large (300)	Medium (20)	High (2,2)	Medium (30)	Cultural layer	Good/Good	Residential site
	13	Vikja I, phase 3	EN/MN	4	1892	Large (300)	High (28)	High (2,35)	Medium (30)	Cultural layer	Good/Good	Residential site
	14	Neset II	EN (early)	5,5	4665	Undetermined	Medium (19)	High (8,12)	Thin (10)	Cultural layer	Good/Good	Residential site
Sogn	15	Masnesberget	EN/MN		434	Undetermined	Medium (25)	Undetermined	Undetermined	Undetermined	Medium/Good	Residential site
Nordhordland	16	Snekkevik 1, phase 3	EN (late)	2	388	Medium (85)	Low (15)	High (1,94)	Medium (15)	Cultural layer	Good/Good	Residential site
	17	Kotedalen phase12	EN (early)	15	4023	Medium (160)	High (30)	High (5,36)	Medium (15)	Cultural layer	Good/Good	Residential site
	18	Kotedalen phase 13	EN (middle)	15	7280	Medium (200)	High (45)	High (9,71)	Medium (20)	Cultural layer	Good/Good	Residential site
Midthordland	19	Torsteinsvik 11, ph. 1	EN (middle)	8	679	Medium (110)	Low (17)	Medium (0,84)	Medium (20)	Cultural layer	Good/Good	Field camp
	20	Flatøy VIII, phase 2	EN (early)	33	622	Medium (150)	Medium (18)	Low (0,19)	Medium (25)	Cultural layer	Good/Good	Field camp
	21	17 Nilsvik	EN (late)	13,75	2408	Medium (72)	Medium (26)	High (1,75)	Medium (35)	Cultural layer	Good/Good	Residential site
	22	4 Nilsvik str. 10	EN (late)	7	818	Large (875)	Medium (25)	Medium (0,78)	Medium (25)	Cultural layer	Good/Good	Residential site
	23	4 Nilsvik str. 29	EN (late)	6	1547	Large (875)	Medium (24)	High (1,72)	Medium (25)	Cultural layer	Good/Good	Residential site
	24	4 Nilsvik str. 30	EN (late)	9	1251	Large (875)	High (29)	Medium (0,92)	Medium (25)	Cultural layer	Good/Good	Residential site
Sunnhordland	25	Austvik III	EN (middle)	4	2525	Medium (175)	Medium (19)	High (6,31)	Thin (10)	Cultural layer	Good/Medium	Field camp
	26	88 Føyno	EN (late)	10	996	Medium (150)	Medium (26)	Medium (1)	Thin (10)	Sand layer	Good/Good	Field camp
	27	115 Føyno	EN (late)	17	1342	Medium (150)	Medium (21)	Medium (0,39)	Medium (20)	Sand layer	Good/Good	Field camp
	28	Sokkamyro 1901-2	EN/MN	10	1675	Large (9000)	High (29)	Medium (0,84)	Medium (20)	Cultural layer	Good/Medium	Residential site
	29	3 Tjernagel	EN/MN	9	2204	Small (40)	Medium (21)	Medium (0,82)	Medium (30)	Gravel layer	Good/Medium	Field camp
Mountain area												
Breheimen	30	Styggvasshelleren, ph. 2	EN/MN	9	1661	Medium (60)	Medium (19)	Medium (1,23)	Medium (20)	Cultural layer	Good/Good	Field camp
Lærdal-Hemsedal	31	Mørkedøla I	LM/EN	8	1929	Medium (60)	Low (15)	High (2,41)	Thin (10)	Sand layer	Good/Good	Field camp
	32	Gyrinos IV	LM/EN/MN	21	2413	Medium (50)	Medium (19)	Medium (0,57)	Medium (20)	Sand layer	Good/Good	Field camp
	33	Blånut IV	EN/MN	20	656	Medium (50)	Low (17)	Low (0,16)	Medium (20)	Sand layer	Good/Good	Field camp
Hardangervidda	34	760 Finnsbergvatn	EN (middle)	9	6301	Medium (50)	Medium (25)	High (2,8)	Medium (20)	Cultural layer	Good/Good	Field camp
	35	526 Nordmannslågen	EN (late)	10	1515	Large (1200)	Medium (21)	Medium (0,75)	Medium (25)	Sand layer	Good/Good	Field camp
	36	1020 Bjornesfjorden	EN (early)	16	1023	Medium (100)	Low (17)	Medium (0,31)	Medium (20)	Cultural layer	Good/Good	Field camp
Haukelifjell	37	Vivik	LM/EN/MN	71	2177	Large (300)	Medium (21)	Medium (0,30)	Thin (10)	Cultural layer	Good/Good	Field camp

***Table 2.** Site data from all excavated sites*

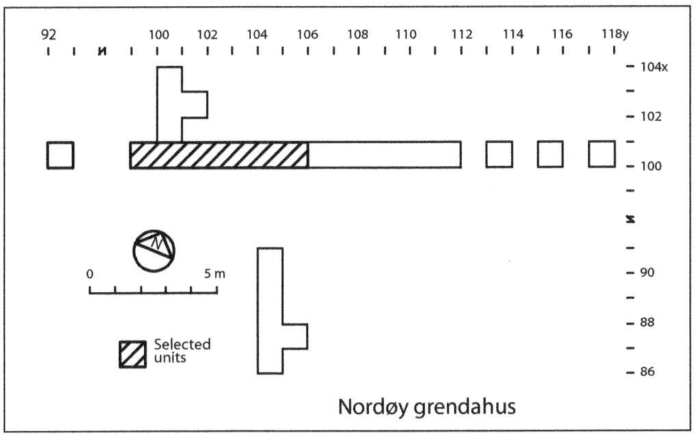

Figure 29. *Nordøy grendahus, Northern Sunnmøre. Plan excavated area. Modified from Søborg (1994).*

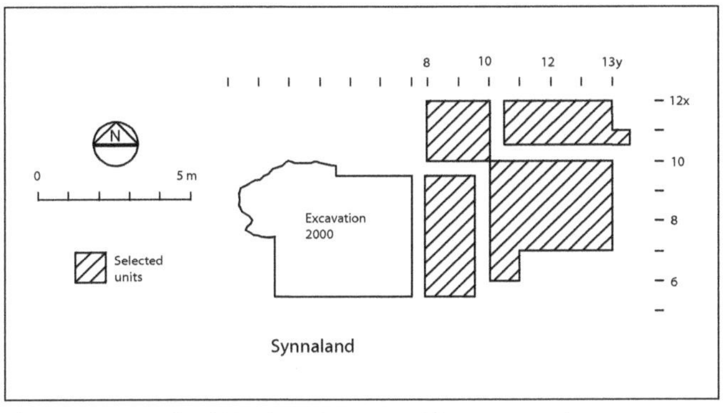

Figure 30. *Synnaland, Northern Sunnmøre. Plan excavated area. Modified from Knudsen (1981).*

Site function: Synnaland faces the broad Harøyfjord to the north. The macro-environment of the site is favourable with regard to sea-hunting and fishing as well as communications. On the micro-level, the harbour conditions are good, but the view from the site itself is less favourable. It is situated on sloping and drained ground close to the shoreline. The site is medium size (50 m^2), the number of artefact forms is low (15), the artefact density is medium (0,59 per litre), the layer thickness is medium (20 cm), and the site has a sand layer, not cultural layer. Seen together, these attributes indicate that the site may have been used on a less permanent basis, perhaps as a field camp or as a short-term residential site.

Valderøya Vest 13, phase 2

The site is situated on the western side of Valderøy, Haram municipality, northern Sunnmøre (Fig. 25). It was excavated by Christopher Prescott, Kjersti Randers and Anne Ågotnes in 1985 and 1986 (Randers & Höglin 1988). Several sites have been investigated at Valderøy. The site was occupied during the late Mesolithic (phase 1, predominantly found in the western part of the excavated area) and early/middle Neolithic (phase 2, predominantly found in the eastern part). The extension of the site is ca 300 m^2, of which 32 m^2 (10 %) were excavated. 17 m^2 were selected for analysis from layers 2A and 2B in squares 98-100x/100-106y (Fig. 31). Altogether 656 lithic tools/debris was analysed.

Dating: Typological/technological elements: late Mesolithic: 4 blades struck by micro blade technique. Early/middle Neolithic: 1 blade tanged point (A-type), 4 Neolithic adzes, 3 slate points, relative dominance of small blades. These elements indicate a mainly early/middle Neolithic date. No radiocarbon samples were dated.

Site-function: The macro-environment of the site is favourable with regard to sea-hunting and fishing. Communications are very good. On the micro-level, the harbour conditions appear to be less favourable, considering that, although the islands Godøy, Giske and Vigra would have guarded against large waves, the site is somewhat exposed towards the North Sea. On the other hand, the view from the site is good, it is situated close to a small river, on level and drained ground close to the shoreline. The site is large (300 m^2 - minimum estimate), the number of artefact forms is medium (19), the artefact density is medium (0,26 per litre), the layer thickness is medium (15 cm), and the site has a gravel layer. Seen together, these attributes are interpreted to indicate that the site was a residential site.

Valderøya vest 26, phase 1

The site is situated on the western side of Valderøy, Haram municipality (Fig. 25). It was excavated by Kjersti Randers in 1986 (Randers & Höglin 1988). It

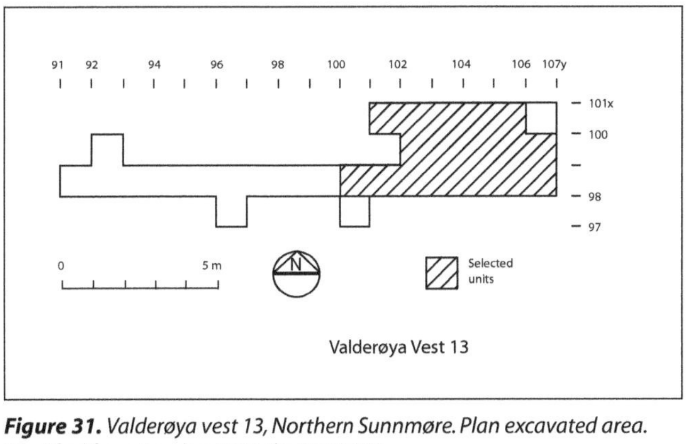

Figure 31. *Valderøya vest 13, Northern Sunnmøre. Plan excavated area. Modified from Randers & Höglin (1988:23).*

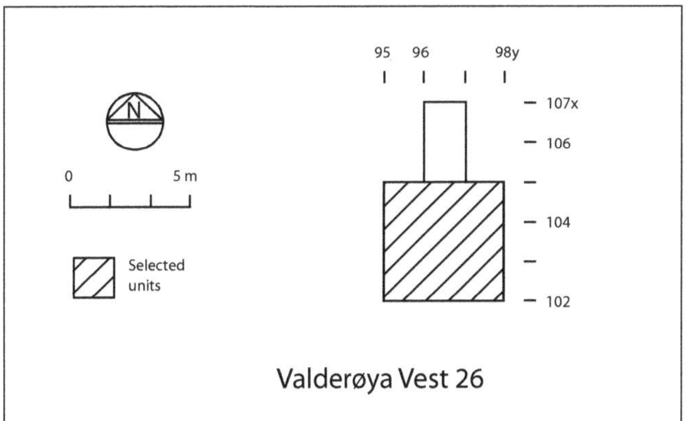

Figure 32. *Valderøya vest 26, Northern Sunnmøre. Plan excavated area. Modified from Randers & Höglin (1988:62).*

was occupied during the early/middle Neolithic (phase 1) and late Neolithic (phase 2). The minimum extension of the site is about 375 m², of which 11 m² (3 %) were excavated. 9 m² were selected for analysis from layers 4-6 in squares 102-105x/95-98y (Fig. 32). Altogether 2400 lithic tools and debris were analysed from the site.

Dating: Typological/technological elements: late Mesolithic: 1 blade produced by microblade technique. Early/middle Neolithic: 1 Vespestad/Vestland adze, 5 slate projectile points, 3 slate point blanks, 2 blade tanged points (A-points), relative dominance of small blades. These elements indicate a mainly early/middle Neolithic date. No radiocarbon samples were dated.

Site-function: The macro-environment is similar to that of 13 Valderøya Vest. On the micro-level, the harbour conditions are good (sheltered bay). The view from the site is good, and it is situated close to a small river, close to the shoreline on level and drained ground. The site is large (375 m²), the number of artefact forms is medium (24), the artefact density is high (1,77 per litre), the layer thickness is medium (35 cm), and the site has a cultural layer. Seen together, these attributes are interpreted to indicate that the site was a residential site.

9 Bustadvika

The site is situated at Djupvik, Ulstein municipality (Fig. 25). It was excavated by David N. Simpson in 1999 (Simpson 2003). One occupational phase from the early Neolithic period can be distinguished at the site. The extension of the site is 200 m², of which 17, 5 m² (8, 5 %) in the eastern part of the site were excavated. The entire excavated area was selected for analysis (Fig. 33). 1847 lithic tools and debris were analysed.

Dating: Typological/technological elements. Late Mesolithic: 1 bi-convex chisel, 1 long and slim struck blank. Early Neolithic: 2 Vespestad adzes. Early/middle Neolithic: 2 blade tanged points (A-type), 1 slate point, 3 ground slate point blanks/preforms. Three radiocarbon samples were dated:

Beta-132227: 4990±80 BP, 105x/43y NØ charcoal layer
Beta-140825: 3750±70 BP, 103x/43y SV layer B1
Beta-140826: 2330±70 BP, 105x/42y SØ, layer B1

The latter two of these gave results which are considerably younger than the archaeological material at the site, and should therefore be disregarded. The result of the Beta-132227, however, is in accordance with the archaeological material, and may indicate a late Mesolithic/early Neolithic date of the site.

Site function: The site is situated in a sheltered bay in the sound Dragsundet between the islands Hareidlandet and

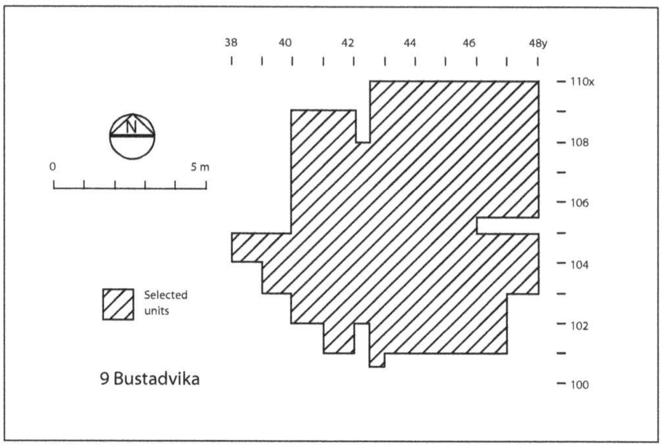

Figure 33. *9 Bustadvika, Southern Sunnmøre. Plan excavated area. Modified from Simpson (2003).*

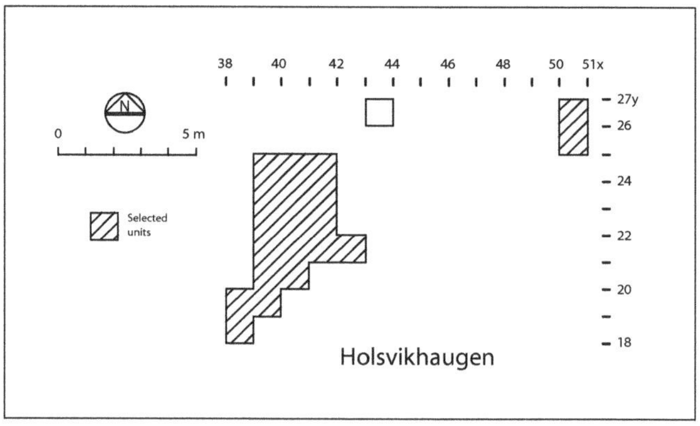

Figure 34. *Holsvikhaugen, Southern Sunnmøre. Plan excavated area. Modified from Olsen (1976a).*

Gurskøy. The macro-environment is therefore favourble with regard to communications as well as for hunting and fishing. The micro environment is characterised by a proximity to the shore, good harbour conditions, relatively good view of the sea, level and drained ground. The site is of medium size (200 m²), the number of artefact forms is high (33), the artefact density is medium (0,58 per litre), the layer thickness is medium (20 cm), and the site has a sand layer, not a cultural layer. Seen together, these attributes indicate that the site may have been used on a less permanent basis, perhaps as a field camp or as a short-term residential site.

Holviksbaugen

The site is situated at Berknes, Volda municipality (Fig. 25). It was excavated by Asle Bruen Olsen in 1976 (A. B. Olsen 1976b). One occupational phase from the early/middle Neolithic period can be distinguished at the site. The extension of the site is 7-800 m², of which 21 m² (3 %) were excavated. 17 m² were selected for analysis from all excavated layers in squares 38 - 42x/19 – 24y (Fig. 34). 507 lithic tools and debris were analysed.

Dating: Typological/technological elements: late Mesolithic: 2 blades produced by micro blade technique were found. Early Neolithic: 2 Vespestad adzes. Early/

middle Neolithic: 1 blade tanged point (A-type), 2 slate points, 1 ground slate point blank, relative dominance of small blades. Middle Neolithic B element: One fragment of a tabular sandstone knife with crescent shape and straight edge. These elements indicate a mainly early/ middle Neolithic date. No radiocarbon samples were dated.

Site function: The site situated at the eastern shore of the broad Vartdalsfjord, at the westernmost point at the headland between Voldafjord and Ørstafjord. The macro-environment is therefore favourable with regard to communications as well as for hunting and fishing. The micro environment is characterised by a proximity to the shore, good harbour conditions, and good view of the sea, level and drained ground. The site is large (700 m^2), the number of artefact forms is medium (20), the artefact density is high (0,2 per litre), the layer thickness is medium (20 cm), and the site has a cultural layer. Seen together, these attributes are interpreted to indicate that the site was a residential site.

17 Havnen, phase 3

The site is situated at Rugsundøy, Skatestraumen, Bremanger municipality (Figs. 3 and 26). At Skatestraumen, a large number of Stone Age sites were surveyed and excavated between 1989 and 1995 (Bergsvik 2002a). 17 Havnen was excavated in 1992, 1993 and 1994.

The site was occupied during 7 phases spanning from the late Mesolithic to the Roman Iron Age, each represented by separate stratigraphical layers. Phase 3 was dated to the early Neolithic. The extension of the site is 300 m^2, of which 89 m^2 (29 %) were excavated. The units selected for analysis are from an area of the site which was excavated with very good stratigraphical control. The phase-correlation of the site distinguished between secure and insecure units (Bergsvik 2002:189). For the current analysis, only the securely correlated layers Ae 1 – 5 and Au 1 and 2 from units 101-103x/51-53y are chosen (Fig. 35). 1635 tools and debris are analysed.

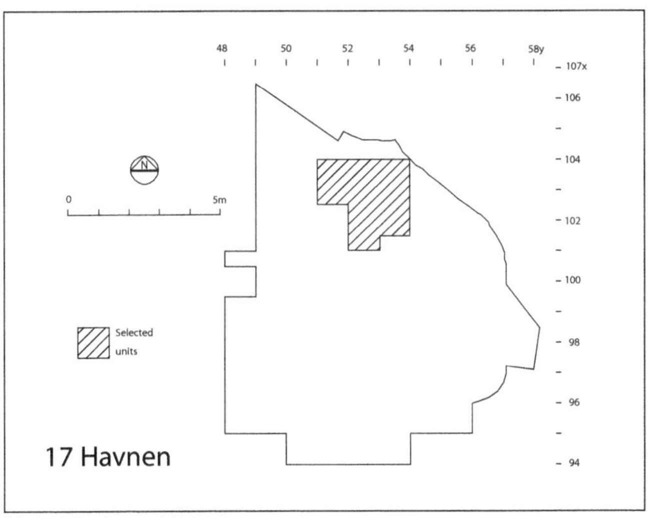

Figure 35. *17 Havnen. Nordfjord. Plan excavated area. Modified from Bergsvik (2002:191).*

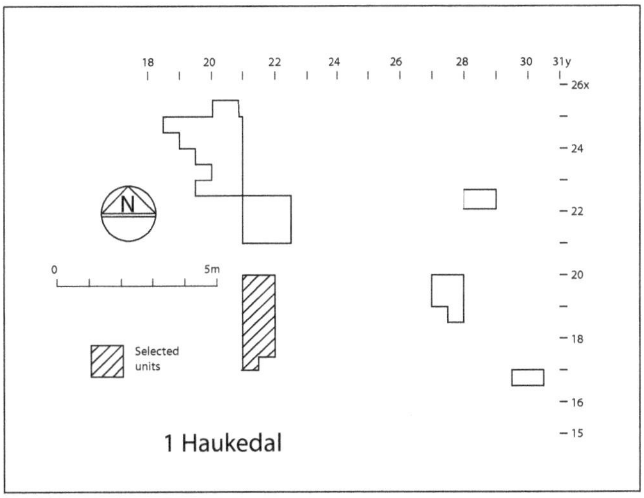

Figure 36. *1 Haukedal. Nordfjord. Plan excavated area. Modified from Bergsvik (2002:111 ff).*

Dating: Typological/technological elements. Late Mesolithic: 2 fragments of tabular sandstone knives. Early/middle Neolithic: 5 tanged blade points (A-type), 1 Vestland chisel, 1 Vespestad/Vestland adze, 5 slate points, 4 slate point blanks, 2 cylindrical cores, relative dominance of small blades. It should be noted that the majority of Neolithic adzes found elsewhere at the site were of the early Neolithic Vespestad type. Altogether 9 charcoal samples were radiocarbon dated from layers associated with phase 3:

- Beta-48164: 4660±80 BP
- Beta-67986: ETH 11612: 4755±55 BP. 102x/52y, NV, layer Ae3
- Beta-67992: 4620±80. 101x/52y S, layer Ae1
- Beta-67993: 5080±140 BP. 101x/52y S, layer Ae2
- Beta-67994: 4820±80 BP. 101x/52y S, Layer Ae 3
- Beta-68245: 4350±60 BP. 102x/53y, NV, Au1
- Beta-78319: 4480±120 BP. 103x/53y SV, layer Ae3
- Beta-78321, CAMS 17544: 2540±70 BP. 101x/55y SV, layer Ae2
- Beta-78324: CAMS 17547: 4560±60 BP. 103x/52y NØ, layer Ae2

If Beta-78321, CAMS 17544 is excluded, these results span about 900 C14-years, from 5200 to 4300 BP. Four of these dates have overlapping results between 4900 and 4500 BP. The temporal spread of the results indicate that phase 3 is the result of several occupational phases during the latter part of the early Neolithic and early part of the middle Neolithic. Some of these occupations were probably brief episodes, while others may have been more substantial. The marked cultural layer in the northern part of the site partly consists of a thick bone midden. This layer, from which the data presented here is collected, was probably accumulated as a result of one longer occupational phase during the latter part of the early Neolithic. A sample AMS dated to 4755±55 BP (Beta-67986-ETH-11612) was collected from this bone midden. On the basis of the above radiocarbon determinations and the technological/typological elements, I assume that phase 3 was occupied during the latter part of the early Neolithic.

Site function: The site is situated on the northern shore of Skatestraumen, just south of the outlet of Nordfjord. Due to the tough weather conditions outside of Bremangerlandet, the main route for traffic of ships along the coast goes through Skatestraumen. The site is situated close to the large rock-art site Vingen (e.g. Bøe 1932). The tidal current channel Skatestraumen is very rich in marine food resources. The macro-environment is therefore favourable with regard to communications, proximity to ritual sites, and for marine hunting and fishing. The micro-environment is also generally good. The site is close to the shoreline, in a good harbour, there is a relatively good view of the sea, and level and drained ground. It is situated close to a small river. The site is large (300 m²), the number of artefact forms is high (27), the artefact density is high (1,55 per litre), the layer thickness is medium (20 cm), and the site has a cultural layer. It should also be mentioned that a house structure (inside which most of the cultural layer has been deposited) is preserved from phase 3. Seen together, these attributes are interpreted to indicate that 17 Havnen was a residential site.

1 Haukedal, phases 2 and 3

The site is situated at Rugsundøy, Skatestraumen, Bremanger municipality (Figs. 3, 4 and 26). It was excavated in 1995 (Bergsvik 2002a). The site was occupied during 7 phases spanning from the late Mesolithic to the Roman Iron Age, each represented by separate stratigraphical layers. Phase 2 and 3 are dated to the early Neolithic. The extension of the site is 150 m², of which 12 m² (8%) have been excavated. The units selected for analysis are from an area of the site which was excavated with very good stratigraphical control. The phase-correlation of the site distinguished between secure and insecure units (Bergsvik 2002:110 ff). For the current analysis, both securely and insecurely correlated layers D 4 – 9/H 1 - 3 from units 17-19x/21y and layers O1-3 from units 18 and 19x/21y were chosen from phase 2 and 3 respectively (Fig. 36). 490 and 629 tools and debris are analysed from the two phases.

Dating: Typological/technological elements. Phase 2: Mesolithic/Neolithic: 1 transverse point. Early/middle Neolithic: 2 blade tanged points (A-points), 4 slate points. Middle Neolithic: 1 Vestland chisel. Phase 3: early/middle Neolithic: 1 Vestland/Vespestad adze, 1 slate point, 1 slate point blank, 1 cylindrical core and relative dominance of small blades.

Four charcoal samples were collected from layers associated with phase 2 (Beta-82833-4), and two were dated from phase 3 (Beta-82832, 82835).

Beta-82834: 5430±60 BP. 23x/20y, SØ, H6,
Beta-82833: 5220±70 BP. 19x/21y, SV, H2
Beta-82835: 5090±70 BP. 24x/20y, SV, P5
Beta-82832: 5080±60 BP. 19x/21y, SV, O2

Beta 82834 is somewhat older than expected. However, the sample may partly consist of charcoal from an older Mesolithic phase 1 at the site (this is more than likely, considering that it was collected relatively close to the bedrock). The three remaining results are in accordance with the typological and technological elements present in the chosen units. On the basis of the above radiocarbon determinations and the technological/typological elements, I assume that phase 2 and 3 were occupied during the early and middle part of the early Neolithic.

Site function: For macro environment see 17 Havnen. The micro environment is generally good. The site is situated close to the shore, there are good harbour conditions, and very good view of the sea, the ground is level and drained. The site is of medium size, (150 m²), the number of artefact forms for phase 2 and 3 are medium (22) and low (17), the artefact densities are medium and high (1,3/2 per litre), the measurements of layer thickness are medium for both phases (15/15 cm), and both phases have cultural layers. Seen together, these attributes are interpreted to indicate that the site was a residential site.

Kleiva, phase 1

The site is situated at Gaddevågen at the south-eastern side of Floralandet, close to the city of Florø, Flora municipality (Fig. 26). It was excavated by Morten Ramstad and Asle Bruen Olsen in 2001 (Ramstad & A. B. Olsen 2002). The site has one early Neolithic phase (1), and was also occupied during the Pre-Roman Iron Age (phase 2). The extension of the site is 150 m², of which 7, 5 m² (5%) were excavated. The units selected for analysis are 7, 5 m² from 50-52x/49-50y (northern area) and 53-54x/50-52y (southern area) from layers C1-4 (Fig. 37). 803 lithic tools and debris were analysed.

Dating: Typological/technological elements. Early Neolithic: 9 Vespestad adzes. Early/middle Neolithic: 2 Vespestad/Vestland adzes. 3 Blade tanged points (A-points). Middle Neolithic B: 2 fragments of tabular sandstone knives. Iron Age: 1 fragment of a soapstone vessel. Two radiocarbon samples were dated from layers associated with phase 3:

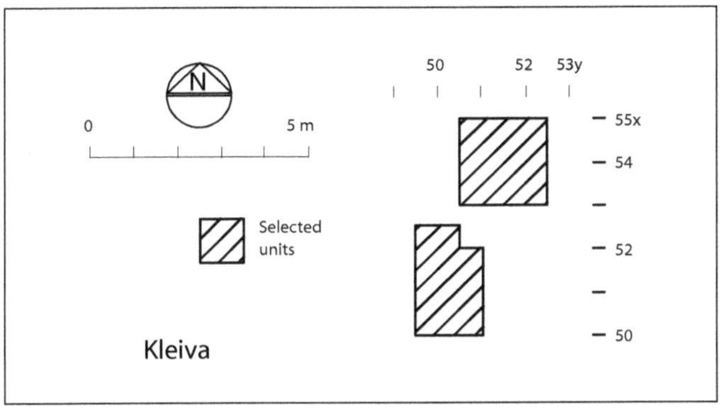

Figure 37. Kleiva, Sunnfjord. Plan excavated area. Modified from Ramstad & Olsen (2002).

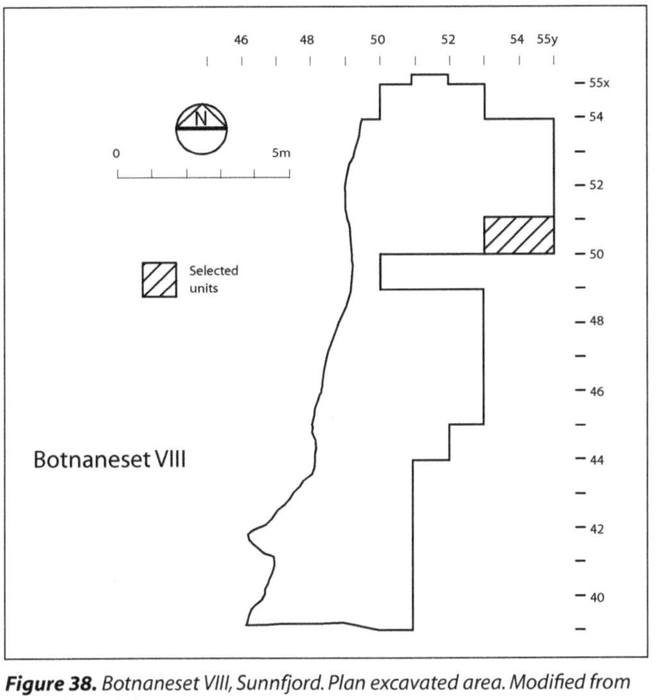

Figure 38. Botnaneset VIII, Sunnfjord. Plan excavated area. Modified from Olsen (1983:86).

Beta-151277: 4980±80 BP. 51x/50y, SØ, structure 4 (pit-hearth)
Beta-156054: 5020±70 BP. 51x/49y, NV, structure 4 (pit-hearth)

The C14-datings and the dominating typological/technological elements indicate that phase 3 was occupied during the middle part of the early Neolithic period.

Site function: On the macro-level, the site has a good location with regard to communication and proximity to food resources. The tidal current channel through Gaddevågen was presumably rich in food resources during the Neolithic (see 1 Vikja for further notes on macro-environment). The micro environment is generally good. The site is situated close to the shore; there are very good harbour conditions, relatively good view of the sea, level and drained ground. The site is of medium size (150 m²), the number of artefact forms is medium (22), the artefact density is medium (0,71 per litre), the layer is of medium thickness (25 cm), and the site has a cultural layer. On the basis of this data, the site is interpreted as a residential site.

Botnaneset VIII, phase 3

The site is situated at Brandsøy, Flora municipality (Fig. 26). It was excavated by Asle Bruen Olsen in 1981 (A. B. Olsen 1983). The site has two late Mesolithic phases (1 and 2) and one early Neolithic phase (3). The extension of the site is 225 m², of which 68 m² (30%) were excavated. The units selected for analysis are 50x/53,54y layer I (Fig. 38). 2068 lithic tools and debris were analysed.

Dating: Typological/technological elements. Late Mesolithic: 1 keeled core, 5 micro blades produced by the microblade technique. Early/middle Neolithic: 2 blade tanged points, 2 cylindrical cores. Micro blades dominate compared to small blades, however, these micro blades are to a large extent made from chert, and was reduced on cylindrical cores, not by the Mesolithic microblade technique or by bipolar technique. Two radiocarbon samples were dated from layers associated with phase 3:

T-4484: 5030±130 BP. 50x/54y lag I
T-4482: 4630±60 BP. 50x/53y lag I

The two C14-datings indicate that phase 3 was occupied during the middle or late part of the early Neolithic period. A. B. Olsen pointed out that T-4484 represents the best sample context of the two dates. T-4482 may have been contaminated by later intrusions. He therefore argued that the oldest date probably was closest to the correct age (A. B. Olsen 1983:91). The typological and technological elements are consistent with this argument. The material was most likely deposited during the early Neolithic period.

Site function: The site is situated south of Norddalsfjorden, at the northern side of the island Brandsøy, about 4 km to the east of Florø (see Vikja I for further notes on macro-environment). The micro environment is generally good. The site is close to the shore, there are very good harbour conditions, relatively good view of the sea, and level and drained ground.

The site is of medium size (225 m²), the number of artefact forms is low (17), the artefact density is high (10,34 per litre), the layer is thin (10 cm), and the site only has a sand layer, not a cultural layer. On the basis of this data, A. B. Olsen (1983) interpreted the site as a result of one, relatively short occupation by a small household. This interpretation is supported here.

Vikja I phases 2 and 3

The site is situated at Brandsøysund, Flora municipality (Fig. 26). It was excavated by Asle Bruen Olsen in 1976 (A. B. Olsen 1976b; 1981:97), in an area where numerous Stone Age sites have been surveyed and excavated. The site has one late Mesolithic phase (1) and two Neolithic phases (2 and 3). The extension of the site is 300 m², of which 18, 5 m² (6%) have been excavated. The units selected for analysis from phase 2 is 7x/3-6y, layers 1a, 1b, 2a, and 3a. Units 1-3x/1,2y layers I, II, III, and IV (Fig. 39). 1767 and 1892 lithic tools and debris were analysed from phase 2 and 3 respectively.

Dating: *Phase 2*: Typological/technological elements. Mesolithic: 1 soapstone net-sinker, 1 adze with bi-convex cross-section, 1 conical microblade core. Early/middle Neolithic: 3 tanged blade points (A-points), 2 slate point blanks, 11 cylindrical cores and relative dominance of small blades. Middle Neolithic: 1 Vestland-adze.

Phase 3: Mesolithic/Neolithic: 1 transverse point. Early/middle Neolithic: 11 Vestland/Vespestad adzes, 3 tanged blade points (A-points), 15 slate points, 21 slate point blanks, 1 cylindrical core and relative dominance of small blades. Middle Neolithic: 3 slate points with hanging shoulders. On the basis of the above elements, I suggest that phase 2 and 3 mainly can be dated to the early/middle Neolithic. No radiocarbon samples were dated.

Site function. The site is situated at the mainland Klavelandet on the eastern shore of the strait Brandsøysund, about 7 km to the east of Florø. The strait is favourable for communications in the north-south direction. Høydalsfjorden and Norddalsfjorden are also easily accessible from Brandsøy. The tidal current channel Brandsøysund is also very rich in marine food resources. Furthermore, the large stone-adze quarry at Stakaneset is situated only 5 km to the east, and Ausevik, the large rock-art site, is located about 10 km to the south-east. The macro-environment is therefore favourable with regard to communications, stone adze acquisition, and rock-art practices as well as for marine hunting and fishing. The micro environment is also generally good. The site is close to the shore, there are very good harbour conditions, relatively good view of the sea, level and drained ground. It is situated close to a small river. The site is large, (300 m²), the number of artefact forms for phase 2 and 3 are medium (20) and high (28), the artefact densities are high (2,2 / 2,35 per litre), the measurements of layer thickness are medium for both phases (30/30 cm), and both phases have cultural layers. Seen together, these attributes are interpreted to indicate that the site was a residential site. It is important to note that production of blanks and adzes of diabase from the quarry at Stakaneset was carried out at the site.

Neset II

The site is situated at Brandsøysund, Flora municipality (Fig. 26). It was excavated by Asle Bruen Olsen in 1978 (A. B. Olsen 1981:98). The site has one early Neolithic phase. The extension of the site is not determined. The units selected for analysis are 50,51x/75y, 52-55x/75,5y, and 57x/81y layers I, II, IIb, and III (Fig. 40). 1767 and 1892 lithic tools and debris were analysed from phase 2 and 3 respectively.

Dating: Typological/technological elements. Late Mesolithic: 4 conical microblade cores, 5 micro blades produced by microblade technique. Early/middle Neolithic: 8 tanged blade points, 1 cylindrical core, and relative dominance of small blades. Two radiocarbon samples were dated from the site:

T-3056: 2450±80 BP. 52x/75y layer I
T-2682: 5230±70 BP. 50x/75y layer I

T-3056 represents an Iron Age occupation at the site that was not documented archaeologically, while the oldest result is clearly in accordance with the typological and technological elements at the site. The significant Mesolithic elements may indicate an independent late Mesolithic occupation. However, it may, as indicated by the radiocarbon date, instead suggest that the entire material has been deposited during the Mesolithic - Neolithic transition.

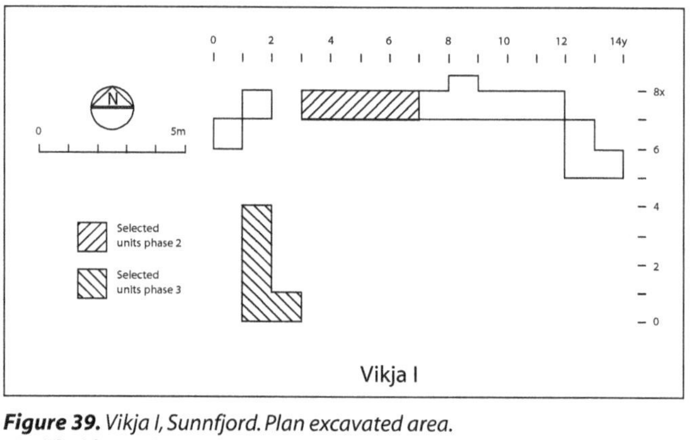

Figure 39. *Vikja I, Sunnfjord. Plan excavated area. Modified from Olsen (1976b).*

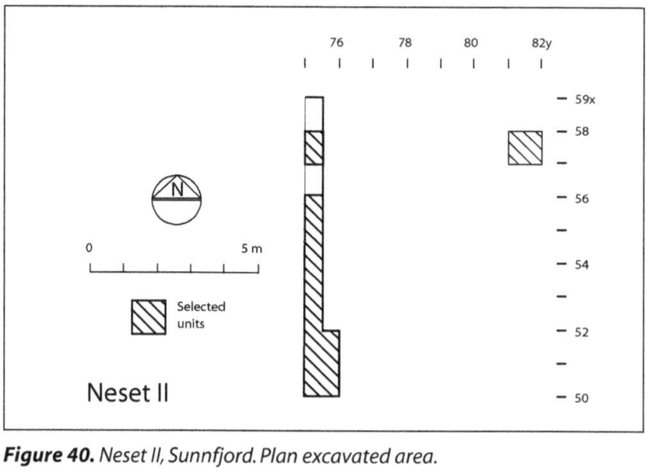

Figure 40. *Neset II, Sunnfjord. Plan excavated area.*

Site function: For macro-environment see Vikja I. The micro environment is good. The site is close to the shore, there are very good harbour conditions, good view of the sea, and level and drained ground. The number of artefact forms is medium (19), the artefact density is high (8,12 per litre – diabase flakes are excluded), the layer thickness is thin (10 cm), and the site has a cultural layer. Seen together, these attributes are interpreted to indicate that the site was a residential site. Production of blanks and adzes of diabase from the quarry at Stakaneset was carried out at the site.

Masnesberget

The site is situated south of the western part of Sognefjorden, just at the outlet of Ikjefjorden, Høyanger municipality (Fig. 27). It has not been excavated. The artefacts were mainly collected by amateurs in the 1920s. The site was occupied during the late Mesolithic and the early Neolithic. The extension of the site is not determined. The material presented here (Museum no. B7811, B8144, B8325, and B8442) was collected from a cultivated field within a small area to the south of a rock outcrop.

Dating: Typological/technological elements. Mesolithic: 12 adzes with round cross-sections, 3 chisels with bi-convex cross-sections, 2 conical cores. Neolithic elements: 7 Vestland/Vespestad adzes, 3 slate, 5 slate point blanks, 4 cylindrical cores and dominance of small blades. Very few artefacts other than adzes can be related to the Mesolithic period. These 15 adzes may, as Trond Klungseth Lødøen have suggested, represent an adze-deposit, not an occupational site (Lødøen 1994:72). The main lithic data indicates an early/middle Neolithic date. No radiocarbon samples were dated.

Site function: On the macro-level, the site is situated about 25 km into the Sognefjord. The location is therefore strategic regarding east-west communications, and less favourable for traffic to the north or south. The Sognefjord offered rich resources in periods of the year, however, compared to the richness and stability on outer coast, the resource basis is less favourable. The micro environment is generally good. The site is close to the shore, there are very good harbour conditions, good view of the sea, level and drained ground. The size of the site, number of artefact forms, artefact density, layer type and layer thickness cannot be evaluated. Lødøen has test-pit surveyed middle Neolithic sites close by. These have cultural layers and score high on the above attributes. On may therefore assume that the current site was of the same type as these, which Lødøen has interpreted as residential sites (Lødøen 1995:63).

Snekkevik 1 phase 3

The site is situated at Fosnstraumen, in Austrheim municipality, facing Kotedalen (below) on the other side of the channel (Figs. 5 and 27). It was excavated in 1988 and 1989 (Bergsvik 1988). The site was occupied during 5 phases spanning from the early Mesolithic to the Iron Age, each represented by separate stratigraphical layers. Phase 3 was dated to the early Neolithic. The spatial extension of phase 3 was 85m², of which 4, 75 m² (ca. 6%) were excavated. The units selected for analysis were 2 m² from 49x/47y, layer C2-3 and D1-3. 51x/49y D1-2, 52x/52y D1-2 and 54x/49y D4 (Fig. 41). 388 lithic tools and debris were analysed.

Dating: Typological/technological elements. Early/middle Neolithic: 3 blade tanged points (A-points), 1 slate point, 3 cylindrical cores and relative dominance

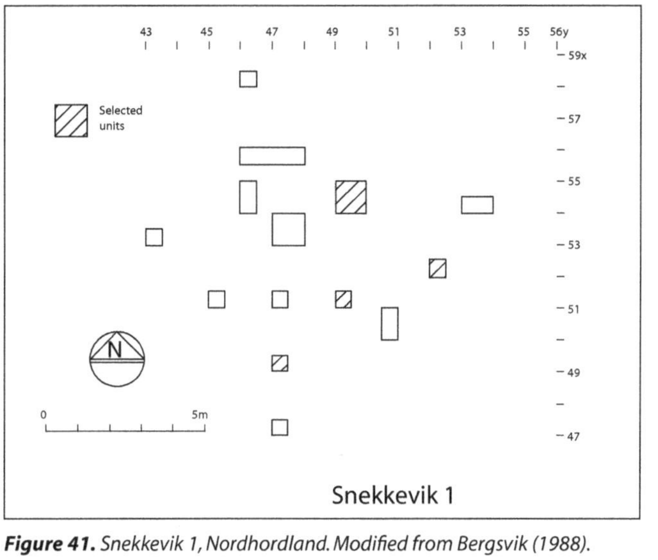

Figure 41. *Snekkevik 1, Nordhordland. Modified from Bergsvik (1988).*

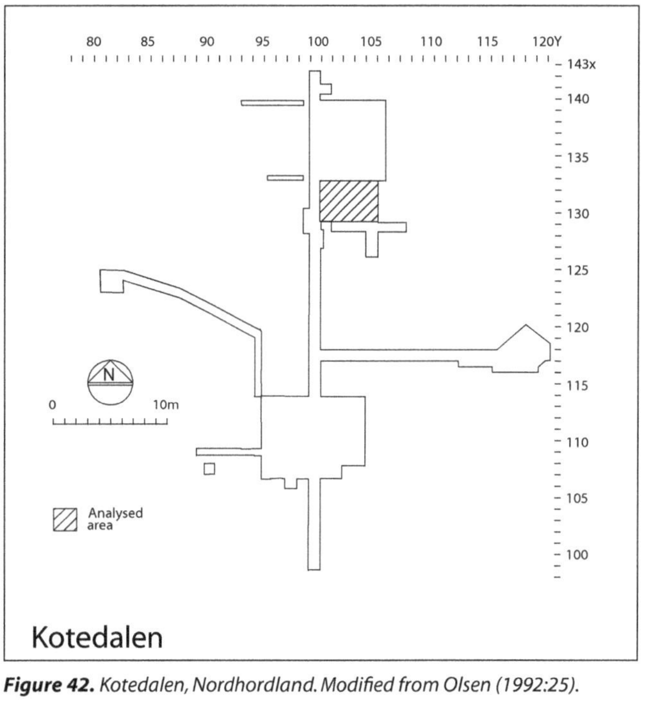

Figure 42. *Kotedalen, Nordhordland. Modified from Olsen (1992:25).*

of small blades. One charcoal sample was radiocarbon dated from layers corresponding to phase 3:

T-7643: 4800±120 BP (3770-3380 BC). 51x/49y layers DI-II

This result is in accordance with the estimated age based on the typological and technological elements. It can, therefore, be argued that phase 3 was accumulated during the latter part of the early Neolithic.

Site function: On the macro-level, Snekkevik has a strategic location with regard to communications north-south along the coast. It is also situated close to the shoreline at Fosnstraumen, a tidal current channel that offered rich and stable marine resources throughout the Stone Age. The micro environment is good. The site is close to the shore, there are very good harbour conditions, good view of the sea, and level and drained ground. The site is of medium size (85 m²), the number of artefact forms is low (15 – but this low number is most likely the result of a low sample-size), the artefact density is high (1,94 per litre), the layer is of medium thickness (15 cm), and the site has a cultural layer. On the basis of this data, the site is interpreted as a residential site.

Kotedalen phases 12 and 13

Kotedalen is situated at Straume, in Radøy municipality (Figs. 5 and 27). It was excavated by Asle Bruen Olsen, David N. Simpson and Jan Elizabeth Warren in 1986 and 87 (A. B. Olsen 1992, Warren 1994, Simpson 1999). As pointed out in Chapter 4, large numbers of sites have been surveyed and excavated at Fosnstraumen, among them the well known site Ramsvikneset (Bakka 1993, Nærøy 1993b). Kotedalen was occupied during 16 phases spanning from the early Mesolithic to the middle Neolithic, each represented by different stratigraphical layers. Phases 12 and 13 were dated to the early Neolithic. The spatial extension of phases 12 and 13 were 160/200 m² (minimum estimates), of which 29/45 m² (ca. 20/22 %) were excavated (A. B. Olsen 1992:36). The units selected for analysis are from area 1B (Fig. 42), which was excavated with very good stratigraphical control. A. B. Olsen and Simpson distinguished between secure and insecure units in their phase-correlation of the site (A. B. Olsen 1992:75, Simpson 1999). For the current analysis, only the securely correlated layers NE2 (phase 12) and NE1 (phase 13) were chosen. Altogether 4023 (phase 12) and 7280 (phase 13) tools and debris are analysed.

Dating: Typological/technological elements. *Phase 12*: Mesolithic typological/technological elements: 1 keeled core, 1 adze with round cross-section. Early/ middle Neolithic: 43 blade tanged points, 14 cylindrical cores, 6 slate points, 1 slate point blank, relative dominance of small blades. *Phase 13*: Mesolithic elements: 1 keeled core. Early Neolithic: Several sherds of TRB-type pottery (with impressions of real cord). Early/middle Neolithic: 3 Vespestad/Vestland adzes, 56 blade tanged points, 27 slate points, 24 slate point blanks, 5 cylindrical cores and a relative dominance of small blades. Middle Neolithic: 3 slate points with hanging shoulders. Several charcoal samples were radiocarbon dated from layers corresponding to phases 12 and 13 at Kotedalen (A. B. Olsen 1992:211, 214):

Phase 12:

T-6235: 5100±60 BP. 127x/99y, layer 8
T-6238: 5050±70 BP. 133x/103y, layer 6
T-7336: 5010±50 BP. 132x/103y, layer NE2I
T-7523: 5010±80 BP. 129x/100y, layer NE2I
T-7524: 5030±80 BP. 129x/100y, layer NE3
T-7519: 5160±120 BP. 121x/88y, layer V9 (bottom)

Phase 13:

T-7052: 4860±60 BP. 1030x/99y, NØ, layer 8e
T-7337: 4920±50 BP. 135x/104y, layer EI
T-7509: 4960±80 BP. 132x/103y, layer NE1II

These results indicate that the layers corresponding to phase 12 and 13 were accumulated during two succeeding periods between 5100 and 4900 BP, i.e. during the early and middle part of the early Neolithic. This is in accordance with the estimated age based on the typological and technological elements that dominate in the relevant layers.

Site function: On the macro-level, Kotedalen has the same favourable location as Snekkevik 1. On the micro-level it rates very high for harbour conditions, view, and level ground. The site is therefore likely to have been chosen for long-term occupations. The site is of medium size (160/200 m² - minimum estimates), the number of artefact forms for both phase 2 and 3 is high (30/45), the artefact densities are high (5,36 / 9,71 per litre), the measurements of layer thickness are medium for both phases (15/20 cm), and both have cultural layers. In contrast to most other open-air sites in western Norway, faunal material was preserved at Kotedalen. A number of ash-filled pit-hearths provided the best preservation conditions. The faunal data indicates that the site was occupied during the summer, autumn and winter, and that the subsistence base was diverse, consisting of a variety of marine and terrestrial species (Hufthammer 1992). Botanical investigations at the site also indicate that, in the Neolithic, a large area behind the site was continuously kept open (Kaland 1992). On the basis of this data, A. B. Olsen (1992) concluded that the site was occupied by sedentary local groups during the early and middle Neolithic phases 12-15. This interpretation was supported by (Nærøy 2000) and myself (Bergsvik 2001b) in more recent reviews of the site. The site should be characterised as a sedentary residential site.

Torsteinsvik 11, phase 1

The site is situated at the northern side of the island Toftøy in Øygarden municipality (Fig. 27). In this area, several Stone Age sites have been surveyed and excavated. 11 Toftøy was excavated by Hein Bjartmann Bjerck in 1984 (Bjerck 1984). The site has been analysed by Arne Johan Nærøy for his chronological work (Nærøy 1993a). Three periods are represented: Phase 1 is dated to the early Neolithic, phase 2 is middle Neolithic A, and phase 3 is middle Neolithic B. The spatial extension the site is 110 m², of which 35m² (ca. 32%) were excavated (Fig. 43). The units selected for analysis are from units 47-49x/46-47y E, F, and G, and 47-48x/49y, layers E and F (8 m²). Altogether 679 tools/debris was analysed.

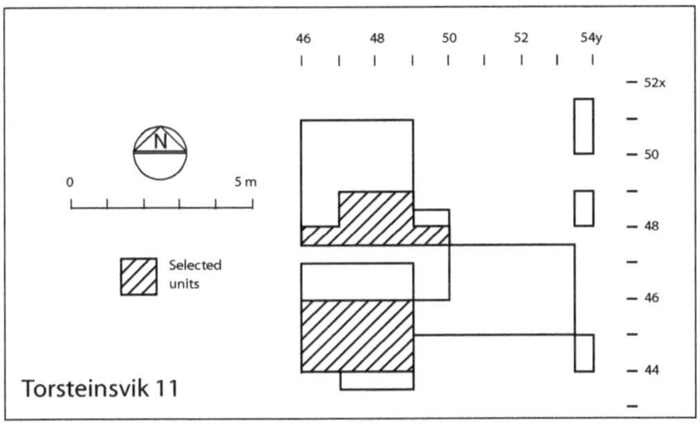

Figure 43. *Torsteinsvik 11, Midthordland. Modified from Bjerck (1984).*

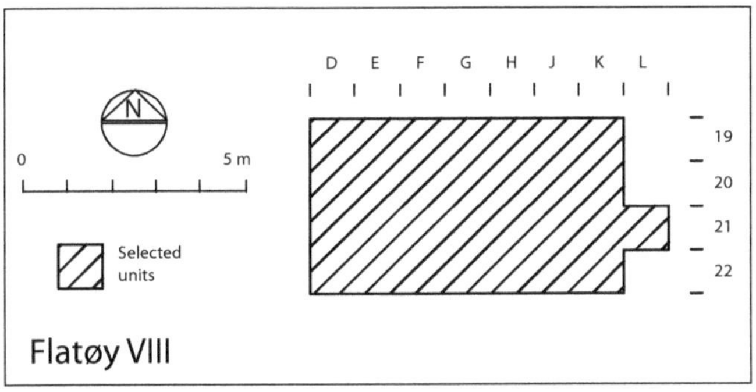

Figure 44. *Flatøy VIII, Midthordland. Modified from Bjørgo (1981).*

Dating: Technological/typological elements. Early/ middle Neolithic: 1 adze of the Vestland/Vespestad type, 5 tanged blade points (A-points), 1 slate point, dominance of small blades. One charcoal sample was radiocarbon dated from layers associated with phase 1:

T-6114: 5040±80 BP. 45x/49y layer 2F

This date is in accordance with the archaeological dating. Phase 1 is therefore dated to the middle part of the early Neolithic.

Site function: 11 Torsteinsvik is situated at the outer coast close to the narrow strait Rongesundet between the two islands Toftøy and Rongøy. Stationary fish species concentrate in Rongesundet. Øygarden is generally rich in marine resources. On the micro-level, the site is close to the shore, there are very good harbour conditions, good view of the sea, and level and drained ground. The site is of medium size (110 m²), the number of artefact forms is low (17), the artefact density is medium (0,84 per litre), the layer is of medium thickness (20 cm), and the site has a cultural layer. The relatively low artefact density and the low number of tool types indicate that the site was in use during a relatively short period of time, perhaps as a short-term residential site or as a field camp.

Flatøy VIII, phase 2

The site is situated at the island Flatøy in Meland municipality (Fig. 27). It was excavated by Tore Bjørgo in 1976 (Bjørgo 1981). Several other sites have been surveyed and excavated at Flatøy (Bjørgo 1981, Simpson 1992). Flatøy VIII is represented by two phases: Phase 1 is dated to the late Mesolithic (layer III) and phase 2 to the early Neolithic (layer 1, 2 and 2A). 29 m² (20%) have been excavated (Fig. 44). The units selected for analysis are layers 2 and 2A from the entire excavated area (29 m²) (Fig. 44). Layer 1 (the modern topsoil) is excluded from the analysis. Layer 3 (Phase 1) is excluded because of its late Mesolithic date. Altogether 622 tools/debris was analysed.

Dating: Typological/technological elements. Late Mesolithic: 4 conical cores, 5 blades produced on microblade cores. Early/middle Neolithic: 1 Vestland/ Vespestad adze, 8 blade tanged points, 5 cylindrical cores and relative dominance of small blades. The Mesolithic elements may have been deposited during phase 1 at the site. Two charcoal samples were radiocarbon dated from the excavated area.

T-2385: 5240±150 BP. Square E20, layer 2A
T-2739: 2550±150 BP. Square H21, layer III

The youngest date (T-2739) is not in accordance with the lithics from layer 2, and should be seen as secondary to the phase. T-2385 is in accordance with the archaeological material. The majority of the data was probably deposited during the early part of the early Neolithic period.

Site function: On the macro-level, Flatøy VIII has a strategic location with regard to communications north-south along the coast and east-west between coast and inland. Furthermore, it is situated close to the tidal current channels (Krossnessundet and Haglesundet), which offered rich and stable marine resources. On the micro-level it rates highly for harbour conditions, view, and level ground. The site (perhaps the entire complex of large sites at Flatøy) may quite likely have been chosen for long-term occupancy. The site is of medium size (150 m²), the number of artefact forms is medium (18), the artefact density is low (0,19 per litre), the layer has medium thickness (25 cm), and the site has a cultural layer. The low artefact density and the relatively low number of tool types indicate that the site was used during a relatively short period of time, perhaps as a short-term residential site or as a field camp.

17 Nilsvik, 4 Nilsvik, structures 10, 29, and 30

The sites are situated on the northern side of the island Bjorøy in Raunefjorden, Fjell municipality (Fig. 27). They were excavated by Kari K. Kristoffersen in 1992 -94 (Kristoffersen 1995). Jan Ivar Trones has analysed parts of 4 Nilsvik (Trones 1996, Trones 1998). A large number of Stone Age sites are surveyed and excavated at Bjorøy. 4 Nilsvik was sporadically occupied during the Mesolithic (phase 1). The early Neolithic (phase 2) and middle Neolithic A and B (phase 3 and 4) are represented by separate stratigraphical layers and structures. 17 Nilsvik is distinguished as a separate site by K. Kristoffersen. This site constitutes, however, the northern natural extension of 4 Nilsvik. The two sites are therefore treated as one unit here. 17 Nilsvik was only occupied during the early Neolithic. Its spatial extension is about 72 m², of which 18 m² (ca. 25%) were excavated (K. Kristoffersen 1995). At 4 Nilsvik, the units selected for analysis are from three different structures: 10, 29, and 30 (Fig. 45). These structures are interpreted as remains of houses by K. Kristoffersen and Trones. The following units are selected: Structure 10: 76x/56y, 75-77x/57-59x (7 m²), layer B 1-3. Structure 29: 72-76x/57-59y (6 m²), layer Ø. Structure 30: 75-78x/53-55y (9 m²), mainly layers X, Y, and Z. Altogether 818 (structure 10), 1547 (structure 29), and 1198 (structure 30) tools/debris were analysed. At 17 Nilsvik, the units 113-114x/58,5-60y and 116-118,5x/58-60y (13,75 m²) layers 2 and 3, were chosen.

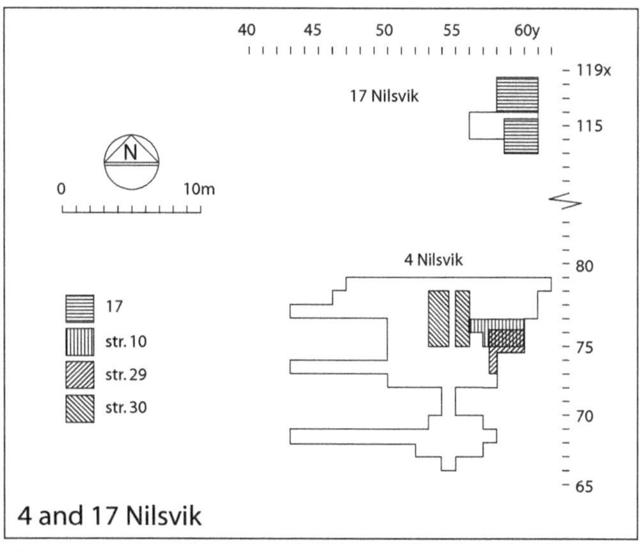

Figure 45. *4 and 17 Nilsvik, Midthordland. Modified from K. Kristoffersen (1995:49, 105).*

Dating: structure 10. Typological/technological elements: early/middle Neolithic: 6 tanged blade points (A-points), 3 slate points, 6 slate point blanks, 2 cylindrical cores, relative dominance of small blades. Middle Neolithic: 1 Vestland chisel. Stratigraphically, structure 10 is partly overlapped by structure 29. Considering that structure 29 probably was used during the early part of the middle Neolithic, structure 10 should be somewhat older. No radiocarbon samples were dated.

Dating: structure 29. Typological/technological elements: early/middle Neolithic: 14 tanged blade points (A-points), 5 slate points, 1 slate point blank, relative dominance of small blades. Two radiocarbon samples were dated from layers associated with structure 29:

> Beta-67789: 4590±100 BP. 73x/57y, hearth, structure 27
> Beta-56847: 3980±60 BP. 74x/59y, NV, layer Ø

Beta-56847 is clearly not in accordance with the archaeological data from the structure. A probable explanation for the late result is that the sample consists of charcoal from middle Neolithic B phase 4, which has a thick cultural deposit stratigraphically above layer Ø. Beta-67789 is in accordance with the typological/ technological elements. On the basis of the above, it can be argued that structure 29 was occupied during the early/middle Neolithic transition.

Dating: structure 30. Typological/technological elements: early/middle Neolithic: 1 Vestland/Vespestad adze, 16 tanged blade points (A-points), 1 slate point, 4 slate point blanks, 3 cylindrical cores, relative dominance of small blades. 53 sherds of pottery were found in structure, 30. On the basis of visual inspection of these sherds, K. Kristoffersen and Trones argue that two of these sherds are similar to the TRB type of sherds found in contemporary phase 13 at Kotedalen (see above), while the others are typical examples of the "local" type which is tempered with coarse quartz. Later visual inspection of these sherds, however, indicates they are all of the "locally produced" type (Leif Inge Åstveit personal communication). Three radiocarbon samples were collected from layers associated with structure 30:

> Beta-67783: 4620±100 BP. 78x/53y, SØ, Z5
> Beta-56843: 4860±60 BP. 72x/52y, NØ, Z4

These results suggest a date to the late part of the early Neolithic. On the basis of the above, I argue that the structure 30 was occupied during the early/middle Neolithic transition.

Dating: 17 Nilsvik. Typological/technological elements: early/middle Neolithic: 3 Vestland/Vespestad adzes, 1 Vestland chisel, 29 tanged blade points (A-points), 2 cylindrical cores, relative dominance of small blades. One conical core indicates that the site was occupied during the late Mesolithic period. Two radiocarbon samples were collected from the site:

Beta-67775: 4690±140 BP. 116x/58y, SØ, layer 5
Beta-67779: 4660±140 BP. 117x/60y, NV, layer 6

These two dates indicate an age at the early/middle Neolithic transition. In sum, the radiocarbon determinations and the typological/technological elements indicate that Nilsvik 17 and the three structures 10, 29, and 30 at Nilsvik 4 were occupied during the early/middle Neolithic transition.

Site function. 4 and 17 Nilsvik are situated close to the tidal current channel Vatlestraumen between Håkonshella and Bjorøy. Vatlestraumen has rich marine resources. Due to the tough weather conditions outside the island of Sotra, the main route for ships along the coast goes through Vatlestraumen. The macro-environment is therefore favourable with regard to communications as well as for marine hunting and fishing. On the micro-level, the site rates high on the location factors harbour conditions, shelter, proximity to fresh water, and level ground (K. Kristoffersen 1995:31). The Nilsvik 4 site-complex is large (875 m² - minimum estimate), while 17 Nilsvik is of medium size (72 m²). The number of artefact forms for 17 Nilsvik and 4 Nilsvik structures 10, 29 and 30 are medium and high (26/25/24/29). The artefact densities are high and medium (1,75/0,78/1,72/0,92 artefacts per litre), the measurements of layer thickness are medium (35/25/25/25 cm), and both units have cultural layers. Trones has suggested that 4 Nilsvik was occupied by sedentary households during the early and middle Neolithic (Trones 1998). On the basis of the above data, his interpretation is supported here for both 4 and 17 Nilsvik.

Austvik III

The site is situated at the northern side of the island Goddo, Bømlo municipality, to the west of the island of Stord, and to the north of Bømlo (Fig. 28). The site was

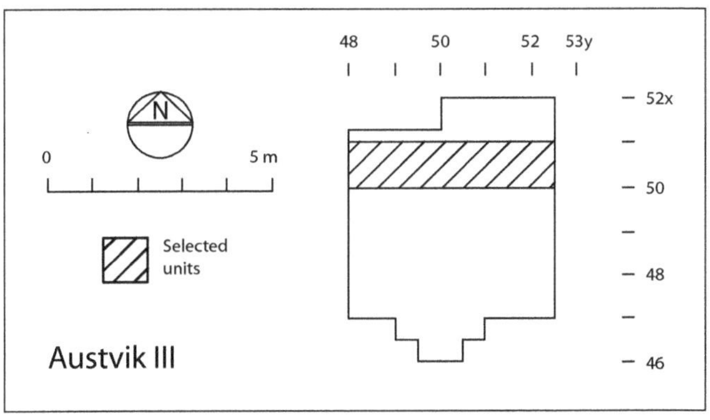

Figure 46. *Austvik III, Sunnhordland. Modified from S. Kristoffersen (1990:27).*

excavated by Siv Kristoffersen in 1988 (S. Kristoffersen 1990, S. Kristoffersen 1995). In this area, several Stone Age sites have been surveyed and excavated. Austvik IV and V are situated a few meters from Austvik III. The spatial extension of the site complex is 175 m², of which 58,5 m² (ca. 33%) were excavated (Fig. 46). At Austvik III, 24 m² were excavated. The site was only occupied during the early Neolithic. The units selected for analysis are from units 50x/48-52y (4 m²), layers 0-5 cm and 5-10 cm. 2525 tools/debris were analysed.

Dating: Typological/technological elements: early Neolithic: 2 Vespestad adzes. Early/middle Neolithic: 24 tanged blade points (A-points), 17 cylindrical cores, relative dominance of small blades. One charcoal sample was radiocarbon dated from the relevant layer:

Beta-48159 (AMS): 5030±60 BP, 48x/50y, hearth.

This result is in accordance with the archaeological dating and the site is therefore likely to have been occupied during the middle part of the early Neolithic.

Site function. Austvik III is situated in an archipelago of small islands (The Fitjar islands) to the north of Bømlo, close to the North Sea. This area is generally rich in marine resources. On the micro-level the site is situated in a narrow bay, close to the contemporary shoreline. The view from the site itself is not good, however, the scores on factors like harbour conditions, shelter and level and drained ground are high. The site is of medium size (175 m²), the number of artefact forms is medium (19), the artefact density is high (6,31 per litre), the layer is thin (20 cm), and the site has a cultural layer. S. Kristoffersen interpreted the site as a result of an occupation by a small household for some period of time (S. Kristoffersen 1995b:169). Her interpretation is supported here.

88 Føyno

The site is situated at the small island Føyno, Stord municipality (Figs. 10 and 28), between the large islands Bømlo and Stord. The site was excavated by Hilde Roland in 1994 and published by Kari K. Kristoffersen (Roland 1994, K. Kristoffersen 2001b). Due to development of this area (The Stord-Bømlo-Sveio highway connection), a number of Stone Age sites have been surveyed and excavated (see also Chapter 4). The spatial extension of the site is ca. 150 m² of which 72 m² (ca. 48%) were excavated (Fig. 47). The site was occupied during the early Neolithic. The units selected for analysis are 52-53x/107, 5-116, 5y (17 m²), layers 1-2. 966 tools/debris were analysed.

Dating: Typological/technological elements: early Neolithic: 1 Vespestad adze. Early/middle Neolithic: 15 tanged blade points (A-points), 6 cylindrical cores, relative dominance of small blades. One radiocarbon

(AMS) date is available from the site (collected from a hearth, structure 7):

Beta-74479: 4810±60 BP, 110x/53y, layer 2.

On the basis of the radiocarbon date and the typological/technological date, it is suggested that the site was occupied during the latter part of the early Neolithic.

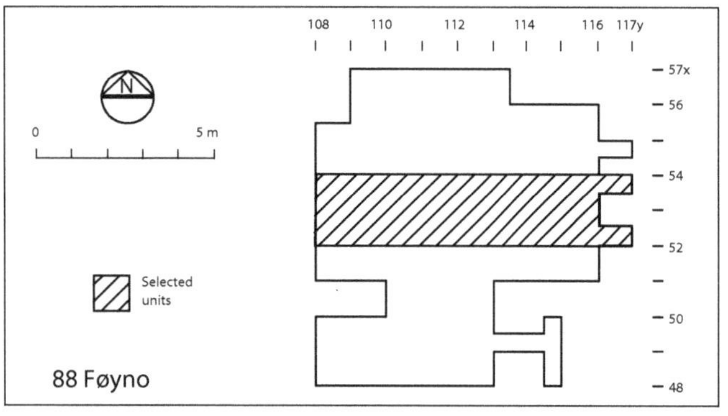

***Figure 47.** 88 Føyno, Sunnhordland. Modified from K. Kristoffersen & Warren (2001:154).*

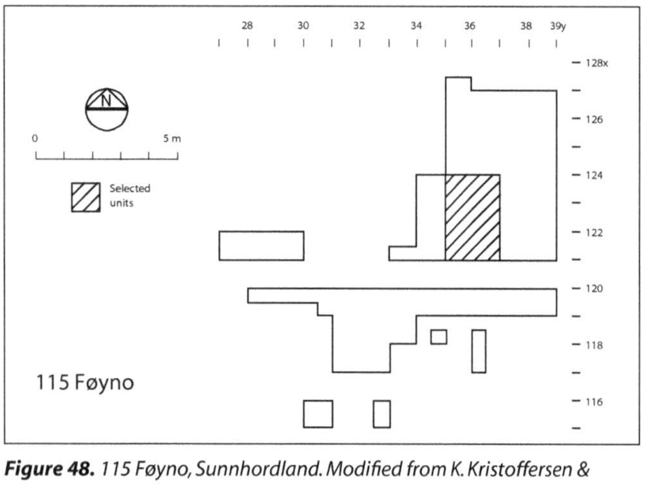

***Figure 48.** 115 Føyno, Sunnhordland. Modified from K. Kristoffersen & Warren (2001:163).*

Site function. 88 Føyno is situated at in Bømlafjorden, between Stord and Bømlo. The macro-environment is favourable with regard to communications as well as for hunting and fishing. On the micro-level, the site scores high on the factors view, harbour conditions, accessible area, and level ground. It scores low on proximity to fresh water and shelter (K. Kristoffersen 2001a). The site is of medium size (150 m²), the number of artefact forms is medium (26), the artefact density is medium (1 per litre), the layer is thin (10 cm), and the site only has a sand layer, not a cultural layer. Although the site location is good, the thin layer and the lack of a cultural layer indicate that the site was used during a relatively short period of time, perhaps as a short-term residential site or as a field camp.

115 Føyno

The site is situated on the small island Føyno, Stord municipality (Figs. 10 and 28), between the large islands Bømlo and Stord. It was excavated by Nils Anfinset in 1994 and published by Kari Kristoffersen (Anfinset 1994, K. Kristoffersen 2001b). Due to development of this area (The Stord-Bømlo-Sveio highway connection), a number of Stone Age sites have been surveyed and excavated. The spatial extension of the site is ca 150 m² of which 50, 25 m² (ca. 33%) were excavated (Fig. 48). The site was occupied during the early Neolithic. The units selected for analysis are 121-123x/35-36y (6 m²), layers 1-2. 1342 tools/debris were analysed.

Dating: Typological/technological elements: early/middle Neolithic: 10 tanged blade points (A-points), 3 cylindrical cores, relative dominance of small blades. It should be noted that 3 Vestland/Vespestad adzes and one Vestland chisel were found outside the analysed area. One conical core indicate that the site have been used during the late Mesolithic. Two radiocarbon dates are available from the site:

Beta-78819:1869±70 BP, 125x/36y, NØ, layer 1
Beta-78820: 4610±70 BP, 121x/37y, layer 1.

Beta-78819 is clearly too young. It probably dates the deforestation of the area. Beta-78820 is in accordance with the archaeological dating. The above elements suggest a date at the early/middle Neolithic transition.

Site function: For macro-environment, see 115 Føyno. On the micro-level, the site scores high on the factors view, harbour conditions, accessible area, and level ground. It scores low on shelter (K. Kristoffersen 2001a). The site is of medium size (150 m²), the number of artefact forms is medium (21), the artefact density is medium (0,39 per litre), the layer is medium (20 cm), and the site only has a sand layer, not a cultural layer. The lack of a cultural layer indicates that the site was in use during a relatively short period of time, perhaps as a short-term residential site or as a field camp.

Sokkamyro 1901-2

The site is situated at Vespestad, southern Bømlo, Bømlo municipality (Fig. 28 and 133). It was excavated by Haakon Schetelig during several periods between 1901 and 1939 (e.g. Schetelig 1901, Shetelig 1922, S. Alsaker 1987). Sokkamyro is mainly Neolithic, but the site area was also occupied during the Mesolithic. The spatial extension of the site complex at Sokkamyro is probably more than 9000 m² (A. B. Olsen 2002), of which ca 80 m² (ca. 0,9%) were excavated (Fig. 49). The units selected for analysis were excavated by Shetelig in 1901-2: units IA-F and IIB, IID-F, mainly from the cultural layer. 3001 tools/debris were analysed. It should be noted

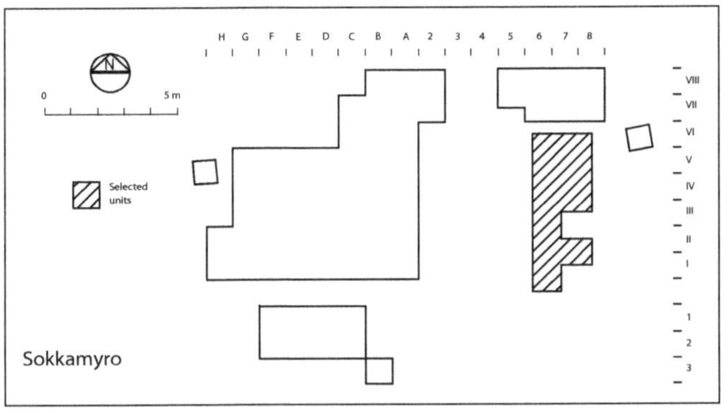

Figure 49. *Sokkamyro, Sunnhordland. Modified from Schetelig (not dated).*

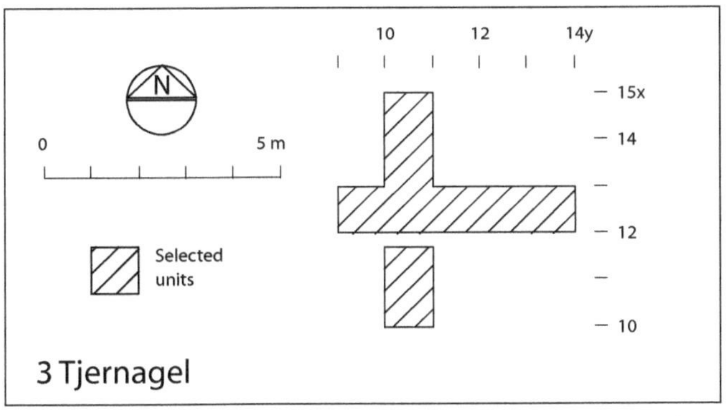

Figure 50. *3 Tjernagel, Sunnhordland. Modified from Bjerck & Ringstad (1985:32).*

that Shetelig did not screen the soil, which resulted in that he missed the smallest artefacts (flakes smaller than 1 cm were not registered at all during my re-analysis of the site). He wrote that most of the greenstone flakes were counted, but were not brought to the museum. Except for a portion of the greenstone debris, however, he noted that the excavation was careful, and that all artefacts were collected and catalogued (Schetelig 1901). If the lack of small flakes is taken into consideration, I find it defendable to apply the re-analysed data from Sokkamyro for my purpose.

Dating: Typological/technological elements: late Mesolithic: 4 adzes with round/oval cross-sections, 2 conical cores, 1 blade produced by the microblade technique. Early Neolithic: 10 Vespestad adzes. Early/middle Neolithic: 3 tanged blade points (A-points), 4 slate points, 1 shard of "locally produced" pottery, 60 cylindrical cores, relative dominance of small blades. A mainly early/middle Neolithic date is also indicated by a large number of short and broad struck greenstone blanks (most likely preforms for Vestland/Vespestad adzes) compared to a low number of long and slim Mesolithic blanks (preforms for adzes with round cross-sections). No radiocarbon determinations are available from the site.

Site function: Sokkamyro is situated at the isthmus between the two deep bays Eidesvåg and Langevåg at southern Bømlo. The site seems to cover the entire isthmus. The main route for ships passes southern Bømlo. The area is generally rich in marine resources. In addition, the stone-adze quarry at Hespriholmen is located only 3 km to the west of Sokkamyro. On the micro-level the site is situated in a narrow bay, close to the contemporary shoreline. The view form the site itself is not the best, but the scores on factors like harbour conditions, shelter and level and drained ground are high. The site is very large (9000 m²), the number of artefact forms is high (29), the artefact density is medium (0.84 per litre – greenstone flakes are excluded), the layer is of medium thickness (20 cm), and the site has a cultural layer. Production of blanks and adzes of greenstone from the quarry at Hespriholmen was carried out at the site. Shetelig (1922) interpreted the site as a combined greenstone adze workshop and residential site. This is supported by the data, which indicates frequent or long-term use by one or several households.

3 Tjernagel

The site is situated at Sveio, Sveio municipality, on the eastern shore of Bømlafjorden (Fig. 28). It was excavated by Åsa Dahlin Hauken in 1983 and published by Hein Bjartmann Bjerck (Hauken 1983, Bjerck & Ringstad 1985). At Tjernagel, several Stone Age sites have been surveyed and excavated. The spatial extension of the site is ca 40 m² of which 9 m² (ca. 23%) were excavated (Fig. 50). The site was occupied during the late Mesolithic and early Neolithic. The units selected for analysis are 10-14x/9-13y (9 m²), layers 2-6. 2204 tools/debris were analysed.

Dating: Typological/technological elements: Mesolithic: 5 micro blades produced by the microblade technique. Early Neolithic: 1 Vespestad adze. Early/middle Neolithic: 9 tanged blade points (A-points), 2 cylindrical cores, relative dominance of small blades. The elements indicate that the site was occupied sporadically during the late Mesolithic, but that it was mainly early/middle Neolithic. No radiocarbon determinations are available from the site.

Site function: Tjernagel 3 is situated at Bømlafjorden. The macro-environment is, therefore, favourable with regard to communications as well as for hunting and fishing. On the micro-level, the site is situated in a narrow bay, relatively far (70 m) from the contemporary shoreline. The view is not good from the site, but there are high scores for harbour conditions, shelter and level and drained ground. The site is small (40 m²), the number of artefact forms is medium (21), the artefact density is medium (0, 82 per litre), the layer is medium (30cm), and the site only has a gravel layer, not a cultural layer. The location of the site, the small area and the lack of a cultural layer indicate that the site was occupied during a relatively short period of time, perhaps as a short-term residential site or as a field camp.

Styggvasshelleren, phase 2

The site is a rock-shelter. It is situated at the northern shore of Styggevatn (1200 m above sea level) in Breheimen, Luster municipality, Sogn og Fjordane (Fig 24). It was excavated by Kjersti Randers in 1982 and 1983 (Randers 1986). Another Stone Age site (Hella) in this area has been excavated. Styggvasshelleren has four phases. Phase 1 is Mesolithic, phase 2 is early Neolithic, phase 3 is late Bronze Age, and phase 4 is from the Migration Period. The extension of the site is about 60 m², of which 47 m² (78%) were excavated. The units selected for analysis are 9 m² from 56-58x/49 from layers 8, 9, and 10 (Fig. 51). 1661 lithic tools and debris were analysed.

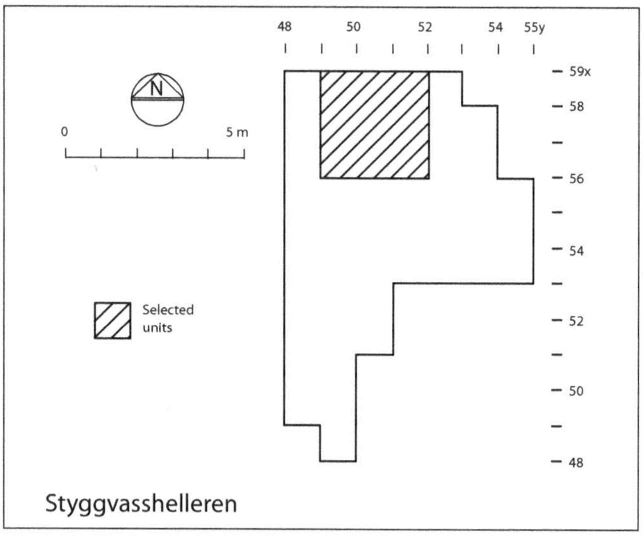

Figure 51. *Styggvasshelleren, Breheimen. Modified from Randers (1986:49).*

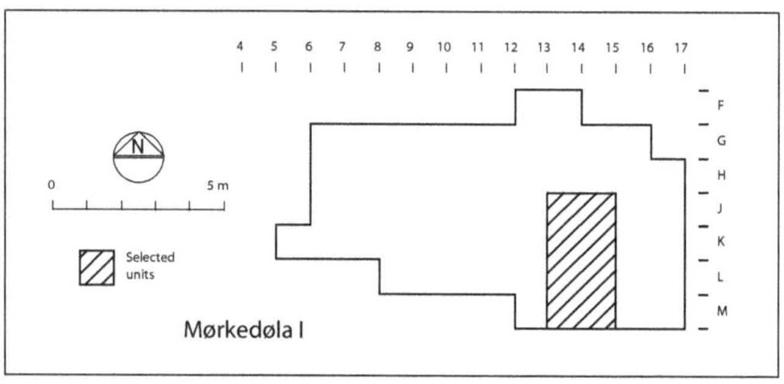

Figure 52. *Mørkedøla I, Lærdal-Hemsedal. Modified from A. B. Johansen (1969:137).*

Dating: Typological/technological elements: early/middle Neolithic: 5 tanged blade projectile points (A-points), 49 slate points, 3 slate point blanks, relative dominance of small blades. It should be noted that a diabase adze of a possible early Neolithic Vespestad type is found in layer 12. One radiocarbon sample was dated from layers associated with phase 2:

T-5519: 4560±80 BP. 57x/50y layer 10

The radiocarbon date indicates an age at the early part of the middle Neolithic. The archaeological elements indicate, however, that it was also occupied during the early Neolithic.

Site function. On the macro-level, the site has a good location with regard to reindeer-hunting. The micro environment is generally good. The site is situated close to the shore of Styggevatn. The rock-shelter guarded effectively against the rapidly changing weather conditions in this area. The site is of medium size (60 m²), the number of artefact forms is medium (19), the artefact density is medium (1,23 per litre), the layer is of medium thickness (20 cm), and the site has a cultural layer. These attributes do not diverge significantly from coastal sites interpreted as residential sites. However, considering that Styggvasshelleren is situated in the alpine zone, it is not likely that people stayed here on an annual basis. Randers interpreted the site as a field camp; the result of frequent occupation by task groups/hunting crews. This interpretation is supported here.

Mørkedøla I

The site is situated at the western shore of Tjørni (1115 m above sea level) in the upper parts of the Lærdal watercourse, Lærdal municipality, Sogn og Fjordane (Fig 24). It was excavated by Odd Espedal and Signe Hvoslef in 1965 and 1967 (Espedal 1965, Hvoslef 1967) and analysed by Arne B. Johansen (1969; 1978). Several Stone Age sites have been surveyed and excavated in this area. Mørkedøla I has been occupied during three phases; phase 1 is late Mesolithic, phase 2 is early/middle Neolithic, phase 3 is late Bronze Age/early Iron Age. The extension of the site is about 60 m², the excavators estimate that about 100 % was excavated. The units selected for analysis are 8 m² (J-M 14, J-M 15) (Fig. 52). 1929 lithic tools and debris were analysed.

Dating: Typological/technological elements: Mesolithic: 3 micro blades produced by microblade technique. Late Mesolithic/Neolithic: 4 tanged flake points (A-points), 1 transverse point. Early/middle Neolithic: 1 slate point (Pyhensilta type), 1 slate point blank. It should be noted that several single-edged points were found at the site. One leaf-shaped bifacial point was also found. One radiocarbon sample was dated from the site:

T-670: 4830±160 BP. J17, layer 1.

The radiocarbon date indicates an early Neolithic age. On the basis of the radiocarbon determination and the typological/technological elements, it can be argued that the site was mainly occupied during the late Mesolithic/early Neolithic transition.

Site function: On the macro-level, the site has a central location with regard to reindeer-hunting. The micro environment is generally good: it lies close to the shoreline, on level and drained ground, and it scores high on view. The site is of medium size (60 m²), the number of artefact forms is low (15), the artefact density is high (2,41 per litre), the layer is thin (10 cm), and the site only has a sand layer, not a cultural layer. The lack of a cultural layer, the low number of artefact forms, and thin layer indicate that the site was occupied during relatively short periods of time, perhaps as a short-term residential site or as a field camp.

Gyrinos IV

The site is situated on the western shore of Gyrinosvatn between Hemsedal and Hallingdal, Ål municipality, Buskerud (Fig 24). It was excavated in 1959 (Martens & Hagen 1961) and re-analysed by Hagen (1963), Indrelid (1973), and Schaller (1984, Schaller-Åhrberg 1990). Along this lake, a number of Stone Age sites have been surveyed and excavated. The extension of the site is about 50 m², of which 21 m² (42%) were excavated. The site was occupied during the late Mesolithic and early Neolithic periods. The artefact material has not been related to the excavated units in the museum collection (all flakes from the site are gathered in one bag); the entire excavated area has therefore been selected for analysis. 2413 tools/debris is analysed.

Dating: Typological/technological elements: Mesolithic: 25 blades produced by microblade technique. Mesolithic/Neolithic: 11 blade/flake tanged points (A-points), 11 single-edged points, and 11 transverse points. Early/middle Neolithic: 1 slate point, 10 blades produced by cylindrical core technique. Late lithic elements: 1 bifacial point. One radiocarbon date is available from the site:

K-711: 5700±120 BP (Indrelid 1973:38).

The provenance of the dated sample is not known. In his discussion of the site, Hagen (1963) dated the site to the early Mesolithic. Later, it has been argued that the site is a mixture of several late Mesolithic and early/middle Neolithic phases of occupation (Indrelid 1973, Schaller-Åhrberg 1990:614). Indrelid's and Schaller-Åhrberg's suggested date is supported here.

Site function: On the macro-level, the site has a good location with regard to reindeer-hunting. The micro environment is also good: it is situated close to the shoreline, on level and drained ground, and it scores high on view. The site is of medium size (50 m²), the number of artefact forms is medium (19), the artefact density is medium (0,57 per litre), the layer is of medium thickness (20 cm), and the site only has a sand layer, no cultural layer. The broad frame of dating and the lack of a cultural layer indicate that the site was used during relatively short periods of time, perhaps as a short-term residential site or as a field camp.

Blånut IV

The site is situated on the southern shore of Gyrinosvatn between Hemsedal and Hallingdal, Ål municipality, Buskerud (Fig 24). It was excavated in 1959 (Martens & Hagen 1961). The extension of the site is about 50 m², of which 20 m² (40%) were excavated (Fig. 53). The site was occupied during the early/middle Neolithic periods, possibly also during the late Mesolithic. The artefact material has not been related to the excavated units in the museum collection (all flakes from the site are gathered in one bag); the entire excavated area has therefore been selected for analysis. 656 tools/debris are analysed.

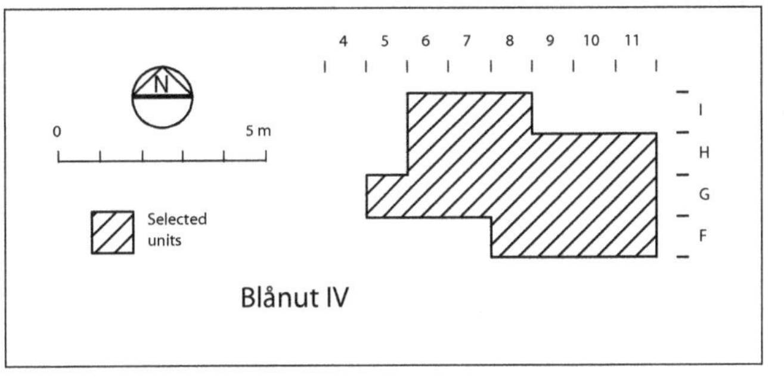

Figure 53. *Blånut IV, Lærdal-Hemsedal. Modified from Martens & Hagen (1961:35).*

Dating: Typological/technological elements: early Neolithic: 1 fragment of a thin-butted, ground flint axe. Early/middle Neolithic: 4 fragments of ground flint axes, 22 fragments of slate points, 10 slate point blanks. One radiocarbon determination is available:

T-256: 6850±50 BP (Hagen 1983:33).

The provenance of the dated sample is not known. The C14-dating may be in accordance with the micro blades found at the site. However, none are produced in the microblade technique. As Martens & Hagen (1961) and Schaller-Åhrberg (1990) point out, the majority of the archaeological material indicates an early Neolithic – possibly a middle Neolithic date instead.

Site function: At both macro-level and micro-level, the location is similar to that of Gyrinos IV. The site is of medium size (50 m²), the number of artefact forms is low (17), the artefact density is low (0,16 per litre), the layer is of medium thickness (20 cm), and the site only has a sand layer, not a cultural layer. The low artefact density, the low number of artefact forms and the lack of a cultural layer indicate that the site was used during relatively short periods of time, perhaps as a short-term residential site or as a field camp.

760 Finnsbergvatn

The site is situated at Sumtangen, at the shore of Finnsbergvatn at the mountain plateau Hardangervidda, Eidfjord municipality, Hordaland (Fig. 24). It was excavated by Svein Indrelid in 1973 (Indrelid 1994). A number of Stone Age sites have been surveyed and excavated at Sumtangen (Bøe 1942, Indrelid 1994:130). The spatial extension the site is ca 50 m² of which 25 m² (ca. 50%) were excavated (Fig. 54). The site is a rectangular house-site which was occupied during the early Neolithic. The units selected for analysis are 10-12x/9-12y (8,5 m²), layers I and II. 6301 tools/debris were analysed.

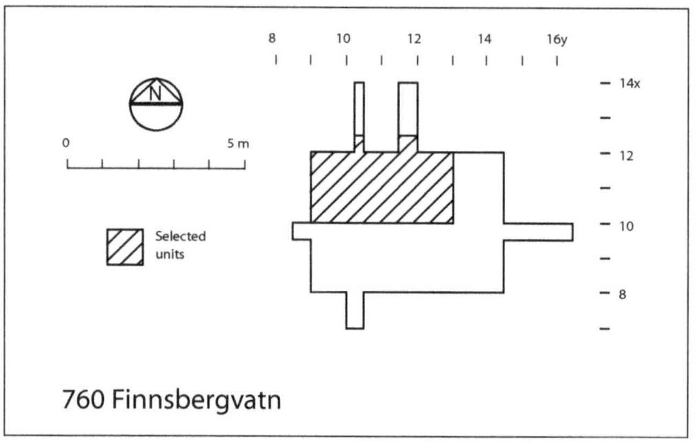

Figure 54. *760 Finnsbergvatn, Hardangervidda. Modified from Indrelid (1994:145).*

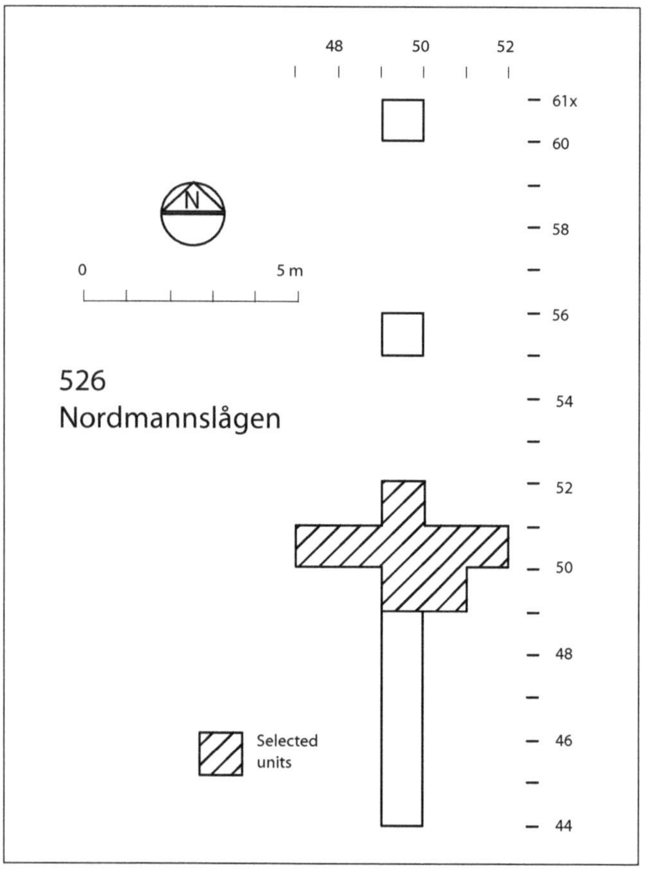

Figure 55. *526 Nordmannslågen, Hardangervidda. Modified from Indrelid (1994:70).*

Dating: Typological/technological elements: late Mesolithic/Neolithic: 12 blade/flake tanged projectile points (A-points), 13 single-edged points, 1 transverse point. Early/middle Neolithic: 5 slate points. Bipolar core technique as well as cylindrical technique is represented by blade points and some small blades (the tanged points and single-edged points were produced from both blades and flakes at the site). Two radiocarbon dates are available from the site:

T-1611: 5100±80 BP, 10x/12y, layer DII.
T-1612: 5080±150 BP, 10x/10y, layer AII.

These two results are in accordance with the typological/technological elements. A mainly early Neolithic date of the site is suggested.

Site function. Most of the sites at Hardangervidda are situated close to major reindeer trekking routes. Site 760 is located less than 2 km from such a route. Furthermore, it is situated at the point where reindeer swim across Finnsbergsvatn. Svein Indrelid (1994:270) argued that reindeer hunting probably was the main cause for the exploitation of Hardangervidda during the Mesolithic and earliest parts of the Neolithic. Similar as many other sites at the mountain plateau, 760 Finnsbergvatn has a central location in the terrain. The area is also generally favourable with regard to livestock-keeping. Pollen data indicates that pasturing took place during the early Neolithic at Hardangervidda (Indrelid & Moe 1983). The pastures are best close to the lakes, where most of the sites are found. The micro environment is characterised by dry and level ground and easy access to fresh water. The site is of medium size (50 m²), the number of artefact forms is medium (25), the artefact density is high (2,8 per litre), the layer is of medium thickness (20 cm), and the site has a cultural layer. These attributes do not diverge significantly from coastal sites interpreted as residential sites. However, considering that 760 Finnsbergvatn is situated in the alpine zone, it is probably not likely that people stayed here on an annual basis. The site should therefore probably be interpreted as a field camp; as the result of frequent reoccupations by task groups/hunting crews hunting reindeer at Hardangervidda. These groups may have practised livestock-keeping.

526 Nordmannslågen

The site is situated on the shore of Nordmannslågen at the mountain plateau Hardangervidda, Eidfjord municipality, Hordaland (Fig 24). It was excavated by Petter Dyrkorn and Denise M. Rich in 1973 (Dyrkorn & Rich 1973) and published by Svein Indrelid (1994). The spatial extension of the site is ca 1200 m² of which 15 m² (ca. 1%) were excavated (Fig. 55). The units selected for analysis are 49-51x/48-51y (10 m²), layers I-III. 1515 tools/debris were analysed.

Dating: Typological/technological elements: late Mesolithic/Neolithic: 23 blade/flake tanged projectile points (A-points), 14 single-edged points, 1 transverse point, 1 slate point blank. Early/middle Neolithic: 2 cylindrical cores, 2 flakes from ground flint axes (1 of the single edged points was made from a ground flint flake), 44 sherds of (fine-tempered) pottery. 10 of the sherds had impressions of real, double-twined cord. Indrelid interpreted the ceramics as a possible TRB-type. This interpretation is confirmed by later inspections of the sherds (personal communication, Einar Østmo). The tanged points and single-edged points were produced from both blades and flakes. One radiocarbon dates is available from the site:

T-1618: 4860±170 BP, 51x/49y, hearth.

This result is in accordance with the typological/ technological elements. An early Neolithic date is therefore suggested.

Site function: Both micro-environment and macro-environment are much the same as for 760 Finnsbergvatn (see above). The site is large (1200 m²), the number of artefact forms is medium (21), the artefact density is medium (0,75 per litre), the layer is of medium thickness (25 cm), and the site has a sand layer. On the basis of the above, the site should probably be interpreted as a field camp left by early Neolithic task groups hunting reindeer. These groups may have practised livestock-keeping.

1020 Bjornesfjorden

The site is situated on eastern the shore of Bjornesfjorden at the mountain plateau Hardangervidda, Uvdal municipality, Buskerud (Fig 24). It was excavated by Svein Indrelid in 1973 (Indrelid 1994:82-84). The spatial extension of the site is ca 100 m² of which 16 m² (ca. 16 %) were excavated (Fig. 56). The entire excavated area was selected for analysis. 1023 tools/debris were analysed.

Dating: Typological/technological elements: late Mesolithic/early Neolithic elements: 4 flake tanged points (A-points), 1 single-edged point, and 12 transverse points. Early/middle Neolithic elements: 3 fragments of ground flint axes. Late Lithic Period: 1 leaf shaped bifacial point. One radiocarbon date is available from the site:

T-1785: 5190±100 BP, 8x/4y, charcoal layer.

This date is confirmed by the typological/technological elements. On the basis of the above, it is argued that the site was mainly used during the transition between the Mesolithic and the Neolithic periods (Indrelid 1994-84). Site function: Both micro-environment and macro-environment is much the same as for 760 Finnsbergvatn (see above). The site is of medium size (100 m²), the number of artefact forms is low (17), the artefact density is medium (0, 31 per litre), the layer is of medium thickness (20 cm), and the site has a cultural layer (the cultural layer is mainly found in an oval depression at the site). On the basis of these attributes, the site may be interpreted as a residential site, possibly as a field camp.

Vivik

The site is situated on eastern the shore of Holmevatn at Haukelifjell, Vinje municipality, Telemark (Fig 24). It was excavated by Knut Odner and Bjørn Myhre in 1962 and 1963 (B. Myhre 1963, Odner 1965; 1968). The spatial extension of the site is ca 300 m² of which 71 m² (ca. 24 %) were excavated (Fig. 57). The artefact material has not been related to the excavated units in the museum collection (all flakes are gathered in one bag); the entire excavated area has therefore been selected for analysis. 2177 tools/debris were analysed.

Dating: Typological/technological elements: late Mesolithic: 3 blades produced by microblade technique. Late Mesolithic/early Neolithic elements: 4 blade/flake tanged points (A-points), 10 single-edged points, and 3 transverse points. Early Neolithic elements: 1 fragment of a thin-butted, ground flint axe. Early/middle Neolithic elements: 4 fragments of ground flint axes, 3 fragments of slate points, 12 slate point blanks. Middle Neolithic

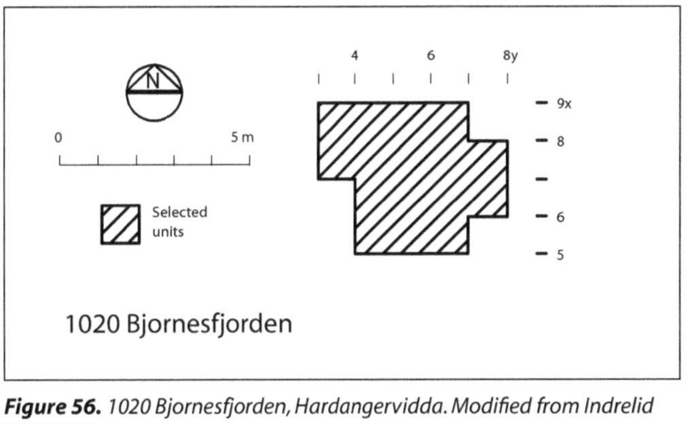

Figure 56. *1020 Bjornesfjorden, Hardangervidda. Modified from Indrelid (1994:82).*

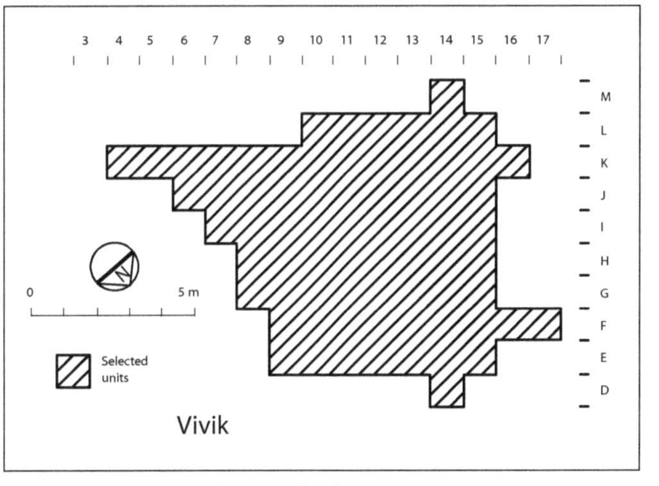

Figure 57. *Vivik, Haukelifjell. Modified from Odner (1965:209).*

elements: 3 B-points. Odner (1965) argued that the site was used during a Mesolithic phase and a middle Neolithic phase. The technological/typological elements indicate, however, that the site was also used during the early Neolithic.[1]

Site function. The macro-environment is favourable with regard to reindeer hunting. At the micro level, there is good view, and the site is situated close to the shoreline, on level and drained ground. The site is large (300 m^2), the number of artefact forms is medium (21), the artefact density is medium (0, 30 per litre), the layer is thin (10 cm), and the site does not have a cultural layer, only a sand layer. The broad frame of dating, the thin layer and the lack of cultural layers indicate that the site was used repeatedly during relatively short periods of time, perhaps as a short-term residential site or as a field camp.

Evaluation of the excavated sites

Dating

Most of the excavated sites are dated on the basis of both archaeological material and radiocarbon determinations. For some of the districts with well-dated sites, it has even been possible to distinguish sites that were used during early, middle and late phases of the early Neolithic period. Many sites are, however, less securely dated, and can be related only generally to the early Neolithic. These differences in resolution imply that it is not possible to discuss changes that have taken place *within* the early Neolithic period to any extent. More problematic, however, is that several sites are only generally dated to the early/middle Neolithic, or generally to the late Mesolithic/early and middle Neolithic. On the other hand, these sites are problematic only to the extent that they make comparisons between the early Neolithic and the preceding and succeeding periods difficult. Assuming that the early Neolithic makes up a significant element at these broadly dated sites, the data may still be important for comparisons with other, more precisely dated sites from this period.

Site functions

Half of the excavated sites (18) presented above have favourable micro and macro site locations. They all have cultural layers. The measured variables site size, number of artefact forms, artefact density and layer thickness have medium or high values (Table 2). In accordance with the initial assumptions, these sites are interpreted as residential sites. Still, there are probably functional differences also within this group of sites. One might, for example, expect that the sites Styggvasshelleren and 760 Finnsbergvatn in the mountain-zone served different purposes than some of the coastal sites, which were most likely used on a year-round basis (A. B. Olsen 1992, Bergsvik 2001b). As Indrelid (1994) and Randers (1986) have pointed out, Hardangervidda and Breheimen were probably occupied on a seasonal basis as field camps. Functional variation may also explain the differences that can be observed within the group of coastal residential sites. These differences may be the results of seasonal vs. year-round occupations, different degrees of reoccupations, or different sizes of the households.

The other half of the sites (19) either do not have cultural layers or score low on one of the measured attributes, or has less favourable site locations. None of them should, however, be interpreted as special purpose camps. If they were, one would expect that they scored low on *more* than one or two of the variables. This means that the main goal of the functional determination – to avoid special purpose sites – has been reached. Nevertheless, as long as the latter sites were mainly the results of shorter stays than the residential sites, it may perhaps be expected that the composition of types, techniques and raw materials is somewhat skewed compared to sites that have been occupied for longer periods of time. In the discussions in the following chapters, these sites will therefore be treated with more caution than the other sites.

Sizes of analysed samples

None of the sites have been excavated totally, and in most cases, the analysed data comes from selected units of the excavated area. These units are taken from the best and most homogenous part of the sites, and where the stratigraphical problems were as few as possible.

Ideally, the sample sizes should have been larger than 1000 artefacts for all of the excavated sites. At most of the sites, relatively large areas were excavated, and sufficiently large samples from good contexts could easily be selected. At some of these sites, large excavated areas made it possible to select only a sufficiently large number of units from the high number of units available. At other sites, the excavated area was small, and only a few units from securely dated contexts were available from the relevant period. In other cases again, the excavated area was large, but had many periods present, which in turn made it difficult to select the early Neolithic units from the stratigraphical sequences. Due to these different circumstances, the numbers of analysed artefacts at the sites vary between 388 and 7280. It should be noted that this difference is partly caused by the different sizes of

1 Håkon Glørstad (2002a:221ff) has showed that Odner's chronological discussion in 1965 was a result of a strong attachment to the developments which had taken place mainly within Swedish and Danish archaeology, and which "forced" much of the early Neolithic of southern Norway into middle Neolithic "pitted ware culture". As I pointed out in Chapter 2 and 5, however, much of the later chronological work (mainly based on radiocarbon determinations), which has been performed in southern Norway has established early Neolithic frameworks independently of Danish and Swedish research.

the analysed units (between 2 m^2 and 20 m^2), and partly the result of different artefact density (between 0, 19 and 10, 34 artefacts per litre).

At the sites with low numbers of analysed artefacts, it follows that the typological, technological and raw material variation is not as large as at the sites with large analysed samples. At the sites with small samples, there is a danger that the quantitative patterns of raw materials, types, and techniques observed in these units might represent different intra-site activities rather than a general trend in the use of these traits. It is probably not coincidental that phase 13 at Kotedalen, which is the site with the highest sample-size, is also the site with the largest number of artefact types. In reverse, the minimally excavated 1 Snekkevik across Fosnstraumen has relatively few artefact forms. Consequently, some of the small totals might not be as representative as the large totals. One might argue that it would be better to leave out the sites with the smallest samples. On the other hand, these sites are often well dated or fill chronological or spatial "gaps", and are important for the overall regional analysis. The sites are not left out, but the problems of low sample sizes have to be kept in mind when the data from the relevant sites is brought into the discussion.

Excavation techniques

The majority of the excavated sites have been excavated during the last 30 years. In this period, stratigraphically complex Stone Age sites in western Norway have been investigated applying a method where stratigraphical layers are divided into 5 cm mechanical shifts (stratigraphical/mechanical method). The soil is water-screened on 4 millimetre mesh. This method allows for a high degree of stratigraphical control, and makes it possible, during the later correlation-work of the sites, to distinguish between securely and less securely dated units (e.g. Simpson 1999). In terms of excavation techniques, the data from these stratigraphically complex excavated sites should be regarded as comparable.

Three of the mountain sites: Gyrinos IV, Blånut IV, and Vivik are problematic because, although the excavations were carried out according to a m^2 grid-system, the lithic material from the different units have later been put together in one bag at the museum. Consequently, all of the collected lithics from the entire excavated area had to be analysed together. If there were several phases at these sites, there are obviously problems in distinguishing between them (see under dating). Two of the coastal sites, Sokkamyro 1901-02 and Masnesberget are also problematic. At Masnesberget, the lithics were collected by non-archaeologists. Artefacts not easily recognised by amateurs, such as small flakes, or quartz and slate flakes, are likely to have been left at the site. Masnesberget is included here, because it represents a district (Sogn) where no other early Neolithic sites have been excavated. At Sokkamyro, which was excavated more than 100 years ago, small flakes were not collected. Otherwise, the data from this site should be regarded as comparable to the other excavated sites.

Regional distribution of sites

For obvious reasons, having site data from all parts of the analysed area was important. This goal was, however, difficult to reach in full. Some districts in western Norway (for example the coast of Hordaland) have been investigated extensively, and this has produced more relevant sites than needed here. In other districts, few excavations have been carried out, and it was more problematic to find good data. Nevertheless, with the exception of outer Sogn, I find that the coastline from Sunnmøre to Sunnhordland have a number of sites in all districts. The problem is much more pressing for the inner parts of western Norway (inner Nordfjord, inner Sogn, Voss and Hardanger (Fig. 58)), in which relevant excavated sites are non-existent. In the mountains, there are several good sites at Hardangervidda and the plateau at Lærdal-Hemsedal, while the areas to the south and north of these districts are relatively poor in terms of sites. This means that comparisons between sites along the coast are largely unproblematic. East-west comparisons, however, will largely have to disregard the inner areas of western Norway and consider only the relation between coastal sites and mountain sites.

Conclusion

Initially, I suggested that the selection of excavated sites would have to satisfy five criteria. According to the above discussions, the goals have largely been met: the main body of the 37 sites are dated to the early Neolithic period, they have been classified as either residential sites or field camps, the sizes of most of the analysed units are sufficiently large, the excavation techniques are fairly comparable, and they have a relatively even distribution throughout the area of analysis.

The raw materials, types, and techniques which have been distinguished at the above sites constitute the most important data in this thesis. In the next chapter each of these categories will be compared in a regional perspective.

Chapter 7

Regional analysis

Introduction

The main purpose of the regional analysis is to describe the quantitative patterns in the distribution of raw materials, types, and techniques that were distinguished in Chapter 5. First, I will investigate if particular elements are found at more than one site or district and how these traits are distributed quantitatively. Second, by means of a correspondence analysis, I will explore the multidimensional variation of these elements. When one raw material, type or technique which is found in similar quantities at two or more sites, systematically is found in larger or smaller quantities at sites in other parts of the region, one may talk about a pattern. The correspondence analysis may uncover whether several patterns vary in similar ways; hopefully, a system of such patterns will constitute the empirical basis for a discussion of boundaries and connections.

Raw materials

Below I will account for the quantitative tendencies, expressed in percentage of the total amount of artefacts at the sites, in the distribution of the different individual raw materials. I will also describe the techniques that have been applied, and the tool-types they have been manufactured into. When relevant, cases of distribution of these individual categories outside of the area of analysis will also be referred to. Most of the raw materials distinguished at the sites will be presented graphically in histograms. A few categories are not presented, because (1) they make up very small percentages (below 1%) and occur at only one site, (2) they are grouped categories that are only minimally represented, or (3) they have been deposited at the sites during preceding periods.[1]

Diabase, greenstone and undetermined basalts: excavated sites

Flakes, blanks and adzes of diabase, greenstone and undetermined basalts have been identified at sites throughout the region (Fig. 59-61). At four of the sites (Kleiva, Neset II, Vikja I phase 2, and Sokkamyro) adze-manufacturing took place, resulting in large amounts of diabase or greenstone. In order to compare the lithic data from these sites with other (non-workshop) residential sites, the diabase flakes and blanks are excluded from the analysis (see site tables in appendix 3 for a presentation of the entire data). The criterion for exclusion is that more than 5% of the debris of these raw materials should be present at the sites.

The majority of the diabase is found in Nordfjord and Sunnfjord. To the north and south of these districts, very little is identified. Greenstone has a wider distribution, and is found at sites from southern Sunnmøre to Sunnhordland, but greenstone is most frequent at the sites in Sunnhordland. Undetermined basalts have been found in the entire coastal area. Except from one basaltic adze/axe found at 1020 Bjornesfjorden, these raw materials or stone adzes made from them were only found at the coastal sites.

The site Masnesberget scores highest for all of the three raw materials. This should probably be explained by the fact that the amateurs that collected lithics at the identified more adzes than other artefacts.

Diabase, greenstone and undetermined basalts: Vespestad adzes

In addition to the raw material classification at the excavated sites, the entire museum collection of Vespestad adzes at Bergen University Museum (from the Sunnmøre district and Sogn og Fjordane and Hordaland counties) has been examined for this work for the purpose of distinguishing between diabase, greenstone, and other basaltic raw materials. In addition, Asle Bruen Olsen's catalogue of Vespestad adzes from Sunnmøre museum (23 specimens) and from the districts Nordmøre and Romsdal (deposited at Museum of Natural history and Archaeology in Trondheim – not inspected by myself) was included (A. B. Olsen 1981: original catalogue deposited at the university library in Bergen).[2] Sigmund Alsaker's catalogue (1982) of Vespestad adzes and chisels from Rogaland (deposited at Archaeological museum in Stavanger – not inspected by me) was also included.[3]

1 Yellow mylonite and soapstone are probably late Mesolithic in the area of analysis (Bergsvik 2002a:283 ff).

2 Note that all of the adzes that A. B. Olsen categorized as "bluish greenstone" (quarry 4 greenstone) is presented as "greenstone" in Fig. 62. See Chapter 8.

3 A. B. Olsen and S. Alsaker performed the surveys during the late 1970's. Vespestad adzes received later than 1978 from Rogaland, Romsdal and Nordmøre are therefore not included in Fig. 62 and 103.

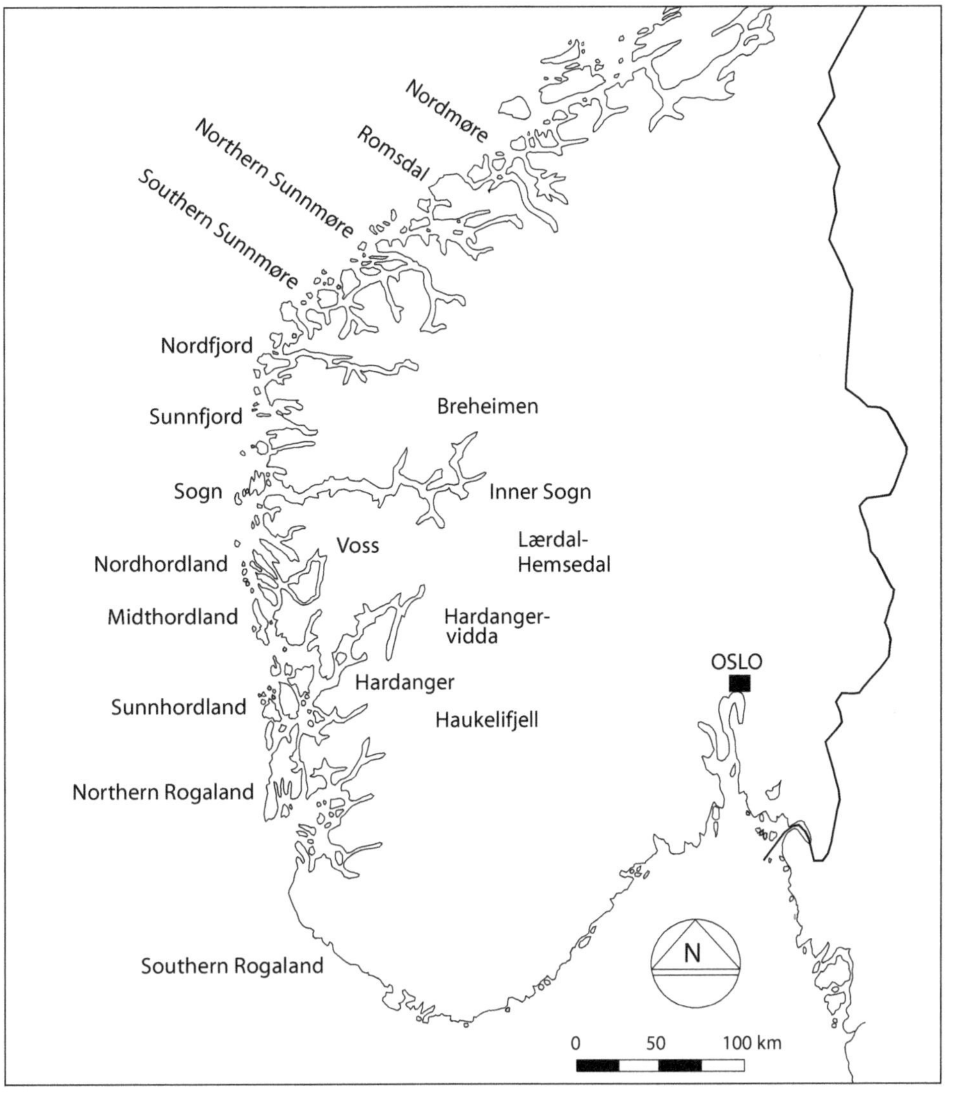

Figure 58. *Southern Norway. Approximate location of districts referred to in the text.*

The new typological and raw material classifications of the adzes diverge significantly from previous determinations of the same data (A. B. Olsen 1981, S. Alsaker 1982, Gjerland 1984).

A. B. Olsen's and S. Alsaker's distinctions between Bømlo greenstone and quarry 4 greenstone are not followed in my visual classification of Vespestad adzes. I only operate with "greenstone" as a grouped term for these two raw materials (see Chapter 8 for a discussion of the geological analysis that led up to this decision). My classifications of diabase, on the other hand, are almost entirely in accordance with A. B. Olsen's determinations (1981).

In Fig. 62, the different raw materials of Vespestad adzes per district are compared (see location of districts Fig. 58. See also map Fig. 103 for a presentation of all Vespestad adzes). In contrast to site data, where the diabase clusters in the Sunnfjord and Nordfjord districts, Fig. 62 shows that diabase adzes have low percentages in Sunnfjord compared to the percentages in the districts further north. Northern Sunnmøre is particularly well supplied with diabase. Considering that the diabase-quarry is situated in Sunnfjord, this different pattern is hard to understand. A. B. Olsen (1981:157, Fig. 54) has argued that the generally large number of adzes found in Sunnmøre should be explained by intensive modern agricultural activity, which has uncovered more archaeological remains than the less intensive agriculture in Nordfjord, Sunnfjord and Sogn (A. B. Olsen 1981:157, see below under Vespestad adzes for further source-criticism). Still, this difference in a total number of adzes should not influence the relative percentage values for each raw material. These percentages clearly show that diabase was more attractive as a raw material for Vespestad adzes at Sunnmøre than in the districts that were situated closer to the quarry. As was demonstrated above, this tendency was also to a certain extent reflected at the residential sites in Nordfjord and Sunnfjord, where greenstone and other basalts are just as well represented as diabase (Fig. 59-61).

The greenstone-adzes are distributed throughout the region. They have low percentages in Nordmøre and Romsdal, but the shares are high in all districts from northern Sunnmøre to southern Rogaland. One obvious reason for this wide distribution is that both greenstone from Bømlo and quarry 4 greenstone (presumably from the northern part of the region) is included in this category. Needless to say, the grouping of these two categories of greenstone disqualifies greenstone as data for discussion of boundaries and connections.

Flint

Flint was used throughout the region, but the percentages vary considerably (Fig. 63). In northern Sunnmøre, it represents between 86% and 95% of the total amount of raw materias, while in southern Sunnmøre, the share is abut 40%. At the sites in the districts between Nordfjord and Midthordland, flint makes up less than 20%. Further south, at most of the sites in Sunnhordland, the percentage of flint increases again. The southernmost sites have percentages about 40%. It should be noted that at Neolithic sites to the south of the analysed area, the percentage of flint is generally higher than 40% (Nygaard 1974, Skjølsvold 1977, Bang-Andersen 1981, Ballin & Jensen 1995). At the mountain sites, the percentages vary between 1% at Mørkedøla 1 to 96% at 1020 Bjornesfjorden.

With regard to the quality of flint, the coastal sites (at 10 of the sites, no quality distinction was made) have largely the same arrangement: medium flint makes up the absolute majority, while fine and coarse flint represent only smaller shares. In contrast, at the mountain sites, fine flint is much more frequent, while coarse flint is not identified at all (see data in Appendix 3).

Flint has been reduced by different techniques in the different districts; in Sunnmøre, the bipolar technique dominates heavily, while other techniques only represent smaller shares. At the mountain sites and at sites to the south of Sunnmøre, some bipolar flint cores are identified, but in these districts, platform cores and other cores are just as frequent. The high amount of blade points and blades in these districts compared to in Sunnmøre indicates that flint was mainly reduced using cylindrical core technique and other platform techniques, rather than bipolar technique.

Three sites deviate from the general pattern in that they have higher flint percentages than most other sites in Sogn and Nordhordland: Masnesberget, Snekkevik 1 and Flatøy VIII. At Masnesberget, problems of representativity (see above) most likely explain the pattern: flint debris is more likely to be collected by amateurs than slate or mylonite. The patterns at Snekkevik and Flatøy may be a result of contamination by material from the preceding Mesolithic phases at these sites, in which flint was more frequent than other raw materials. Two sites in Sunnhordland, Sokkamyro and Tjernagel 3 may also have too high percentages, essentially for the same reasons as at Flatøy and Snekkevik. Still, Sokkamyro and Tjernagel 3 may also be part of a general trend to the south of the analysed area (Rogaland and Agder counties). In Rogaland, flint was frequently used during the late Mesolithic and also during the early and middle Neolithic periods (Skjølsvold 1977; Bang-Andersen 1981; Skjelstad 2003).

Fine, green quartzite 1

Large amounts of fine, green quartzite are found at the mountain sites Mørkedøla 1, Blånut IV and Gyrinos IV. A few specimens are also identified at mountain sites Styggvasshelleren to the north and 760 Finnsbergvatn to the south. It has also been found at five of the coastal sites (Fig. 64).

The raw material has been prepared by platform technique, resulting in some blade points, some blades, and a few platform cores. It has, however, mainly been reduced by simple flake techniques.

Fine, brown quartzite 1

This raw material is only identified at two sites, both at the mountain-plateau: Mørkedøla I (3%) and Gyrinos IV (41%) (Fig. 65). One blade point and a handful of blades and platform cores may indicate that cylindrical core technique may have been practiced, although no cylindrical cores have been identified. The large amount of debris indicates that it has mainly been reduced by simple flake technique.

Fine, black quartzite 1

This raw material represents as much as 35% of the total amount of raw materials at Masnesberget, Sogn. Outside of Sogn, it is only minimally present at the sites (Fig. 66).

At Masnesberget, it was prepared by cylindrical core technique, represented by a number of blades and platform cores. The blade and the cylindrical core of fine, black quartzite 1 at 17 Havnen and 1 Haukedal could indicate such a technique in Nordfjord as well. To the south of Sogn, only flakes were found.

Considering that Masnesberget was collected by amateurs, fine, black quartzite 1 may be somewhat over-represented at this site at the expense of other raw materials that were identified as artefacts. On the other hand, this raw material was much in use at other sites in Sogn. Trond Klungseth Lødøen has identified fine, black quartzite 1 at no less than 17 surveyed sites in this district. At many of these sites, early Neolithic elements were present (Lødøen 1995:135-145).

Fine blue quartzite 1

Small amounts (less than 0, 5%) of fine blue quartzite 1 have been identified at Kotedalen, Flatøy VIII, and Nilsvik 4 at the coast (Fig. 67). Somewhat larger amounts are found at the mountain site Styggvasshelleren. It was prepared by different core techniques and manufactured into small blades and blade points.

Fine blue quartzite 7

Very small amounts of this characteristic raw material have been identified at several of the coastal sites at Sunnmøre, Nordfjord and Sunnfjord. It has also been found at Sokkamyro in Sunnhordland (Fig. 68). The small blades found may indicate that it was prepared by cylindrical core technique.

Fine, blue quartzite 8

Fine, blue quartzite 8 has high percentages at two sites: Nilsvik 17 in Midthordland and Tjernagel 3 in Sunnhordland. A few specimens were found at Austvik III (Fig. 69).

The raw material was reduced by cylindrical core technique at all of these sites. This technique is represented by blade points, blades and platform cores.

Fine blue quartzite 8 is found at sites that are situated in the main distribution area of rhyolite. Considering that it also was treated with cylindrical core technique, it probably supplemented the more common rhyolite. At Nilsvik 17, where rhyolite dominates, it can probably be interpreted as such. At Tjernagel 3, on the other hand, rhyolite is only minimally present. One explanation may be, as noted under rhyolite, that the site was used as a field camp during a short period, resulting in a skewed raw-material profile compared to the general trend. Alternatively, the site was occupied before rhyolite became the dominating raw material. The high elevation of the site indicates an early date (Bjerck & Ringstad 1985:38).

Fine, grey quartzite 2

Fine grey quartzite 2 constitutes about 22% at 17 Havnen in Nordfjord. At 1 Haukedal it has only small shares. Outside of Nordfjord a few specimens has been identified at two sites: Kotedalen phase 12 and at the mountain site Styggvasshelleren (Fig. 70).

At 17 Havnen, the raw material has been reduced by cylindrical core technique only. This technique is mainly represented by a large amount of small blades, however, one blade point and one cylindrical core was also found. The blade point at Kotedalen indicates that this technique was also applied on fine grey quartzite in Nordhordland. At Styggvasshelleren, however, the raw material was reduced by bipolar technique.

Similar to chert, the raw material is intensively used at one site, and only minimally present at others. 17 Havnen phase 3 has been interpreted as a residential site. The tendency should therefore not be interpreted as skewed compared to the general trend in Nordfjord. It may rather reflect that fine, grey quartzite 2 was only utilised intensively in a short period during the latter part of the early Neolithic.

Fine, grey quartzite 3

At the mountain sites 760 Finnsbergvatn and 526 Nordmannslågen, fine, grey quartzite 3 constitutes 34% and 1% respectively. It has not been found at other sites (Fig. 71). The raw material has been prepared using platform technique, as well as bipolar technique and flake technique.

Fine grey quartzite 11

Small percentages of this raw material have been found at Austvik III and Sokkamyro in Sunnhordland (Fig. 72). It has been prepared by cylindrical core technique.

Fine white quartzite 16

Small amounts of fine white quartzite have been found throughout the area of analysis (Fig. 73). It has been treated by flake technique and has been manufactured into scraping tools and probably flake knives.

Medium grey quartzite 1

This raw material is only identified in Sunnfjord and Sogn. At the sites in Sunnfjord, the percentages are high, and vary between 11% and 35 % of the total (Fig. 74).

Medium grey quartzite 1 was first of all reduced by cylindrical core technique, represented by one blade point and a number of small blades and platform cores.

The site Kleiva diverges from the rest of the Sunnfjord assemblages by a complete lack of medium, grey quartzite 1. However, Kleiva may have been occupied only briefly, and may therefore not be representative for the general trend in Sunnfjord.

Medium grey quartzite 9

This raw material constitutes small shares at sites between northern Sunnmøre and Midthordland at the coast. At Styggvasshelleren in the mountains it has about 3% (Fig. 75). Small blades and one tanged point indicate that it was prepared by using cylindrical core technique, but it was also struck by using bipolar technique.

Medium grey quartzite 10 and 11

These raw materials are only present at the sites Nilsvik 4/17 (Midthordland) and Austvik III (Sunnhordland), where they make up between 2 and 8% of the total (Fig. 76). They were only treated by flake technique at these sites, were used as scraping instruments and probably as knives.

Coarse quartzites

Small percentages of different colour variations of coarse quartzites have been identified at the coast as well as at the mountain sites (Fig. 77). They have generally been treated by flake technique and have mainly been manufactured into knives.

Quartz crystal

Quartz crystal is present at almost all of the coastal and mountain sites, but only a few sites have more than 5%. One coastal site, Bustadvika, has 23% and two mountain sites, 526 Nordmannslågen and Vivik have 20% and 59% respectively (Fig. 78).

The bipolar technique has been used almost exclusively. A few blades and cores indicate a slight element of platform technique.

The site Bustadvika diverges from the rest of the coastal assemblages by a high percentage of quartz crystal. The divergence may be explained by the fact that that the site may have been a field camp, and is not representative for the general trend. Considering that quartz crystal was mainly used during the late Mesolithic on the coast (A. B. Olsen 1992, Bergsvik 2002a:283), a more relevant explanation may be that the quartz crystal first of all has been deposited during the Mesolithic occupation at the site. An earlier date could explain the high percentages at Vivik and 526 Nordmannslågen, however, at these and other mountain sites, possibly Neolithic tanged projectile points were, in contrast to the west coast, also made of quartz crystal. A significant percentage of Neolithic quartz crystal should therefore not be seen as problematic in the mountains.

Fine, white quartz

At the mountain sites Styggvasshelleren, 526 Nordmannslågen and 760 Finnsbergvatn, and at the coastal site 9 Bustadvika, fine white quartz represents high percentages. The raw material is present at almost all other sites, but it generally makes up less than 10% of the total (Fig. 79).

In Sunnmøre, bipolar technique clearly dominates. Elsewhere other core-types are found, indicating that it was also frequently reduced by platform techniques and flake techniques.

Medium and coarse quartz

The percentages of medium and coarse quartz vary considerably across space. At some sites, it represents as much as 35% and at others it has not been identified. On the coast, the percentages are higher in Nordfjord and Sunnhordland than in Sunnmøre, Sunnfjord and Nordhordland. Only the mountain sites Vivik and Styggvasshelleren have significant amounts of medium and coarse quartz (Fig. 80).

In Sunnmøre, bipolar technique was used to prepare the raw materials, while it was reduced using platform techniques and flake techniques in other districts.

Coarse and medium grey mylonite

This raw material is represented at all of the sites in Sunnfjord. It makes up as much as 23% of the total at Kleiva, 13% at Botnaneset VIII and 10% at Vikja I, phase 2. It is also present at sites in Kotedalen in Nordhordland and 17 Havnen in Nordfjord (Fig. 81).

It appears to have been reduced by platform core technique, mainly represented by blade points, small blades, and platform cores.

Some of the sites within the main distribution area in Sunnfjord have low figures, and some have very high percentages. As noted below under blue mylonite, At Kleiva and Botnaneset the percentages are high, and may be the result of these sites being short-term camps, and that the raw material profiles are skewed.

Blue mylonite

Blue mylonite dominates in Sunnfjord. To the north and south of this district, there is a fall-off, but it still has high percentages at the sites 17 Havnen and 1 Haukedal in Nordfjord and Kotedalen in Nordhordland. Small quantities are present in Sunnmøre and Midthordland/ Sunnhordland, which indicates a wide geographical distribution of this raw material. It has not been identified at the mountain sites (Fig. 82).

Cylindrical core technique has mainly been used to prepare blue mylonite. This is clearly indicated by the presence of blade points and the relative dominance of small blades compared to other blades, and also by the presence of cylindrical cores and platform cores. Bipolar technique has only been applied to a limited extent.

Some of the sites within the main distribution area lack blue mylonite, some have low figures, and some have very high percentages. The absence at Masnesberget may at least partly be a result of selective and unsystematic sampling at this site. At Kleiva the percentage is very high, and is probably the result of this site being a short-term camp, as well as a skewed profile compared to the general early Neolithic trend. Botnaneset VIII, which has very little blue mylonite and very much chert (see below) can be explained is a similar fashion, only that the tendency is the reverse from that at Kleiva. A reduced importance of blue mylonite through time might also explain some of the low numbers, for example at the relatively late Snekkevik 1.

Chert

Chert dominates at two sites, 1 Haukedal, phase 3 in Nordfjord and Botnaneset VIII, phase 3 in Sunnfjord. At other sites it has low shares. Chert has a relatively wide geographical spread. A few specimens are identified at Flatøy VIII in Midthordland to the south (Flatøy VIII) and at Valderøya Vest, northern Sunnmøre to the north (Fig. 83).

The significant presence of blade points, small blades and platform cores at the sites indicates that chert was mainly reduced by means of cylindrical core technique in the entire distribution area.

With the exception of 1 Haukedal phase 3 and Botnaneset VIII, the use of chert is modest. The question is whether these sites are representative of a general trend of significant chert-use in Nordfjord-Sunnfjord, or whether they are not typical. Botnaneset VIII has been characterised as a field camp. One relevant interpretation is therefore that the inhabitants of this site had acquired a few blocks of chert before they arrived and used them before leaving. Thus the raw-material profile at Botnaneset VIII may be skewed compared to the general trend of the early Neolithic. This explanation is however less probable for the residential site 1 Haukedal. An alternative explanation, which is supported by the concurring radiocarbon determinations from these two sites, is that chert was utilised intensively only during a short time during the middle part of the early Neolithic. Before and after 5000 BP (3750 BC), chert may have been less important.

Rhyolite

Rhyolite dominates at the coastal sites in Sunnhordland, Midthordland and Nordhordland (main distribution area), where it generally constitutes a very high percentage of the total amount of raw materials in use. Within these three districts, it is not possible to distinguish any marked fall-off in any direction: the sites that are situated in Nordhordland are almost equally well supplied with rhyolite as the sites in Sunnhordland. To the north of Nordhordland, however, there is a marked fall-off. Rhyolite makes up no more than 1-4 % of the total at the sites between Sunnfjord and southern Sunnmøre. In northern Sunnmøre, rhyolite has not been found at the sites in this analysis. A few specimens of rhyolite (less than 0,5 %) have been found at three of the mountain sites (Fig. 84).

It should be noted that to the south of the analysed area, rhyolite is frequent at excavated sites at Karmøy, Rogaland (Nygaard 1974). It represents a small percentage at Eigerøy, Rogaland (Bang-Andersen 1981:21). A few specimens were also found at Lista, Vest-Agder (Ballin & Jensen 1995:189). To the north of the analysed area, a few artefacts of rhyolite were collected at Korsnes, Romsdal and at Gjevilvatnet, Sør-Trøndelag (S. Alsaker 1987:59).

Cylindrical core technique has been applied for rhyolite in the entire region. This technique is represented by several cylindrical cores/platform cores and particularly by blade points (A-points) and a large amount of small blades.

It has not been reduced by the Mesolithic microblade technique[4]. Bipolar technique has only been used on this raw material to a minor degree. As is noted by S. Alsaker, rhyolite was carried from the quarry at Siggjo as big blocks, and these were reduced further at the residential sites. One such block was found at Garnes, Bergen, about 80 km from Siggjo (S. Alsaker 1987:53, Bergsvik 2002b:16). David Simpson also refitted a rhyolite-block at Krossnes 6, in Nordhordland (Simpson 1996).

Chronological changes within the early Neolithic period and functional differences of sites might partly explain some of the minor quantitative divergences between the different sites in the main distribution area of rhyolite. A possible decreasing tendency in the use of rhyolite through time probably explains the difference between Kotedalen phase 12 and 13. Some of the differences might also be a result of functional differences or length of occupation. The low percentage of rhyolite at Tjernagel 3 is probably a result of sporadic and short-term use.

Slate

Most of the coastal sites in Nordfjord, Sunnfjord and Nordhordland have relatively high percentages of slate. At some sites it represents more than 20% of the total. In Sunnmøre (except for 9 Bustadvika), Midthordland and Sunnhordland, however, the percentages are significantly lower. About half of the mountain sites have high portions of slate (Fig. 85-90). The colour variations are distributed as follows: *Red slate* constitutes share of about 5% at the sites 17 Havnen in Nordfjord and Vikja I in Sunnfjord. A few specimens have been identified at sites at 13 Valderøya Vest, Sunnmøre, Kotedalen, Nordhordland and Mørkedøla I in the mountains. *Green slate* is found at many of the sites in Nordhordland, Sunfjord and Nordfjord, where there are significant amounts of green slate. Sites at Sunnmøre and Sunnhordland, on the other hand, generally have low figures. Styggvasshelleren, Mørkedøla I and Blånut VI are the only mountain sites with green slate. *Grey slate* makes up approximately 2-4% of the total in Nordfjord, while the site Bustadvika in Sunnmøre has 6, 5%. At other coastal sites, only a few specimens have been found. In the mountains, Vivik has 7%, Blånut 1%, while the rest have a few specimens only. *Brown slate* is only present at the sites in Nordfjord and at Kotedalen. *Blue slate 1* is only found as a few specimens on the coast (Synnaland and Sokkamyro), while it makes up 2 and 3 % at the mountain sites Blånut IV and Vivik respectively. Points have been found at Styggvasshelleren and Gyrinos IV. *Blue slate 2* has only been identified at one site, 760 Finnsbergvatn.

4 To my knowledge, the only convincing rhyolite blade produced by microblade technique in the region is found at Flatøy XII by Tore Bjørgo (1981:92)

Projectile points are the only tool type made of slate at the excavated sites. According to the above, slate is generally well represented at coastal sites in Nordfjord, Sunnfjord and Nordhordland. There are, however, some sites in these districts that have very little or no slate. The absence at Botnaneset VIII and the small percentages at Kleiva may be explained by the brief occupations at these sites. Another reason may be that some of these sites were occupied during the earliest phases of the early Neolithic (for a similar trend in Sunnfjord, see blue mylonite).

Anorthosite

Anorthosite is identified at Kotedalen and Snekkevik in Nordhordland. At these sites, a few points and a fairly large number of flakes were found. A few specimens were found at Torsteinsvik in Midthordland, 17 Havnen and 1 Haukedal phase 2 in Nordfjord and at the mountain site Styggvasshelleren (Fig. 91). Ground projectile points are the only tool type made of anorthosite at the excavated sites.

Sandstone

The largest percentages of sandstone are found at sites in Nordfjord, Sunnfjord and Sogn. Sites to the north and south of these districts have generally less of this raw material. Sandstone has only been found at one mountain site: 1020 Bjornesfjorden (Fig. 92). It was exclusively used for grinding-slabs.

Pumice

Pumice is identified at the majority of the coastal sites (Fig. 93) (only pumice with grooves are presented here in order to exclude specimens that may have been deposited naturally at the sites. See appendix 3 for all pumice at the sites). It generally makes up quite small percentages. Pumice has not been identified at the mountain assemblages analysed here, but one piece was identified in later layers 6+7 at Styggvasshelleren (Randers 1986:63).

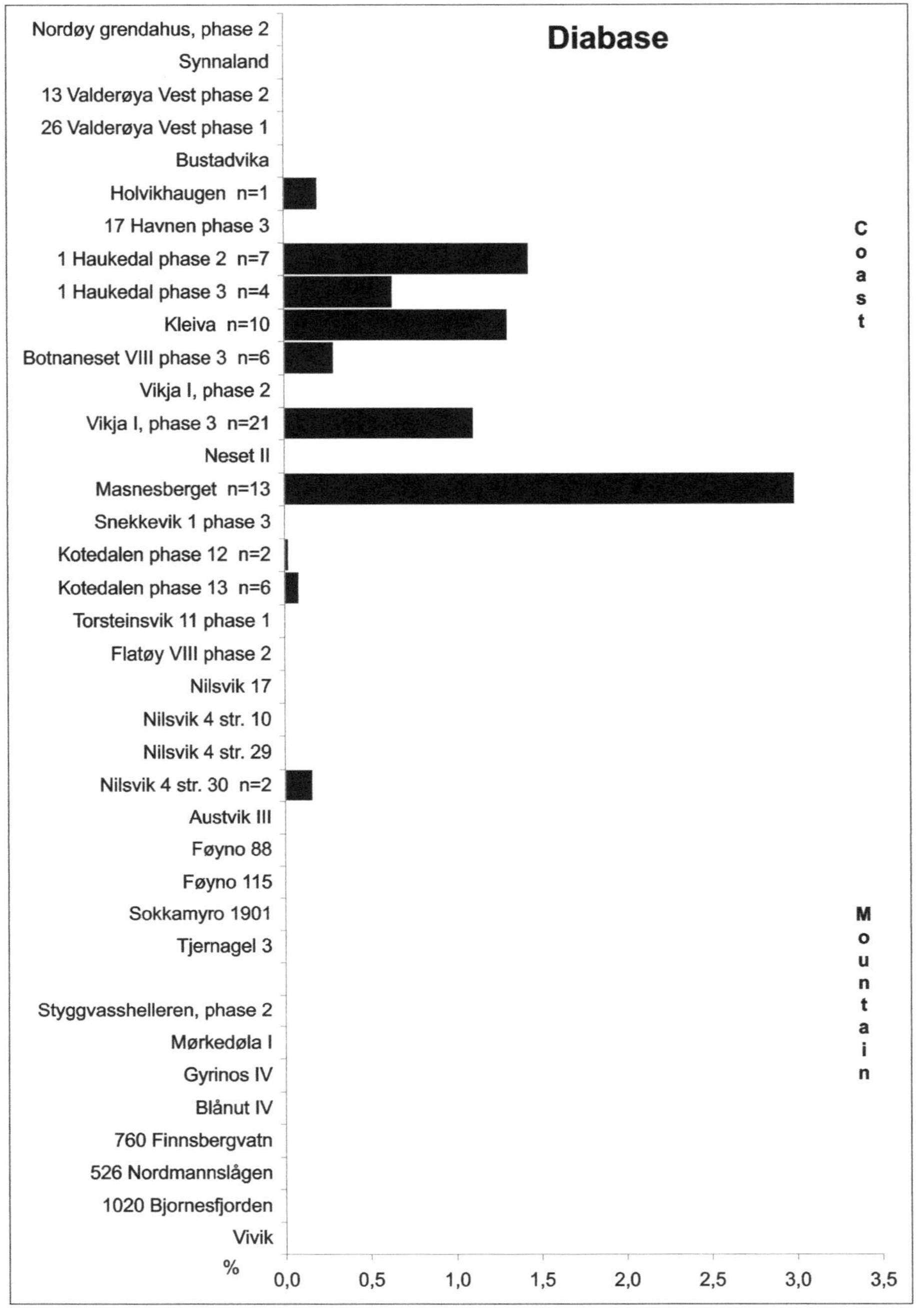

Figure 59. Percentages of diabase at excavated sites.

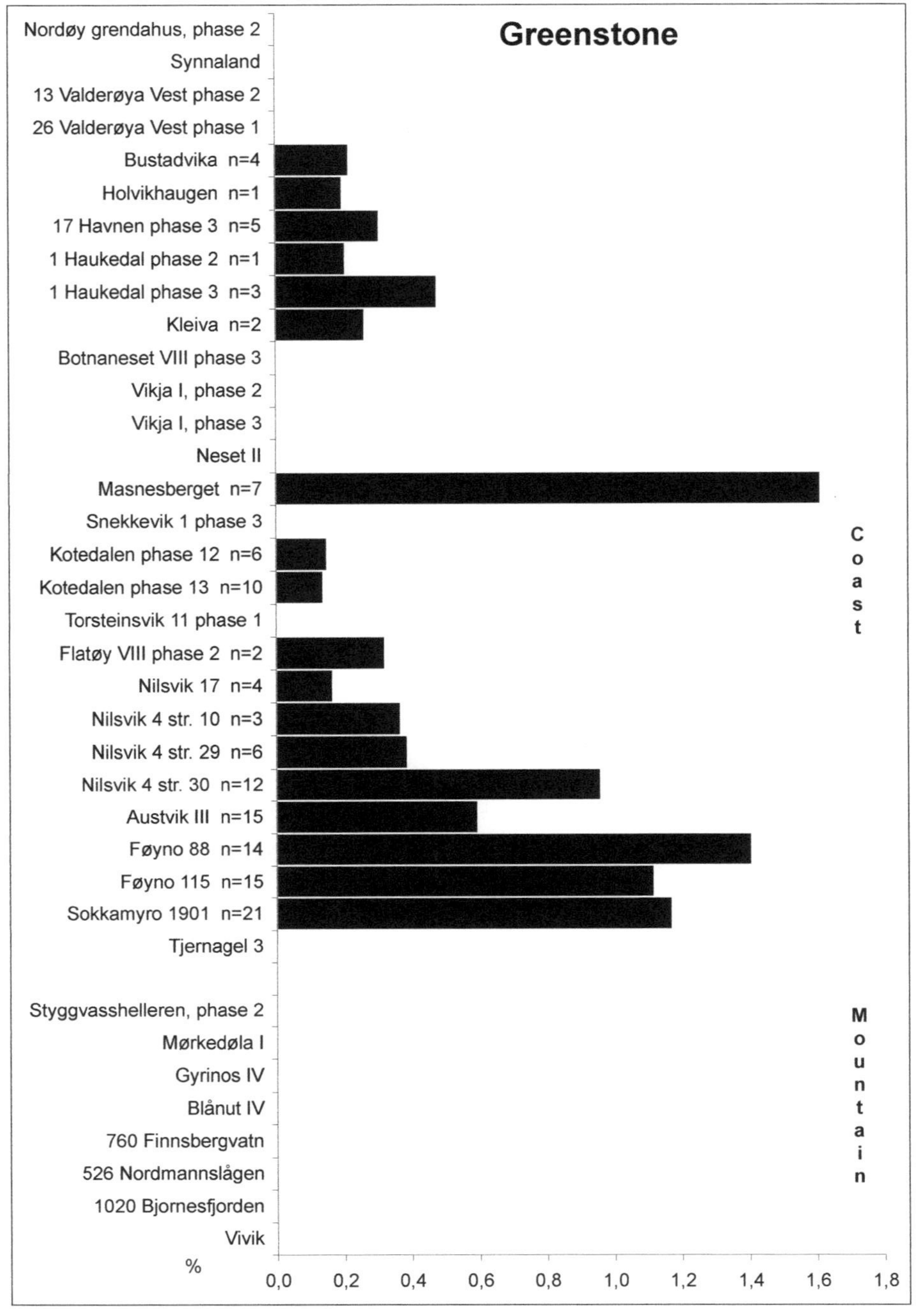

Figure 60. Percentages of greenstone at excavated sites.

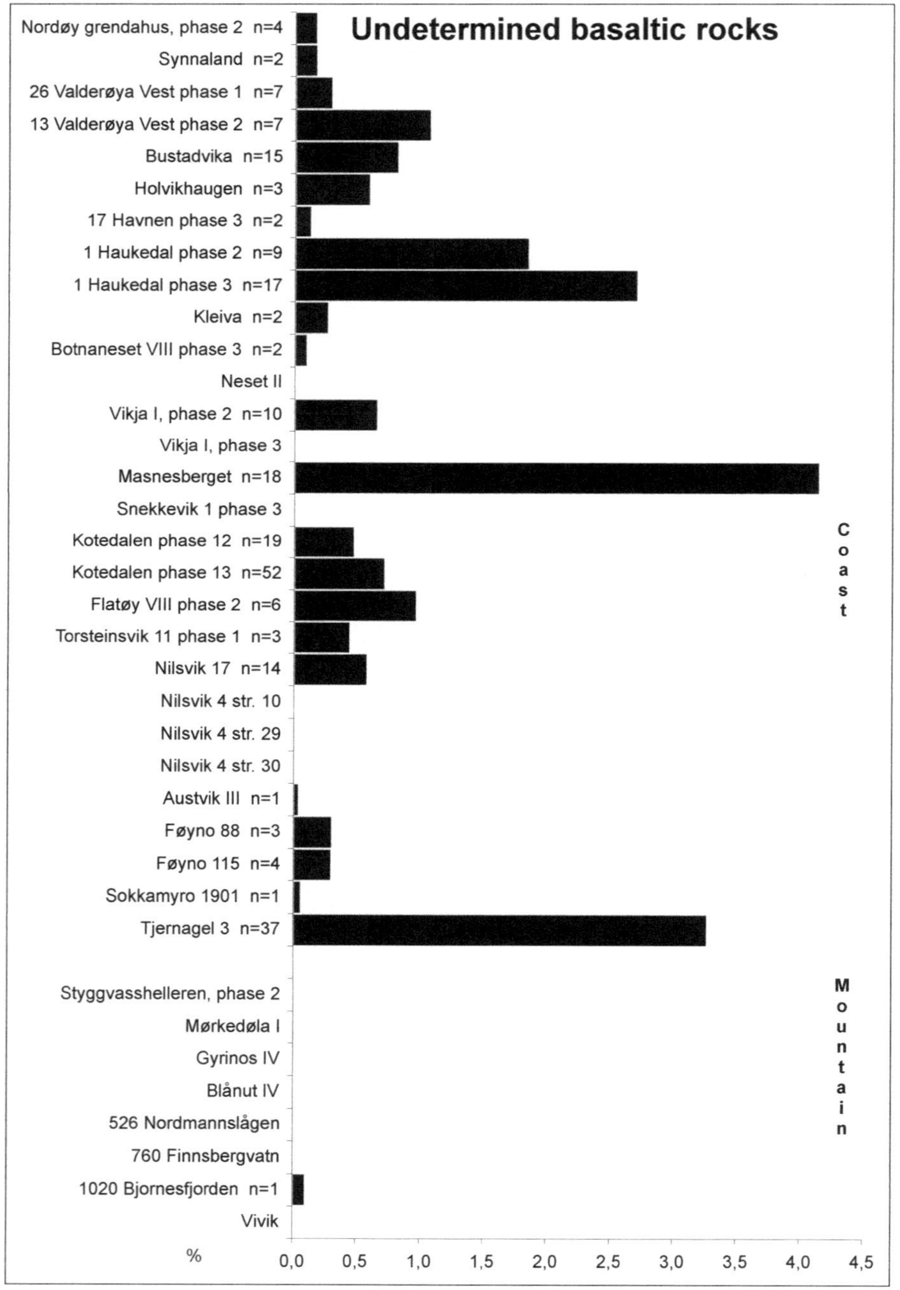

Figure 61. *Percentages of undetermined basaltic rocks at excavated sites.*

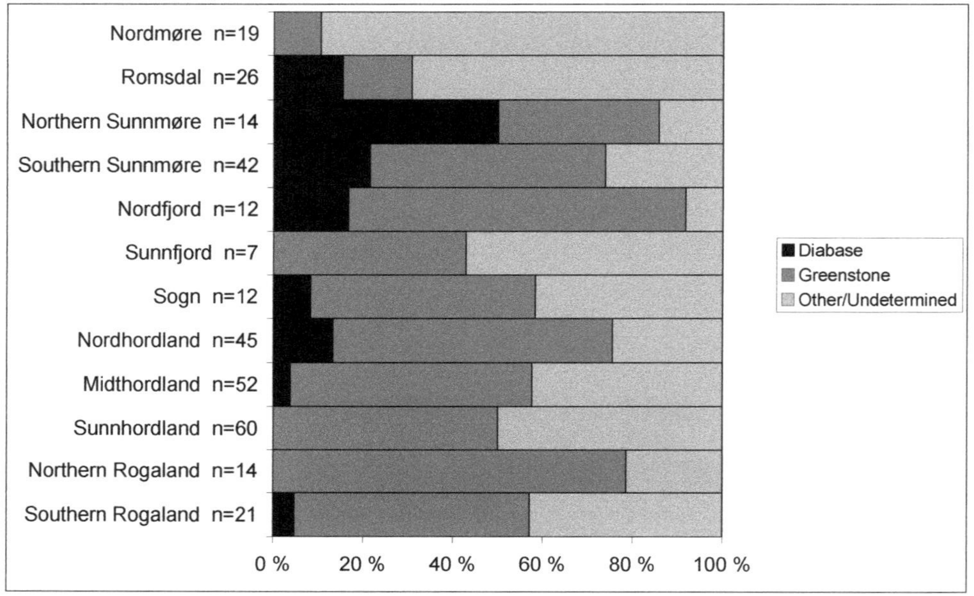

Figure 62. *Percentages of diabase, greenstone and undetermined basaltic Vespestad adzes found in the coastal districts. Stray finds as well as adzes from excavations are included. Data from Sunnmøre in the north to Sunnhordland in the south has been analysed for the purpose of the present work. Data from Nordmøre and Romsdal are extracted from Olsen (1981). Data from northern and southern Rogaland are extracted from Alsaker (1982).*

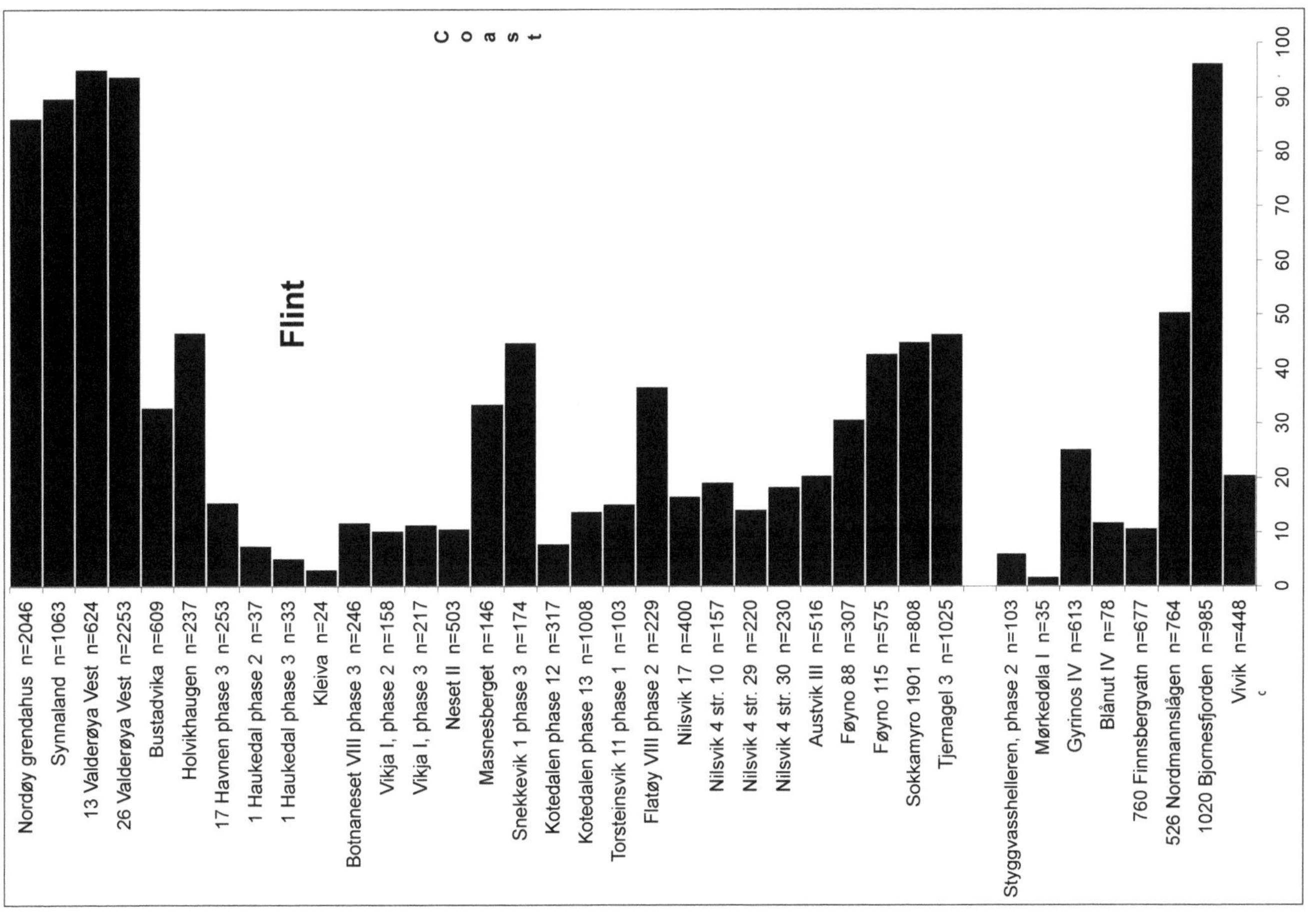

Figure 63. Percentages of flint at excavated sites.

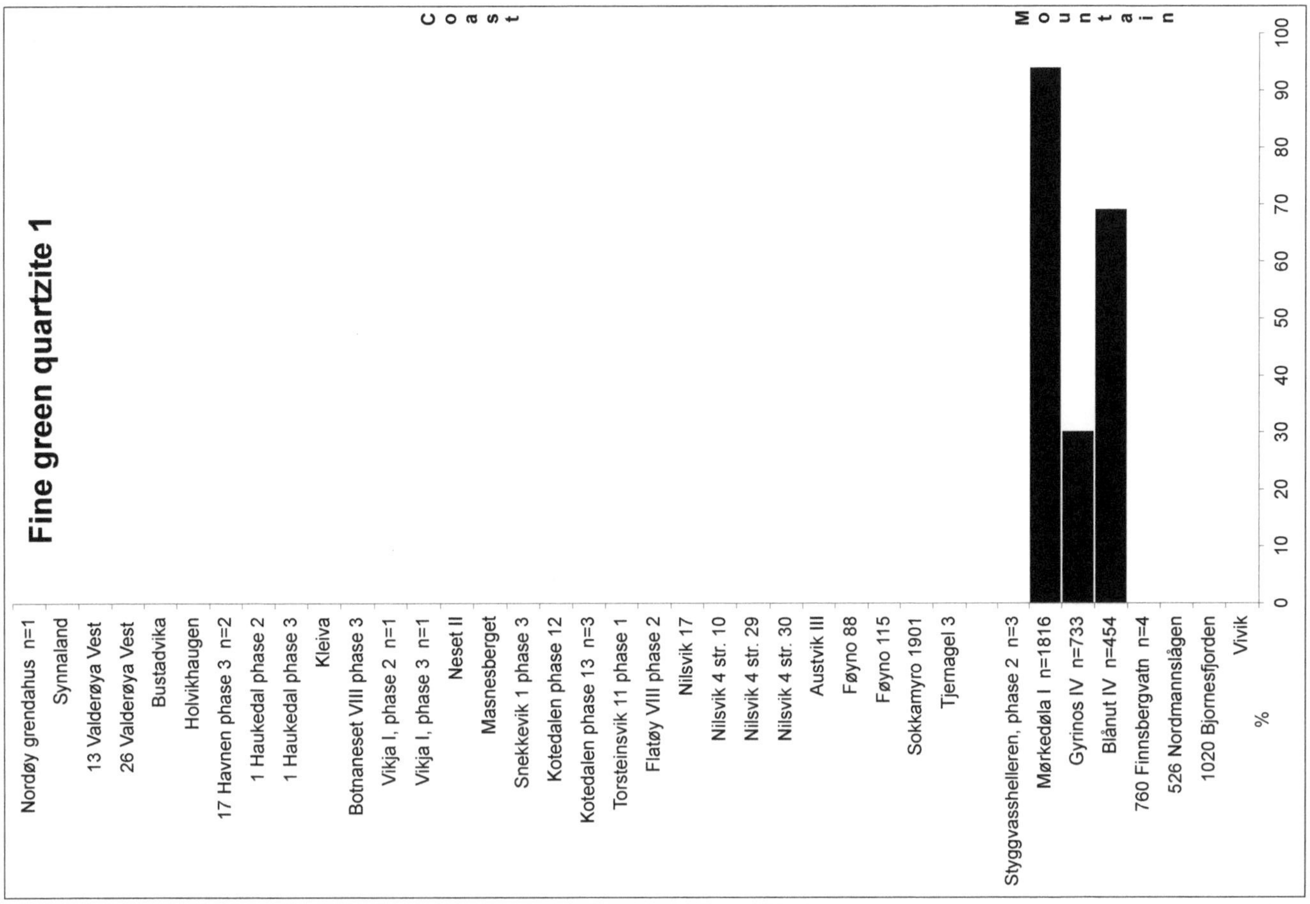

Figure 64. Percentages of fine green quartzite 1 at excavated sites.

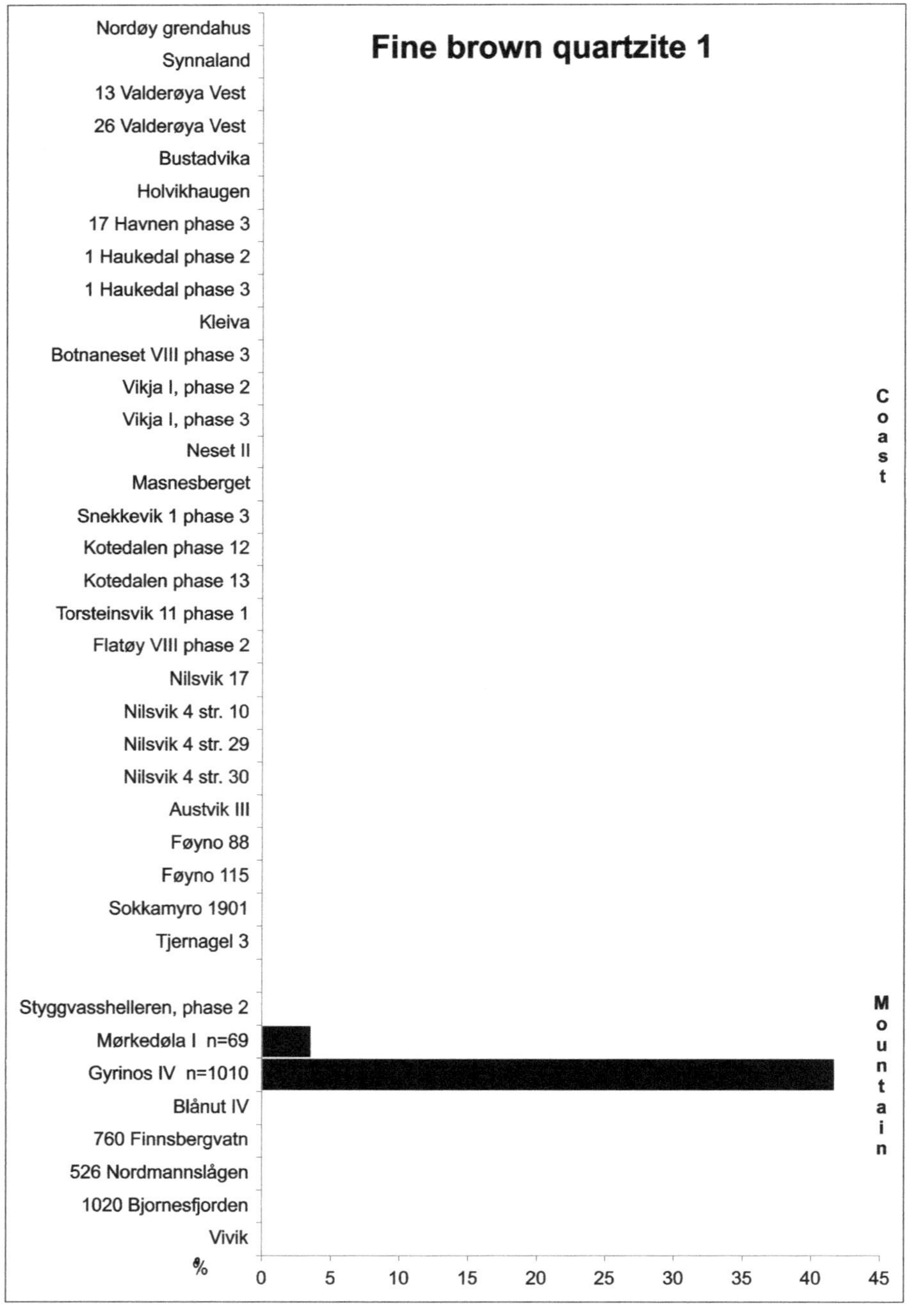

Figure 65. Percentages of fine brown quartzite 1 at excavated sites.

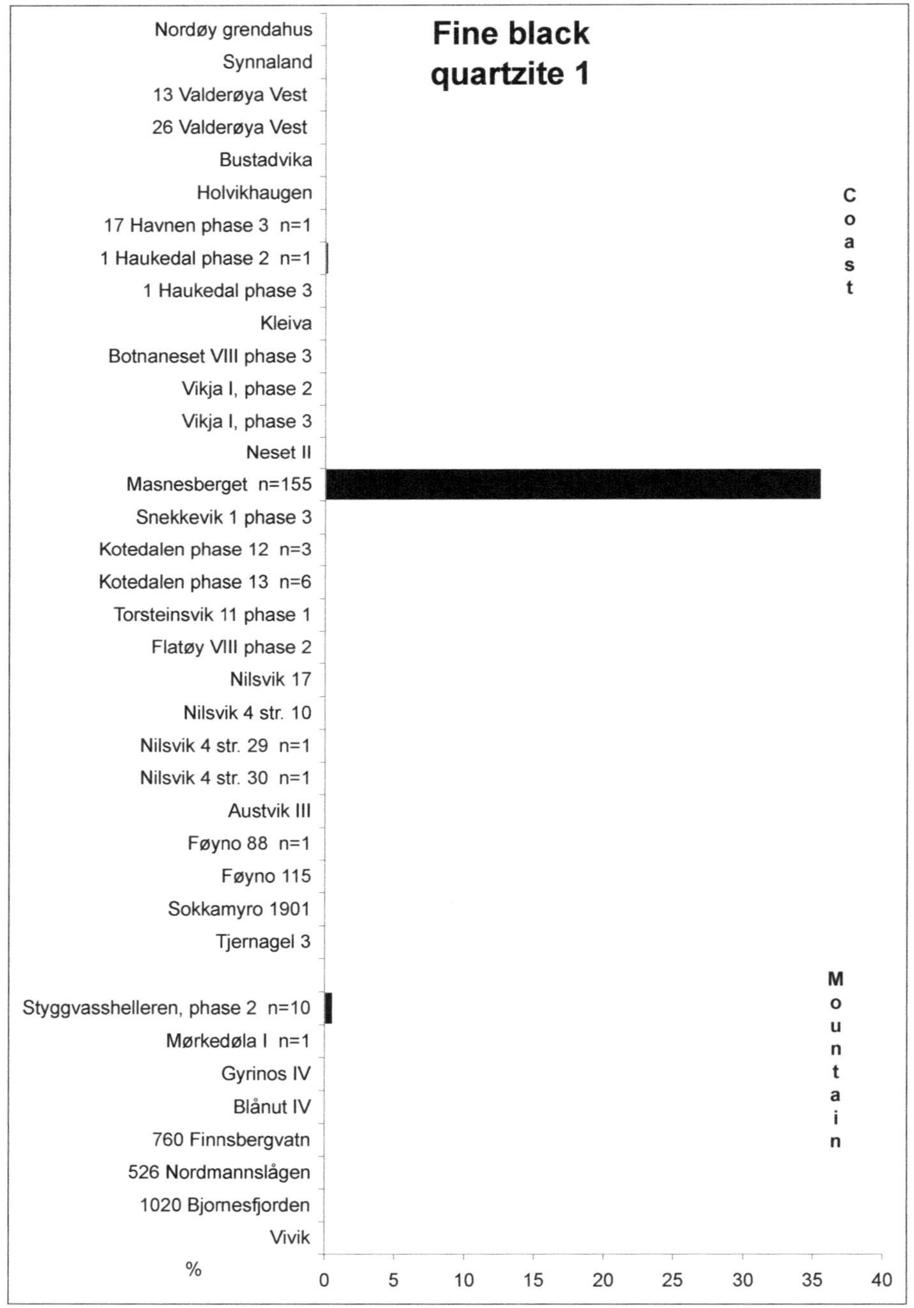

Figure 66. Percentages of fine black quartzite 1 at excavated sites.

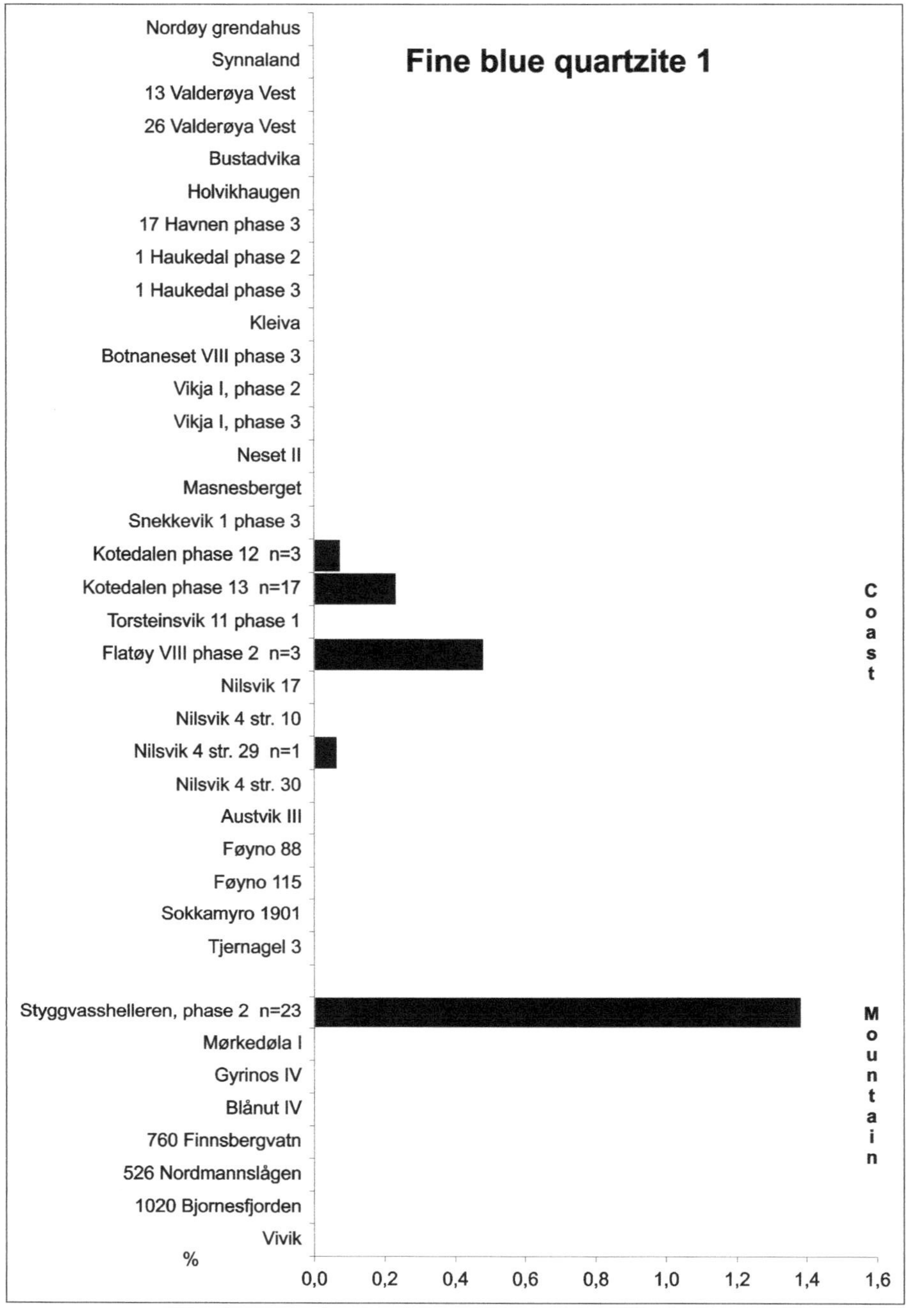

Figure 67. *Percentages of fine blue quartzite 1 at excavated sites.*

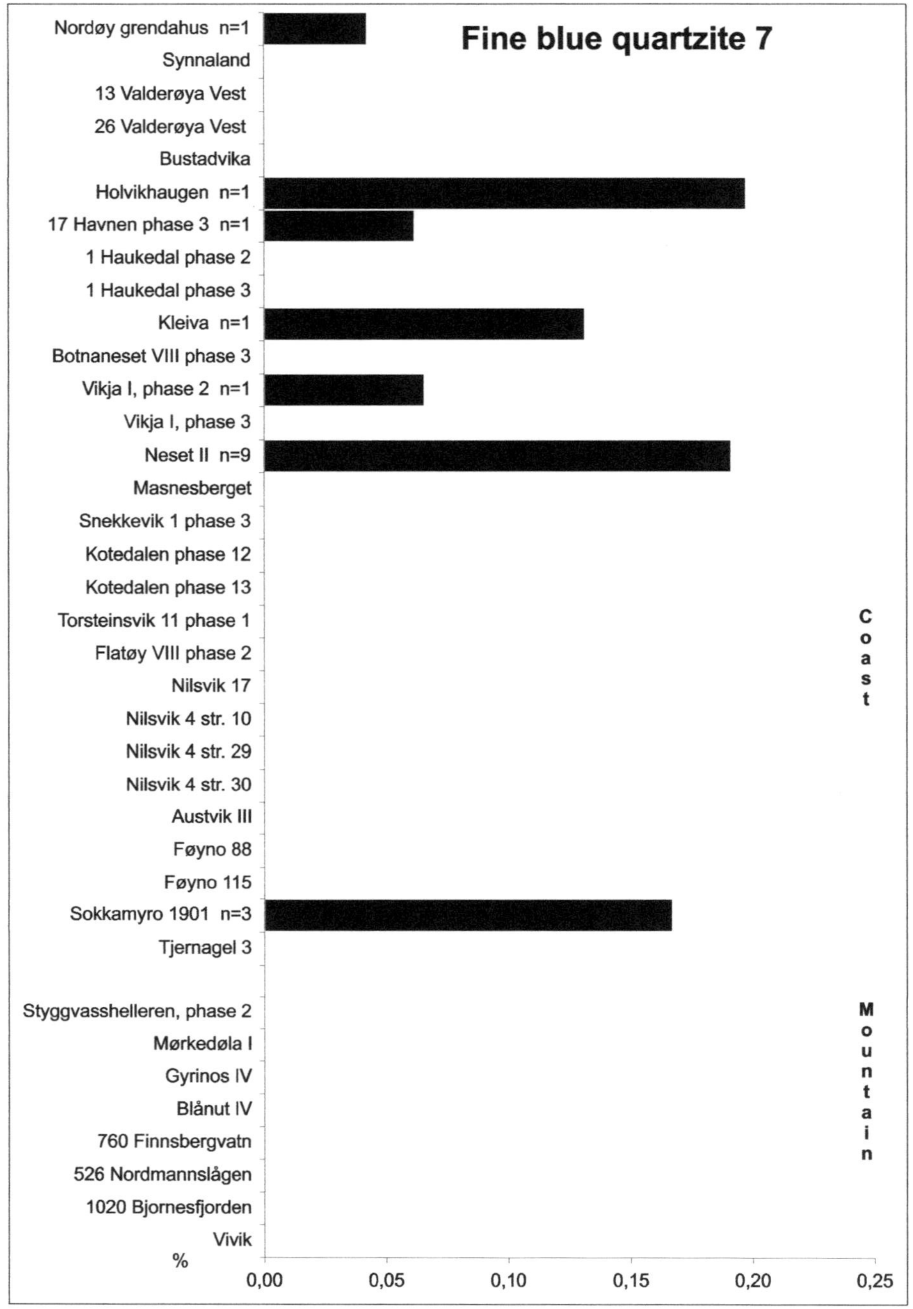

Figure 68. *Percentages of fine blue quartzite 7 at excavated sites.*

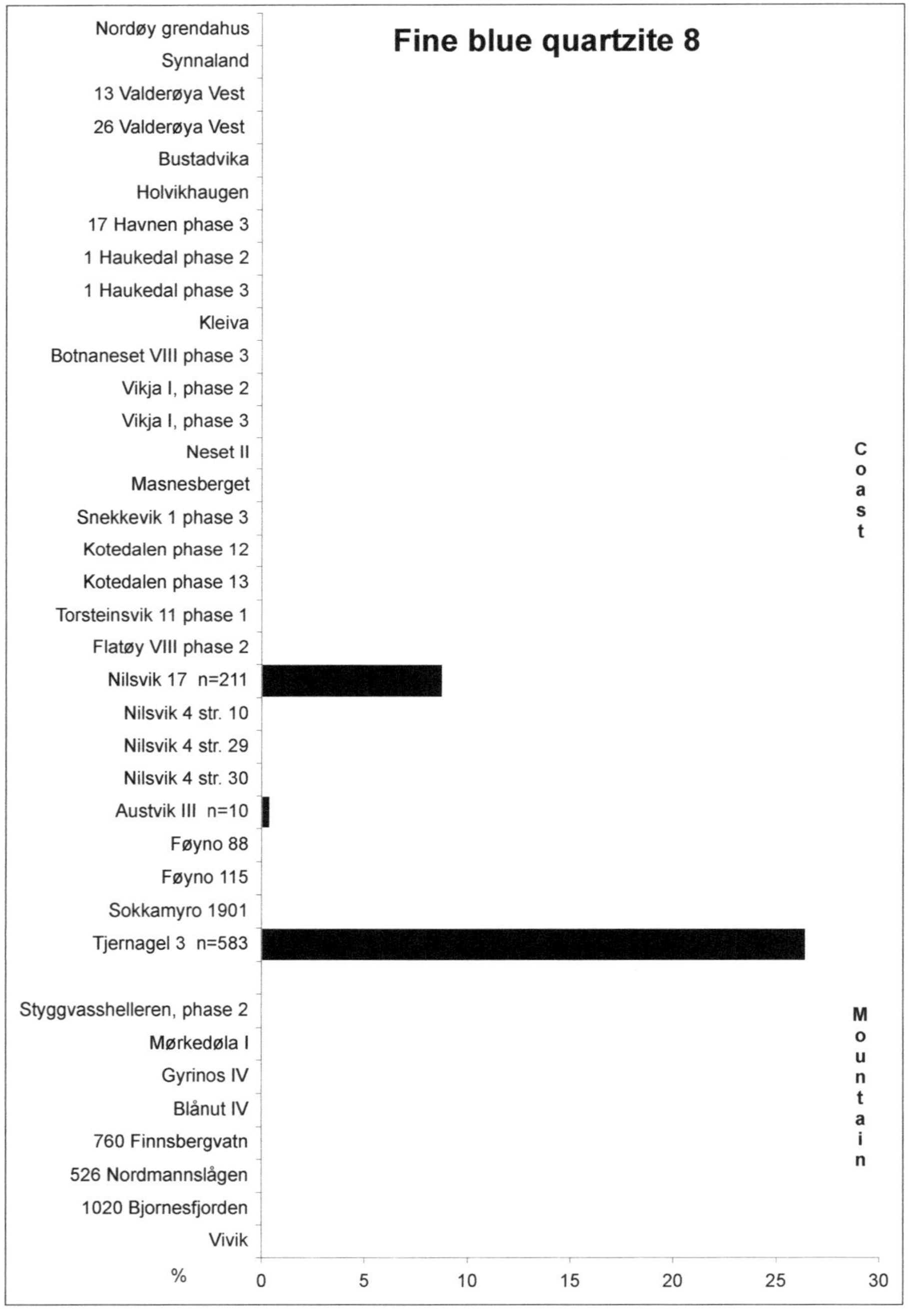

Figure 69. *Percentages of fine blue quartzite 8 at excavated sites.*

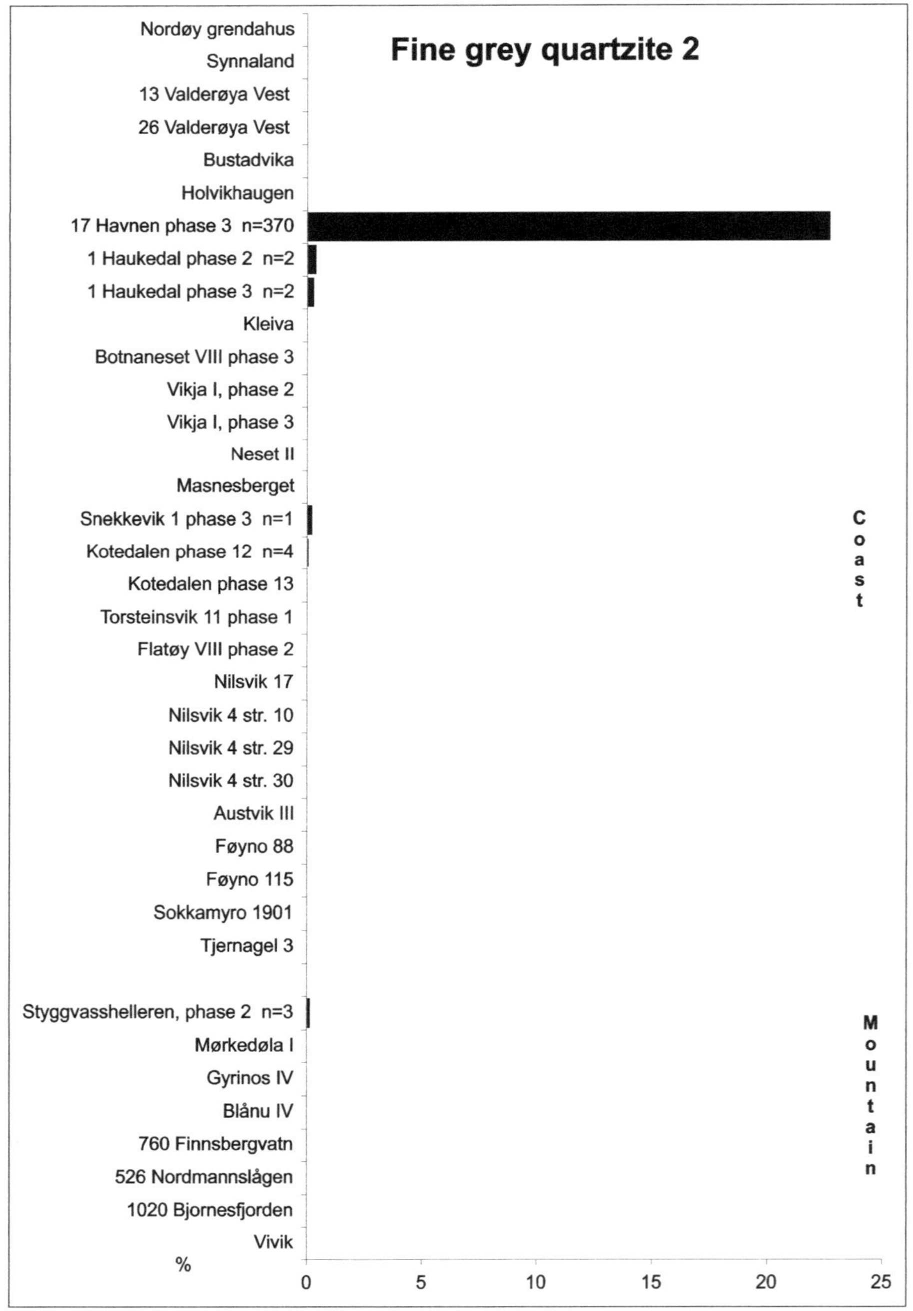

Figure 70. *Percentages of fine grey quartzite 2 at excavated sites.*

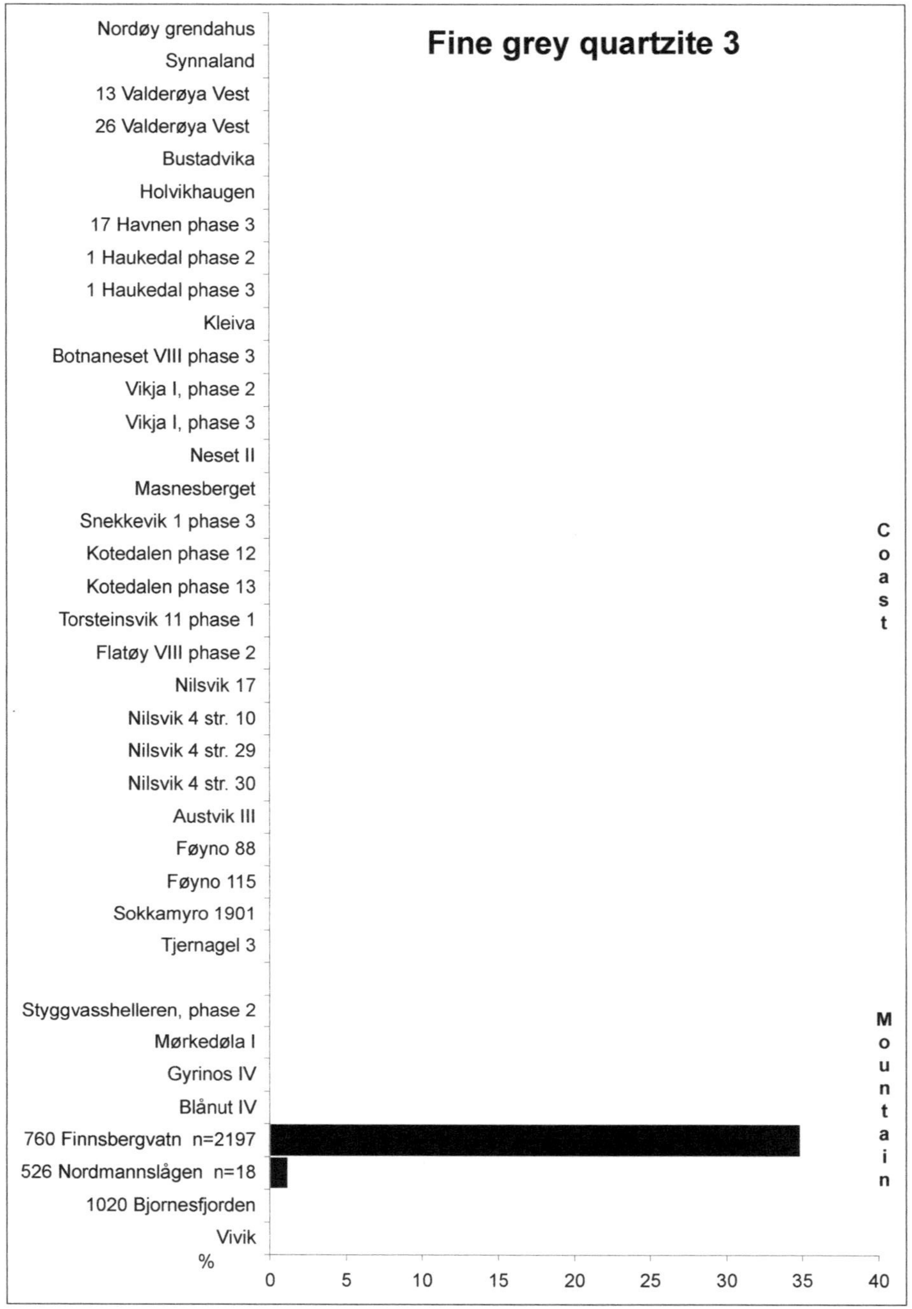

Figure 71. *Percentages of fine grey quartzite 3 at excavated sites.*

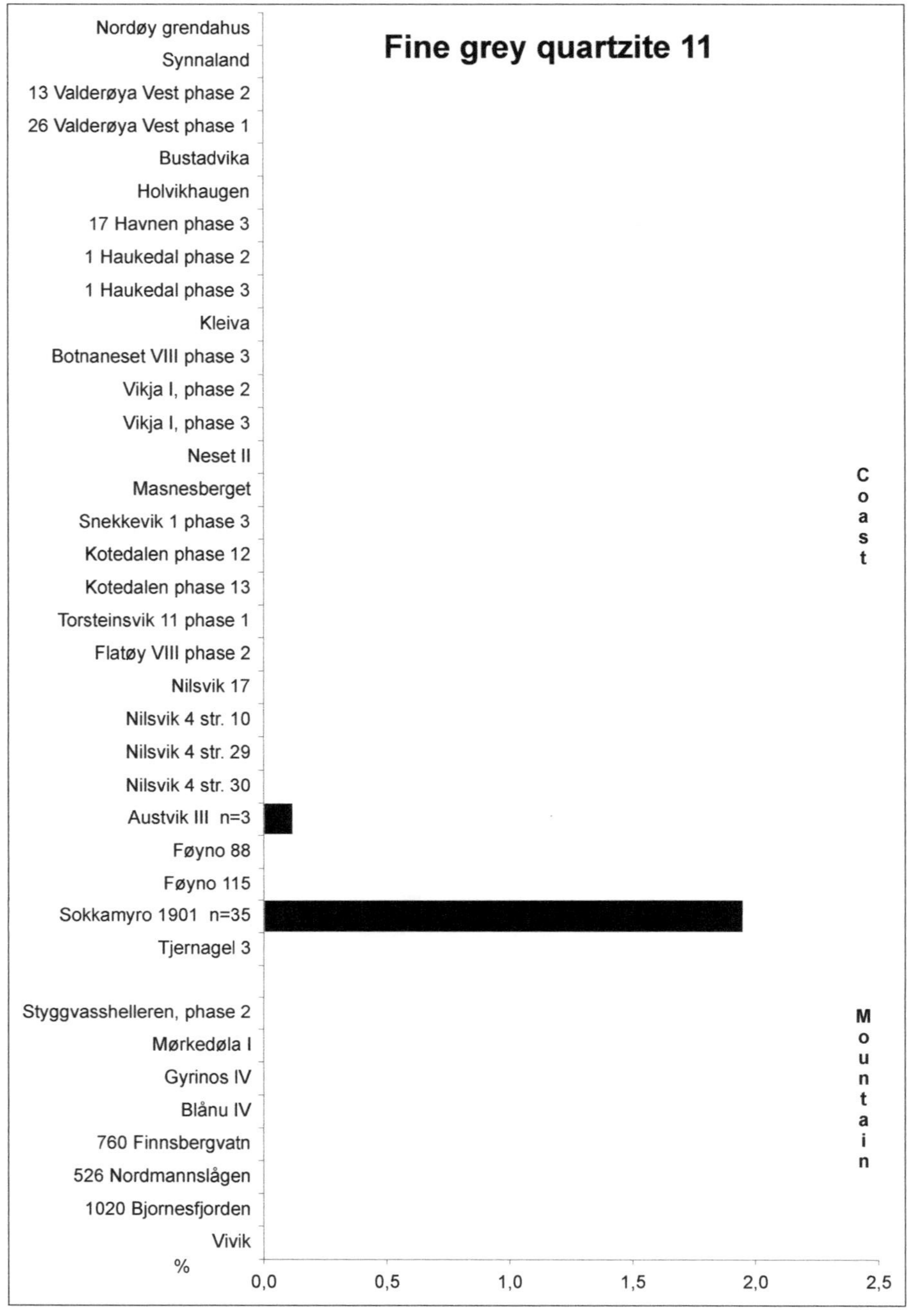

Figure 72. *Percentages of fine grey quartzite 11 at excavated sites.*

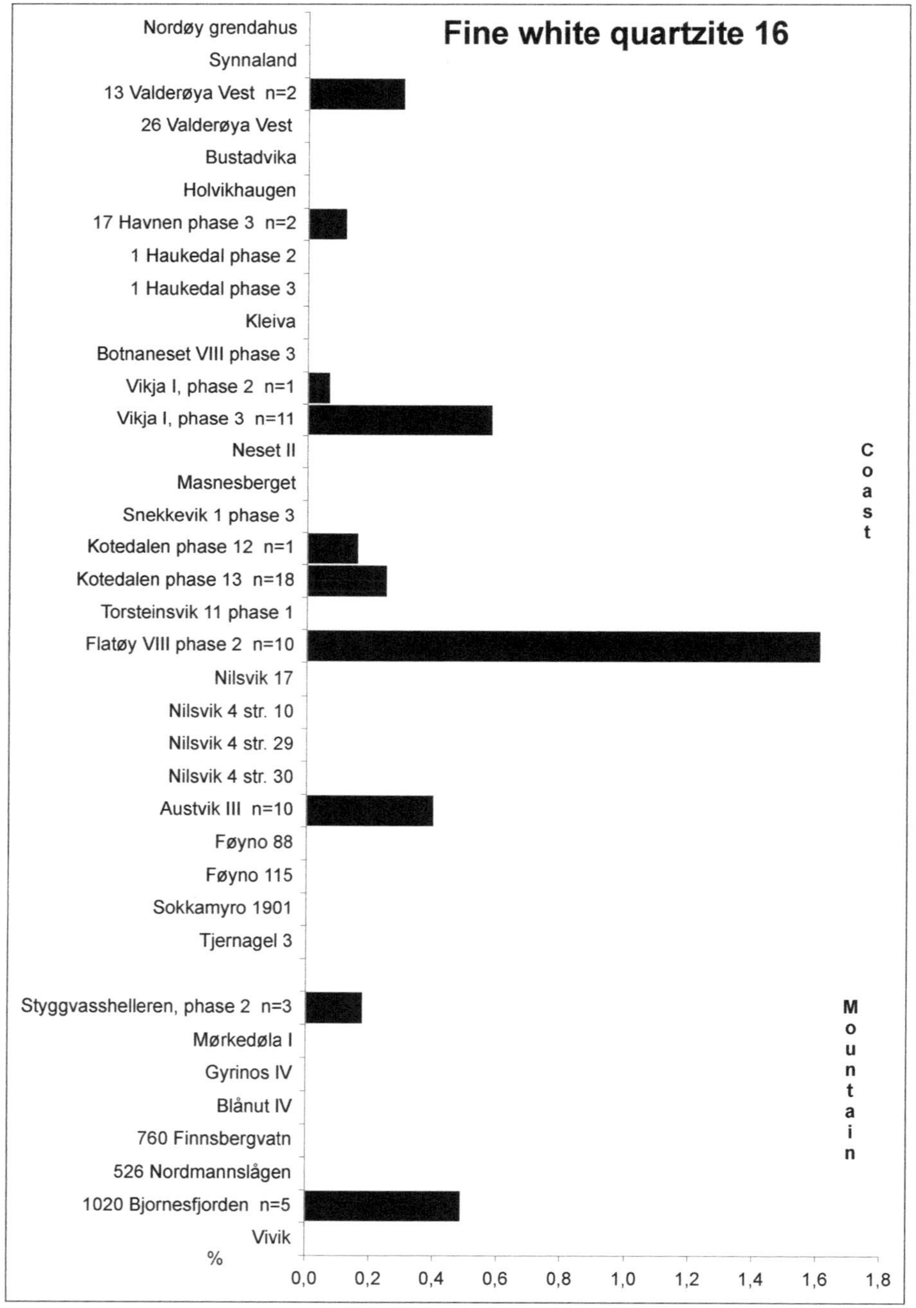

Figure 73. Percentages of fine white quartzite 16 at excavated sites.

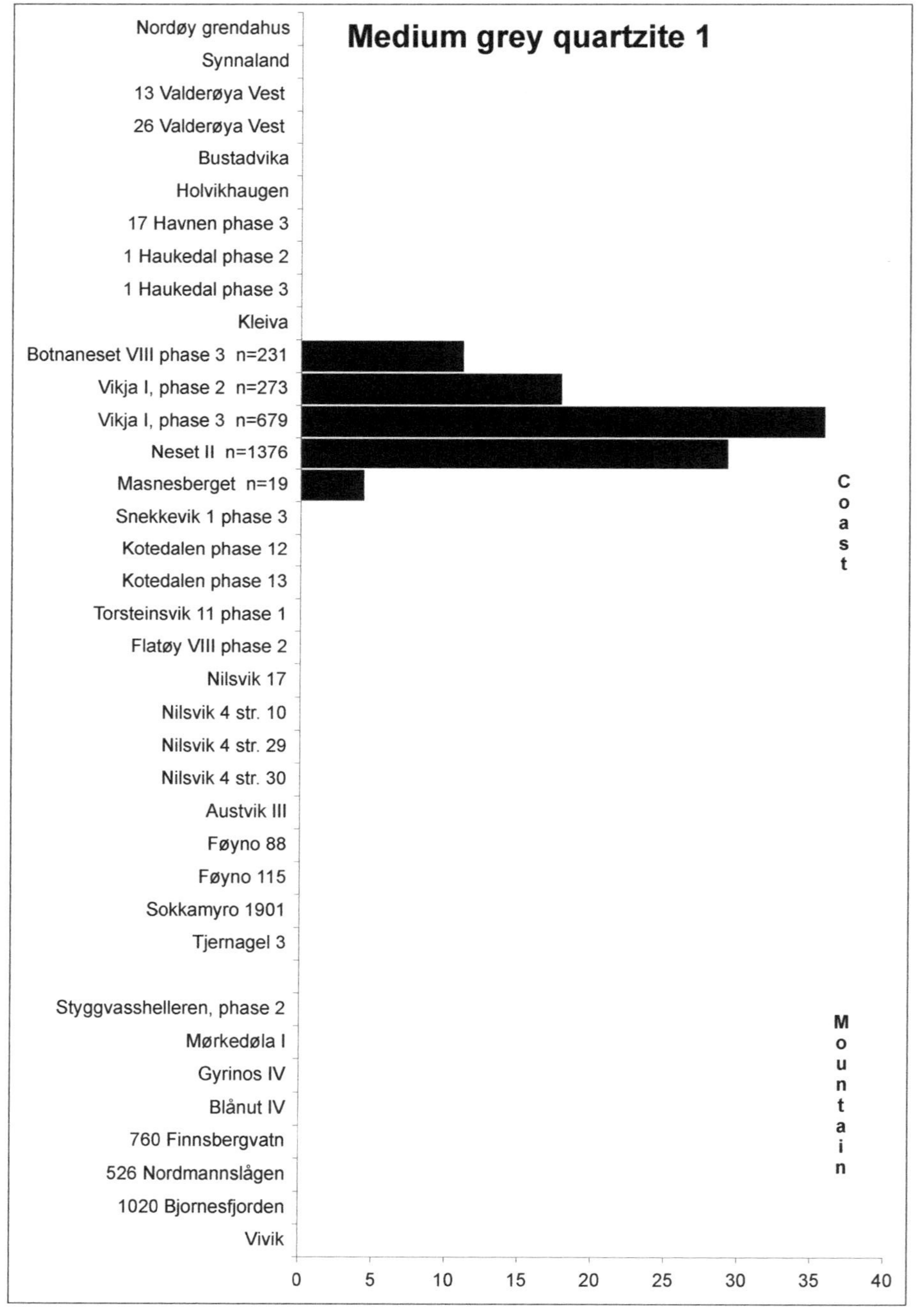

Figure 74. Percentages of medium grey quartzite 1 at excavated sites.

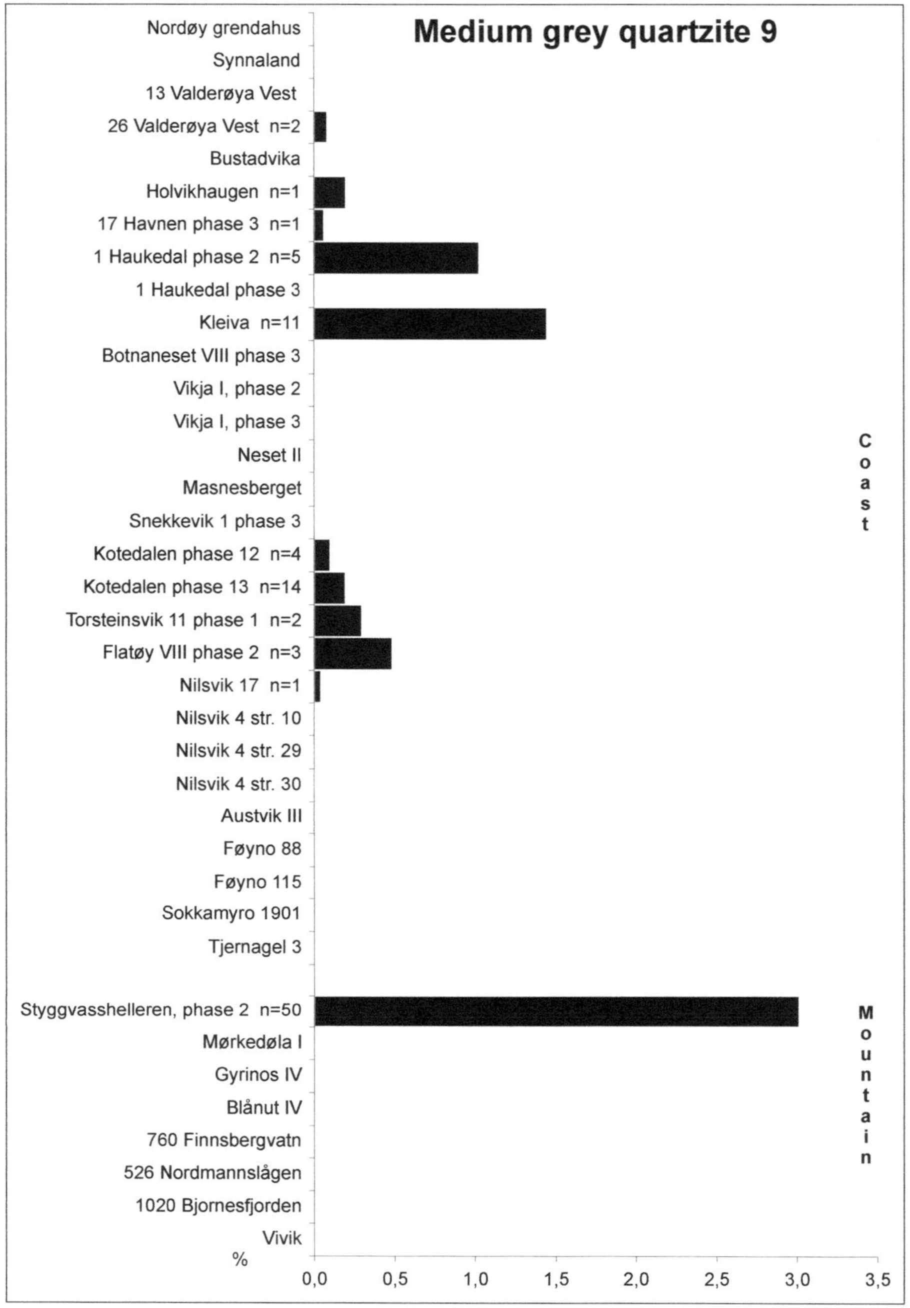

Figure 75. *Percentages of medium grey quartzite 9 at excavated sites.*

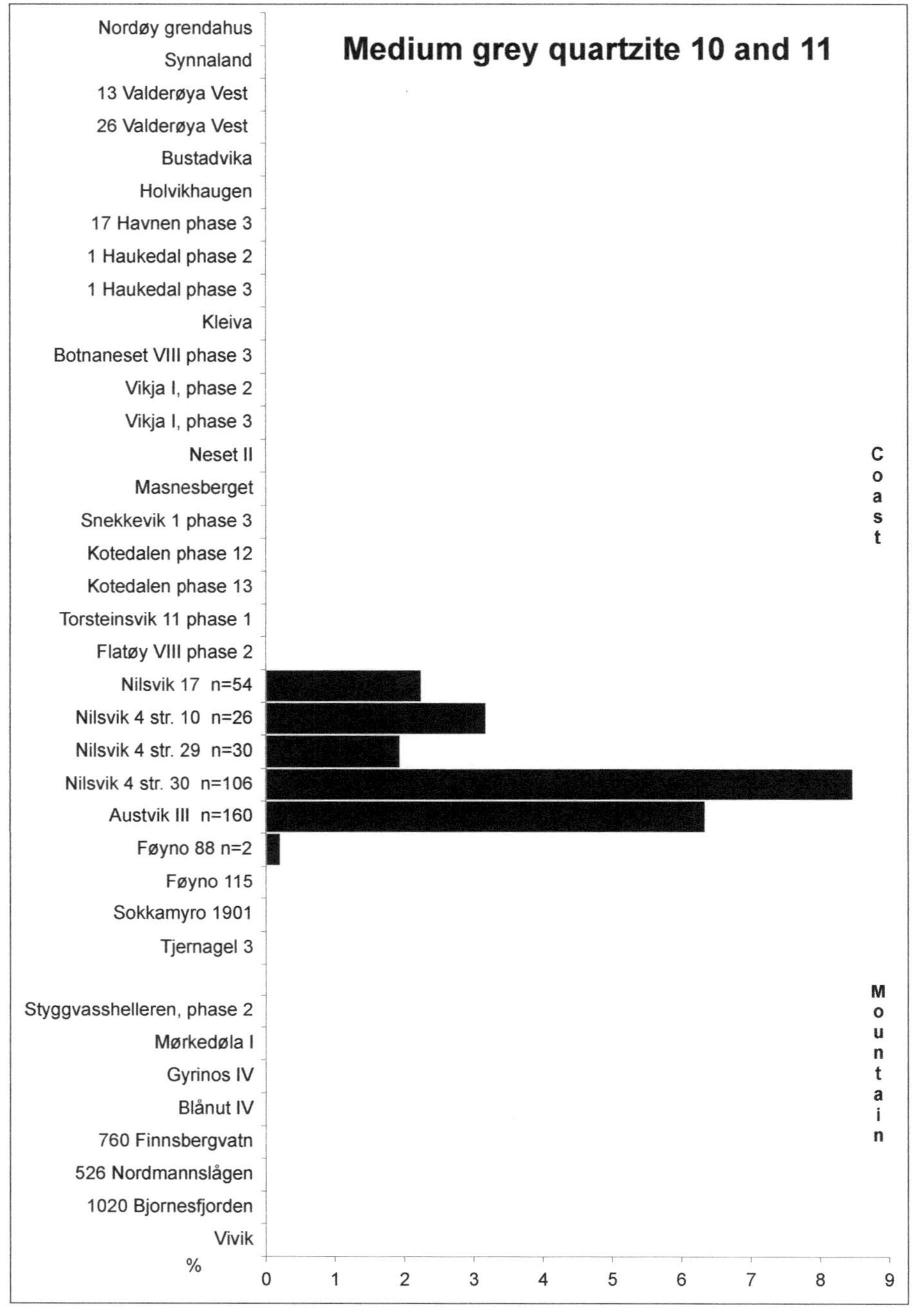

Figure 76. *Percentages of medium grey quartzite 10 and 11 at excavated sites.*

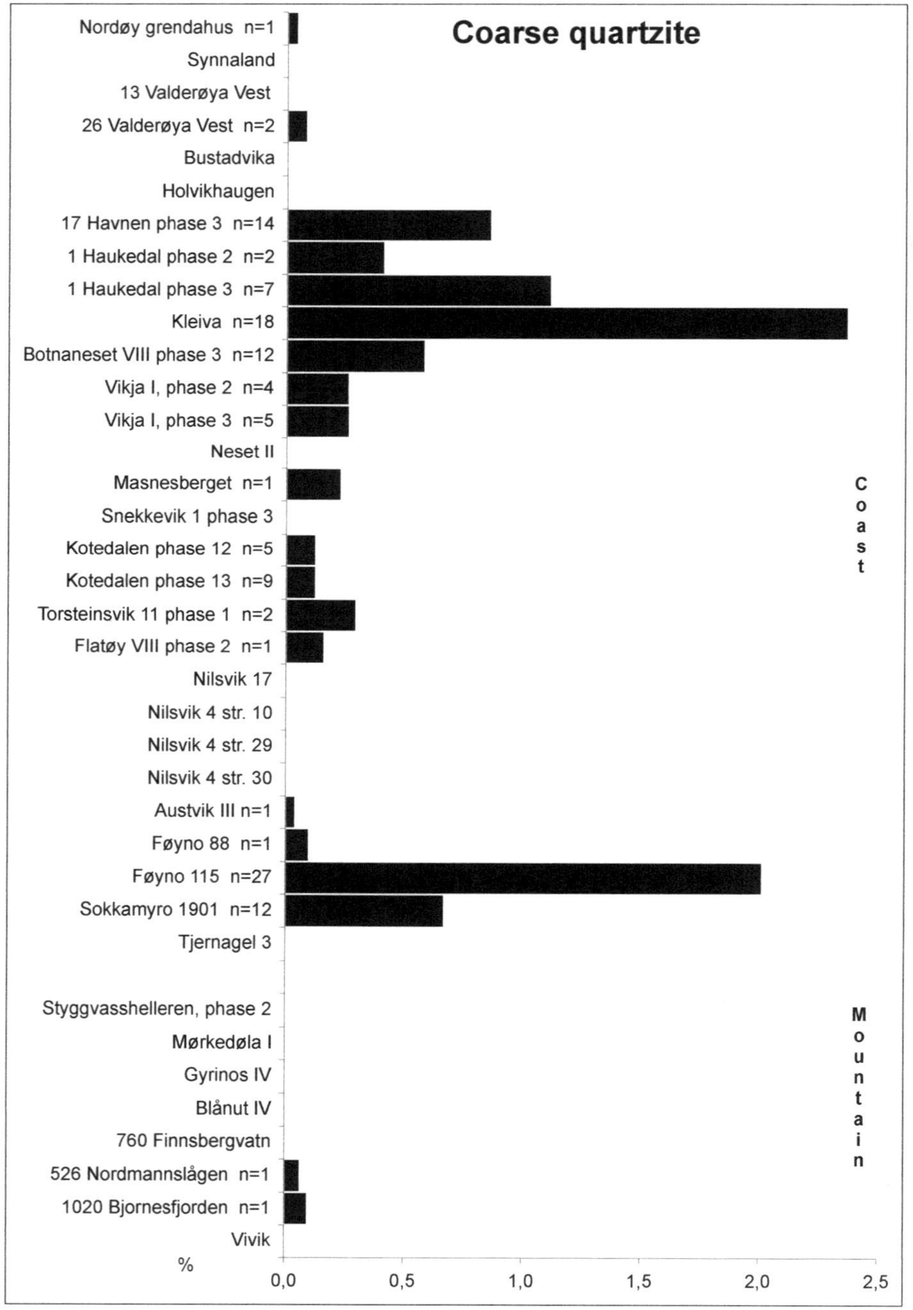

Figure 77. *Percentages of coarse quartzite at excavated sites.*

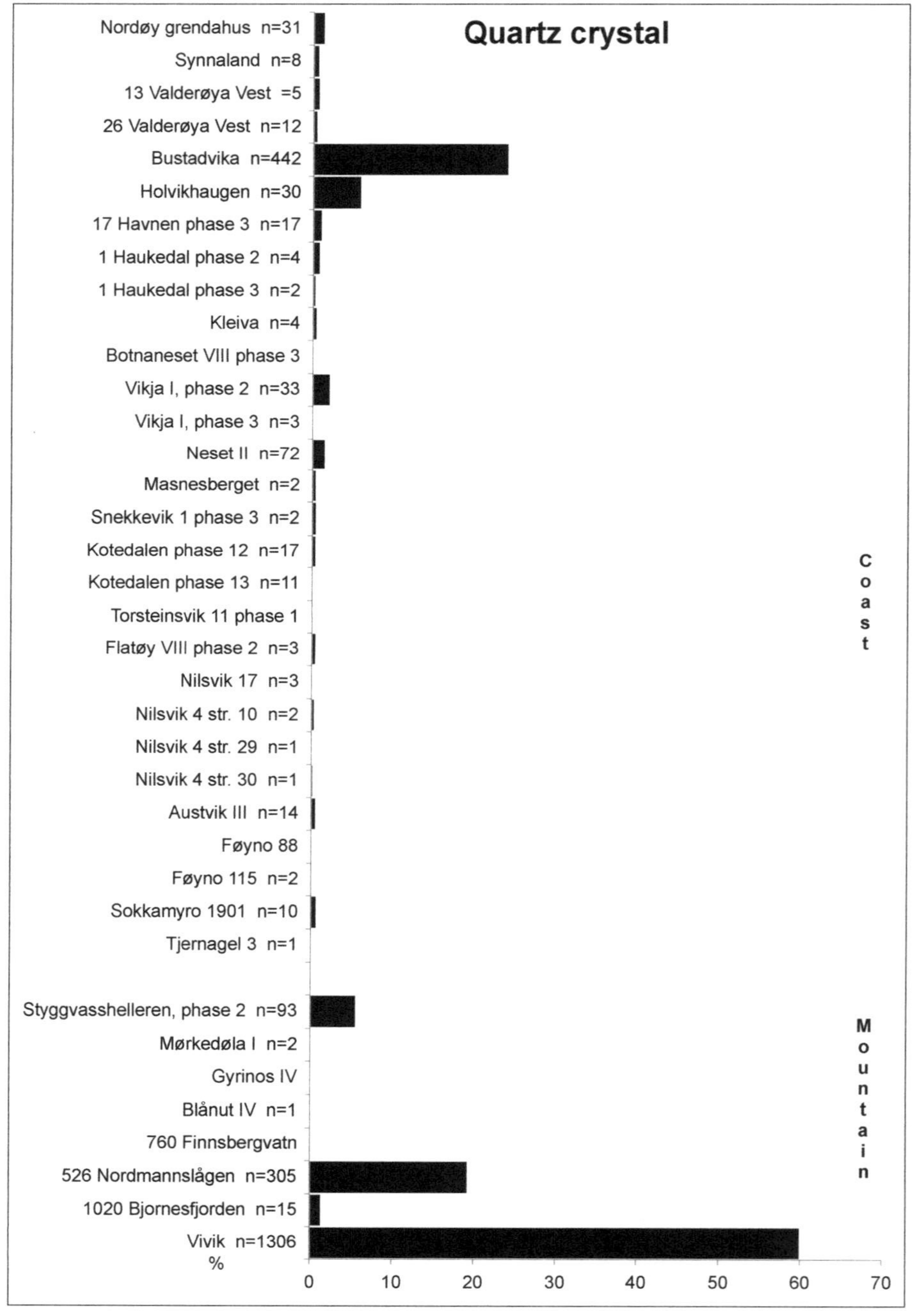

Figure 78. *Percentages of quartz crystal at excavated sites.*

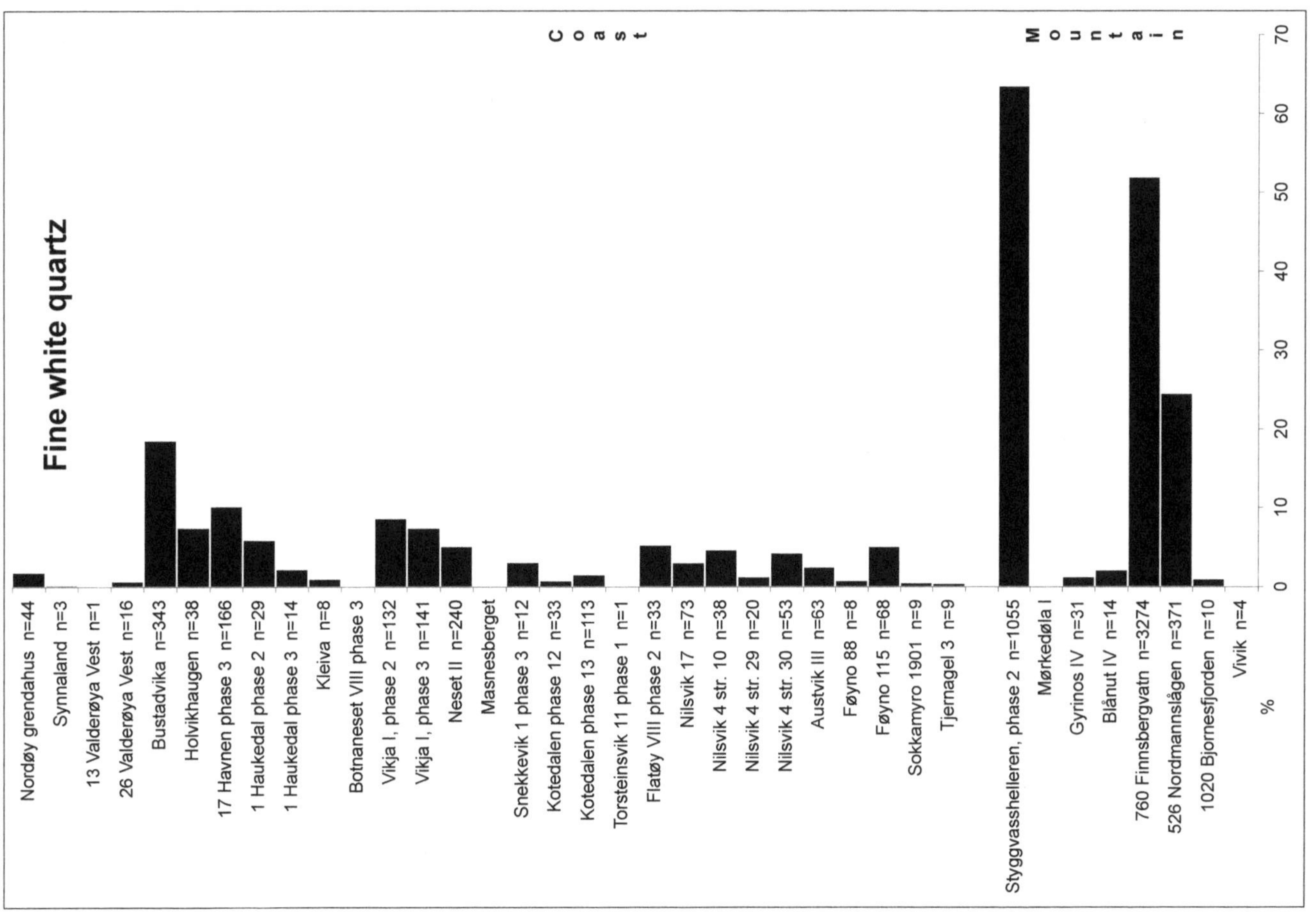

Figure 79. Percentages of fine white quartz at excavated sites.

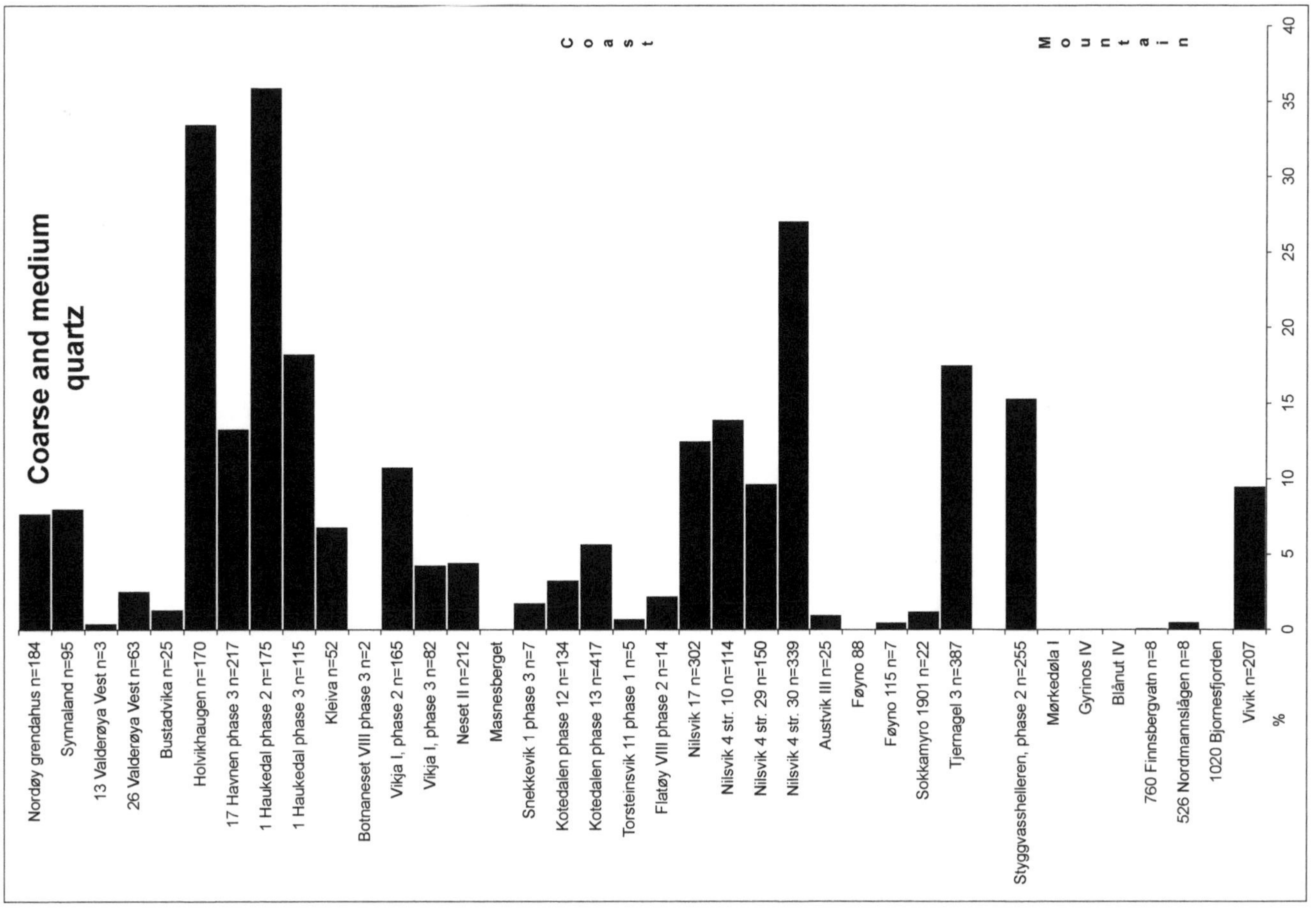

Figure 80. Percentages of coarse and medium quartz at excavated sites.

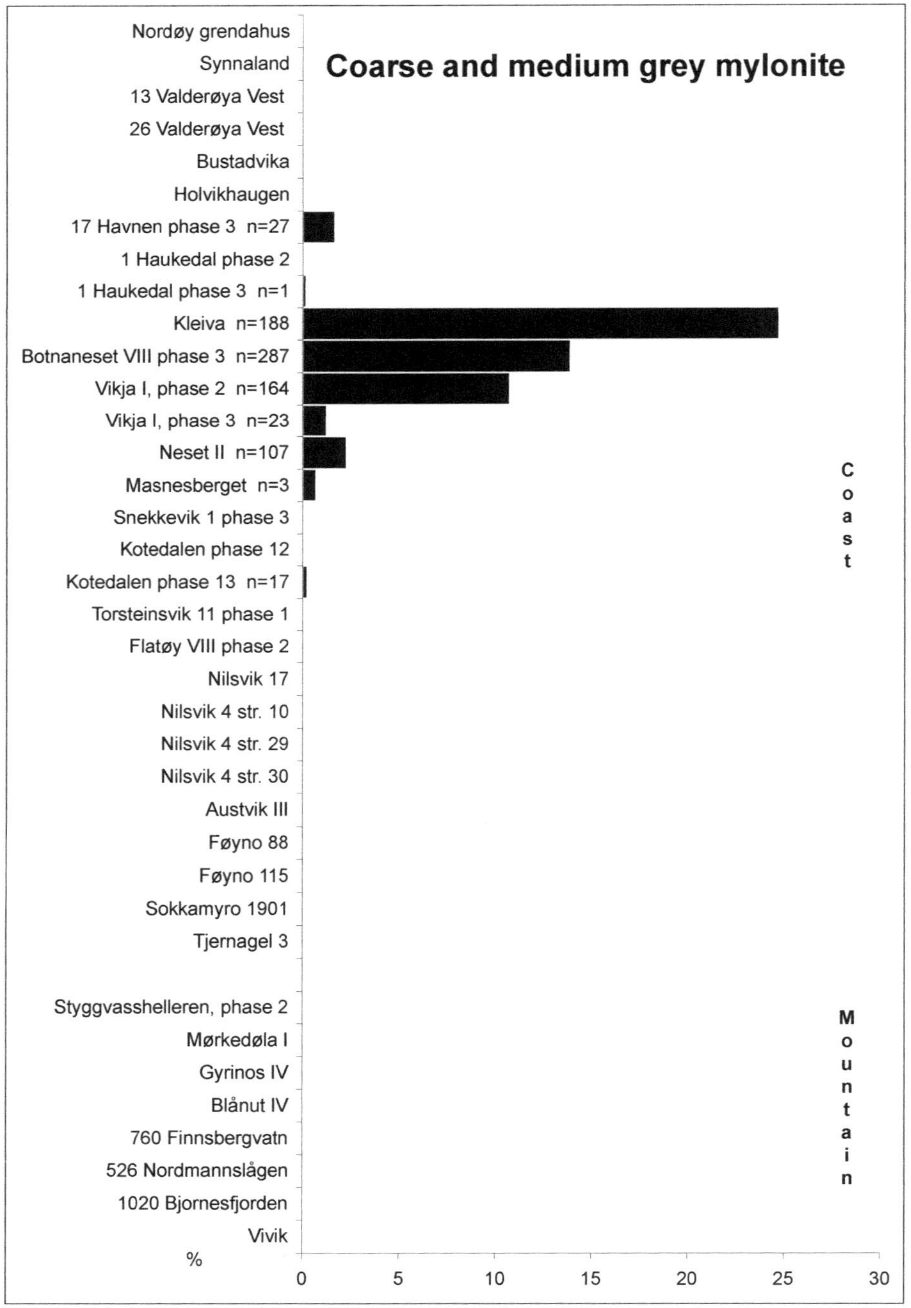

Figure 81. Percentages of coarse and medium grey mylonite at excavated sites.

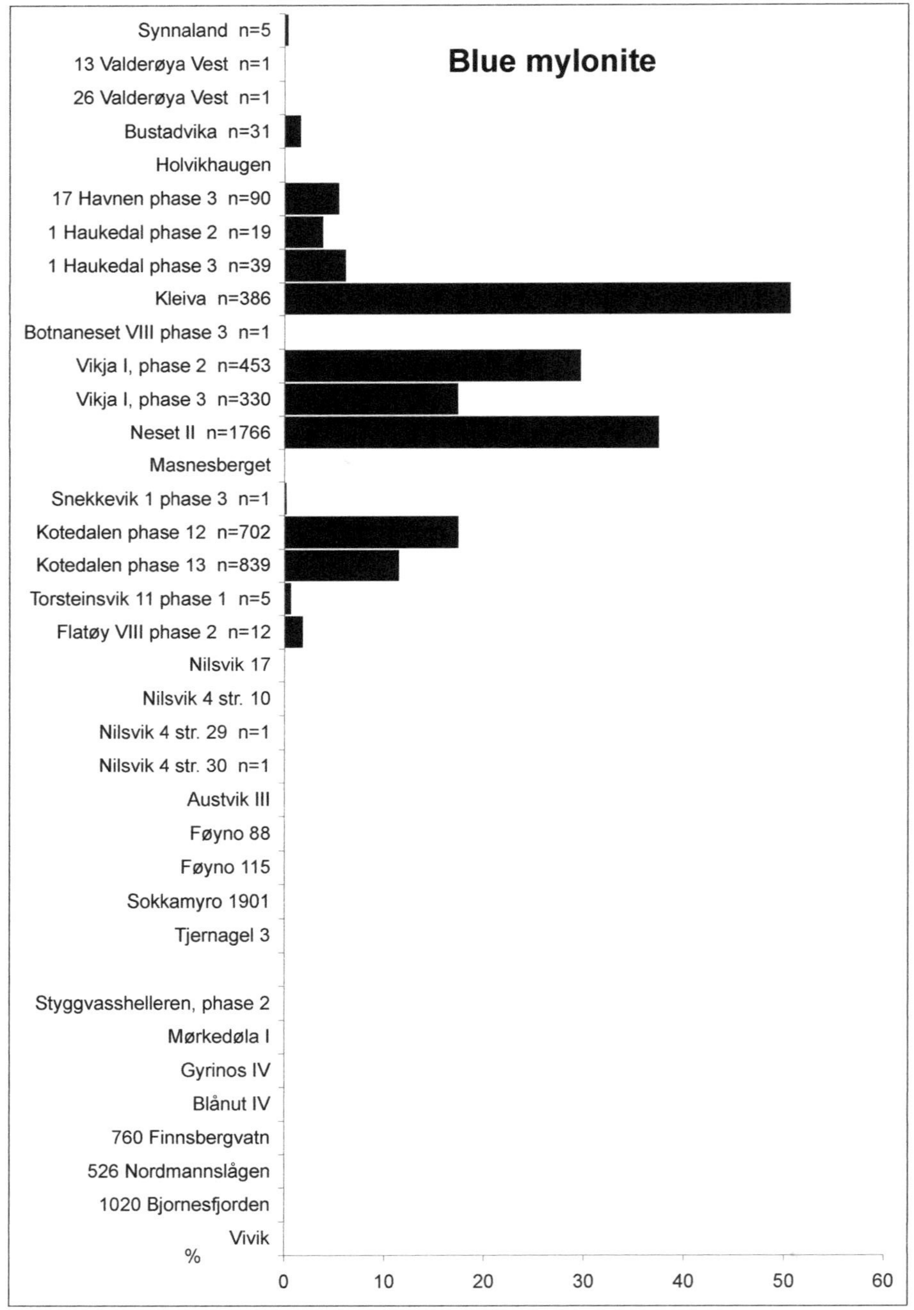

Figure 82. Percentages of blue mylonite at excavated sites.

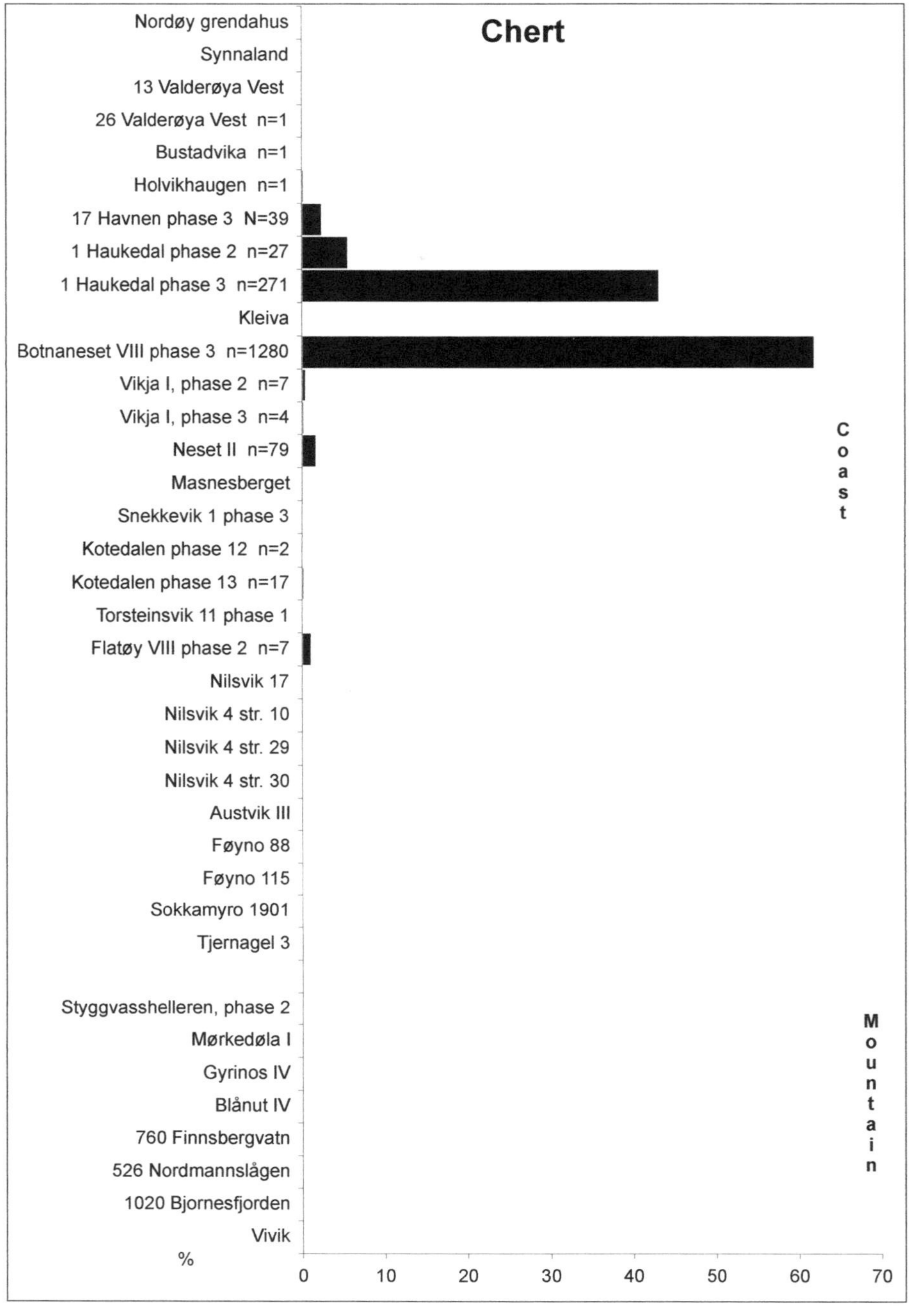

Figure 83. *Percentages of chert at excavated sites.*

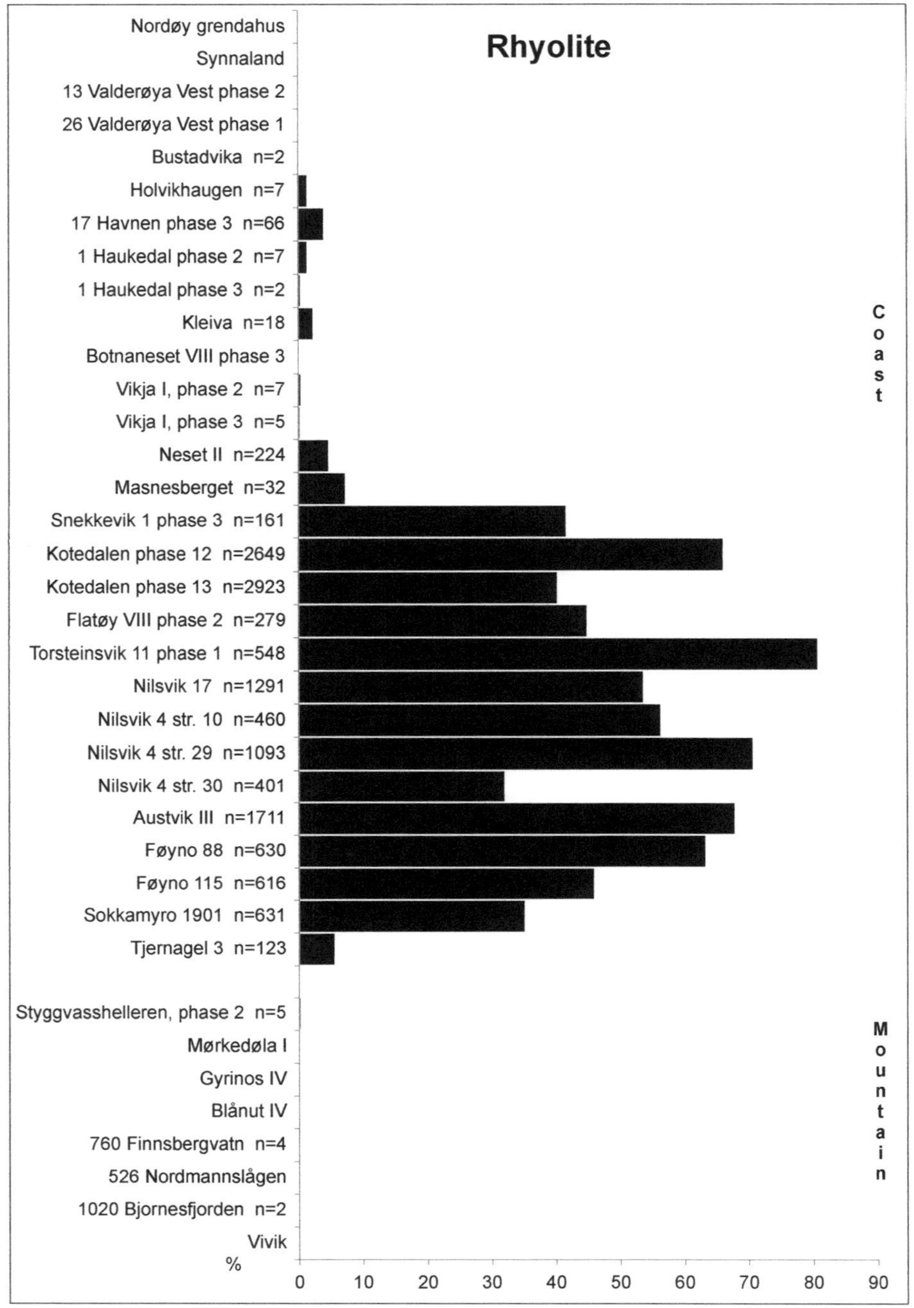

Figure 84. *Percentages of rhyolite at excavated sites.*

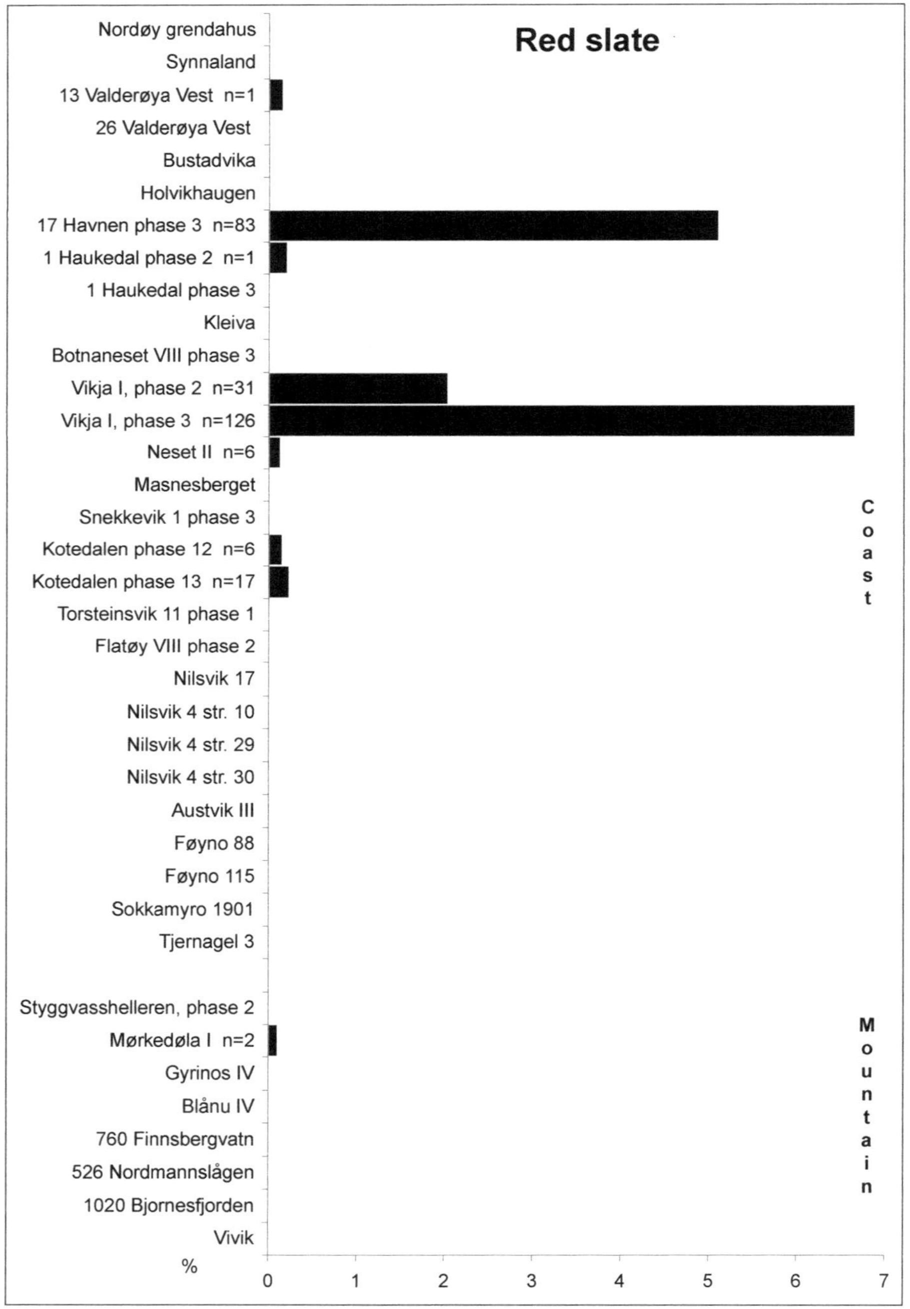

Figure 85. Percentages of red slate at excavated sites.

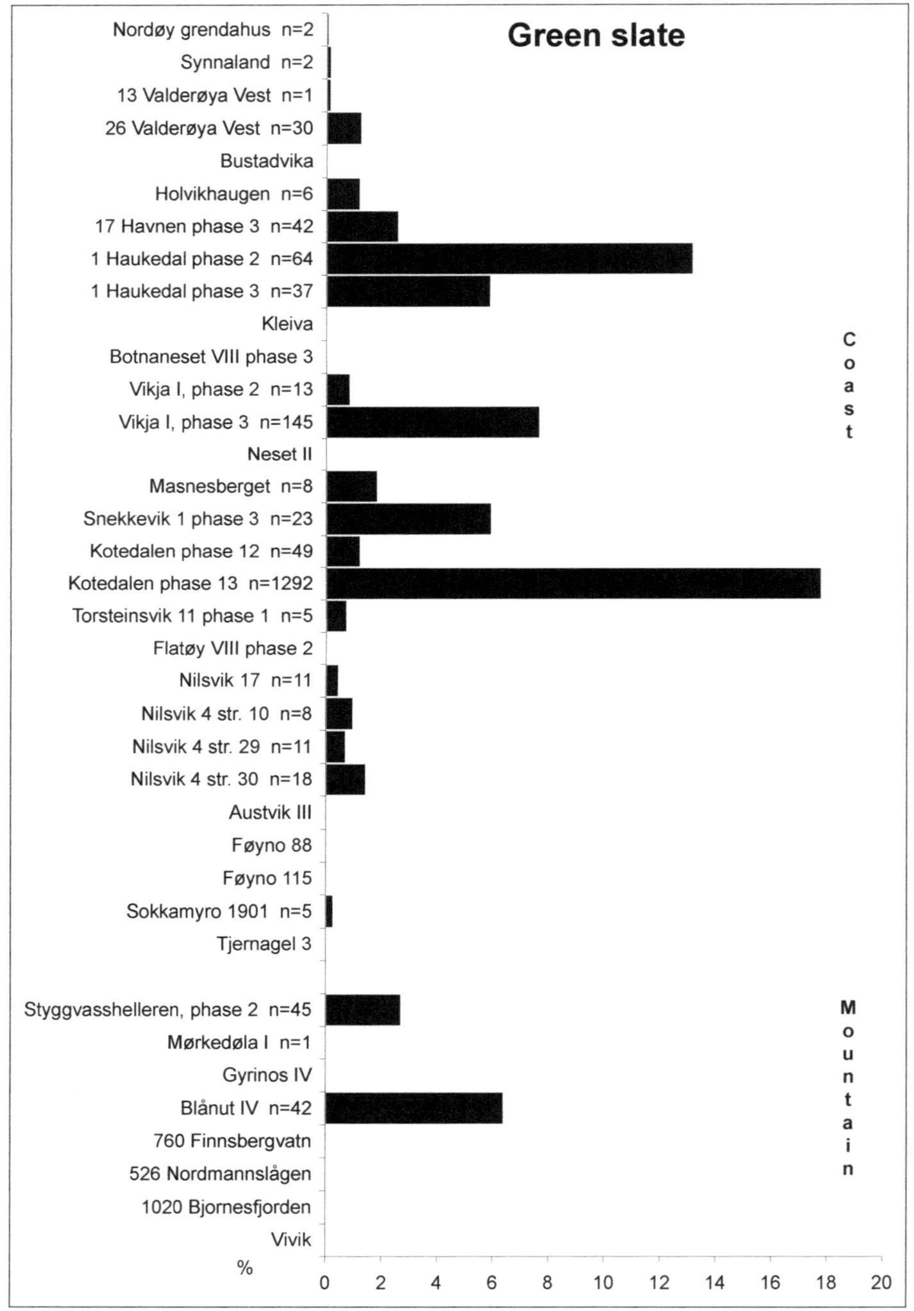

Figure 86. Percentages of green slate at excavated sites.

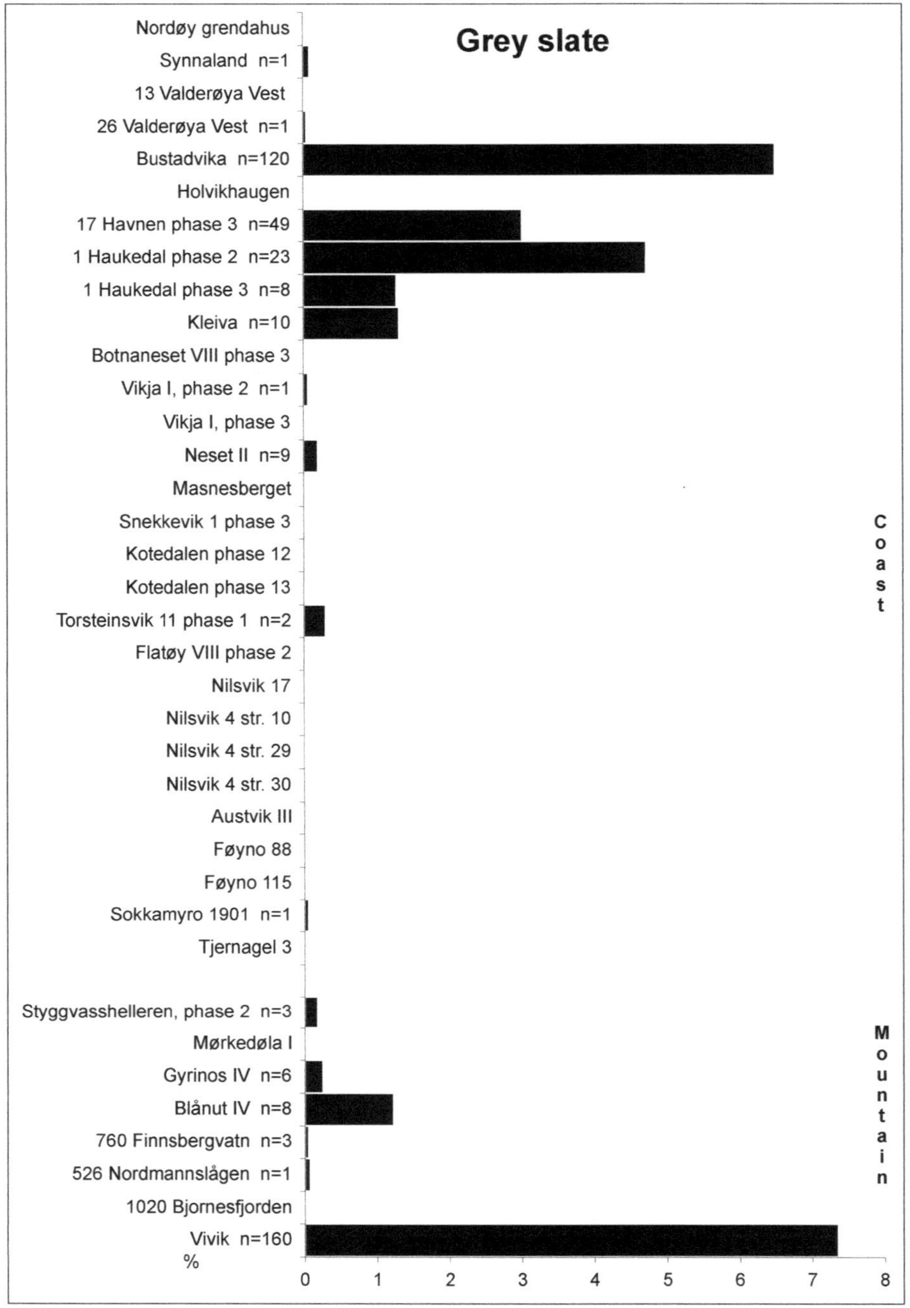

Figure 87. Percentages of grey slate at excavated sites.

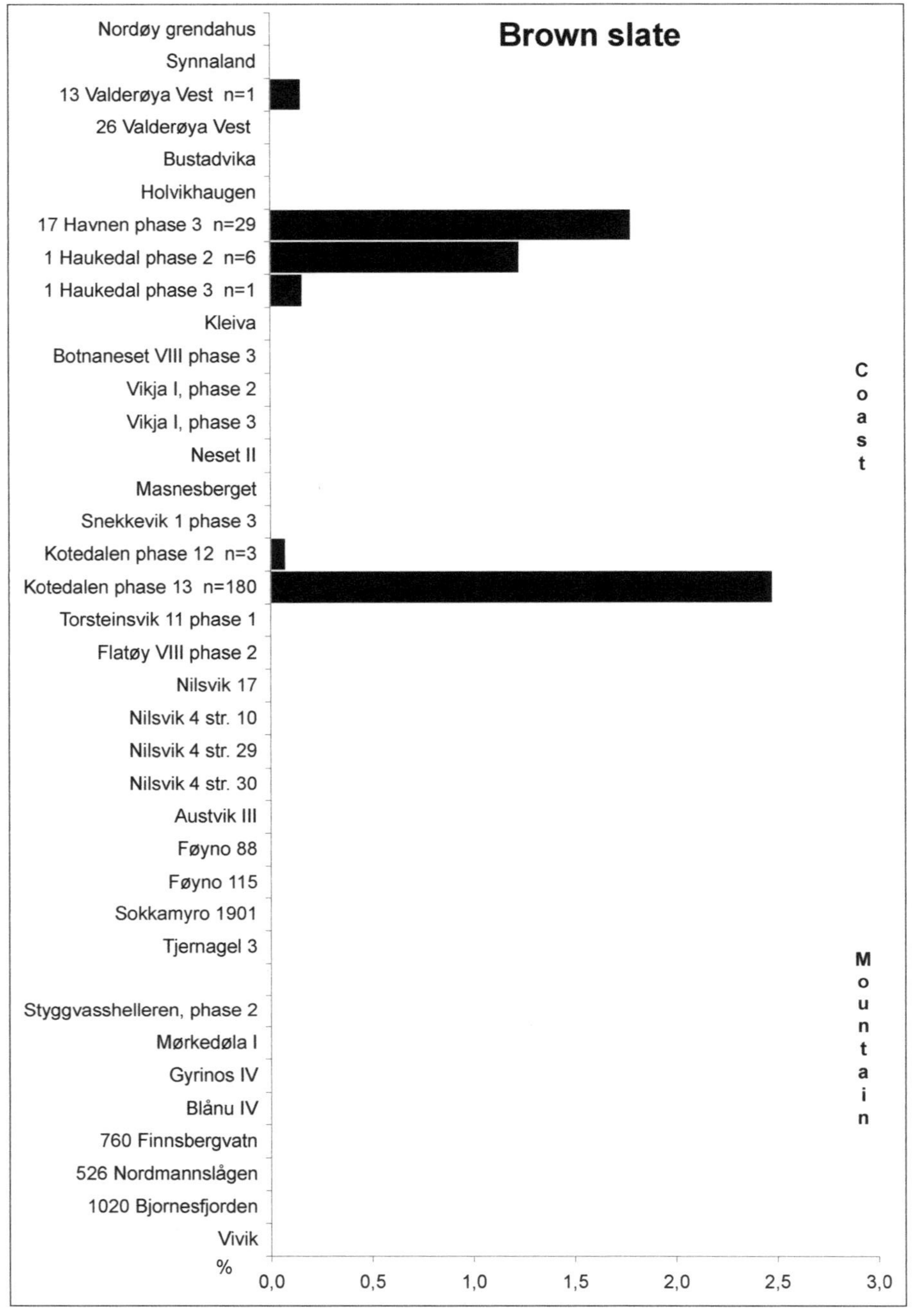

Figure 88. Percentages of brown slate at excavated sites.

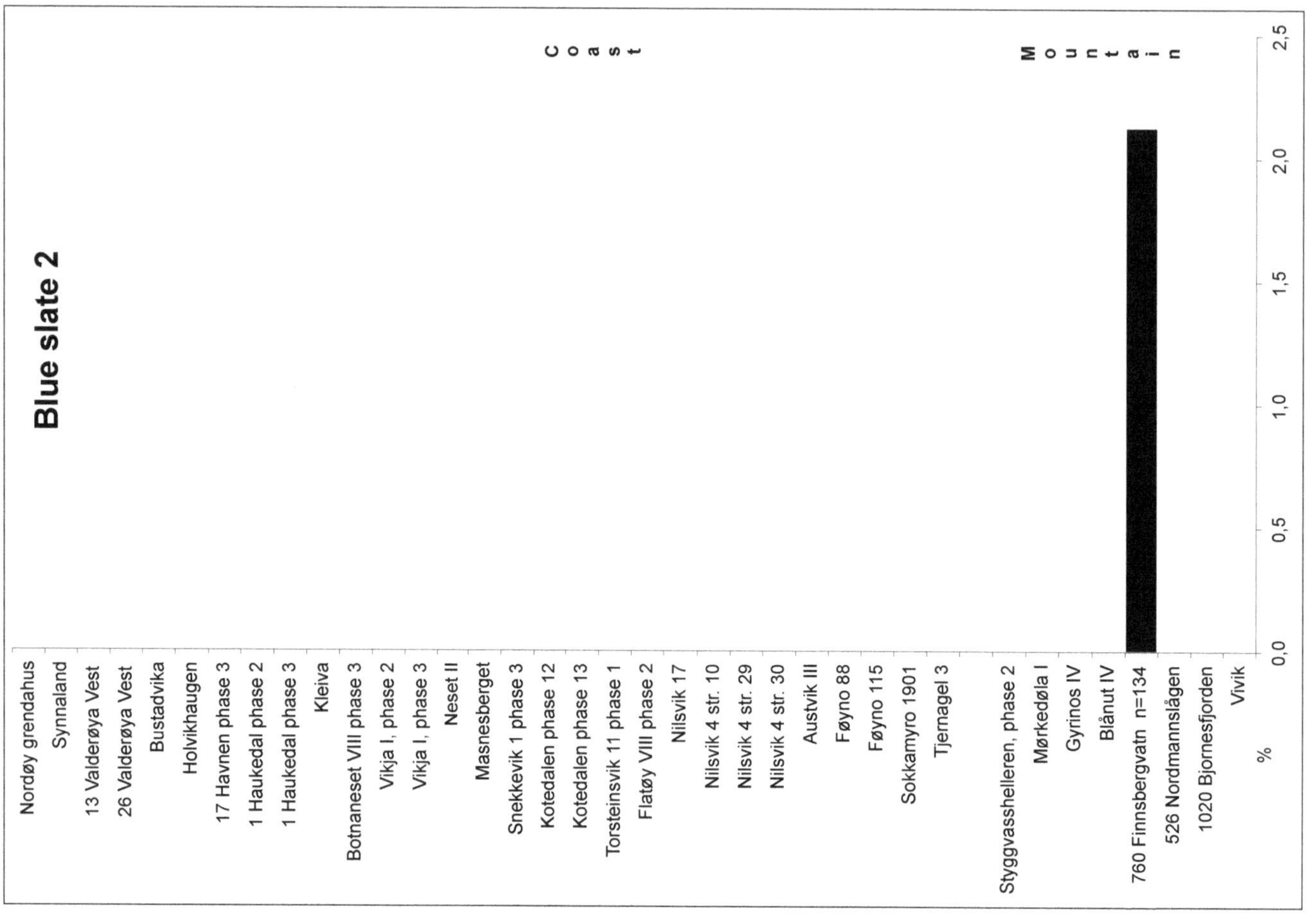

Figure 90. Percentages of blue slate 2 at excavated sites.

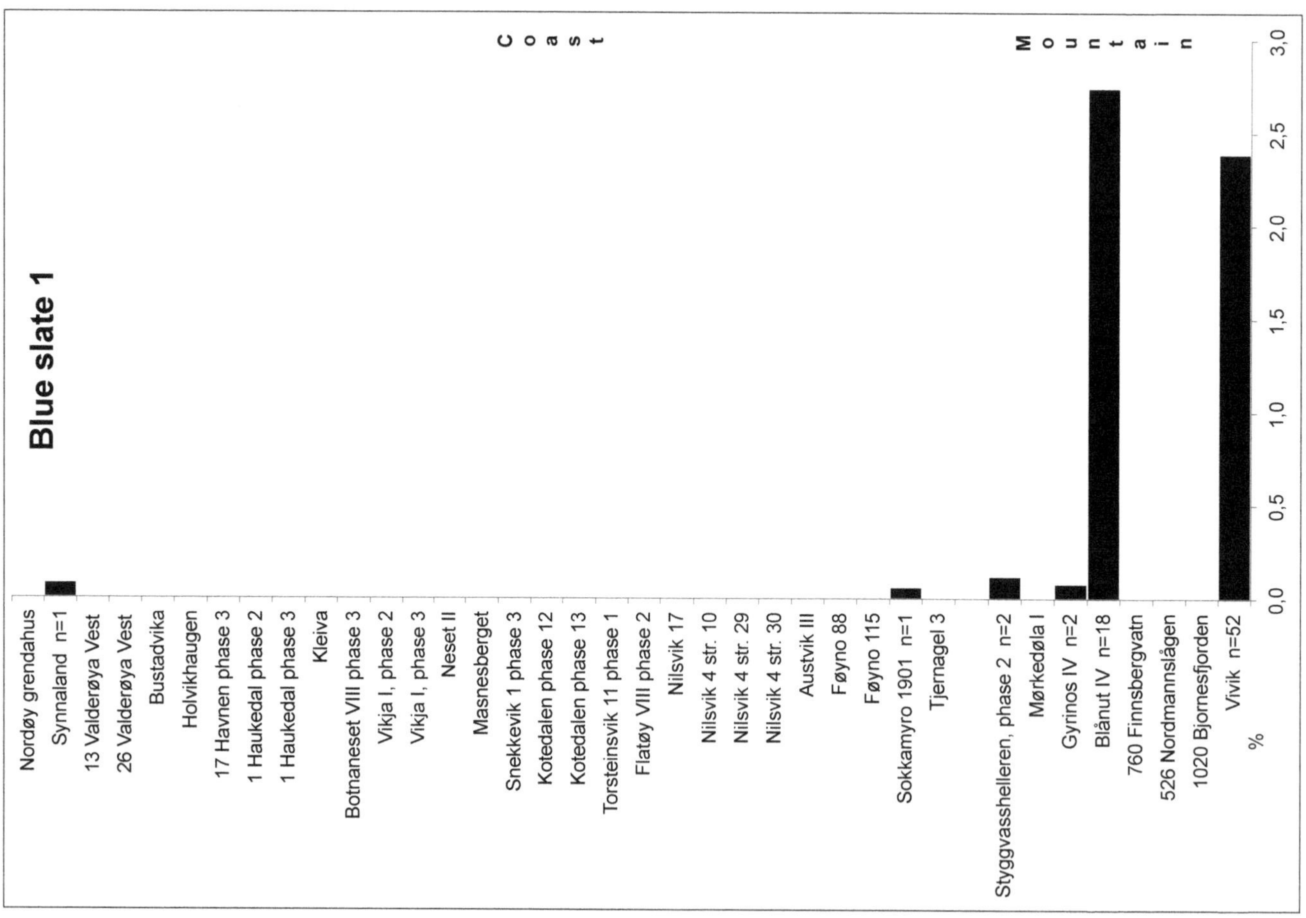

Figure 89. Percentages of blue slate 1 at excavated sites.

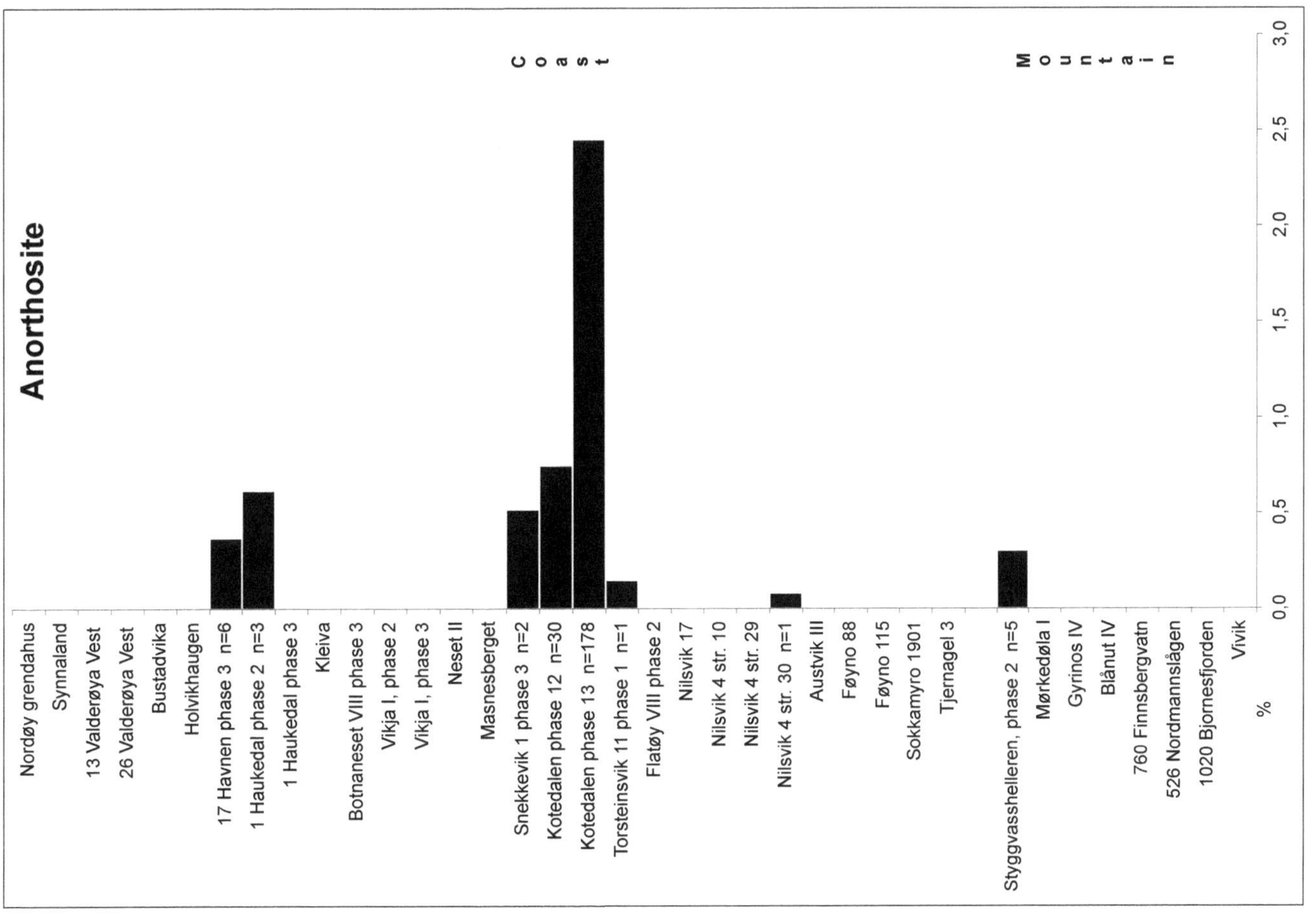

Figure 91. Percentages of anorthosite at excavated sites.

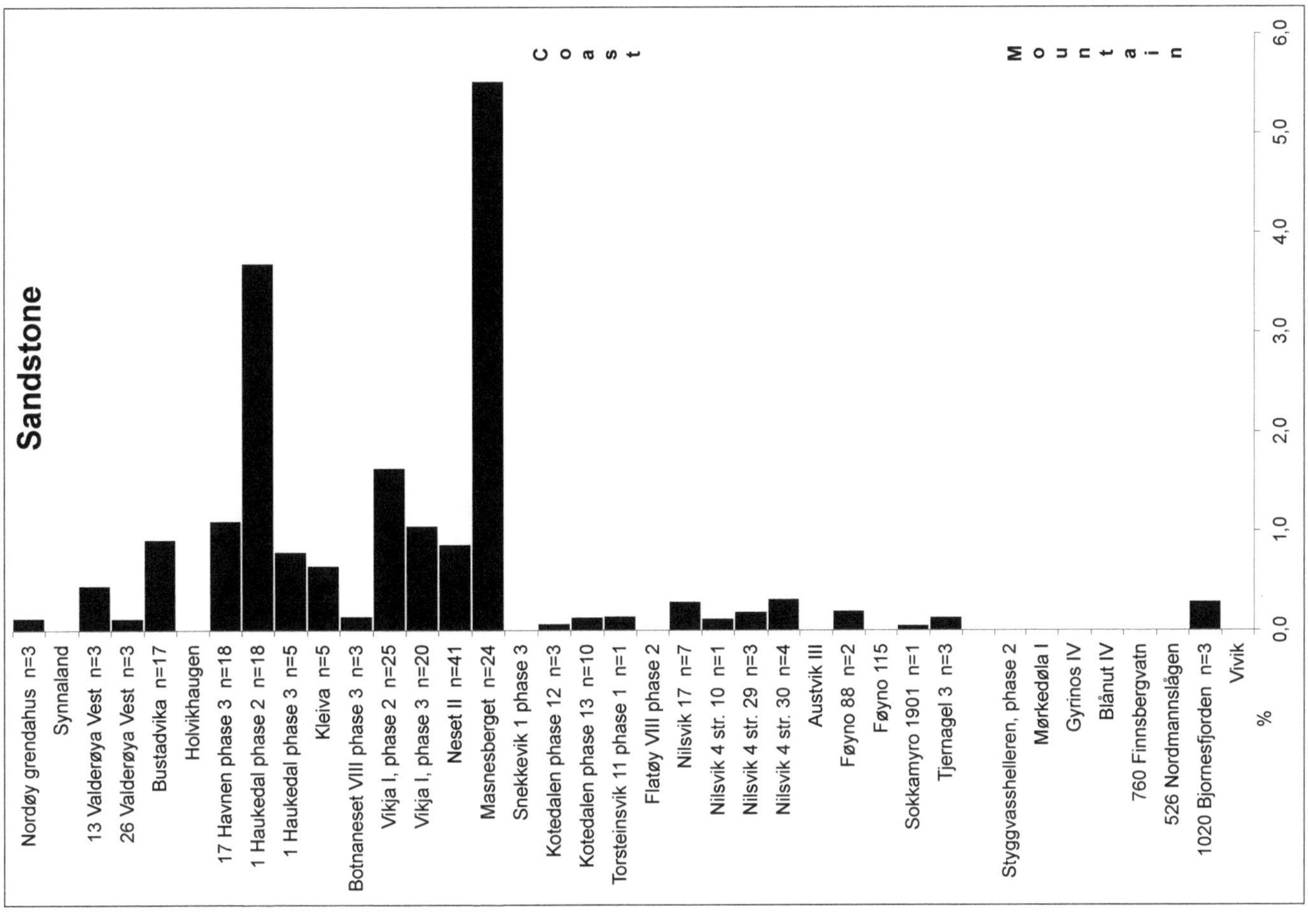

Figure 92. Percentages of sandstone at excavated sites.

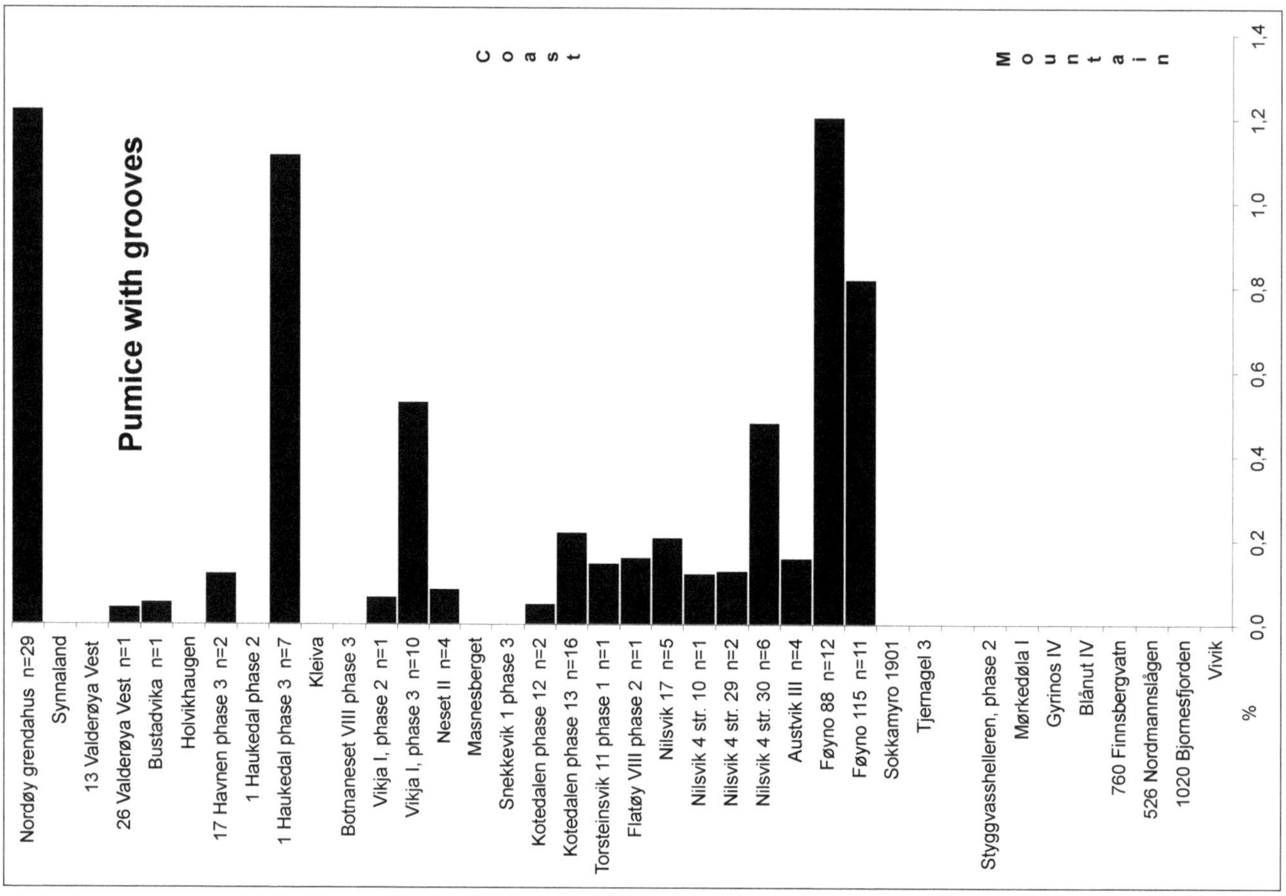

Figure 93. *Percentages of pumice with grooves at excavated sites.*

Types

Below, I will compare the distribution patterns of projectile points, retouched flakes/blades, small round smooth stones, grindstone slabs, adzes and axes, and slate knives. This will be expressed in percentage of the total amount of artefacts at the sites. I will mainly describe the tendencies within the area of analysis; however, since most of the types in question are also found outside of this area, I will refer to, and in several cases also apply data from adjacent regions. Some types found at the excavated sites were probably deposited during periods before or after the early Neolithic.[5] These will not be presented here. Locally produced pottery has been identified at a few of the excavated sites. As argued in Chapter 5, this type of pottery is held to be mainly middle Neolithic, and will not be considered here.

Several of the (presumably Neolithic) adzes are undetermined types that have not been sufficiently described for the purpose of comparison.[6] These will not be presented. Neither will hammer stones, anvil stones, and flint nodules, which have been unsystematically collected at sites and are therefore not suitable for comparative analyses.

Blade points and flake points (A-points)

A-points are found throughout the analysed area. There are, however, quantitative differences between the districts. A-points make up an average of about 1% in the districts Nordhordland, Midthordland and Sunnhordland, while these points generally represent less than 0,5% to the north of these districts and at the mountain sites (Fig. 94).

In Chapter 5 it was noted that A-points appear as blade points and flake points. Unfortunately, I was not aware of this technological distinction until most of the coastal sites had been examined. The distinction was therefore only made in the mountain assemblages where flake points are generally well represented compared to blade points (Styggvasshelleren excluded). However, from my examination of the coastal sites, I had the impression that flake points were only minimally represented on the coast. In order to document this possible difference, I investigated once again the A-points at three of the coastal

5 The Mesolithic types that are not presented include micro blades struck by microblade technique, conical microblade cores, keeled cores, flake drills, soapstone net-sinkers, tabular sandstone knives, adzes/chisels with bi-convex and round/oval cross-sections, entire crystals. Middle Neolithic: Vestland adzes and chisels, locally produced pottery, tabular sandstone knives (crescent shape), B-points. Late lithic Period: bifacial points, soapstone vessels.

6 These include: ground adzes with rectangular cross sections (Vestland/Vespestad adzes), other four-sided adzes/chisels, other struck/ground adzes/chisels, flakes struck from ground stone artefacts, struck blanks.

sites: Austvik III, Kotedalen phase 13 and 17 Havnen.[7] Assuming that these three sites are representative of the coastal region, the result of this inquiry is interesting (Fig. 95).

The histogram clearly shows a marked difference between the coast and the mountain: A-points are predominantly made of blades on the coast, while flake points are much more common at the mountain sites.

Outside of the area of analysis, A-points from Slettabø, Holeheia and Nygård in southern Rogaland have been termed "blade points" by Arne Skjølsvold (1977:137 ff; 1980a:14; 1980b:19), and this classification is supported by the illustrated specimen. There may, however, still be a significant amount of flake points among the early Neolithic A-points at these sites. In his thesis about Gjedlestadvige R20 in southern Rogaland, Sveinung Bang-Andersen observed that the majority of the A-points were made from flakes (Bang-Andersen 1981:28). The drawn specimens are, however, mainly blade points. Further south, at the site Lundevågen, R18 in Vest-Agder, Torben Bjarke Ballin and Ole Lass Jensen noted that 40% of the A-points were made from flakes, while 52% were made from blades (Ballin & Jensen 1995:189). Judging from illustrations of A-points from eastern Norway, Mesolithic phase 4/early Neolithic points seem to have been made from both flakes and blades (Østmo 1976:46, Glørstad 1999:74).

On the basis of the above, one may conclude that A-points constitute fairly small amounts at sites to the north of Nordfjord, and that they dominate significantly in Hordaland to the south of Masnesberget. Within the coastal assemblages, blade points clearly dominate, while flake points and blade points make up almost equal shares at the mountain sites (except Styggvasshelleren, where blade points dominate). From southern Rogaland and eastwards along the coast, flake points constitute significant percentages at early Neolithic sites.

Single-edged points

This type of projectile point has not been identified at all in the coastal assemblages, while they are frequent in the mountain assemlages (Fig. 96). Single-edged points occur frequently in Mesolithic phase 4 and Neolithic assemblages in eastern Norway (Mikkelsen 1975, Østmo 1985:76, Glørstad 1998). They are also found at several coastal sites in Rogaland, to the south of the analysed area (e.g. Skjølsvold 1977:320, Skjølsvold 1980b, Bang-Andersen 1981:29), and also in Agder (Ballin & Jensen 1995:188).

7 The units that were investigated at these three sites are somewhat larger than the units used in the first examination. The number of A-points is therefore different from the ones presented in appendix 3.

Transverse points

Only two transverse points were found at the coastal sites, while this projectile-type occurs frequently at the mountain sites (Fig. 97). Transverse points are standard in the Mesolithic phase 4 and Neolithic lithic assemblages in eastern Norway and Agder (e.g. Mikkelsen 1975, Ballin & Jensen 1995:188, Glørstad 1999). They also occur in Rogaland (Skjølsvold 1980b:24).

Slate projectile points

Slate points are found throughout the area of analysis, but they are not evenly distributed (Fig. 98). On the coast, they are numerous at sites in the districts Nordfjord, Sogn, Nordhordland and Midthordland, and occur less frequently in Sunnmøre and Sunnhordland. The lack of points at Kleiva, Botnaneset VIII and Neset II in Sunnfjord may indicate that slate industry was not practised in this district, but considering that slate flakes were found at these sites (see Chapter 10), and the fact that that Vikja I has numerous points, this absence should probably be seen as coincidental rather than reflecting a particular trend in Sunnfjord.

Slate points/blanks are found at all of the mountain sites except at 1020 Bjornesfjoren. The sites Styggvasshelleren and Blånut IV are particularly well supplied.

To the north of the area of analysis, slate points from all periods are well represented (e.g. Bjørn 1921a; 1921b, Hinsch 1948, Narmo 1993, Ramstad 1999). They also occur as stray finds as well as at excavated sites in Rogaland, Agder and lowland eastern Norway (Nummedal & Bjørn 1930, Hagen 1946, Ingstad 1970, Skjølsvold 1977, Lindblom 1980, Bang-Andersen 1981, Østmo 1985, Boaz 1997). The large majority of the points found to the south and east of the area of analysis are, however, either identified at sites dated to the middle Neolithic, or they are points with hanging shoulders. This may indicate that in these regions (inland eastern Norway excluded), slate industry was not introduced before the middle Neolithic.

Retouched flakes and blades

For the purpose of this analysis, flakes and blades with retouch are presented together. They are found at all of the excavated sites. There is a relative dominance of retouched artefacts at sites in Hordaland to the south of Masnesberget, while they are clearly less frequent from Neset II and northwards. Retouched flakes and blades are fairly frequent at the mountain sites (Fig. 99).

Small round smooth stones

These stones have been collected on beaches or in rivers, and have probably not been modified further. They are assumed to have been used as toys or as amulets (Mortensen 1991). Small round stones have probably been collected somewhat unsystematically by the excavators. Quantitative changes or absence of these items should therefore be treated with more caution than other categories. Their wide distribution (Fig. 100), however, indicates that these stones were an element in the material culture throughout the coastal area of analysis.

Grindstone slabs

Grindstone slabs and fragments of slabs are found at almost all of the coastal sites. This artefact type is best represented at sites in Sunnfjord and Nordfjord, while there are fewer in Sunnmøre to the north of Holvikshaugen and in Hordaland to the south of Masnesberget. Two sites, Masnesberget and Sokkamyro probably have too many grindstone slabs because of problems of representativity at these sites (see Chapter 6). At the mountain sites, this artefact type is present only at 1020 Bjornesfjorden (Fig. 101).

TRB pottery

Sherds of fine-tempered pottery with thin walls, several of which are decorated with impressions of real cords, have been identified at the sites Kotedalen phase 13 and 526 Nordmannslågen (Fig. 102). Asle Bruen Olsen and Svein Indrelid, who have published these sites, both have suggested that these sherds strongly resemble early Neolithic TRB "C" and "B" groups respectively (A. B. Olsen 1992, Indrelid 1994:254). Professor Einar Østmo from the University of Oslo, who inspected these sherds during my work with the sites, also believes that they resemble early TRB groups. Early Neolithic pottery has also been identified at Ramsvikneset, Nordhordland (Bakka 1993), Sumtangen, Hardangervidda (Bøe 1942), Gjellestadvige, Rogaland (Bang-Andersen 1981),[8] and possibly at R3 Lundevågen, Vest-Agder (Ballin & Jensen 1995), and Rødsmyr and Lundevoll, Østfold (Østmo 1988, but see K. B. Johansen 2000).

Vespestad adzes

Vespestad adzes are only identified at the coastal excavated sites Holvikshaugen (2), Bustadvika (2), Masnesberget (3), Føyno 88 (1), Sokkamyro (9) and Tjernagel 3 (1). In order to have a more representative distribution pattern of this adze-type, I examined all of the adzes in the area of analysis (from excavated sites as well as stray finds). Altogether 251 Vespestad adzes from Bergen University Museum and Sunnmøre Museum were identified in this area.[9] In addition, Sigmund Alsaker's data from southern Rogaland and northern Rogaland (S. Alsaker 1982) and Asle Bruen Olsen's data from Romsdal and Nordmøre (A. B. Olsen 1981) are presented. S. Alsaker and A. B. Olsen have identified altogether 81 Vespestad adzes from these districts (complete list of the Vespestad adzes are found in appendix 5).

The distribution pattern (Fig. 103) indicates that Vespestad adzes were common in all of the coastal districts from Nordmøre to southern Rogaland, but there are also differences within this region. Considering that the adzes come partly from excavated sites and are partly stray finds delivered by non-professionals, however, a few source-critical comments will appear below.

Stray finds are first of all collected by farmers, but farming does not have the same character and intensity everywhere. In areas that have many cultivated fields, such as Romsdal, Sunnmøre, and Rogaland, the chances of finding stone adzes are greater than in areas where there are only pastures. The coastal districts Sunnfjord and outer Sogn are dominated by pastures and this probably explain the relative lack of Vespestad adzes from these districts. Cultivation is, on the other hand, common in the fjord districts inner Sogn, inner Nordfjord, Voss and Hardanger. One would consequently expect that if adzes had been frequently deposited in these areas, they would have been collected by the farmers. As adzes have not been acquired from farmers in these districts, one may therefore conclude that fewer adzes have been deposited there than in the coastal districts.

The frequency of excavations in the different districts also has to be taken into account. Outer Nordfjord is dominated by pastures, but in this district, there have been a large number of excavations during later years, and these have produced quite a few Vespestad adzes. The coast of Hordaland is perhaps the area most intensively investigated by archaeologists in western Norway. This is probably the main reason for the large amount of adzes from Nordhordland, Midthordland, and Sunnhordland.

If these two factors are considered, one may conclude that, within the region of western Norway, Vespestad adzes have been quite evenly distributed along the coast. In the fjords further east, however, adzes have been less common.

The middle Norwegian collection of stone adzes at the Museum of Natural History and Archaeology in Trondheim has only been surveyed for the purpose of distinguishing Vespestad adzes in Romsdal and Nordmøre (A. B. Olsen 1981). Data from the counties further north have not yet been subjected to such an analysis. However, according to Olsen, who has inspected the entire Stone Adze collection from Middle Norway, saw very few adzes of this kind in the three relevant counties Sør-Trøndelag, Nord-Trøndelag and Nordland (personal communication 2003). This may indicate that there is a marked northern fall-off in the use of Vespestad adzes approximately at the boundary between Nordmøre and Sør-Trøndelag.

8 For Gjellestadvige, Bang-Andersen has not himself characterized the ceramic sherds as TRB pottery, but this has been done by A. B. Olsen in a review of the site (A. B. Olsen 1992:144) where he has argued that most of the site material is early Neolithic. Olsen's position on this matter is supported by later authors (Åstveit 1999, Amundsen 2000).

9 The Vespestad adzes which are deposited at Sunnmøre Museum have been identified by Asle Bruen Olsen (1981). They have not been inspected by me.

When the data from western and eastern Norway is compared, the contrasts are also very marked. In 1906, A. W. Brøgger distinguished 24 adzes of "western Norwegian forms" in eastern Norway, which was interpreted as indicating a low degree of communication between the different parts of the country (Brøgger 1906:89). More recent surveys have not changed this impression. Berit Gjerland's survey of the adze-collections from the districts Ringerike, Drammen and Larvik in eastern Norway did not produce any Vespestad adzes (Gjerland 1984:134). A survey by Ballin and Jensen (1995) did not result in Vespestad adzes either. In a recent survey of the adze/axe collection at Museum of Cultural History in Oslo, Håkon Glørstad and Espen Uleberg distinguished 27 Vespestad adzes. These were later inspected for the present work. On the basis of the typological criteria used in this thesis, I could only characterise 4 Vespestad adzes from these 27 specimens.[10] The remaining 23 were either Vestland adzes or atypical adzes. It is therefore likely that Glørstad and Uleberg have been too generous and inclusive. On the other hand, the generosity may be seen as reassuring, because they have included all of the possible types, not only the secure adzes. The four remaining adzes distinguished by myself may thus be very close to the actual number of Vespestad adzes deposited in the collection at the Museum of Cultural History in Oslo. If this is correct, Brøgger's statement that the Vespestad adze was a western Norwegian type is still relevant. In fact, the result indicates a more marked difference between the early Neolithic eastern and western Norwegian assemblages than observed by Brøgger in 1906.

Undetermined stone adzes and stone adze flakes

A number of stone adzes of undetermined types and stone adze flakes have been identified at the sites. The first category (Fig. 104) shows a fairly even distribution in the (coastal) area of analysis. Ground stone adze flakes (Fig. 105) are also found at almost all of the sites, but they are quite numerous in Nordfjord and Nordhordland compared to the other districts.

Thin-butted axes

No thin-butted axes have been identified at the coastal excavated sites in the area of analysis. To the best of my knowledge, such axes or flakes from such axes have not been found at any site dated to the early Neolithic in western Norway. At the mountain sites, however, ground flakes from flint axes are found at Blånut IV, 526 Nordmannslågen, 1026 Bjornesfjorden and Vivik. Two flakes at Blånut IV and Vivik could positively be identified as fragments of thin-butted types (Fig. 106).

Previously, several scholars have surveyed the thin-butted axes in western Norway (Brøgger 1906, Hinsch 1955). The last work was performed by Evy Berg, who concluded, in line with her predecessors, that these axes are infrequent in this region (Berg 1986). A problem is that she only considered the flint implements, and disregarded the stone axes. In order to have a more complete database of the thin-butted axes in the area of analysis, I therefore surveyed the collection at Bergen Museum once again together with Professor Einar Østmo from University of Oslo. Our survey resulted in altogether three thin-butted flint axes and 12 thin-butted stone[11] axes (Fig. 107, listed in appendix 6). There is a theoretical possibility that some of the ground flakes of basaltic rocks at the coastal sites (see above) may be fragments of stone axes, but considering that this type of axes has not been found at any habitation site, this option is hardly probable.

Although the number of axes is larger than Hinsch's and Berg's results, there is still a major difference between our result and the large number of axes found in eastern Norway, particularly when compared to the area around the Oslo fjord (Fig. 108) (Hinsch 1955, Østmo 1988, Mikkelsen 1989) but also to the interior valleys (Indrelid 1994) and Rogaland (Berg 1986). In Trøndelag and Nordmøre/Romsdal, on the other hand, the pattern appears to be similar to that found in western Norway (Ramstad 1999, Østmo 2000).

On the basis of the above, it is apparent that, compared to eastern Norway and Rogaland, Hinsch and Berg were generally correct: thin-butted axes are fairly sparse in coastal western Norway. Important to note here is that the mountain sites in this analysis are situated within the main distribution area of the axes. Another important point is that while thin-butted axes found in eastern Norway are made of both flint and stone (equal shares) (Hinsch 1955:52), the western Norwegian implements are predominantly made of stone.

Slate knives

The distribution of slate knives has not been investigated for the present work. A survey of the entire museum collections of slate knives in Norway south of the Arctic Circle has been performed by Hans Christian Søborg (1986; 1988). The knives were mainly stray finds, but some were also from excavated sites. In his distribution study (Fig. 109), Søborg found that slate knives were only sporadically found south of Sunnmøre (Stad), while they were more frequent in Sunnmøre, Romsdal, and Nordmøre. They were particularly numerous on the coast of Sør-Trøndelag and northwards. Stray finds and excavation reports received by Bergen University Museum after Søborg completed his work have not significantly altered his distribution patterns.

10 Ingrid Fuglestvedt has published a possible Vespestad adze from Svevollen, Hedmark (Fuglestvedt 1995:108). The edge is, however, completely crushed. A secure typological determination cannot be established.

11 The raw material of all of these axes can be classified as "undetermined basaltic rock". None of them resemble the diabase or greenstone in the reference collection.

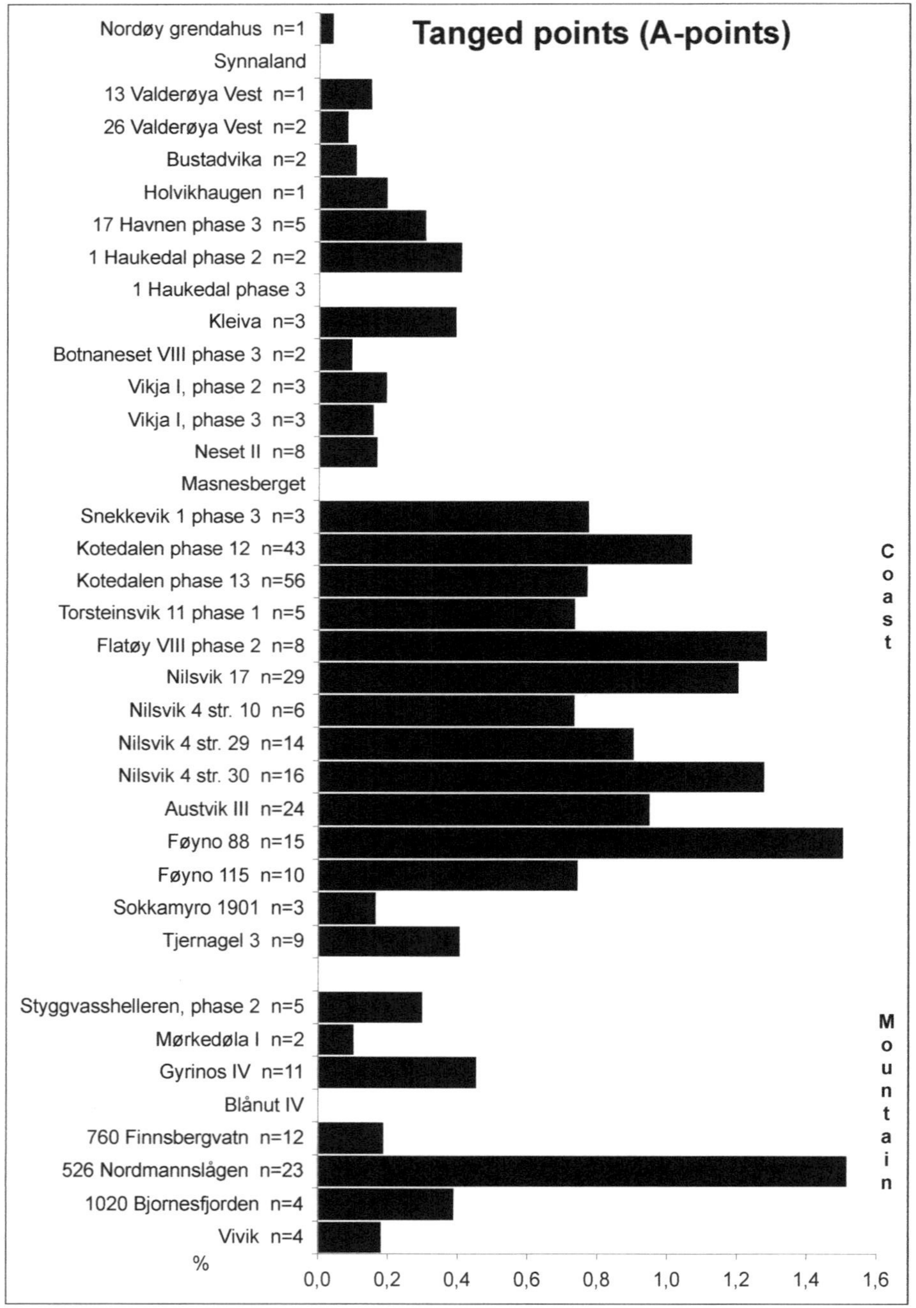

Figure 94. *Percentages of tanged points (A-points) at excavated sites.*

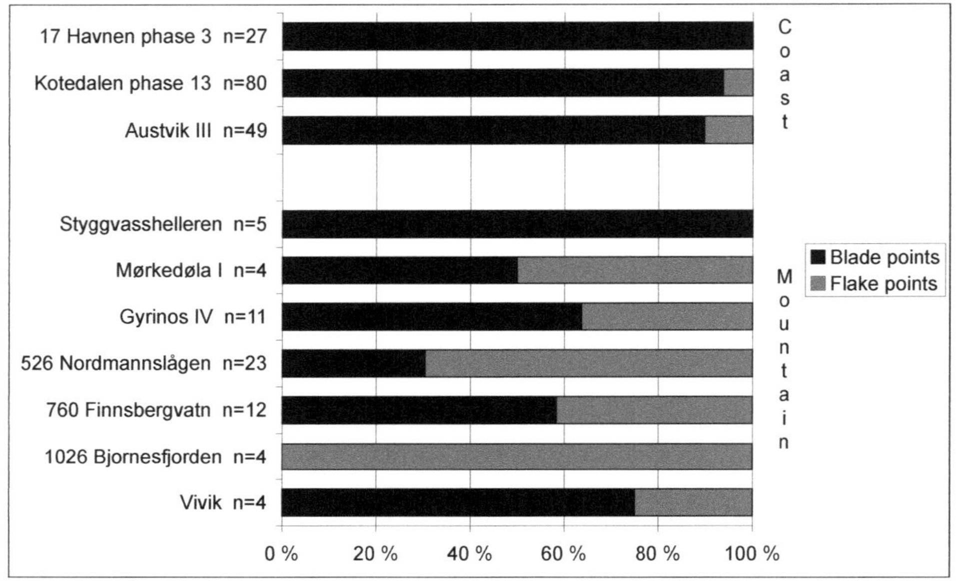

Figure 95. *Percentages of blade points and flake points at selected excavated coastal and mountain sites.*

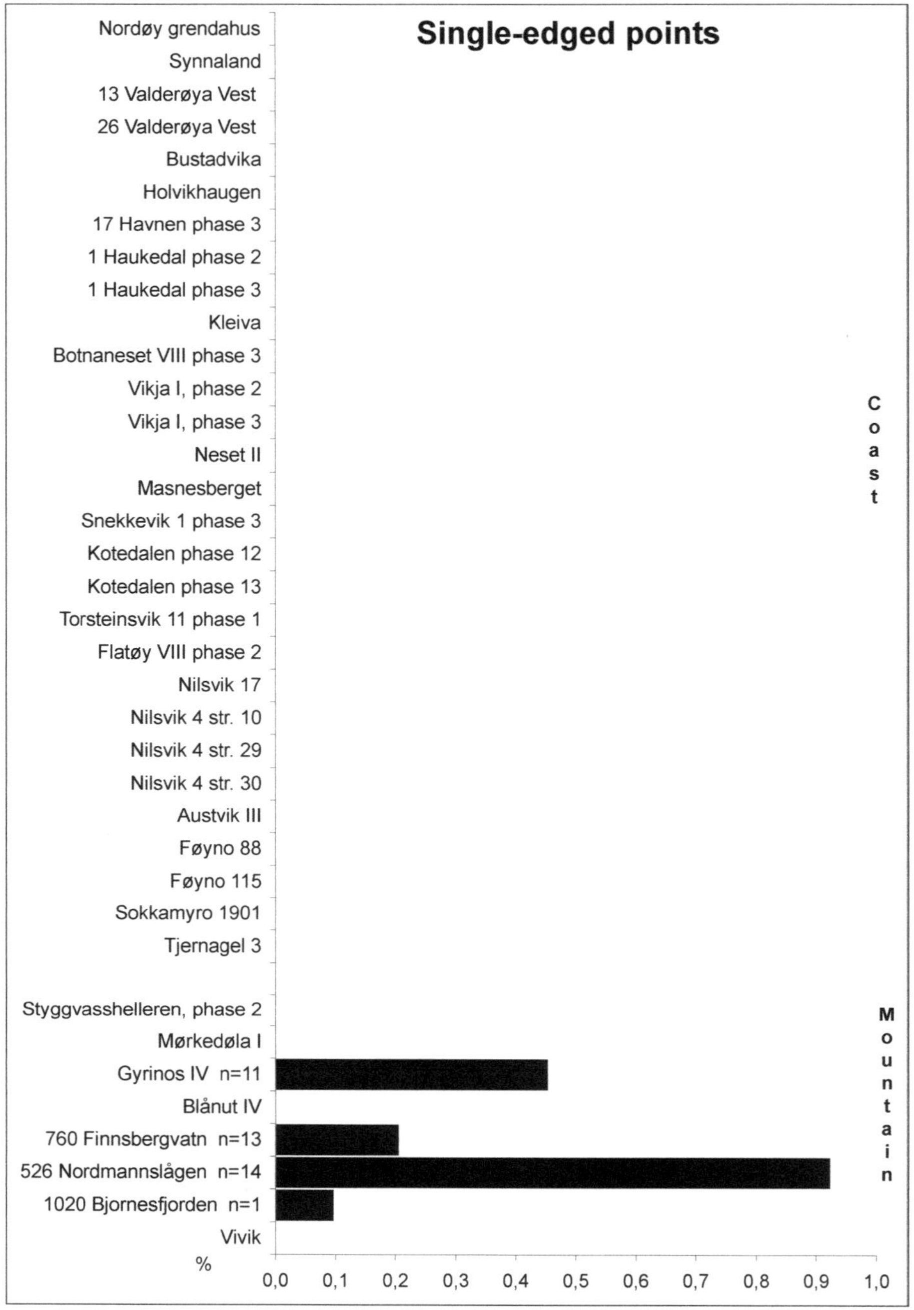

Figure 96. *Percentages of single-edged points at excavated sites.*

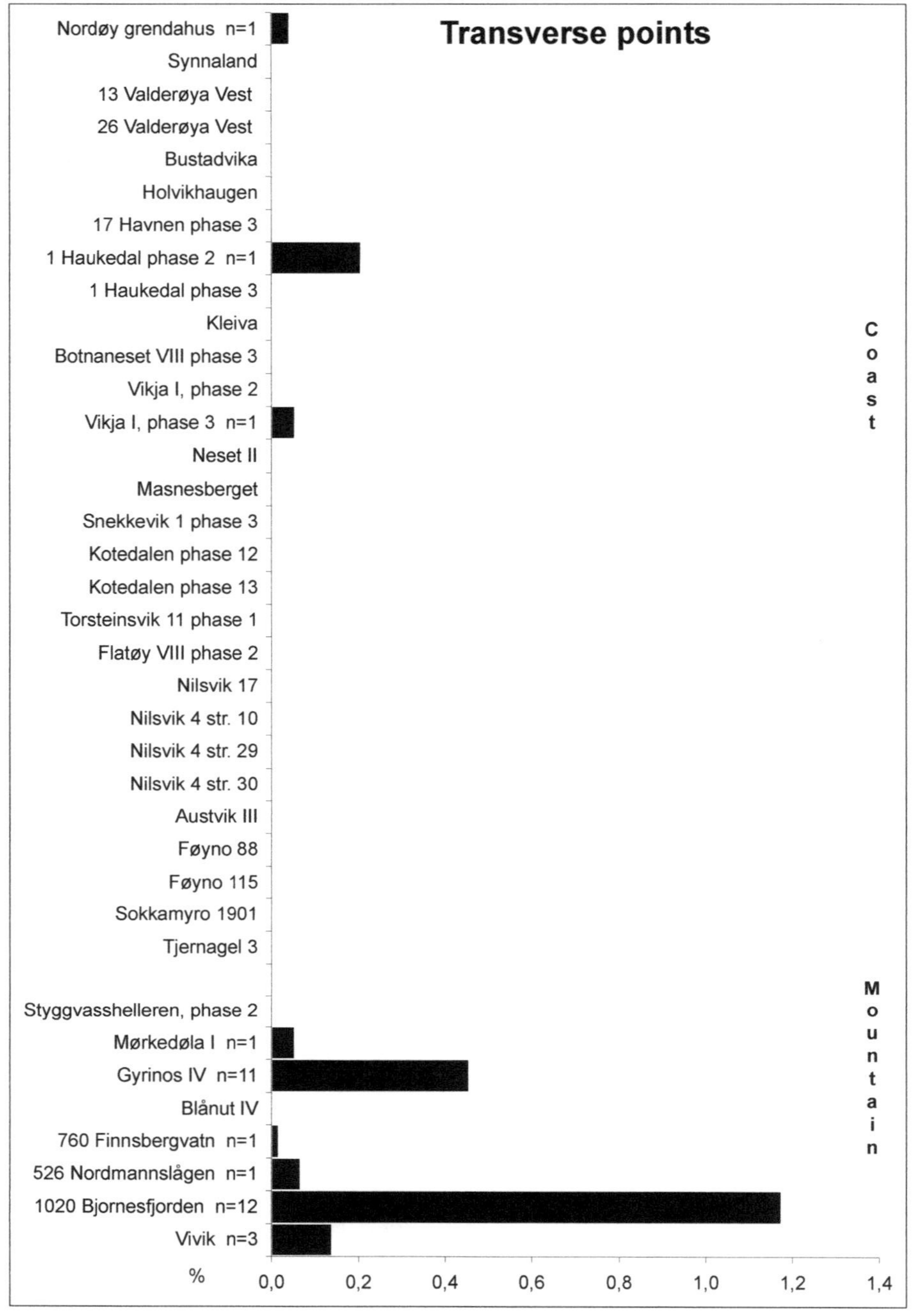

Figure 97. *Percentages of transverse points at excavated sites.*

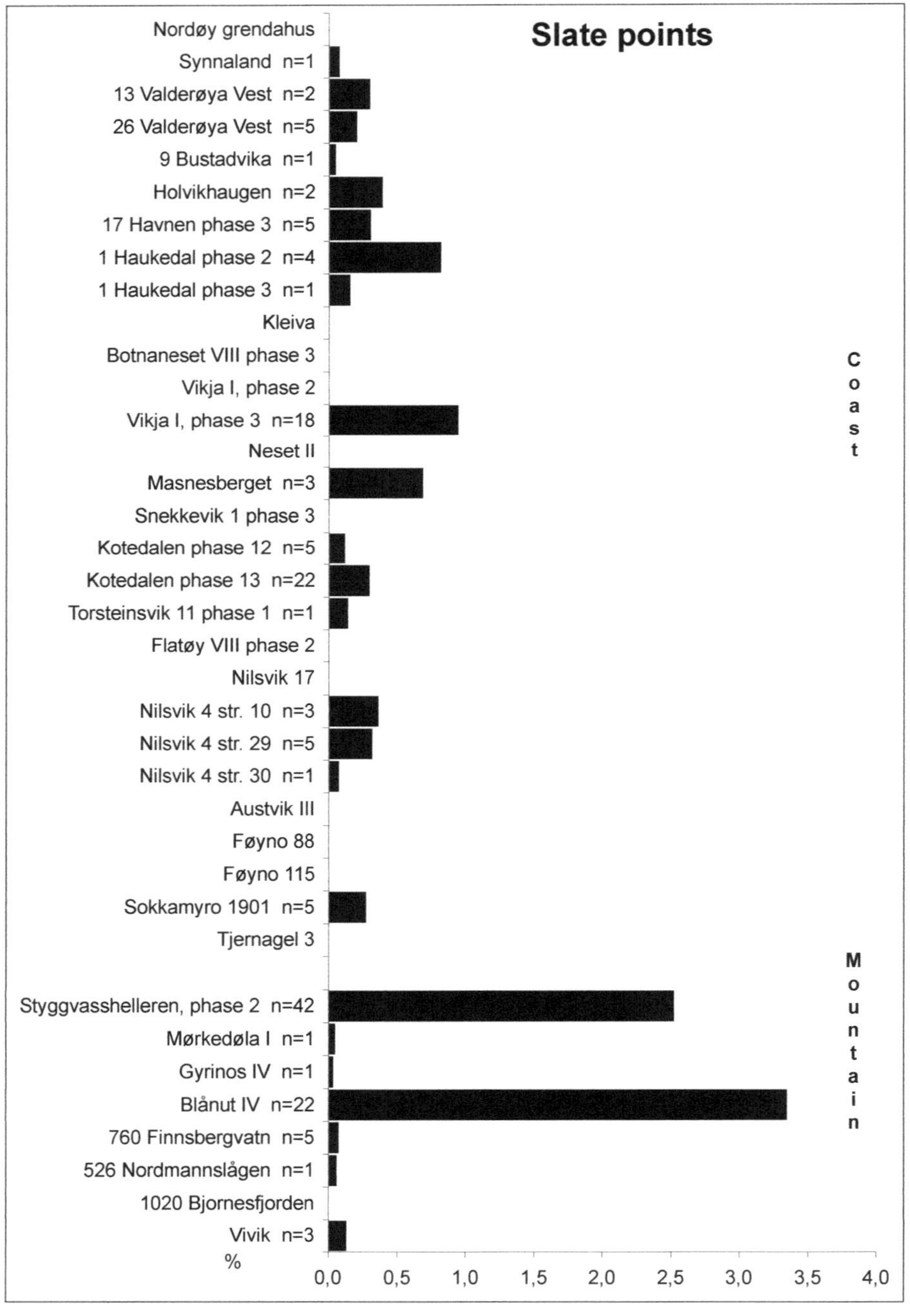

Figure 98. *Percentages of slate points at excavated sites.*

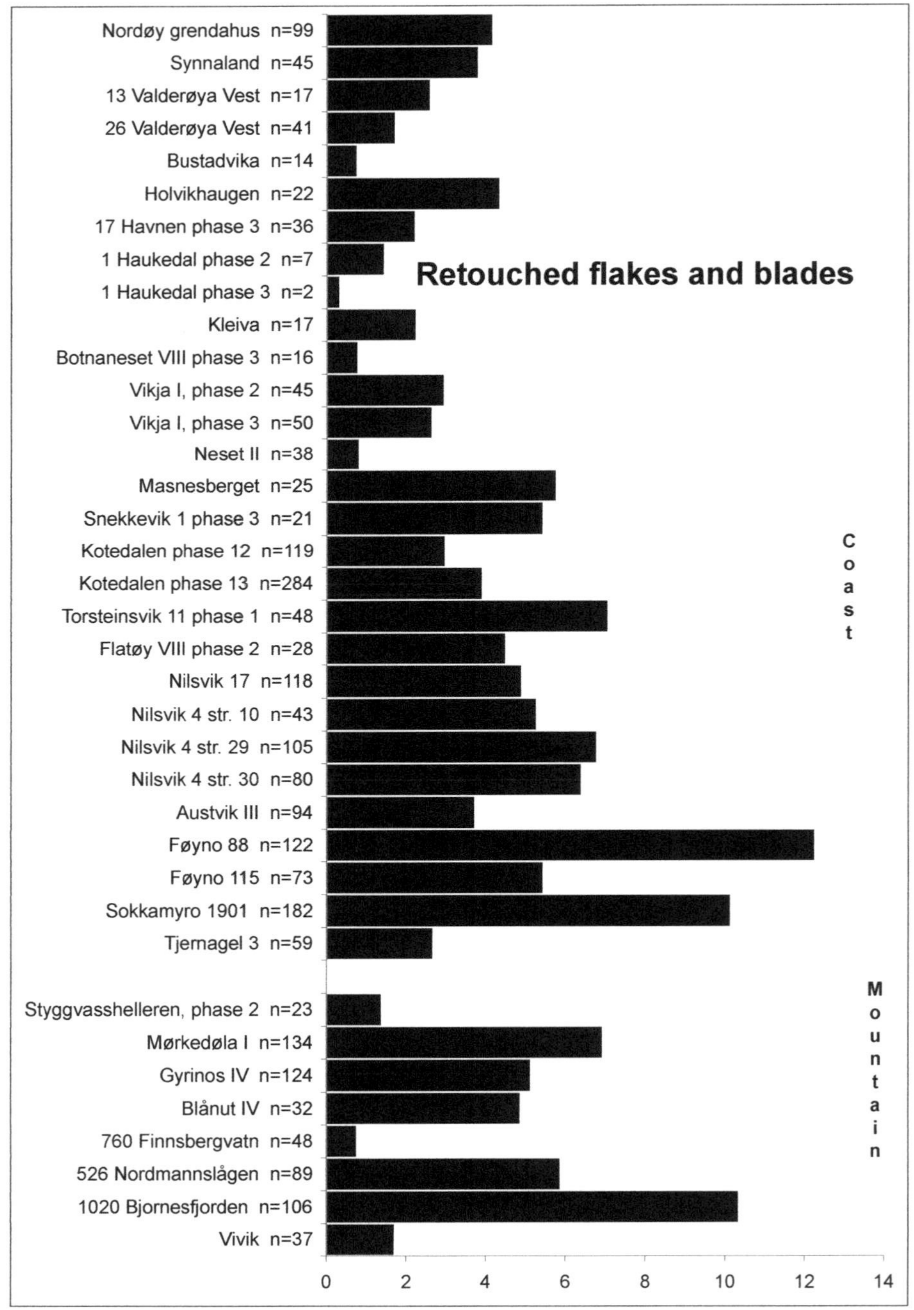

Figure 99. *Percentages of retouched flakes and blades at excavated sites.*

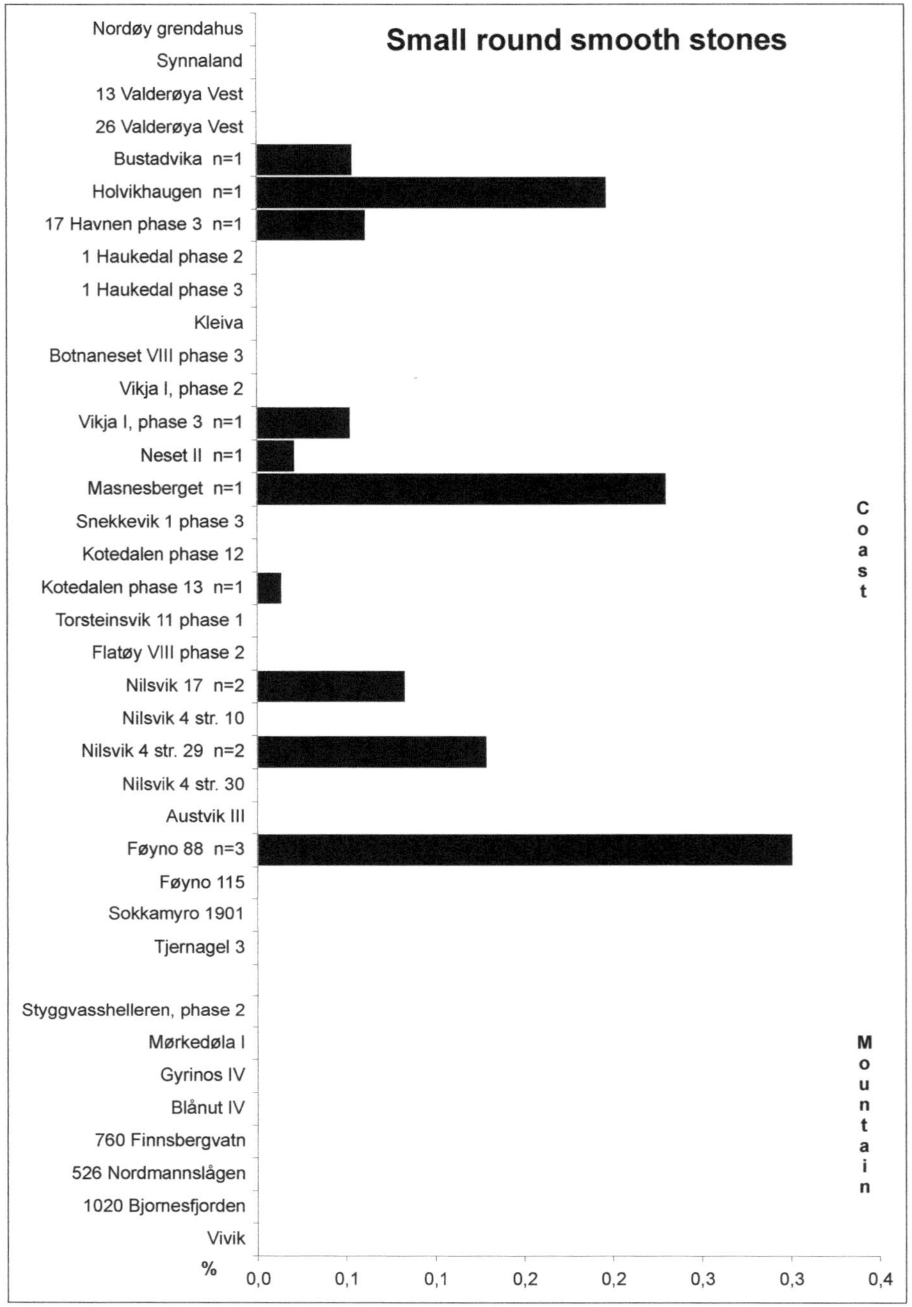

Figure 100. Percentages of small round smooth stones at excavated sites.

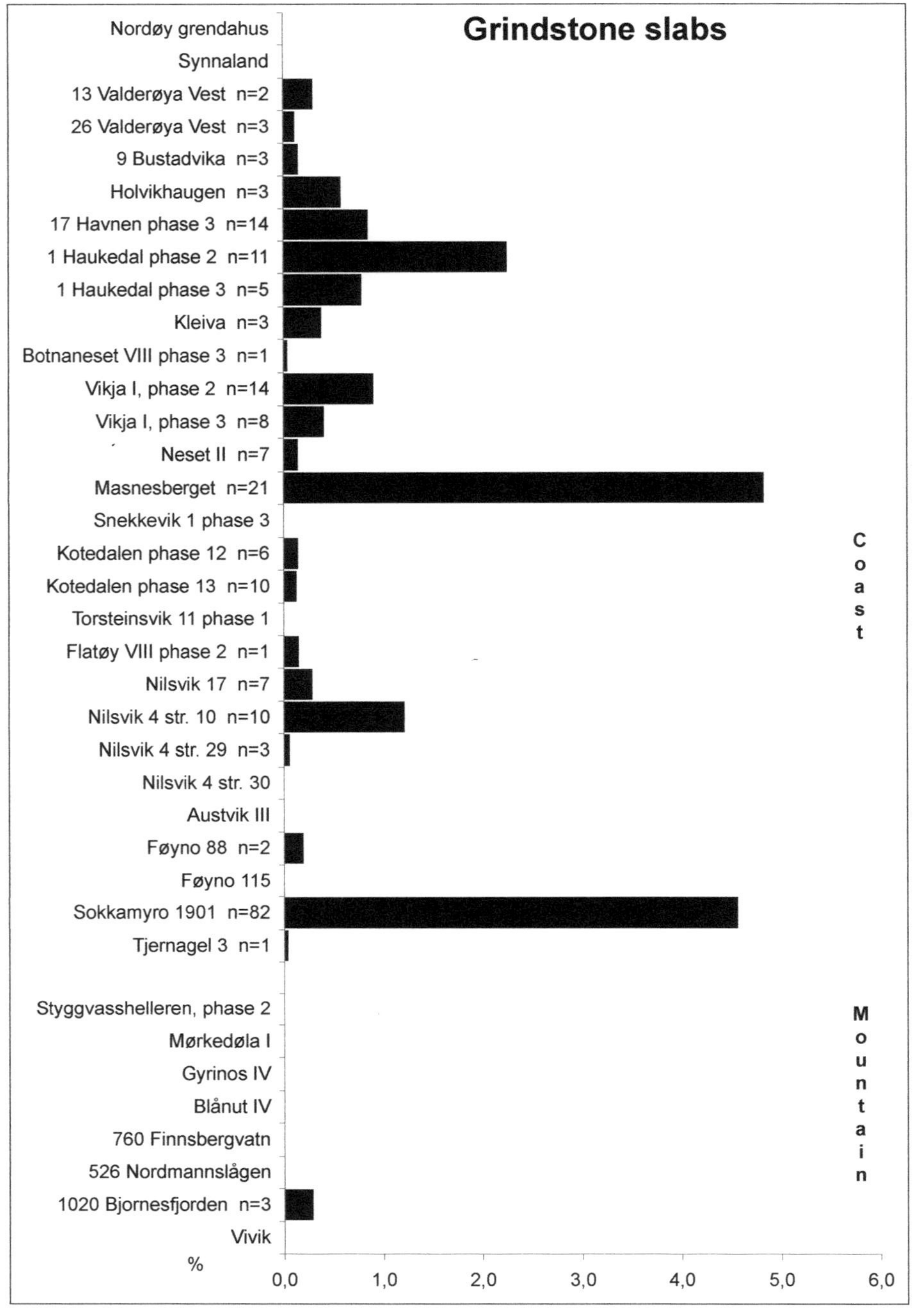

Figure 101. Percentages of grindstone slabs at excavated sites.

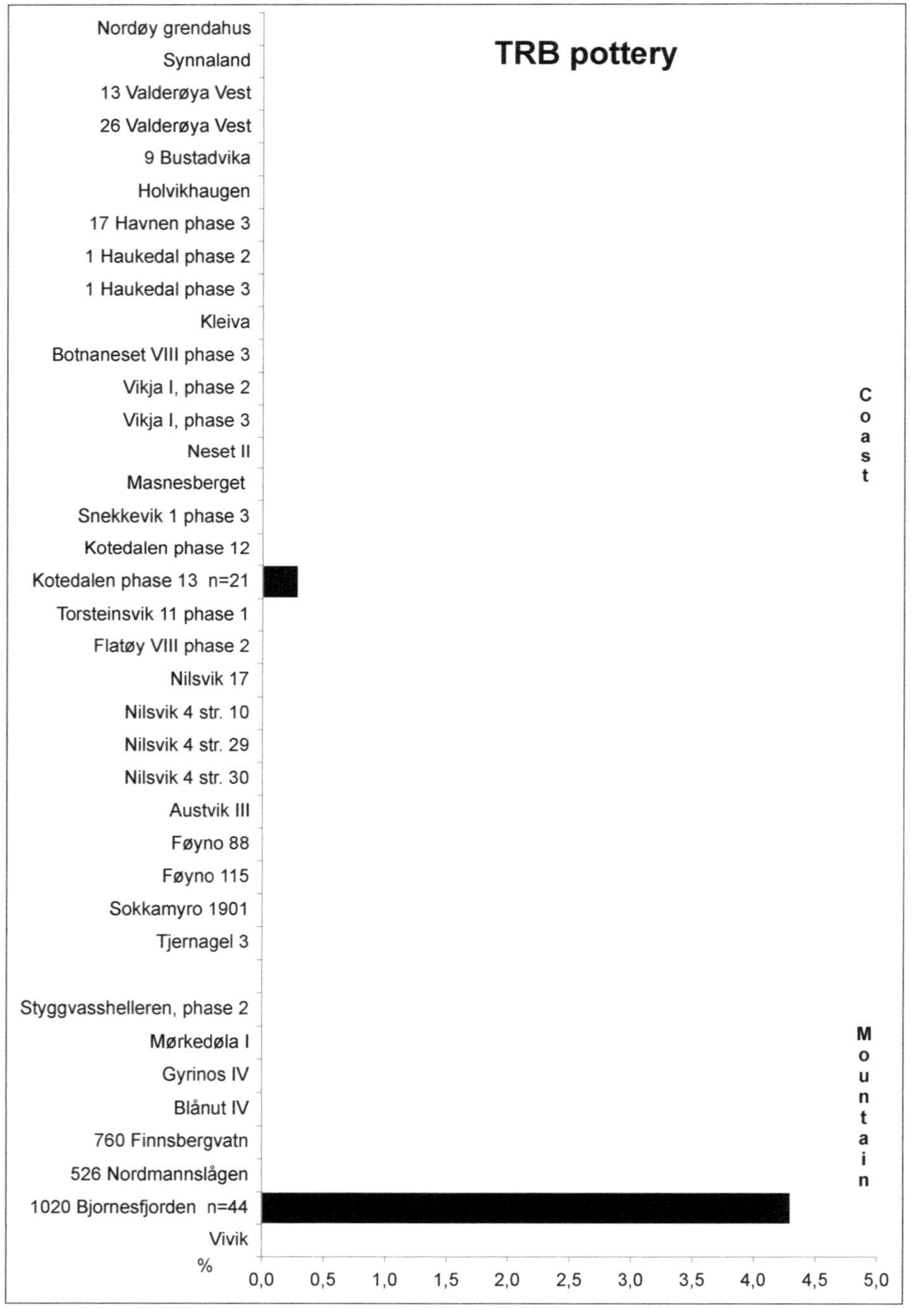

Figure 102. *Percentages of TRB-pottery sherds at excavated sites.*

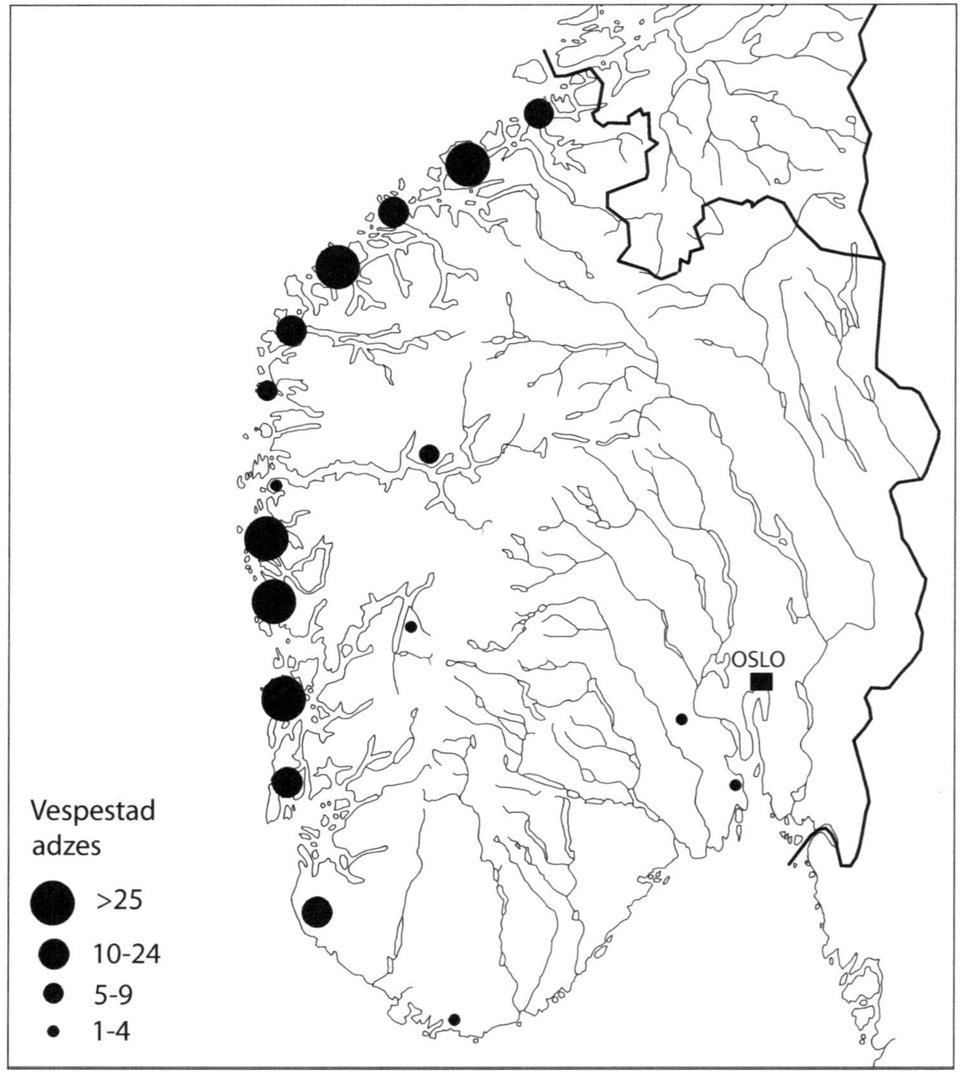

Figure 103. *Distribution of vespestad adzes in southern Norway to the south of Trøndelag. Data from Nordmøre and Romsdal are extracted from Olsen (1981). Data from northern and southern Rogaland are extracted from Alsaker (1982). Data the remaining districts has been analysed for the purpose of this thesis.*

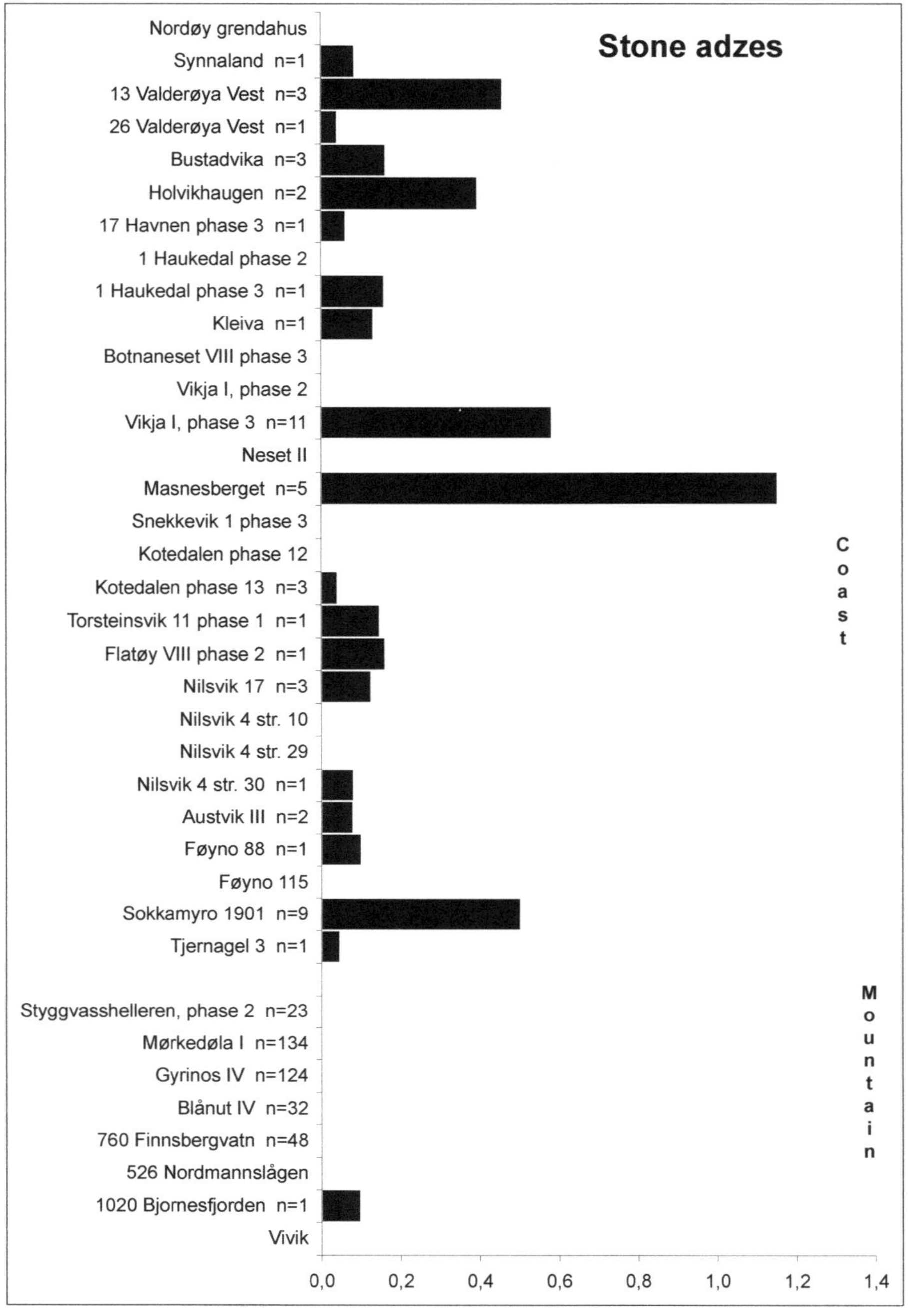

Figure 104. *Percentages of stone adzes at excavated sites.*

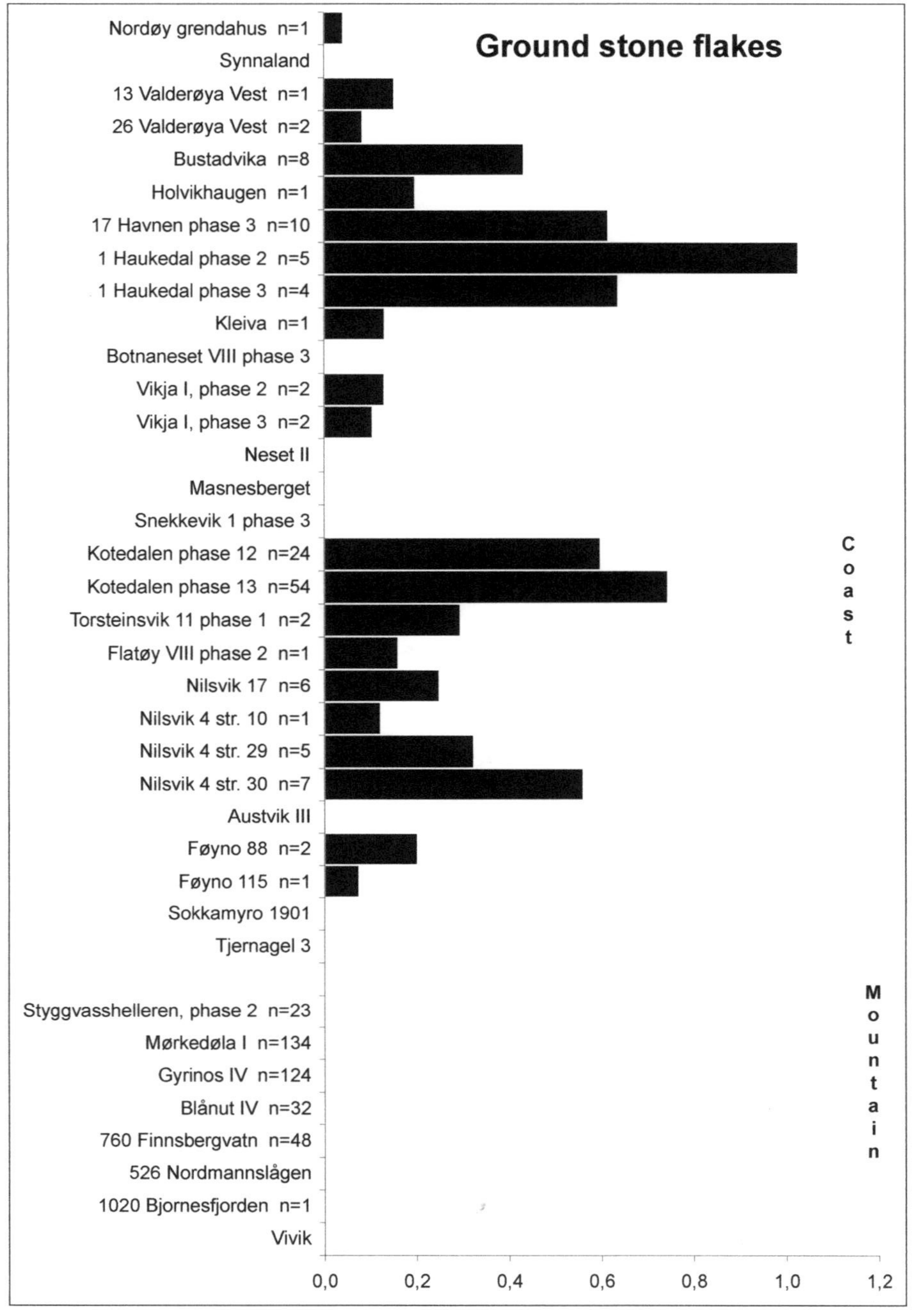

Figure 105. *Percentages of ground stone flakes at excavated sites.*

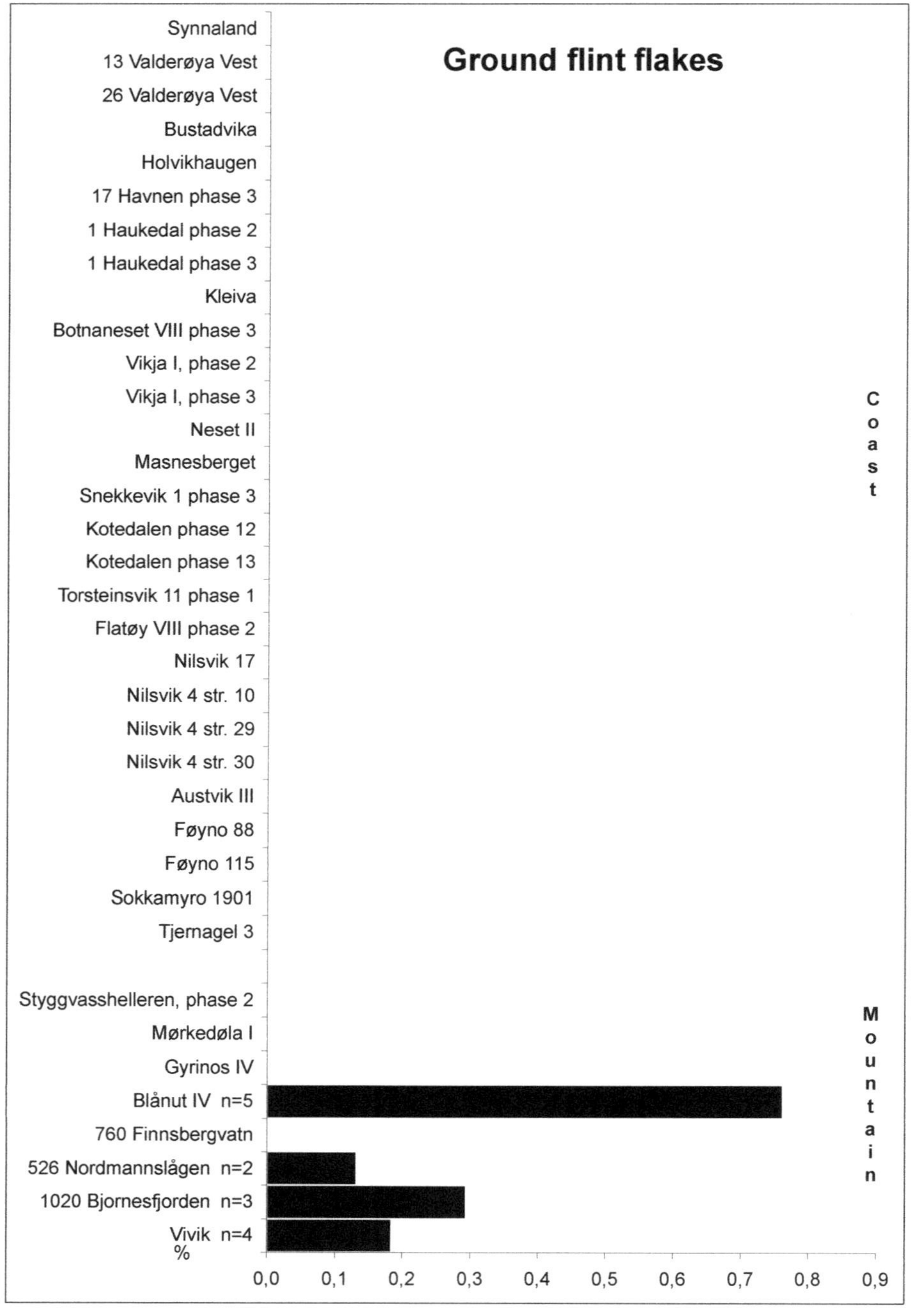

Figure 106. Percentages of ground flint flakes at excavated sites.

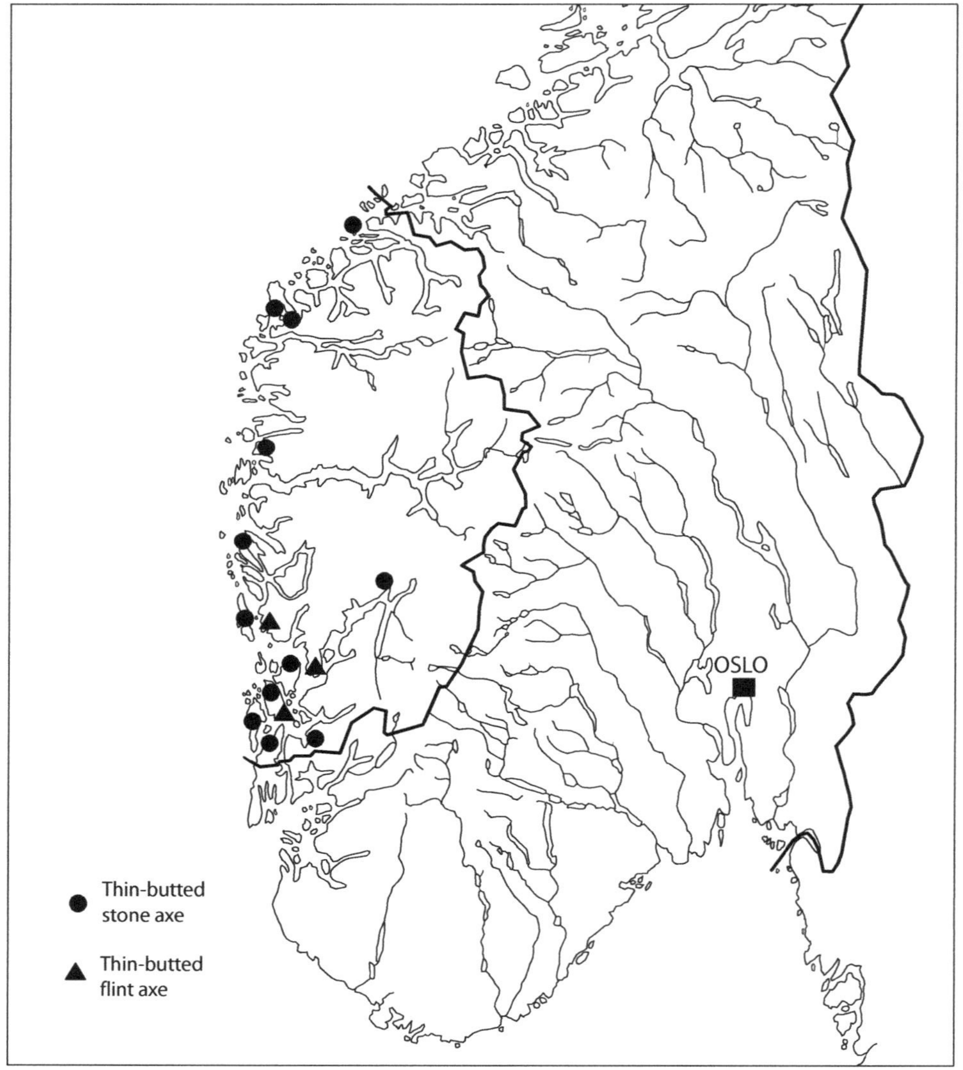

Figure 107. Distribution of thin-butted flint axes and thin-butted stone adzes in the districts northern and southern Sunnmøre and the counties Sogn og Fjordane and Hordaland.

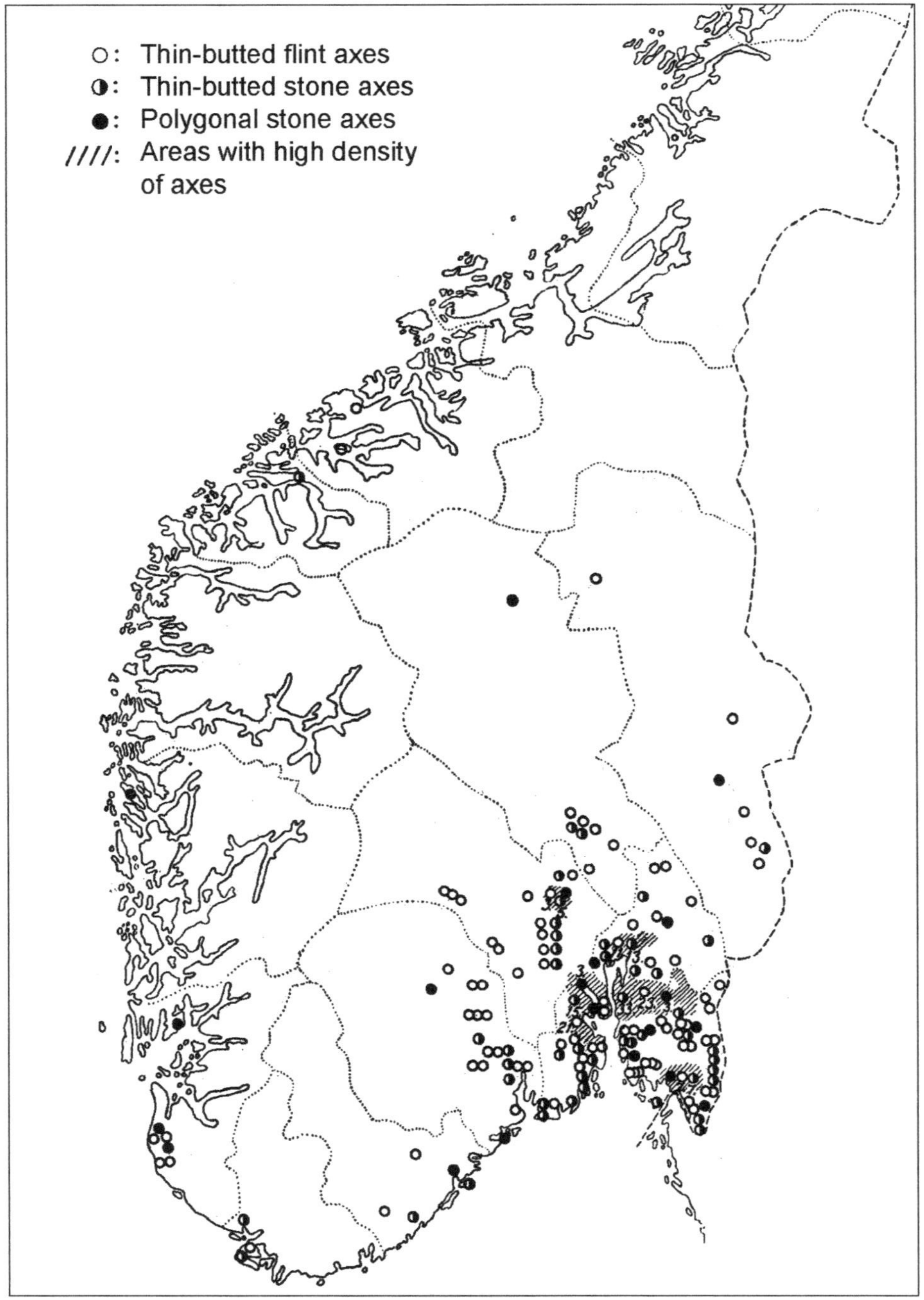

Figure 108. Distribution of thin-butted axes and polygonal axes in southern Norway. Modified from Hinsch (1955:52).

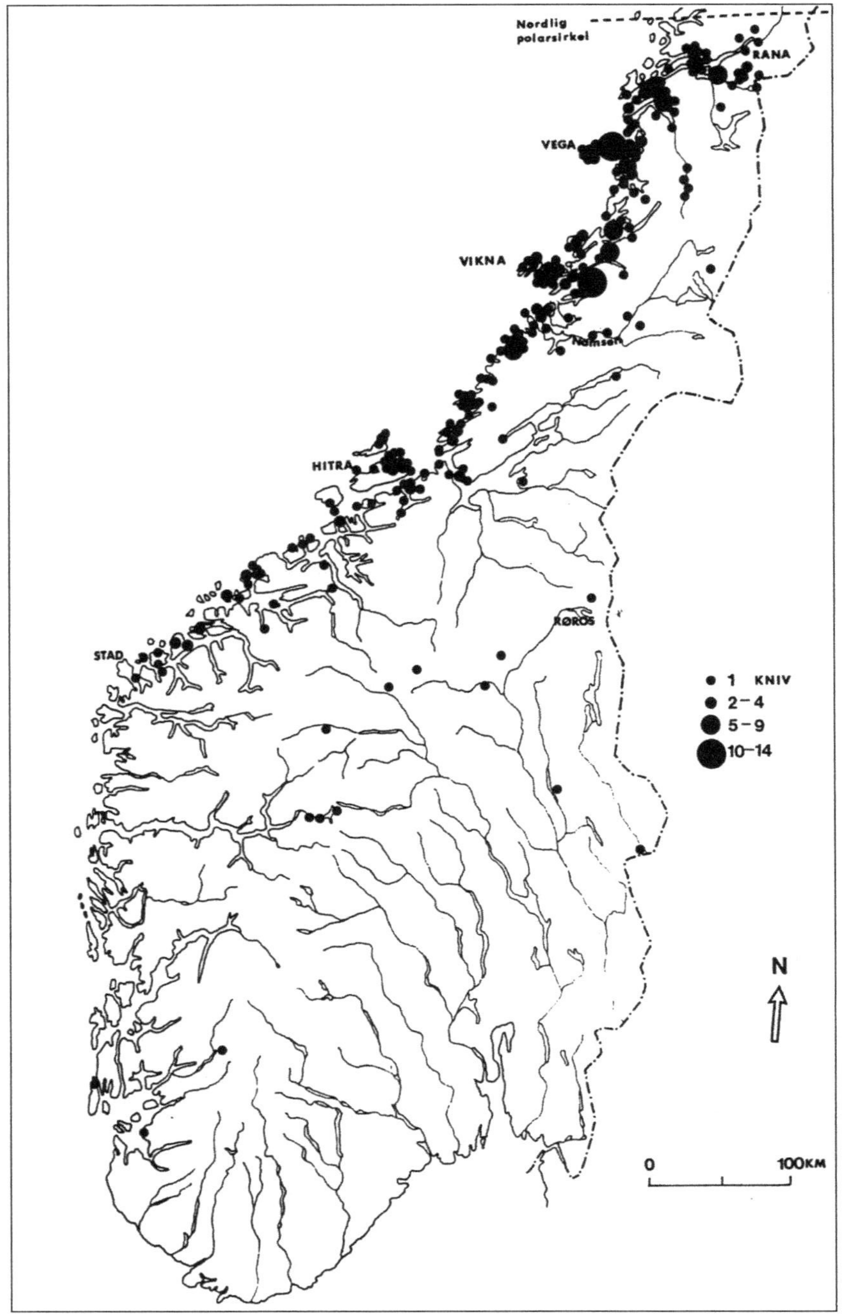

Figure 109. Distribution of slate knives in Norway to the south of the Polar Circle. Reproduced from Søborg (1988:26).

Techniques

Three main lithic techniques are distinguished in the region: cylindrical core technique, bipolar technique, and direct percussion/anvil percussion. All three techniques appear throughout the region, but there are marked differences in percentages between the districts. Below, these differences will be expressed in percentages of the total amount of artecats a the sites.

Cylindrical core technique

The frequency of cylindrical core technique is difficult to measure accurately, because the cores are often reduced further at the sites. Thus, a comparison of the number of cores of this type may not be very indicative of the relative importance of this technique. However, the technique may also be measured by the number of small blades at the sites, because this type of blades was mainly produced using cylindrical core technique. A comparison of the percentages of small blades and A-points in addition to the percentages of cylindrical cores at the sites should, therefore, be indicative of the frequency of this technique.

Cylindrical cores are particularly frequent at sites in Nordhordland, Midthordland, and Sunnhordland. In Sunnfjord and Nordfjord they are less common, they are seldom found in Sunnmøre and they are rare at mountain sites (Fig. 110). The percentages of blades and blade points largely confirm this pattern: at most of the sites in Hordaland, these products are relatively frequent, and constitute about 5%. There are fewer small blades at the sites in Sunnfjord and Nordfjord to the north of Masnesberget. They are infrequent in northern Sunnmøre (about 1-2% of the total amount of artefacts). At the mountain sites, small blades are generally infrequent (Fig. 111). On the coast, A-points (blade points) should be seen as a product of cylindrical core technique. It appears that the relative distribution of these points is almost identical to that of cylindrical cores and small blades (Fig. 94). The relatively high percentage of A-points at the mountain sites diverges from this pattern, but may be explained by the fact that a large portion were made from flakes, not blades (see above).

From the above, it is apparent that the cylindrical core technique dominates at the sites in Hordaland between Masnesberget and Tjernagel 3. However, the sites in Midthordland, Nilsvik 4 and 17, are not entirely in accordance with this. One likely explanation for the quite low percentages of small blades and cylindrical cores (not A-points) at these sites may be that they are dated to the transition to the middle Neolithic. According to Nærøy, the middle Neolithic is a period of decline for the cylindrical core technique in Hordaland (Nærøy 1993a:89).

Early Neolithic cylindrical core technique has been identified at a number of sites outside the area of analysis, first of all in Rogaland and Vest-Agder (Nygaard 1974, Skjølsvold 1977, Skjølsvold 1980a, Bang-Andersen 1981, Ballin & Jensen 1995). Further east, cylindrical cores are rare at sites from this period, but they are numerous at middle Neolithic sites in eastern Norway (e.g. Nummedal & Bjørn 1930, Hinsch 1955, Ingstad 1965, Østmo 1988, Mikkelsen 1989).

Bipolar technique

Bipolar technique is only identified by bipolar cores (Fig. 112). Cores of this type dominate in northern and southern Sunnmøre, where they constitute 12% or more of the total amount of artefacts. To the south of Sunnmøre, however, there is a marked change in core techniques. At sites from 17 Havnen in Nordfjord and southwards along the coast, bipolar cores generally represent less than 2% of the total. Bipolar cores are generally few or absent at the mountain sites, except for Styggvasshelleren, which has a large quantity of cores of this type.

The distribution of bipolar cores to the south and to the east of the area of analysis is difficult to evaluate, because they have been subject to systematic classification at the sites only recently. They are, for example, almost non-existent at sites excavated in Rogaland during the 1970's (e.g. Skjølsvold 1977:59, Bang-Andersen 1981:36), while quite a few are classified at Neolithic sites excavated twenty years later (Ballin & Jensen 1995:202). Recent excavations at Svinesund in Østfold, eastern Norway may, however, indicate that bipolar cores are rare at early Neolithic sites in this area (Tørhaug 2002:81).

Freehand percussion/anvil percussion

The frequency of the flake techniques freehand percussion and anvil percussion is difficult to measure, but a significant amount of the cores under the categories "platform cores" and "other cores" may be a result of these techniques (Figs. 113 and 114).

"Other cores" are found at almost all of the excavated sites, but there are marked differences between the districts. In Sunnmøre, there are generally few of this category, while they are more frequent at sites in Nordfjord and Sogn, between Holvikshaugen and Masnesberget. Few have been found on the coast south of Masnesberget. There are several "other cores" at the mountain sites, particularly at Blånut IV and Vivik.

"Platform cores" include four different types: other cores with one platform, other cores with two platforms, one-sided cores with one platform, and one-sided cores with two platforms. These appear in small quantities at sites at Sunnmøre and Nordfjord until 17 Havnen. They are quite numerous in Sunnhordland and Nordhordland between Botnaneset VIII and Kotedalen. Only a few platform cores are found at the coast to the south of Kotedalen. They occur in small quantities at the mountain sites.

Blade technique and flake technique

I Chapter 5, I argued that the cylindrical core technique was a blade technique, identified by cylindrical cores and small blades. Bipolar percussion, anvil percussion/ freehand percussion was mainly argued to be flake techniques, identified by bipolar cores and other cores/ platform cores.

For cylindrical cores and bipolar cores, these assumptions are probably still valid. For platform cores and other cores, however, there are problems. At the coastal sites, particularly in Nordfjord and Sunnfjord between Holvikshaugen and Masnesberget, and at Gyrinos IV and Mørkedøla I in the mountains, a number of small blades are made from the same raw materials as the other cores. The same is true for the fairly large amount of platform cores that are found at the coastal sites between Botnaneset VIII and Kotedalen and the mountain sites Gyrinos IV and Mørkedøla I. Other cores and platform cores are, therefore, most likely the results of both blade techniques and flake techniques. This implies that high percentages of other cores and platform cores cannot be used as indicators of flake technique being particularly frequent at the sites.

As a result of the above, the data may indicate that different combinations of techniques were practised in different areas. (1) Bipolar flake technique was particularly common at sites in Sunnmøre north of 17 Havnen, and the blade technique was hardly practised at these sites. (2) At sites between 17 Havnen and Masnesberget, the blade technique was common, but flake freehand techniques and anvil techniques may also have been practised on the same raw materials. (3) To the south of Masnesberget, blade technique dominated, while other techniques were less common. (4) Blade technique and bipolar flake technique did not occur at most of the mountain sites, which may indicate that freehand techniques and anvil techniques dominated at these sites.

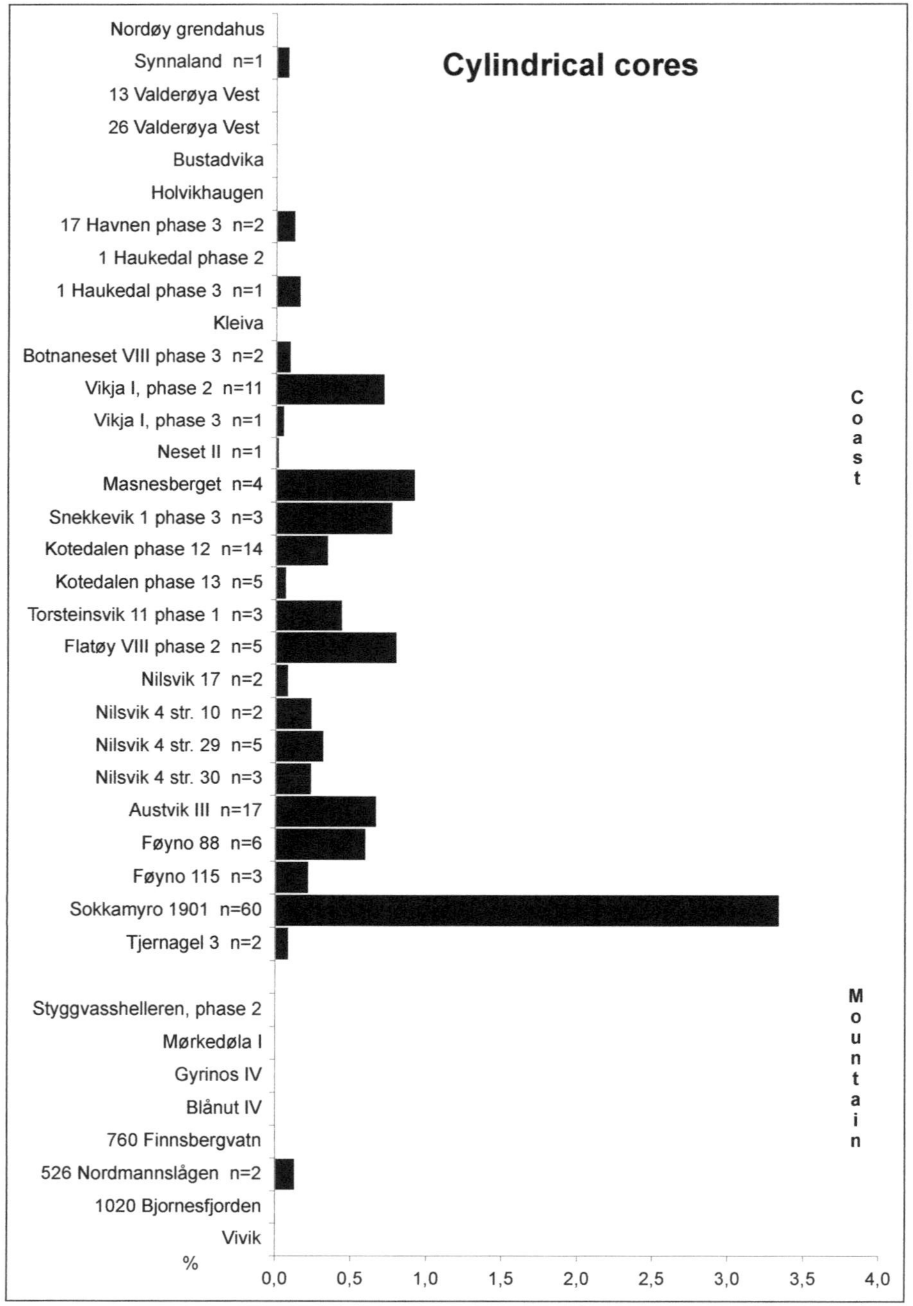

Figure 110. Percentages of cylindrical cores at excavated sites.

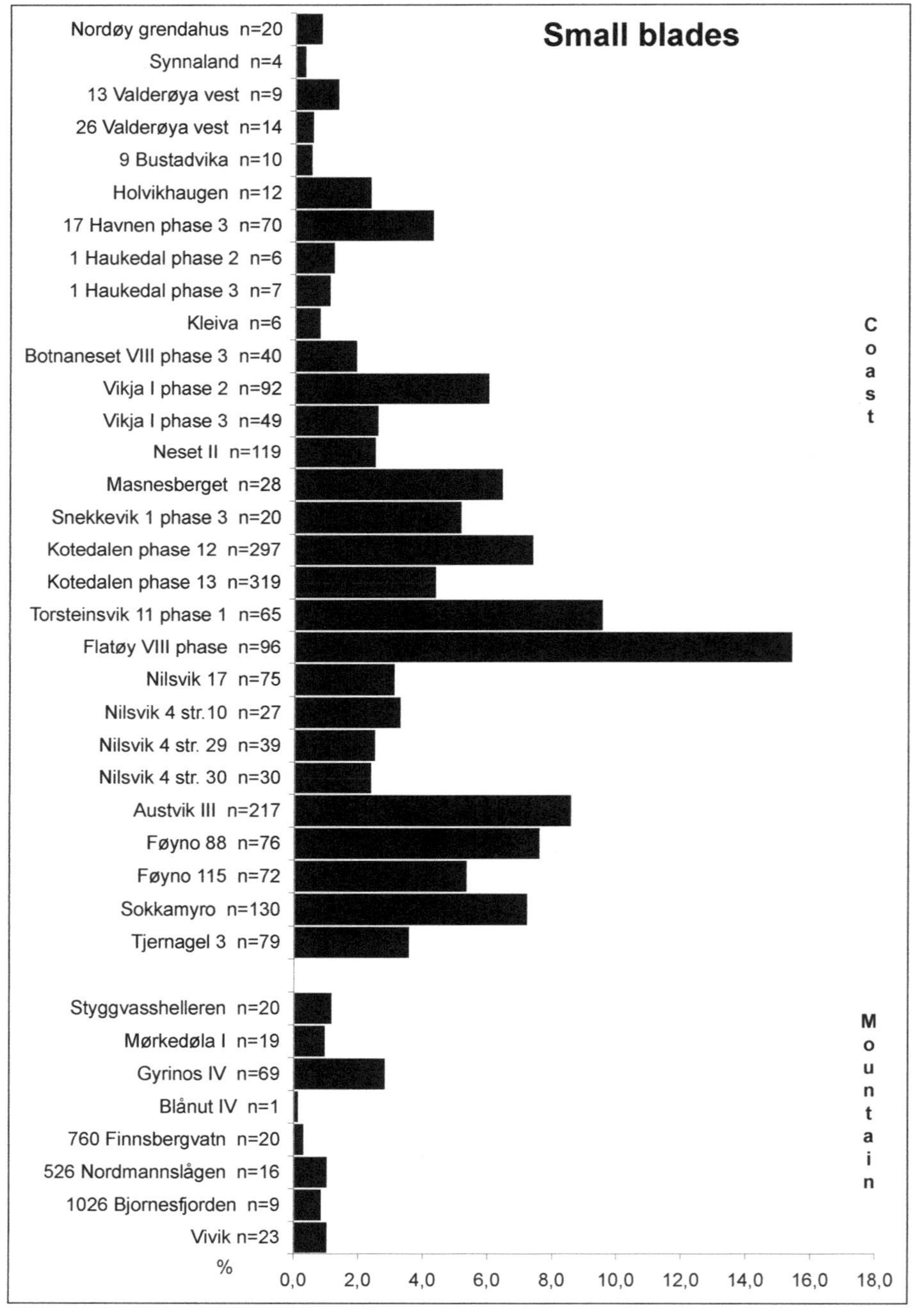

Figure 111. Percentages of small blades at excavated sites.

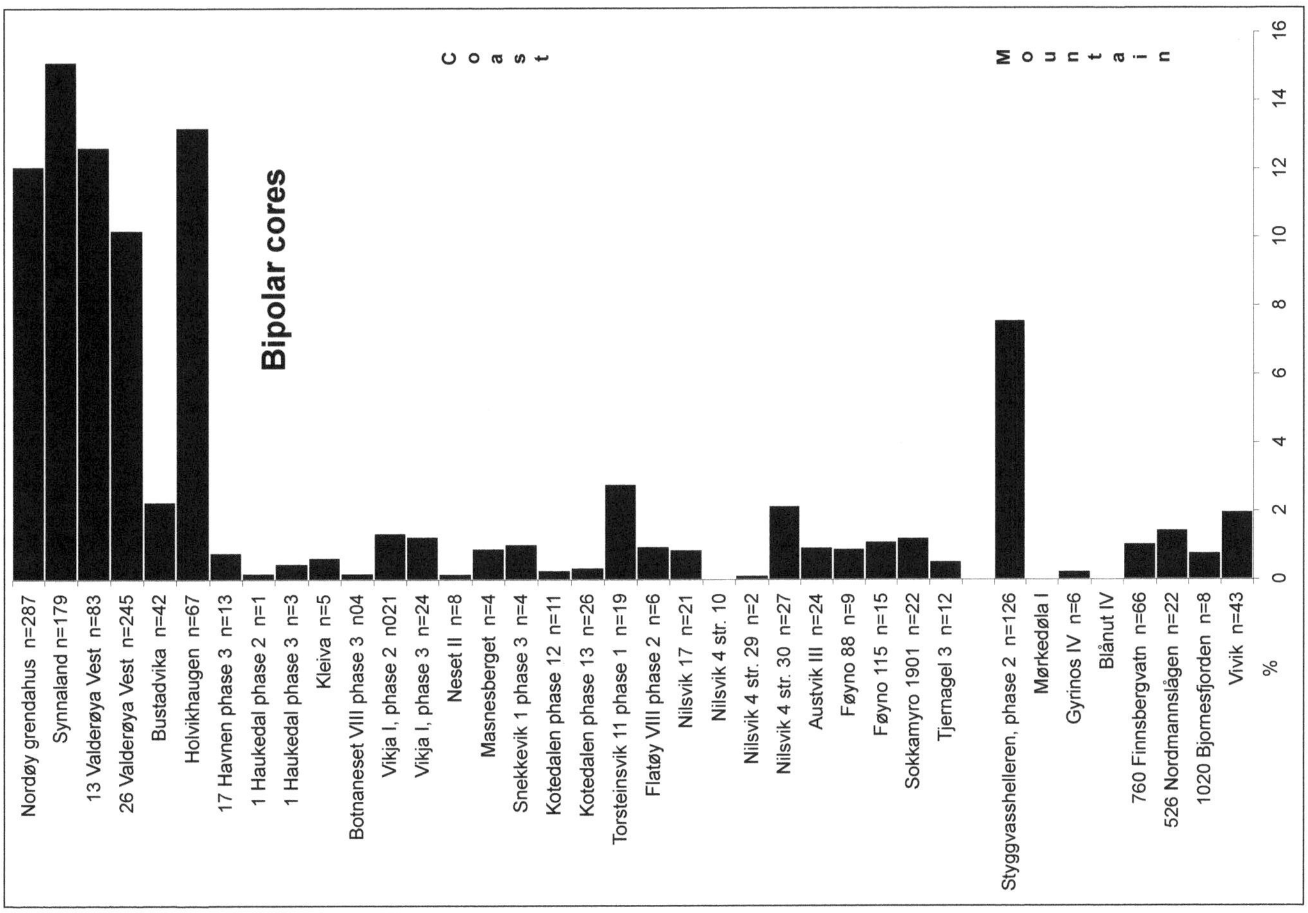

Figure 112. Percentages of bipolar cores at excavated sites.

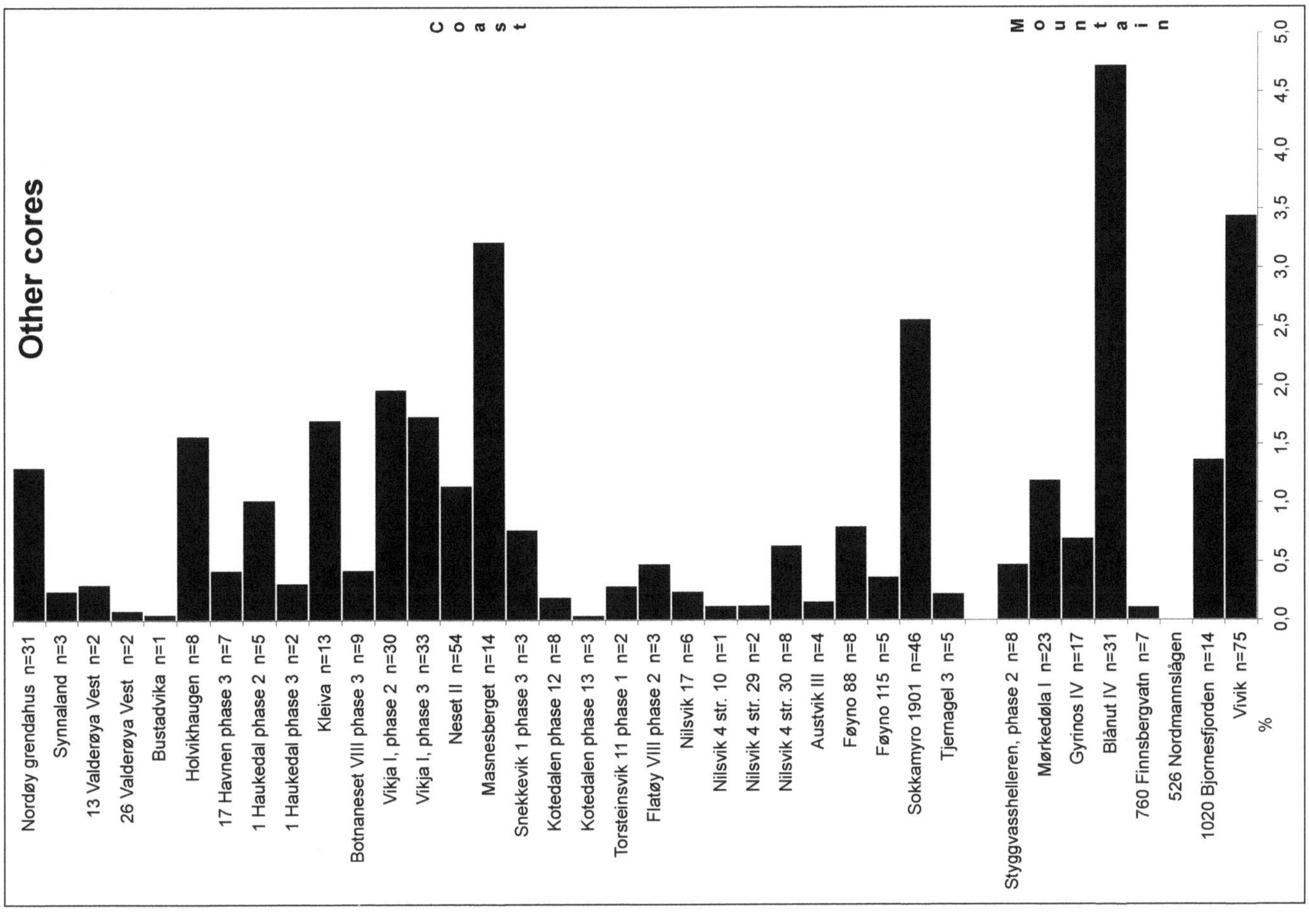

Figure 113. Percentages of other cores at excavated sites.

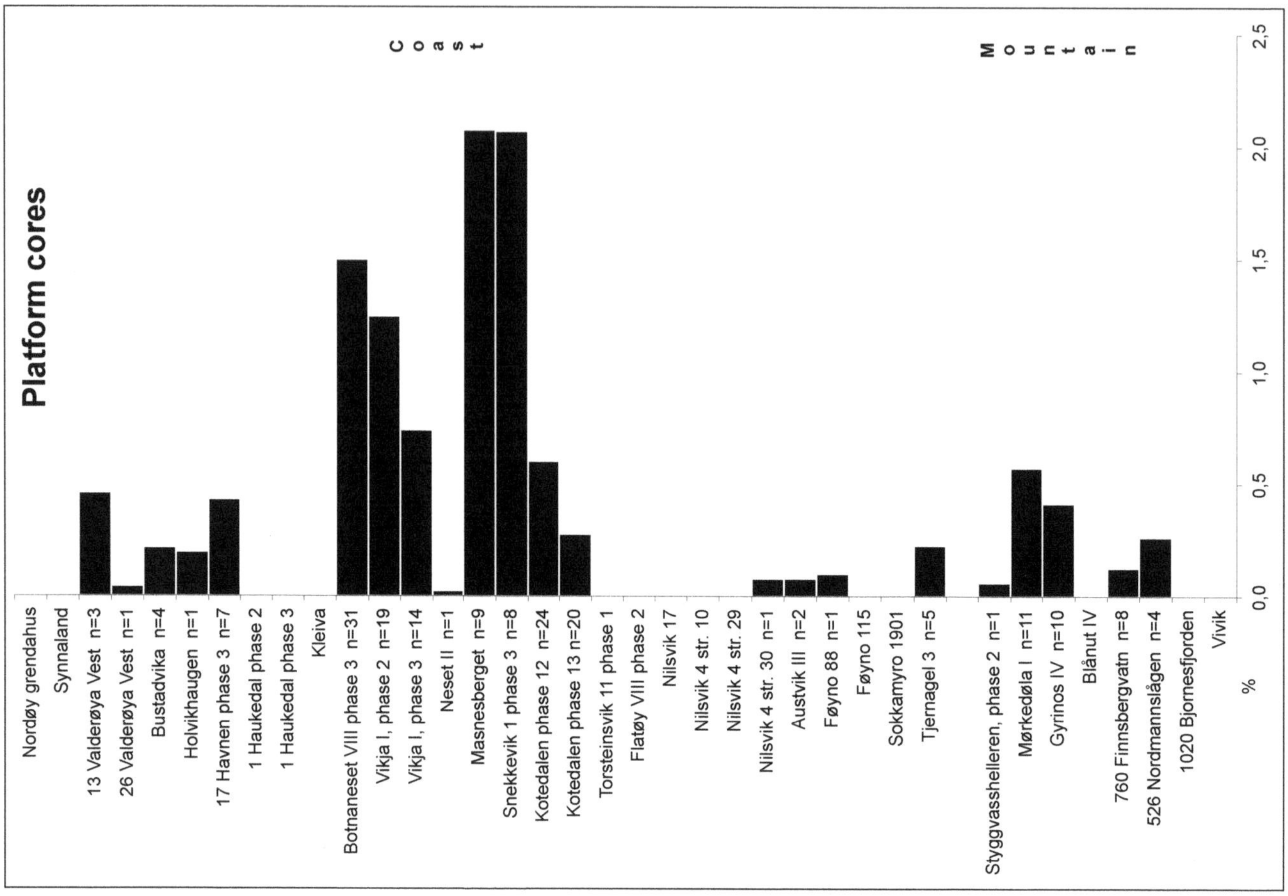

Figure 114. *Percentages of platform cores at excavated sites.*

Correspondence analysis

In order to explore the multidimensional variation of individual raw materials, types and techniques, I performed a correspondence analysis. This is a method for summarising the variation in frequency or presence/absence data and characterising its main patterns on a small number of new axes. It has the advantage of being based on the original data values (Greenachre 1993, Shennan 1997, Hjellebrekke 1999). Raw materials and types/techniques were analysed separately. Cross-tables of both values and a binary indicator matrix (presence-absence) were analysed for types/techniques, while for raw materials, only cross-tables of sites and values were analysed (see appendix 7 for matrices used as basis for the analysis and co-ordinates for the first three axis). The data was processed toghther with Professor Lars Forsberg in the program EXCELSTAT, and analysed in MACSPIN. The results will be discussed below.

Types and techniques

Two analyses were performed on types and techniques. The first was based on a binary indicator matrix (presence/absence of elements), the second was based on a cross-table of values.

One problem with the presence/absence analysis in this context is that categories that are minimally present at the sites will influence the graph just as much as elements that are more significant or dominating. This may in some cases be confusing. On the other hand, if there is a systematic absence of categories in some areas, this will produce a marked tendency in the plot. In the present case, the presence/absence analysis (Appendix 10, Colour Figure 4) shows that the mountain sites (black and light and dark green) are significantly different from the main bulk of the coastal sites. The main reason for this is the presence of ground flint flakes, single edged points and transverse points. These items are almost not found at the coastal sites. These mountain sites are also generally characterised by a lack of grindstone slabs and stone adzes/stone adze flakes. For the coastal sites, there are no variations that systematically follow the different districts. Most sites concentrate close to origo. Only individual sites are found further away, such as Kotedalen phase 13 (light blue), which is drawn towards TRB pottery, and Nordøy Grendahus (yellow), which is affected by the presence of struck blanks and a transverse point. Struck blanks and transverse points also influence other sites in Sunnhordland (pink), Sunnfjord (brown), and Nordfjord (red).

Two main conclusions can be drawn from this analysis. Firstly, the majority of the mountain sites diverge significantly from the coastal assemblages. Secondly, except for minor divergences, the graph does not display any significant differences in the composition of types and techniques between the coastal sites.

The graph based on values (Appendix 10, Colour Figure 5) is clearly different. As for the presence/absence diagram, the elements single-edged points and transverse points influence several of the mountain sites (dark and light green), although one of the Hardangervidda sites (760 Finnsbergvatn) is drawn towards bipolar cores. This is also the case with Styggvasshelleren (black). In contrast to the presence/absence diagram, the graph also shows that there are systematic differences between sites in the different districts. All sites in northern and southern Sunnmøre (orange and yellow) are clearly characterised by the large amount of bipolar cores. For the remaining coastal sites, there are also differences, although they are not as marked as towards Sunnmøre. Sites in Nordfjord (red) and Sunnfjord/Sogn (brown) are found together to the lower left of origo, mainly affected by other cores, grindstone slabs and slate points. Sites in Nordhordland, Midthordland and Sunnhordland, on the other hand, concentrate above these northern sites, mainly influenced by a higher number of A-points, small blades, and retouched flakes and blades.

This graph thus indicates that there are marked differences in the relative composition of types and techniques between four main areas. The majority of the mountain sites also diverge from the main bulk of the coastal sites. Three main areas can be distinguished at the coast: The composition at the Sunnmøre sites is the most distinctive. There are also marked differences between the sites in Sogn og Fjordane and Hordaland.

Raw materials

The analysis of the raw materials was performed in two steps. During the first step, all sites were included. During the second step, only coastal sites were included. Step 1 is presented in Appendix 10, Colour Figure 6. The graph is strongly influenced by the raw materials blue slate 2, fine grey quartzite 3, fine brown quartzite 1 and blue slate 1. As is evident from the graphs for the distribution of the individual raw materials, these categories are found almost only at mountain sites. This influences strongly the plots of 760 Finnsbergvatn, 526 Nordmannslågen, Blånut IV, Mørkedøla I, and Gyrinos IV, and indicates that the raw material profiles at these five mountain sites are significantly different from those at the remaining sites, which again confirms what has already been noted for the individual analysis, namely that most of the mountain sites have very different raw material composition from the sites on the coast, and that they also are distinctly different from each other. The remaining mountain sites resemble the coastal assemblages. Styggvasshelleren is quite similar to the sites in Nordfjord and Sunnfjord, 1020 Bjornesfjorden is first of all characterised by the high percentage of flint. As a result, it plots similar to the Sunnmøre sites. The composition at Vivik is diverse, and no raw materials used exclusively in the mountains distinguished it from the coastal assemblages.

Having established the main difference between the coastal and the mountain sites, it is necessary to move to the second step in the analysis, which is to summarise the raw material values at the coastal sites only. Such a step is important to take, because in Appendix 10, colour Figure 6, the coastal sites and raw materials concentrate close to origo in the graph due to the marked influences of the mountain sites noted above. The plot of the coastal values is presented in Appendix 10, colour Figure 7. The plots form a disc-like, relatively flat surface, and the graph is rotated in order to face this surface. The sites in the different districts have been given different colour codes (see text Appendix 10, Colour Figure 6). It is apparent that sites in the same districts are mainly found together in the graph. The sites at northern Sunnmøre (dark blue) concentrate at the upper left side, mainly affected by the heavy dominance of flint. The same is true for southern Sunnmøre (light blue), even if 9 Bustadvika is drawn downwards somewhat due to the significant presence of grey slate and quartz crystal at this site. The sites in Nordfjord (dark green) are more spread due to the different composition of raw materials at these three sites. 17 Havnen is affected by fine grey quartzite 2, while 1 Haukedal phase 3 is drawn towards chert. These sites nevertheless make up a group close to origo in the diagram. Four of the Sunnfjord/Sogn sites (light green) are found close together at the bottom of the graph, mainly affected by the marked presence of the mylonites and medium grey quartzite 1. The remaining two sites plot quite differently. Botnaneset VIII is drawn towards chert, while Masnesberget is found close to fine black quartzite, which dominates at this site. As noted above, the composition of raw materials at Masnesberget is influenced by the fact that this site has been collected by amateurs. The remaining sites plot quite close together, probably due to the dominance of rhyolite. There are, however, differences which can be related to variations between the districts. Two of the Nordhordland (red) sites, Kotedalen phase 12 and 13 are drawn towards green and brown slate, anorthosite, and probably also blue mylonite. Snekkevik 1, on the other hand, while largely lacking these raw materials, is found together with the remaining sites in Hordaland. It is, however, likely that the result from Snekkevik 1 is largely due to the small sample size compared to Kotedalen. Most of the sites in Midthordland and Sunnhordland (orange and yellow) plot close together. Only one site, 3 Tjernagel, stands out because of a relative lack of rhyolite and a marked presence of flint and fine grey quartzite 8.

In the analysis of the distribution of the individual raw materials earlier in this chapter, it was apparent that there was a marked fall-off and distribution boundaries of several raw materials in the same areas. This impression is confirmed by the correspondence analysis, which shows that there are marked differences between the different districts with regard to the composition of the raw materials. Although it does not clearly show how the fall-off looks like, it is apparent from this analysis that the composition of raw materials at the sites shows a large degree of variation across space. Firstly, it can be concluded that there are major differences between most of the sites in the mountain districts and the coastal sites in general. Secondly, within the individual mountain sites, there are large differences, even though the sites at Lærdal/Hemsedal and Hardangervidda largely plot in two different main areas in the graph. Thirdly, there are marked differences between the coastal sites in Sunnmøre, Nordfjord, Sunnfjord, Sogn, Nordhordland and Midthordland/Sunnhordland.

Summary of the correspondence analysis

Conclusively, the correspondence analysis of types and techniques shows that four different areas can be distinguished in the presence/absence graphs and the value graph: Mountain, Sunnmøre, Sogn og Fjordane and Hordaland. The raw material graphs with values show that the area of analysis is subdivided even more. In these diagrams, the Lærdal/Hemsedal sites are clearly distinguishable from the sites at Hardangervidda, and there are also marked differences generally between coastal and mountain sites. Within the coastal region, there are marked differences between sites in the districts northern/southern Sunnmøre, Nordfjord, Sunnfjord/Sogn, Nordhordland and Midthordland/Sunnhordland.

Source critical factors

In this chapter, possible cultural differences or breakage in communication is the theme. I assume that the general tendency in early Neolithic raw material use, types, and techniques for each district or region would best reflect such differences. I am, therefore, not particularly interested in variations that, for example, are the results of differences in activity patterns within the same group.

The ideal situation for the above analysis would have been to only have compared sites that were contemporary, of the same type, excavated by the same methods and that had identical sample sizes. As pointed out in Chapter 6, however, there is significant variation in the data for all of these factors, and it is quite likely that they are responsible for a number of the patterns that were observed above. The factors discussed for each of the variables above, and can be summarised as follows:

(1) Site types. It was expected in Chapter 6 that the sites characterised as field camps would have a less varied or different composition of artefacts compared to the residential sites. The difference in site type may explain some of the very high or very low raw material percentages at the field camps Kleiva, Botnaneset VIII, Torsteinsvik 11, Flatøy VIII, and 3 Tjernagel. This factor may also explain some of the profound differences between the mountain sites.

(2) Differences in excavation methods and selective collection of artefacts. Sokkamyro and Masnesberget in particular were expected to be problematic in this respect. It is likely that some of the high percentages of some raw materials and absence of others at these sites are results of selective and inconsistent collection of artefacts.

(3) Differences in sample sizes. Some of the sites in this analysis have quite small sample sizes, particularly Holvikshaugen, 1 Haukedal, Masnesberget and Snekkevik 1. It is possible that larger sizes of the samples from these sites would have adjusted the somewhat high percentages at some of these sites.

(4) Differences in dating. Some of the sites have significant elements from other periods, but it is also apparent that the sites in this analysis cannot all be dated to the same part of the early Neolithic. Some of the observed differences have come as a result of changes that took place during the period. Such changes are, of course, important for evaluating the problem at hand in a small scale diachronic perspective, even if this problem will not be particularly important in this thesis.

Conclusions

Initially, I pointed out that in order to be able to discuss ethnic boundaries, one would have to find systematic, concurring patterns in the distribution of raw materials, types and techniques. The above analyses show that there are significant, systematic differences in the distribution of different traits in the area of analysis.

At the coast, five districts may be distinguished tentatively on the basis of the above descriptions:

(1) Northern and southern Sunnmøre
(2) Nordfjord
(3) Sunnfjord/Sogn
(4) Nordhordland
(5) Midthordland/Sunnhordland

Most of the mountains assemblages diverge profoundly from the coastal assemblages. There are also major differences between the different areas in the mountains. In Chapter 9 I will discuss the boundaries between the different districts in a context of ethnicity. In the next chapter, however, I will investigate the possible provenance of the raw materials that has been distinguished in this thesis. This will partly be done on the basis of geological analyses.

Chapter 8

Raw material sources

Introduction

The majority of the raw materials which have been documented by applying the reference-system were probably quarried in specific bedrock outcrops. Some of these quarries are known (A. B. Johansen 1978, A. B. Olsen & S. Alsaker 1984, S. Alsaker 1987) (Fig. 115). In this chapter, I will review these known sources and perform a supplementary geological analyses of quarries as well as artefacts. For the remaining raw materials, I will attempt to localise possible source areas on the basis of new geological data and distribution patterns.

For obvious reasons, geological analyses can only be performed on a very small portion of the raw materials that have been classified in this thesis. Testing the validity of the reference system is consequently another important aim of the geological analysis.

Procedure for detecting the sources

Provenance studies in this region are hampered by a number of difficulties. One of the most serious is that western Norway is geologically very complex, and only a few areas have sufficiently detailed geological maps. Therefore, even if the raw materials are split into individual types and they are geologically determined, one can still not expect to relate them all to specific sources.

In order to characterise the raw materials and to detect their provenance, I followed a basic procedure, mainly based on Earle and Ericsson (1977:5): (1) Raw-material samples from the archaeological sites are shown to geologists (both amateurs and professionals), with specific knowledge of different areas or possible sources. A raw material and a region for research is selected based on their information. (2) Geological surveys of possible sources are performed and samples taken from these sources. (3) The samples are analysed using different techniques (mineralogical, geochemical, or trace-element analysis) in order to isolate the characteristic element composition for each source. (4) Samples of artefact material are collected from the archaeological sites and compared to the data from the sources in order to identify the specific sources for the artefacts.

For the current work, step 1 in this procedure was followed for all of the raw materials in the analysis. The reference collection was showed to several geologists

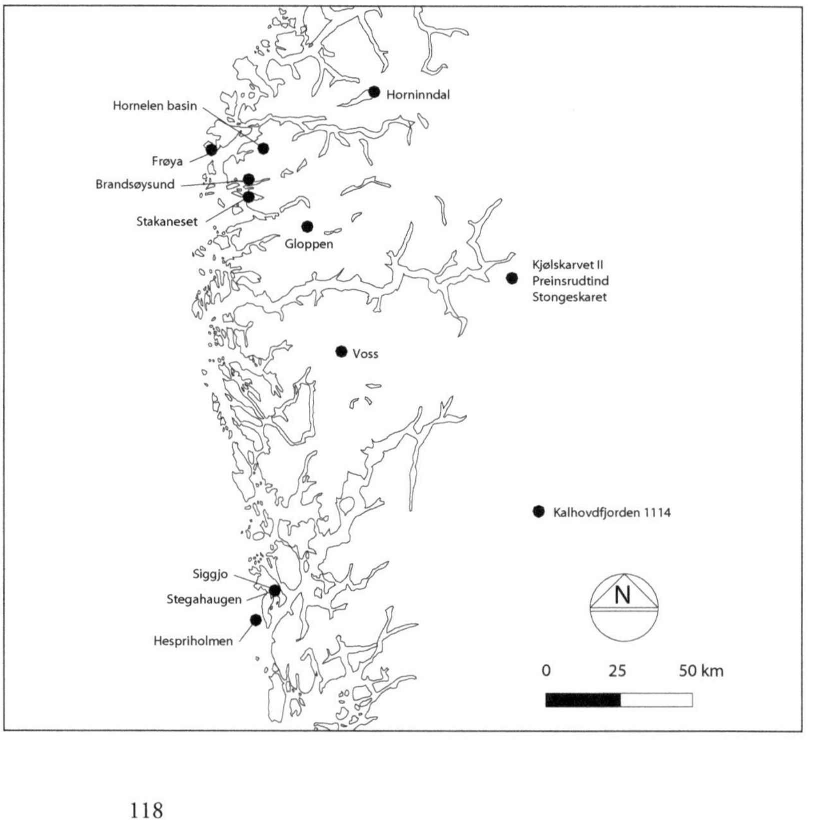

Figure 115. *Lithic bedrock sources and quarries in western Norway. For references to the individual sites, see text.*

who are particularly well acquainted with the area of analysis: Eystein Grimstad, Norwgian Geotechnical Institute, Inge Bryhni, Geological Museum, University of Oslo, Harald Furnes, and Øystein Jansen, both from the Dept. of Earth Science, University of Bergen, and Bjørn Russenes from the administration of the county of Sogn og Fjordane. Part of the collection has been characterised (based only on visual inspection) by Einar Alsaker, Statoil (E. Alsaker 2002). In addition, I showed parts of the collection to the amateur geologists Odd Naustheller, Førde, Svein Brandsøy, Florø and Ståle Edvardsen, Florø.

On the basis of discussions with these geologists and amateur geologists, it became clear that the chances for relating individual raw materials to specific sources or source-areas by geological methods were limited for the majority of the quartzes and quartzites, because they were expected to consist of few characteristic elements and because possible source areas had been located for very few of them. For fine green quartzite 1, chert, mylonites, and anorthosite, the possible source areas have been located, and geological analyses should ideally have been performed, but it could not be done within the current project. For these raw materials, only step 1 of the procedure has been carried out. Nevertheless, I find that one may still produce convincing arguments for possible source-areas. Below, I will apply the qualitative and quantitative distribution pattern of the individual raw materials in Chapter 7 as arguments for possible provenance.

Step 2–4 in the procedure were carried out on greenstone, diabase, rhyolite, sandstone, and slate. The geological analyses have been performed by Professor Rolf Birger Pedersen, Dept. of Earth Science, University of Bergen.

Greenstone

Previous research

During Haakon Schetelig's excavations of Sokkamyro in 1902 (Fig. 134), large amounts of greenstone debris were found, and this led him to conclude that a quarry was situated nearby (Schetelig 1901). In 1923 the quarry was found by Johs. Bøe and the geologist Kolderup at Hespriholmen, a small islet just off the coast of southern Bømlo (Kolderup 1925). However, the quarry was not investigated until the late 1970's, by Sigmund Alsaker (A. B. Olsen & S. Alsaker 1984, S. Alsaker 1987).

Outcrops of greenstone are known at several locations along the coast (Furnes et al. 1980). One of the largest is situated in the Bømlo-Stord district in Sunnhordland. Hespriholmen is located in the northern part of two basic sub aerial lava belts which cross Southern Bømlo in a SW-NW direction. Greenstone has been extracted at three separate quarries, the largest of which appear as two "mines" into the western part of the islet. The quarries cover about 250 m^2. S. Alsaker estimated that about 427 m^3 have been quarried at Hespriholmen.

On the basis of data from nearby workshop sites and the presence of typologically different greenstone adzes, S. Alsaker argued that the extraction of greenstone possibly commenced as early as about 9500 BP (8800 BC) and lasted until the end of the late Neolithic, about 3500 BP (1800 BC) (S. Alsaker 1987:81ff). Two charcoal samples were radiocarbon dated from an excavated test-square in the quarry-area. Both samples were collected from the bottom layer (S. Alsaker 1987:22):

T-3205: 5280±250 BP (calibrated date is found in appendix 4)
T-3206: 4620±110 BP

The primary reduction sequences into blocks took place in the quarry-area at Hespriholmen, while the secondary reduction of the blocks into blanks was done at workshop sites up to 10 km away from the quarry. The final grinding/pecking and polishing were performed at residential sites (A. B. Olsen & S. Alsaker 1984:82).

20 km to the north of Hespriholmen, S. Alsaker also surveyed a small greenstone-quarry at Stegahaugen. According to S. Alsaker, this quarry is the source of less than 5% of the greenstone adzes manufactured at Bømlo. In addition, S. Alsaker suggested that outcrops of greenstone may have been utilised close to Røyksund to the south of Stegahaugen (S. Alsaker 1987:32 ff).

The greenstone at Hespriholmen and Stegahaugen has been geologically described by the geologist Harald Furnes, on the basis of visual inspection as well as geochemical trace element analyses. Furnes maintained that the greenstone has a massive, fine grained texture lacking slate structure or macroscopically visible crystals (Furnes referred in A. B. Olsen & S. Alsaker 1984). The Stegahaugen outcrop is chemically identical to that of Hespriholmen, but the weathering patinas of the two greenstones are somewhat different.

On the basis of Furnes's results, S. Alsaker argued that three of the following criteria should be satisfied in order to classify greenstone quarried at Hespriholmen macroscopically (visually): It should be (1) homogenous without vesicular parts (parts with bladders), (2) completely aphyrical (with no fennocrystals), (3) close to Munsell 42/1 – olive grey in colour in fresh fractures, and (4) hair-thin yellowish-white continuous epidote bands or remains of such bands on cleavage faces. According to S. Alsaker, application of these criteria makes it easy to recognise greenstone from Hespriholmen by visual inspection (S. Alsaker 1987:33).

From a total of 2209 stone adzes of western Norwegian types in the museum collections in Norway, S. Alsaker and A. B. Olsen classified 736 as greenstone from the quarries at Bømlo. The majority of these specimens were concentrated to the districts from Rogaland to Nordhordland. To the north of Nordhordland, there was a marked fall-off in the distribution pattern (A. B. Olsen & S. Alsaker 1984:84). In order to test these determinations based on visual inspections, Furnes and S. Alsaker compared geochemical samples from the quarry and from workshops close to the quarry with samples from 40 distributed adzes which had been visually identified as greenstone by S. Alsaker and A. B. Olsen. The geochemical analysis (XRF trace element plotting in Ti/Y/Zr discrimination diagrams) indicated, according to S. Alsaker, a geochemical affinity for all of the adzes to the greenstone sources at Bømlo. This led him to conclude that the criteria for visual classification were valid (S. Alsaker 1987:55-58).

Asle Bruen Olsen have suggested that a third greenstone-quarry is probably situated further north at the coast of western Norway. The existence of such a quarry is indicated by the fact that a significant portion of the Neolithic adzes in the districts to the north of Nordhordland are made from a "bluish greenstone" (hereafter termed quarry 4 greenstone). Based on the analyses of a thin-section of one of the adzes of this raw material, the geologist Finn Skjerlie concluded that it was a mylonitizised and re-crystalised greenstone with slate structure, which probably originate on the coast of Møre og Romsdal or Nordfjord. On the basis of visual inspection only, A. B. Olsen classified altogether 188 adzes as made of quarry 4 greenstone. The main distribution area of adzes of this raw material was between Romsdal and Sunnfjord (A. B. Olsen 1981).

Geological investigations of greenstone for the present work

In their analyses, A. B. Olsen and S. Alsaker distinguished the Vespestad adze as a separate category and classified the raw materials of all adzes of this type. In theory, I might have applied their results directly for the current thesis. Still, I chose to inspect collection of Vespestad adzes from the Bergen University Museum district myself. One reason for performing this new inspection was that the adze collection at the museum has grown considerably since A. B. Olsen and S. Alsaker completed their survey in the late 1970's. Another reason was that I personally found it very problematic to distinguish between the different categories of greenstone based only on visual inspection. When I applied S. Alsaker's four criteria for visual classification of greenstone from Hespriholmen (see above), I found it very hard to distinguish specimens that S. Alsaker had visually classified as coming from Hespriholmen from the ones that had been classified as quarry 4 greenstone by A. B. Olsen. I therefore developed the working hypothesis that the majority of the adzes that had been determined as quarry 4 greenstone in reality had been quarried at Hespriholmen and distributed to the northern districts as a supplement to the diabase.

In order to test this hypothesis, I arranged a blind-test together with A. B. Olsen (he did not know where they had been found). All of the 230 Vespestad adzes from the collection at Bergen University Museum (appendix 5, except the ones deposited at Sunnmøre Museum) were put on a table. None of them had information about find context. He was asked to distinguish between diabase, quarry 4 greenstone, greenstone from Hespriholmen, and undetermined rocks. With regard to the two greenstone-types, he classified 29 as quarry 4 greenstone, and 34 as greenstone from Hespriholmen. Afterwards, I checked his classifications against the find contexts, and this showed an almost identical distribution pattern for the two supposedly different greenstone-types. If these adzes had been taken out at Hespriholmen and a quarry in the northern part of western Norway, an identical distribution pattern would not be expected. Thus, the result strengthened my hypothesis about a common origin of these adzes at Hespriholmen.

The problem was investigated further, by selecting five adzes that had been classified as quarry 4 greenstone by A. B. Olsen in his thesis from 1981 (no. 44-48) and three new adzes selected by myself (no. 49-51) for geological isotope analysis (Table 3). Among these was the adze (no. 47) that had already been described by Skjerlie on the basis of a thin-section (see above). The eight adzes had all been found far from Hespriholmen, in Sunnmøre and Nordfjord. Furthermore, I selected four adzes that had been visually and geochemically classified as greenstone from Bømlo by S. Alsaker and Furnes (no. 53-56). All of these 12 specimens satisfied S. Alsaker's criteria for visual classification of greenstone from Bømlo. The adzes were compared to five samples that had been collected at the quarry at Hespriholmen (no. 57-61). If all of the adzes plotted in the same area, I would regard my working hypothesis as sufficiently tested.

The method of isotope analysis was chosen firstly because less test material was needed than for trace element analysis (it was important not to destroy the artefacts in the museum collections). Secondly, the isotope analysis is more diagnostic than trace element analysis, because of greater precision. The isotope-variation for the relevant species of rocks (greenstones, diabase, and rhyolite) in western Norway is generally large (e.g. Pedersen & Dunning 1997). Significant differences in the composition of isotopes between the different sources were expected. The method that was applied here is frequently used by geologists for provenance studies in western Norway and elsewhere, and I find that describing the method and the relevant isotopes in full detail in this contribution is unnecessary (but see text Fig. 116 and Table 3). A presentation of the method and the relevant isotope-variation is found in Pedersen & Dunning (1997).

Rolf Birger Pedersen ran the tests at the Dept. of Earth Science. His results did not support my working hypothesis at all (Table 3, Fig. 116). Only two of the four chosen adzes that had been geologically tested by S. Alsaker and Furnes plotted in the same area as Hespriholmen. They had been found in Nordhordland (no. 54), and Sogn (no. 53) respectively. The other two, which had both been found in Sunnhordland, close to Bømlo (no. 55, 56), plotted quite differently in two different areas. The remaining eight adzes from the northern districts also plotted close together.

The relatively large number of adzes that plotted close together indicated, as A. B. Olsen already had suggested (A. B. Olsen 1981:161), the presence of an unidentified, significantly utilised greenstone quarry (quarry 4) somewhere in western Norway, most likely in the northern part of this region. One adze that plotted by itself (no. 55) indicates the presence of a greenstone quarry 5 as well. However, considering that the adzes had only been checked against the samples from Hespriholmen, there was still a theoretical possibility that the nine adzes may have been quarried at the already known greenstone and diabase quarries at Stegahaugen, Bømlo, or at the diabase quarry at Stakaneset. Samples were therefore collected from these sites and tested in a similar fashion by Pedersen (no. 88-93). The nine adzes did not originate at any of these sources. The greenstone samples from Stegahaugen plotted similar to the samples from Hespriholmen (no. 90, 91), while the four diabase-samples plotted in two mutually different areas of the diagram.

These results show that the working hypothesis was wrong; a large portion of the greenstone Vespestad adzes have not been quarried at Hespriholmen or at any of the other quarries known today. 10 out of 12 greenstone adzes originate at other quarries. As many as nine of these come from the same, unidentified quarry 4 greenstone, and this indicates that this raw material must have represented a significant portion of the adzes during the early Neolithic. Consequently, I found it necessary, once again, to try to distinguish between Bømlo greenstone and quarry 4 greenstone visually. This time the test was done together with Rolf Birger Pedersen. The 12 analysed adzes were used as references, and the rest of the 111 Vespestad adzes that already had been classified as greenstone were blind tested against these analysed items. We specifically looked for slate structure as a way of distinguishing the quarry 4 greenstone from the Bømlo greenstone. About ten adzes were distinguished as quarry 4 greenstone. When these were checked against the data on find-context, the majority of these had been found at southern Bømlo and Sveio, close to Hespriholmen. I find it unlikely that our classification is reliable, because if these adzes had been taken out at a northern quarry 4 greenstone, one would not expect so many of them to be found at Bømlo. It is much more likely that these adzes originate at Hespriholmen or Stegahaugen, considering the proximity to these quarries. The majority of these ten specimens were, most likely, classified wrongly by us as quarry 4 greenstone.

As a result of these tests, it can be argued that S. Alsaker's criteria for distinction of greenstone from Bømlo are problematic, because firstly, in our isotope analysis, only two out of 12 adzes, which all satisfied S. Alsaker's visual criteria, matched the samples from Hespriholmen and Stegahaugen. Secondly, out of four adzes (out of 40), which had trace elements that had affinity to the Bømlo greenstone in S. Alsaker's and Furnes' analysis, two did not match the isotope analysis on samples from Hespriholmen and Stegahaugen. These results indicate that the geological support that is claimed by S. Alsaker[1] for his visual criteria is only partly supported by our new geological analysis. This result obviously has serious implications for S. Alsaker's classification of 736 adzes as greenstone from Bømlo. It is likely that he has been too inclusive, and that the real number is much lower.

A question that is relevant for the current work is whether the above conclusion undermines S. Alsaker's and A. B. Olsen's work on the regional distribution of Vespestad adzes of quarry 4 greenstone and Bømlo greenstone. I believe that it is still possible to argue that the greenstone from Bømlo dominates in the southern districts and that quarry 4 greenstone (together with diabase) dominates in the north. Such a dualistic distribution pattern is indicated by the fact that eight out of nine adzes that were identified as quarry 4 greenstone in our analysis have been found in Nordfjord or in Sunnmøre. None of the analysed adzes from this area were from Bømlo. It is also likely that, although if the number of Vespestad adzes made of greenstone from Bømlo is probably lower than S. Alsaker's estimation, Hespriholmen or Stegahaugen still supplied a significant share of these adzes in the southern districts. This is indicated by the different geological analyses, which, after all, has a considerable match between adzes and the quarries. It is

1 A close inspection of S. Alsaker's data from the geological analysis (S. Alsaker 1987:115-117) uncovers serious problems. Firstly, the scientific designations of the trace elements are not correctly written, and this makes it difficult to find out which elements that the values actually refer to. Secondly, the values themselves (expressed in ppm) are much higher than one would expect for the presumed elements. It is therefore difficult to evaluate his interpretations against the data.

Samples from known quarries							
No	**District**	**Artefact**	**Rock type**	**143Nd/144Nd**	**+/-**	**87Sr/86Sr**	**+/-**
58	Hespriholmen	Adze blank	Greenst.	0.512181	7	0.714959	9
57	Hespriholmen	Flake	Greenst.	0.512183	7	0.715626	9
61	Hespriholmen	Adze blank	Greenst.	0.512183	7	0.711313	9
60	Hespriholmen	Flake	Greenst.	0.512189	7	0.713334	9
59	Hespriholmen	Adze blank	Greenst.	0.512197	6	0.716966	9
88	Stakaneset	Flake	Diabase	0.512520	6	0.708348	9
89	Stakaneset	Adze blank	Diabase	0.512526	6	0.708642	8
90	Stegahaugen A	Flake	Greenst.	0.512188	6	0.716179	9
91	Stegahaugen A	Flake	Greenst.	0.512181	6	0.715480	9
92	Stegahaugen B	Flake	Diabase	0.512127	6	0.738125	9
93	Stegahaugen B	Flake	Diabase	0.512141	5	0.731435	9
94	Siggjo	Block	Rhyolite	0.511963	5	0.794791	9
95	Siggjo	Block	Rhyolite	0.511971	5	0.773930	9

Lithic artefacts from residential sites							
No	**District**	**Artefact**	**Rock type**	**143Nd/144Nd**	**+/-**	**87Sr/86Sr**	**+/-**
54	Outer Sogn	Vesp. Adze	Greenst.	0.512140	7	0.714812	9
53	Nordhordland	Vesp. Adze	Greenst.	0.512181	5	0.718090	12
55	Sunnhordland	Vesp. Adze	Greenst.	0.512386	6	0.709197	10
50	Nordfjord	Vesp. Adze	Greenst.	0.513084	7	0.704319	8
49	Nordfjord	Vesp. Adze	Greenst.	0.513087	7	0.703301	8
56	Sunnhordland	Vesp. Adze	Greenst.	0.513090	7	0.704137	9
47	S. Sunnmøre	Vesp. Adze	Greenst.	0.513092	6	0.702857	8
46	S. Sunnmøre	Vesp. Adze	Greenst.	0.513092	5	0.703020	9
51	Nordfjord	Vesp. Adze	Greenst.	0.513093	7	0.703443	9
44	S. Sunnmøre	Vesp. Adze	Greenst.	0.513093	7	0.703997	9
48	Nordfjord	Vesp. Adze	Greenst.	0.513095	7	0.703072	9
45	S. Sunnmøre	Vesp. Adze	Greenst.	0.513101	7	0.703393	8
10	Midthordland	Flake	Rhyolite	0.511955	8	0.809101	8
16	S. Sunnmøre	Flake	Rhyolite	0.511962	6	0.817901	9
22	Breheimen	Blade	Rhyolite	0.511963	6	0.810615	9
26	S. Sunnmøre	Core	Rhyolite	0.511996	4	0.801184	9
37	Nordhordland	Core	Rhyolite	0.511958	4	0.818613	9
42	Nordfjord	Flake	Rhyolite	0.511950	6	0.790223	9

Table 3. *Nd and Sr isotopic data of lithic artefacts and samples from known quarries. The following analytical procedures were used: 8 mm diameter cores were drilled from the artefact and milled in an agat mortar. The subsequent chemical processing of the powdered samples were carried out in a clean room, using reagents purified in two bottles Teflon stills. Samples were dissolved in a mixture of HF and HNO_3. Sr and rare earth elements (REE) were separated by specific extraction chromatography, and Nd were subsequently separated using a low-pressure ion-exchange chromatographic set-up with phosphate-coated Teflon powder as the ion-exchange resin. Sr and Nd were loaded on double Re-filaments and analysed in static and multi-dynamic mode on a Finnigan 262 thermal ionisation mass spectrometer. Sr and Nd-isotopic ratios were corrected for mass fractionation using a $^{88}Sr/^{86}Sr$ value of 8.375209 and $^{146}Nd/^{144}Nd$ ratio of 0.7219 respectively. Analytical errors (+/-) are given on the 2-sigma level, and refer to the last digits in the reported isotopic ratios. See appendix 8 for more information on context of samples.*

also supported by the size of the quarry at Hespriholmen itself, by the radiocarbon results from the production site at Hespriholmen, and by the large amount of debris from adze production that was identified at the early Neolithic workshop at Sokkamyro.

Nevertheless, it must be stressed that, although southern and northern distribution areas of greenstone may have been a reality, the concrete boundaries and fall-off for these distributions as presented by S. Alsaker and A. B. Olsen (1984:84) cannot be convincingly supported by the visual classifications for the present work. These patterns are only indicated by the few geological tests that have been performed.

As for the remaining, visually classified adzes in the current analysis, I have chosen to apply the grouped term "greenstone" as a designation for both quarry 4 greenstone and greenstone from Bømlo. Admittedly, this represents an unforeseen, regrettable drawback for the analysis, because by taking this restrictive position, a discussion of social boundaries on the basis of the greenstone data is impossible.

If a more reliable data-set is to be established, an extensive programme on geological analysis of greenstone should be initiated. Such a programme should also include finding the unidentified quarries, as a definite aim. It is most pressing to locate the source of the quarry

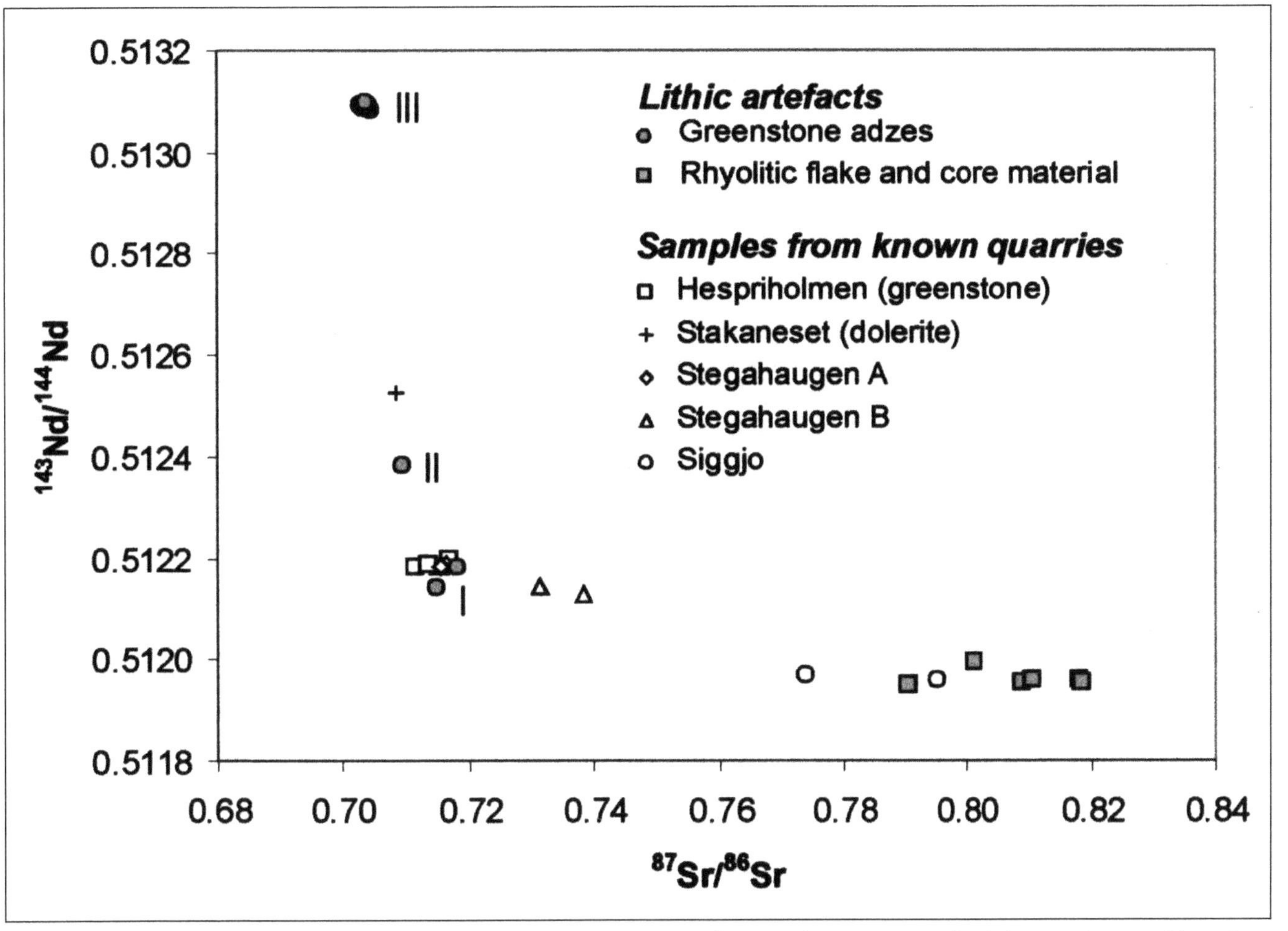

Figure 116. *Diagram where the 87Sr/86Sr and 143Nd/144Nd isotopic compositions of lithic artefacts are compared with the isotopic compositions of potential sources (known Stone Age quarries). The isotopic data show that the analysed greenstone adzes stem from three different sources marked I, II and III on the diagram. Hespriholmen and Stegahaugen A have similar isotopic compositions as Group I adzes, and is therefore the likely source of these adzes. Group III adzes have "depleted mantle type" isotopic compositions, which limits the source of these adzes to greenstone sequences associated with ophiolite complexes.*

Rhyolitic artefacts show very low 143Nd/144Nd, and very high 87Sr/86Sr isotopic ratios. Rhyolites from Bømlo (Siggjo) and Stord (Kattenakken) are the only volcanic sequences in western Norway that are known to exhibit such isotopic compositions. Two samples from the Siggjo quarry show 87Sr/86Sr compositions that differ beyond the analytical uncertainty of the individual analyses. This suggests that the quarry is inhomogeneous with regards to Sr-isotopic composition, which is compatible with the relatively large range in Sr-isotopic compositions seen in the artefacts.

The diagram illustrates furthermore that diabase samples from Stakaneset and Stegahaugen B have distinctly different isotopic compositions, and that artefacts from these quarries therefore also may be well distinguished by using the Sr-Nd isotopic tracer method.

4 greenstone, which, according to both Skjerlie and Pedersen, is situated somewhere along coastline between Sogn and Nordmøre.

Diabase

Prior to the 1970's, diabase adzes were not distinguished from greenstone, and one generally held that they had been quarried at Bømlo. During his excavations in Flora in Sunnfjord, however, Asle Bruen Olsen found large amounts of flakes from adze production, and assumed that there was a large quarry close by (A. B. Olsen 1977). The quarry was found by A. B. Olsen and the amateur archaeologist Svein Brandsøy in a diabase dyke at Stakaneset in Flora in 1978, and was soon investigated and published (A. B. Olsen 1981, A. B. Olsen & S. Alsaker 1984).

The dyke is two m wide and 200 m long, and diabase has been extracted at five different quarries. Due to heavy erosion of the dyke, it is not possible to evaluate precisely the amount of raw material that has been taken out, but A. B. Olsen estimated that more than 100 m^3 and less than 800 m^3 diabase were quarried (A. B. Olsen & S. Alsaker 1984:73). The primary and secondary reduction sequences took place in the same manner as for the greenstone at Hespriholmen.

A. B. Olsen argued that the extraction of diabase possibly commenced as early as about 9500 BP and lasted until the end of the late Neolithic, about 3500 BP. Nevertheless, the site appears to have been more intensively used during the Mesolithic than later. This is indicated by the fairly low number of distributed Neolithic adzes made of diabase compared to the Mesolithic (A. B. Olsen & S. Alsaker 1984:96), and also by the radiocarbon determinations from excavated test-squares from quarry I, II and III: (A. B. Olsen 1981:102,103, A. B. Olsen & S. Alsaker 1984:79):

Quarry I

T-3052: Layer III: 3940±110 BP

T-3053: Layer VI: 5840±100 BP

T-3238: Layer IX: 6050±90 BP

T-3239: Layer XII: 6300±160 BP

Quarry II

T-3055: Middle layer: 6620±100 BP

Quarry III

T-3439: Upper layer: 5590±100 BP

T-3440: Bottom layer: 6330±130 BP

The diabase at Stakaneset has been geologically described by Finn Skjerlie on the basis of thin sections as well as geochemical analyses. According to Skjerlie, the dykes in Flora generally have a relative uniform mineralogy, dominated by plagioclase, clinoporyxene, and titanomagnetite. The dyke at Stakaneset is characteristic because of its white and opaque sericitized phenocrysts in green to brownish-green matrix. These phenocrysts makes it easy to identify artefacts made of Stakanes-diabase by visual inspection (Skjerlie referred in A. B. Olsen & S. Alsaker 1984:71, 72). Skjerlie compared thin sections of samples from the quarry with sections from 19 distributed adzes which had been visually identified as diabase by A. B. Olsen. Skjerlie found that the adzes were identical to the samples from the quarry. In addition, five adzes selected for trace-element analysis all fell within the diabase Flora field (A. B. Olsen & S. Alsaker 1984:85, 86). As a result of these geological tests, it is argued here that visual identification is a reliable method for the purpose of identifying Stakanes-diabase in the present work. During his examination of the museum collections, A. B. Olsen identified altogether 330 adzes as diabase.

Several other diabase dykes are known from Sunnfjord, but none of these appear to have been quarried during the Stone Age (A. B. Olsen & S. Alsaker 1984:72). Diabase also occurs at several locations in Norway, among these several dykes in Sunnhordland (Skjerlie & A. B. Olsen 1981). One of these, at Stegahaugen B, Bømlo, have been utilised during the Stone Age. It has a distinct mineralogical composition and is visually very different from the diabase from Stakaneset (S. Alsaker 1987:33). Two samples from this source are tested by isotope analysis for the present work by Rolf Birger Pedersen. The results confirm that it is different from diabase as well as from the variety of greenstones (Table 3, Fig. 116).

In order to distinguish diabase from Stakaneset from greenstone, and from diabase from Bømlo, isotope analysis was performed at two samples from the quarry at Stakaneset (see above) (Table 3, Fig. 116). The result shows a distinguishable composition of isotopes. Ideally, isotope analysis should also have performed on the adze material. Such analysis could, however, not be carried out within the current project. Instead, I chose to apply A. B. Olsen's results (A. B. Olsen 1981). In order to include in the current analysis the Vespestad diabase adzes that has been received by Bergen University Museum after 1981, and also to test A. B. Olsen's visual classification of diabase from 1981, I performed a blind-test on A. B. Olsen (see also under greenstone). Our classification of diabase matched completely the adzes that A. B. Olsen had characterised as diabase in 1981. I therefore find it safe to apply only visual classification of diabase in the present work.

Rhyolite

Previous research

The first to identify rhyolite was Haakon Schetelig, who termed it "dark quartz with light veins" in the report from the first excavations at Sokkamyro. He suggested that it was quarried locally at Bømlo (Schetelig 1901). The raw material, which was later termed "årekvarts" (veined quartz), was commonly found in large quantities at excavated sites in western Norway in the years to come, but it was not until the late 1970's that a more precise geological description was provided and the quarry was located.

Although several outcrops of rhyolite exist in western Norway, only one quarry is known. It is situated at the top of the mountain Siggjo at Bømlo in Sunnhordland, and it was located and published by Sigmund Alsaker (1987). The outcrop of rhyolite (zone II) covers approximately 10.000 m². Small quarries and workshops were found in an area of about 6000 m² (quarry-area). Three charcoal samples from an excavated test-square in the quarry-area were radiocarbon dated (S. Alsaker 1987:22):

T-3578: 4930±80 BP. Layer III (top layer)

T-3577: 4890±60 BP. Layer IV (middle layer):

T-3576: 4940±90 BP. Layer VI (bottom layer)

These dates suggest that the main extraction of rhyolite took place during the early Neolithic. In the quarry-area,

S. Alsaker mainly found remains of the quarrying itself (large pieces without bulbs of percussion). Few traces of secondary treatment of the raw material (flakes from core preparation) could be detected. This indicates that rhyolite was carried down from Siggjo as large blocks, not as prepared cores (S. Alsaker 1987:51).

Rhyolite is easily distinguishable visually because of a homogenous faint black-bluish texture and by a web of white (eroded) veins that leap across it. When heavily eroded because of chemical reactions, the texture tends to become grey or off-white. However, it is still easy to recognise because of the characteristic veins. Microscopical studies of rhyolite from the source show that it consists of an often indistinguishable aggregate of quartz and feldspar. Geochemically, it has a complex and rare composition (Nordås 1987). Until now, no rhyolite implements from archaeological sites have been compared geologically with samples from the source. They have only been identified by visual inspection.

Geological analysis for the present work

In order to test the visual determinations, I selected seven artefacts which had been determined as rhyolite for isotope analysis (for choice of method see under greenstone). No. 26 and 16 had been found at the at excavated sites Bustadvika and Holvikshaugen, Sunnmøre, no. 41 and 42 at 17 Havnen, Nordfjord, no. 22 at Styggvasshelleren, Breheimen, no. 37 and 38 at Kotedalen, Nordhordland, and no. 10 at 4 Nilsvik, Midthordland. These specimens from different parts of the area of analysis were compared to two blocks that had been collected at the quarry (no. 94 and 95).

All specimens from the excavated sites plotted very similar to the samples from the quarries (Table 3, Fig. 116). There is a slight spread of the plot because of differences in the values of strontium isotopes. According to Pedersen, these differences are most likely a result of different degrees of chemical erosion on the individual samples.

Due to a quite a similar geochemical composition of the rhyolite at Siggjo and a rhyolite outcrop at Kattenakken, Stord, just north of Bømlo, it is not possible to exclude Stord as a possible source for the seven specimens from the sites. However, until now, no quarry has been found at Stord. Considering that there is a large quarry-area at Siggjo, and that the radiocarbon dating from the quarry are clearly coincidental with the dating at the excavated sites with large quantities of rhyolite, I find it likely that all of the analysed samples have been quarried at Siggjo. On the basis of this unequivocal result, it is possible to conclude that the visual classification of rhyolite on the basis of the reference-system is reliable.

Sandstone and slate

Previous research

Several scholars have discussed the possible origin of the slate points and knives found in western Norway, but most of these studies have only been based on visual classifications. Mineralogical analyses of slate and sandstone have only been performed by Tore Bjørgo. Prior to the 1960's, it was generally believed that that much of the slate found in this region was imported from Middle and Northern Norway (e.g. Gjessing 1945:210). In his presentation of the site Ramsvikneset, Nordhordland, Egil Bakka supported this hypothesis for the red points and the specimens with differently coloured bands. For the points and debris of grey and green slate found at this site, however, he rather pointed to the un-metamorphosed middle Devonian sequences e.g. (hereafter termed Devonian sequences) of sandstone and siltstone (hereafter termed slate) at the coast between the Sognefjord and Nordfjord (Steel et al. 1985) as a more likely source (Bakka 1964, Bakka 1993:62).

Inge Lindblom had 58 slate points from Rogaland inspected by geologists, who concluded that as many as 35 points were not made of slate, but were rather quarried from erratic boulders or from the local bedrock. For the remaining 24, the majority have probably been made of locally available slates, while the Devonian sequences is suggested as a possible source for two red slate points (Lindblom 1980:14-20).

The geologist Finn Skjerlie analysed thin sections of three samples of slate and sandstone from his sites at Flatøy, Midthordland, for Tore Bjørgo. Skjerlie concluded that they were all from the Devonian sequences (Bjørgo 1981:148-149).

In his work on the slate knives, Hans Christian Søborg had a selection of red knives with green bands visually inspected by several geologists. It was pointed to a number of possible source areas in middle and northern Norway and Sweden, particularly along the "erosion boundary" of the Caledonian mountain range. The Devonian sequences further south were not considered by Søborg as relevant as a source, because of their coarse nature, and a lack of "false slate", which was a characteristic element of many of the knives in his analysis (Søborg 1988:231-33). The grey qualities, he argued, are difficult to relate to specific sources.

At Hardangervidda, no quarries have yet been located, but Svein Indrelid has pointed out that many of the slate points and the debris that occur at some of the sites were similar to the slate from the Kambro-Silurian sediments in the north-western part of the mountain massif (Indrelid 1994:251).

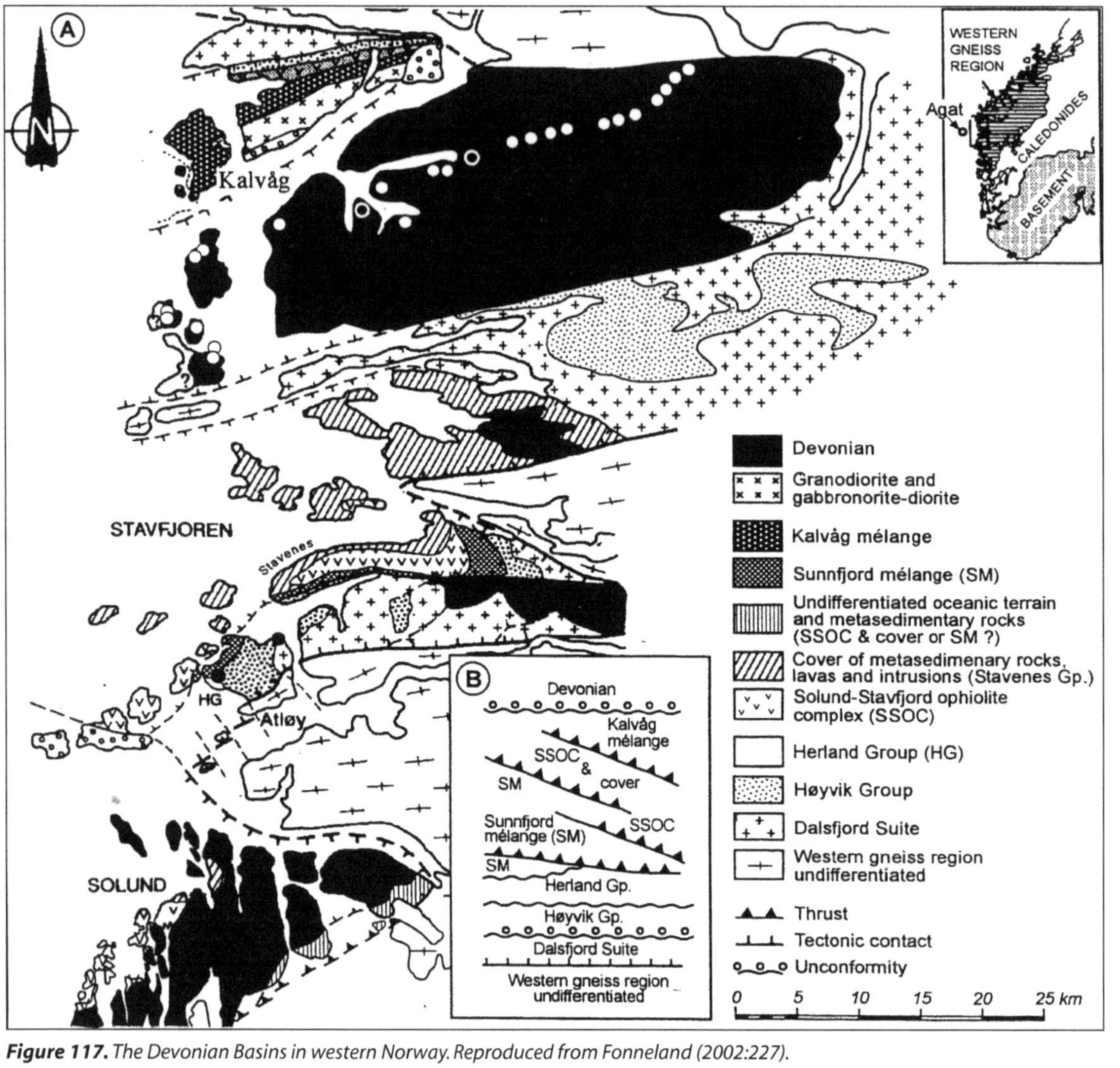

Figure 117. *The Devonian Basins in western Norway. Reproduced from Fonneland (2002:227).*

During the excavation project at Skatestraumen, Nordfjord (Bergsvik 2002a) samples of sandstone and different colour variations of slate were visually inspected by several geologists. In his geological report from the project, Einar Alsaker pointed out that the Devonian sequences of western Norway may have been likely sources for several of them, but they may also stem from other sedimentary sequences along the coast. More detailed mineralogical studies were needed in order to establish a more secure database for the general category of slate and sandstone (E. Alsaker 2002:350-52).

Geological investigations for the present work

Although there is some disagreement on the topic, most of the authors mentioned above agree that the Devonian sequences are probable raw material sources for slate implements found in western Norway. For the purpose of the present analysis, I surveyed the largest of the Devonian sequences in Sunnfjord (the Hornelen basin) (Fig. 117) in order to locate outcrops of relevant sandstone and slate (Bergsvik 1998). During this survey, I found several areas of sandstone as well as green, grey, and red slate that closely resembled the samples in the reference collection. Several samples were taken from the slate outcrops in the Hornelen basin. Samples of sandstone had already been collected from this area by geologists from the University of Bergen.

Until now, no prehistoric quarries have been found in the Devonian sequences. The sources of sandstone and slate may therefore theoretically be situated at all of the large basins Hornelen, Kvamshesten, Håsteinen, and Solund (one slight possibility may also be the Devonian sequences in the outer Trondheim region to the north). The Hornelen basin is, however, definitely the largest and also the most promising, because there are many suitable sandstones and slates. Kvamshesten has both sandstone and slate, but these are located at rather high altitudes (far away from the shorelines), and may not have been relevant for that reason. (On the other hand, the rhyolite quarry at Siggjo clearly shows that altitude was not necessarily a problem). Håsteinen and Solund mainly consist of unsuitable conglomerates (Steel et al. 1985). The most probable sources of sandstone and slate are therefore situated in the Hornelen basin.

Geological references					
No.	**District**	**Site**	**Raw material category**	**Rock type**	**Provenance**
Several geol. samples	Sunnfjord	Several sites	Sandstones	Sandstones	References Devonian sequences
A	Sunnfjord	Hovden	Red slate	Siltstone	Reference Devonian sequences
B	Sunnfjord	Kalvåg	Red slate	Siltstone	Reference Devonian sequences
C	Sunnfjord	Botnen	Green slate	Siltstone	Reference Devonian sequences
D	Sunnfjord	Botnen	Green slate	Siltstone	Reference Devonian sequences

Samples from excavated sites					
No.	**District**	**Site**	**Raw material category**	**Rock type**	**Provenance**
5	Midthordland	4 Nilsvik, str 10	Brown slate	Siltstone	Devonian sequences
6	Midthordland	4 Nilsvik, str 29	Grey slate	Siltstone	Devonian sequences
7	Midthordland	4 Nilsvik, str 29	Sandstone	Sandstone	Devonian sequences
11	N. Sunnmøre	Valderøya Vest 13	Sandstone	Sandstone	Devonian sequences
13	N. Sunnmøre	Valderøya Vest 26	Sandstone	Sandstone	Devonian sequences
14	N. Sunnmøre	Valderøya Vest 26	Green slate	Siltstone	Devonian sequences
15	N. Sunnmøre	Valderøya Vest 26	Green slate	Siltstone	Devonian sequences
18	S. Sunnmøre	Holvikshaugen	Sandstone	Sandstone	Devonian sequences
19	S. Sunnmøre	Holvikshaugen	Green slate	Siltstone	Devonian sequences
20	S. Sunnmøre	Holvikshaugen	Green slate	Siltstone	Devonian sequences
21	Sunnhordland	88 Føyno	Sandstone	Sandstone	Devonian sequences
29	Nordhordland	Kotedalen	Sandstone	Sandstone	Devonian sequences
30	Nordhordland	Kotedalen	Sandstone	Sandstone	Devonian sequences
31	Nordhordland	Kotedalen	Sandstone	Sandstone	Devonian sequences
32	Nordhordland	Kotedalen	Green slate	Siltstone	Devonian sequences
52	Inner Sogn	Mørkedøla 1	Green slate	Siltstone	Devonian sequences
69	Nordhordland	Kotedalen	Green slate	Siltstone	Devonian sequences
70	Nordhordland	Kotedalen	Green slate	Insecure	Possibly from Devonian sequences
71	Sunnfjord	Vikja I	Green slate	Siltstone	Devonian sequences
72	Sunnfjord	Vikja I	Red slate	Siltstone	Devonian sequences
73	Nordfjord	1 Haukedal	Green slate	Insecure	Possibly from Devonian sequences
74	Nordfjord	1 Haukedal	Green slate	Siltstone	Devonian sequences
75	Nordfjord	17 Havnen	Green slate	Siltstone	Devonian sequences
76	Nordfjord	17 Havnen	Green slate	Siltstone	Devonian sequences
77	Breheimen	Styggvasshelleren	Green slate	Siltstone	Devonian sequences
78	Hardangervidda	760 Finnsbergvatn	Blue slate 2	Indetermined	Unknown
80	Hardangervidda	760 Finnsbergvatn	Blue slate 2	Indetermined	Unknown
81	Sunnfjord	Vikja I	Sandstone	Sandstone	Devonian sequences
82	Nordfjord	17 Havnen	Sandstone	Sandstone	Devonian sequences
83	Nordfjord	17 Havnen	Sandstone	Sandstone	Devonian sequences
84	Sunnhordland	Sokkamyro	Green slate	Indetermined	Unknown
85	Sunnhordland	Sokkamyro	Green slate	Siltstone	Devonian sequences
86	Hemsedal	Blånut IV	Blue slate 1	Indetermined	Unknown
87	Hemsedal	Blånut IV	Blue slate 1	Indetermined	Unknown

***Table 4.** Geological analysis. Geological references from the Hornelen basin in the Devonian sequences and thin-sectioned and microscopically examined samples of artefacts from excavated sites. See appendix 8 for more information on context of samples.*

The Devonian basins have a low metamorphic grade and are the youngest sedimentary rocks on land in southern Norway. Distinguishing them from other rocks by means of isotope analysis or trace element analysis is problematic. Their mineralogical composition and textures are, however, quite characteristic, and a mineralogical study was performed. Thin-sections were taken from sandstone and slate artefacts from the sites and compared microscopically to thin sections from samples of sandstone and slate that had been collected from the Devonian sequences in Sunnfjord. Altogether 33 samples from excavated sites throughout the area of analysis were analysed (Table 4): 11 samples of sandstone and 22 samples of red slate (1), green slate (15) grey slate (1), brown slate (1), blue slate 1 (2), and blue slate 2 (2). The analysis was carried out by Rolf Birger Pedersen.

Similar to Skjerlie's and Bjørgo's earlier results from Flatøy (see above), all of the 11 samples of sandstone from the excavated sites had a mineralogical composition that was almost identical to the samples of sandstone that had been collected from the Devonian sequences (Colour

Figure 8, Appendix 10). This indicates strongly that the visual classifications of sandstone have been consistent, and it is likely that a large majority of the items classified as sandstone in the above analysis have been extracted in the Devonian sequences.

16 of the 22 samples of slate from the excavated sites also show a mineralogical composition identical to the samples from the Devonian sequences. These include the red, brown, grey, and most of the green samples. Two samples of green slate (no. 70 and 73) are possibly from the Devonian sequences, and the remaining five (no. 78, 80, 84, 86 and 87) were unidentified slates (not from the Devonian sequences). The two possibly Devonian slates and one of the unidentified slates had been visually classified as green slate, while 12 green slates were identified as Devonian.

This result is important for the evaluation of the visual classifications of slate based on the reference collection. It means that the visually classified red, brown and grey slates are most likely from the Devonian sequences. Nevertheless, considering that only one sample has been analysed of each of these variants, this conclusion is still tentative. One might, for example, question whether it is likely that the grey slate found at Vivik was transported from the coast. For evaluation of the visually classified green slates at the sites, the result is more reliable. Only a small percentage probably has other sources than the Devonian sequences. The only green slate that has definitely not Devonian origin has been found at Sokkamyro, one of the sites that are located furthest away from Devonian slates. It is also possible that the green slate at Blånut IV also has other sources, because the two analysed samples of blue slate 1 at this site do not originate in the Devonian sequences. The two samples of blue slate 2 at 760 Finnsbergvatn are also unidentified rocks. As noted above, Indrelid has suggested local sources for much of the slate that was used at Hardangervidda. This is not contradicted by the above results.

These results from the mineralogical investigations, which indicate that much of the slate and sandstones originate in the middle Devonian sedimentary sequences in Sunnfjord, are also supported by the distribution patterns for the sandstones and slates (Fig. 85-90, 92), which show that the frequencies of sandstone and red, green, brown, and grey slate are highest in Sunnfjord and Nordfjord and decrease with increasing distance from these districts. Some of the green and grey slates that are found far from the Devonian sequences, probably originate at other sources. This is also the case for the blue slates.

Chert

Chert was first distinguished during Asle Bruen Olsen's excavations of the site Botnaneset VIII in Sunnfjord (termed "grey quartzite with dark lines") (A. B. Olsen 1983:123). It was later identified as chert during the investigations at Skatestraumen, Nordfjord (Bergsvik 2002a:283).

Chert is only known at one outcrop in western Norway. The geologists Inge Bryhni and K. Lyse (1985) surveyed this deposit at the island of Frøya, Bremanger. It has also been studied by Rodmar Ravnås and Harald Furnes (1995). Due to its flint-like character, the chert at Frøya is well suited for production of struck stone tools, however, despite extensive searching, no quarries have yet been located at Frøya. On a fresh cut surface, the chert appears homogeneous and faint dark grey, with crystallised thin lines. On natural surfaces or on eroded specimens it is light grey with dark grey patterns. Quartz makes up 85-95% of this rock, while iron oxides are accessory minerals. Secondary minerals are biotite, garnet, chlorite and feldspar (Ravnås & Furnes 1995).

There is a striking visual similarity between the chert at Frøya and the raw material characterised as chert at the excavated sites. Another indication of Frøya as a source is the distribution patterns of chert (Fig. 83). The two sites with the highest percentages of chert, 1 Haukedal phase 2 and Botnaneset VIII, are located at almost equal distances from Frøya in the northern and southern direction respectively. At excavated sites further to the north and south, chert is only minimally represented.

Anorthosite

Ground projectile points of anorthosite (also termed "labradorite" in the archaeological literature) were first identified at the site Ramsvikneset in Nordhordland. Egil Bakka, who excavated this site, believed that it was probably of local origin. He pointed out that anorthosite was common in the Bergen arches of geological formation (Bakka 1964; 1993:65). Bakka's view on this matter is supported by Rolf Birger Pedersen, who has inspected several specimens of anorthosite from the excavated sites. The distribution pattern of this raw material, which shows that it was most frequently used at sites in Nordhordland (Fig. 91), and less frequently in adjacent districts, indicates that this assumption is correct.

Flint and pumice

Flint and pumice do not occur in the bedrock in Norway. Flint was probably found as nodules dropped by sea-ice along the outer coast and is therefore not traceable to specific sources. (E. Johansen 1957, Pettersen 1986, Berg-

Hansen 1999). Deposits of pumice are found along the ancient shorelines along the outer coast of Norway. They probably originate from a series of volcanic eruptions on Iceland, and have drifted to Norway across the North Sea (Noe-Nygaard 1979:129).

According to Kristian Pettersen (1986), a large number of icebergs carrying flint are likely to have drifted ashore on the long stretches of beaches at the large islands in the northern part of western Norway and in Sør-Trøndelag. Flint deposits may therefore have been more numerous in this area compared to further south, where the beaches were smaller and less frequent. The same argument may apply for pumice.

Quartzites

Quartzite occurs less frequently than quartz, but the outcrops are generally larger. Similar to quartz, it is visually and geologically difficult to distinguish, but in addition to silicates, it sometimes consists of characteristic minerals, which may be distinguishable. 15 different quartzites have been distinguished visually at the sites, but only one of them, fine green quartzite 1, can be convincingly related to a specific quarry.

Fine green quartzite 1 was first distinguished at Gyrinos (Martens & Hagen 1961) and later at sites in the mountains of Lærdal (A. B. Johansen 1969; 1978). It has often been termed "Lærdalskvartsitt" (quartzite from Lærdal) in the literature. One quartzite-outcrop, Kjølskarvet, quarry II, which is situated in the alpine-zone above the inner part of the Sognefjord (Lærdal), has extensive traces of prehistoric quarrying (A. B. Johansen 1978:143-151). Samples from the source are geologically described by geologist Finn Skjerlie on the basis of microscopically studied thin-sections as a "mylonitizised quartzite" formed as a part of the Jotun complex. It consists of 80-85% quartz, 15-20% epidote and small amounts of sericite and amfibolite (Bjørgo 1981:146). The Jotun-complex is also represented at the west coast along the Sognefjord (Milnes et al. 1988). Theoretically, the fine green quartzite may originate there. However, thin sections of three samples from coastal sites (Ramsvikneset, Flatøy III and Nordeide) also show similar mineralogical composition as the samples from the quarry (Bjørgo 1981:148). The samples from the reference collection also show significant visual resemblance to samples from Kjølskarvet II. The raw material has been identified in large quantities at sites close to the quarry and less on the coast (Fig. 64). One may therefore argue that the fine green quartzite 1 in the present work originate at Kjølskarvet quarry II.

The remaining quartzites can only be related to possible source-areas on the basis of the quantitative regional distribution patterns in Chapter 7 (Fig. 65-77):

Fine brown quartzite 1 has only been identified at the sites Mørkedøla I and Gyrinos IV (Fig. 65). It is most likely quarried in the mountains of Lærdal and Hemsedal.

Fine black quartzite 1 dominates at Masnesberget, and is only minimally represented at other sites (Fig. 66). The source is probably somewhere in Sogn.

Fine blue quartzite 1 has been identified at Kotedalen, Flatøy VIII, and Nilsvik 4. It is also found at Styggvasshelleren (Fig. 67). The fairly large amount at Styggvasshelleren may indicate that the source is located in Breheimen. It may, however, also be situated in Nordhordland/Midthordland.

Fine blue quartzite 7 is only found in small quantities, mainly at sites to the north of Sogn (Fig. 68). Its source is most likely situated somewhere along the northern part of the coast.

Fine blue quartzite 8 has been identified at three sites, Tjernagel 3, Austvik III and Nilsvik 17 (Fig.69). Its source is probably located in Midthordland or Sunnhordland.

Fine grey quartzite 2 is represented with almost 25% of the total at 17 Havnen, and is only sparsely found at other sites (Fig. 70). It is therefore likely that its source lies somewhere in Nordfjord.

Fine grey quartzite 3 has only been documented at 760 Finnsbergvatn and the adjacent site 525 Nordmannslågen (Fig. 71). It was probably quarried at Hardangervidda.

Fine grey quartzite 11 is only identified at the sites Austvik III and Sokkamyro (Fig. 72), and may have been taken out in Sunnhordland.

Fine white quartzite 16 is found in small quantities throughout the area of analysis (Fig. 73). It is not possible to relate it to any potential source areas.

Medium grey quartzite 1 is only found at sites in Sunnfjord and Sogn (Fig. 74). It is therefore probably quarried in one of these districts.

Medium grey quartzite 9 is found at sites between Nilsvik 17 in Midthordland and 26 Valderøya vest in northern Sunnmøre (Fig. 75). A relative dominance at Kleiva and 1 Haukedal may indicate that the source may be situated in Nordfjord or Sunnfjord. Another possibility is Breheimen, considering the quite large amount at Styggvasshelleren.

Medium grey quartzite 10 and 11 are only found at sites in Midthordland and Sunnhordland (Fig. 76). The sources are most likely found in one of these districts.

Coarse quartzites are found at a large majority of the sites, and the percentages vary (Fig. 77). Considering that it is a grouped category, it is not possible to evaluate a possible source area.

Quartz, quartz crystal and chalcedony

The sources of quartz, crystal quartz and chalcedony are difficult to trace. First, the mineralogy and geochemical composition of the different quartzes are very similar. Second, they are extremely frequent and poorly mapped. Third, there seem to be large variations both in colour and grain-size within each quartz outcrop. Fourth, the quartz-outcrops are generally very small. This means that prehistoric quarrying is likely to have low archaeological visibility. Within the area of analysis, six quarried quartz outcrops are known (Fig. 115): Horninndal, Nordfjord (chalcedony) (Kaldhol 1947), Gloppen, Nordfjord (B13550), Stongeskaret, Lærdal mountains (B 15989), Preinsrudtind, mountains of Hemsedal (A. B. Johansen 1978:70), Voss, Hordaland (Gustafson 1983), and Kalhovdfjorden, lok. 1114, Telemark (Indrelid 1994:264) (C 34609 and C35215). None of the samples of quartz in the reference collection can be convincingly related to any of these quarries.

Quartz crystal, fine white quartz and coarse and medium quartz are all grouped categories. They are represented at almost all of the sites and they are found in varying quantities (Fig. 78-80). Possible source areas can not be evaluated on the basis of this data.

Mylonites

Mylonites were first distinguished by Asle Bruen Olsen during his surveys and excavations in Flora (A.B.Olsen 1981; 1983), and have later been distinguished at a number of Mesolithic and Neolithic sites at the coast of western Norway (Bergsvik 1999). Mylonites and cataclasites occur in fault-zones in most of the region, but they are poorly mapped. Surveyed outcrops of mylonites are particularly frequent in the area to the east of Florø. These generally have a flint-like quality, and are excellent for production of struck stone tools. One outcrop of fine, green cataclasite (mylonite) was found by Asle Bruen Olsen near to a Stone Age site at Leversundet in 1979 (B-no. 13929). Later I have also surveyed a number of similar outcrops in this local area, which is geologically distinguished as "charnockithic gneiss, often mylonitic" (Kildal 1970). Close by, on the western shore of Brandsøysundet, one outcrop of blue mylonite has been located by the amateur archaeologist Svein Brandsøy. Despite of extensive surveying no quarries have yet been located.

The mylonites have not been geologically tested. However, the blue, green, and grey mylonites in the reference collection strongly resemble the samples from the outcrops in Sunnfjord. This may indicate that the mylonites were mainly collected or quarried in this district. The distribution pattern of the mylonites shows that they were frequently used in Sunnfjord and that there was a gradual fall-off towards the north and south (Fig. 81, 82). This also supports the theory that the mylonites originate in Sunnfjord.

Conclusions

The aim of this chapter was to perform supplementary investigations of the previously known lithic raw material sources and to localise new sources or source areas for other raw materials. This was done partly based on new geological analyses, partly based on regional distribution patterns. Three levels of precision can be distinguished.

(1) Undetermined. Several of the of the raw materials distinguished are only grouped categories which, most likely, have been quarried or collected at a number of different locations. Greenstone, flint, pumice, fine white quartzite 16, coarse quartzite, and the different varieties of quartz belong to this category. Greenstone has been subjected to geological tests for the present work in order to distinguish between adzes that have been quarried at Bømlo and an (for the time being) unidentified quarry 4. The geological isotope analysis did not match the visual classification of these adzes, contrary to S. Alsaker's result (1987). Consequently, Bømlo greenstone and quarry 4 greenstone are grouped in the visual classifications.

(2) Based on distribution patterns. The majority of the raw materials can be related to source areas on the basis of visual classification based on the reference-system and distribution patterns. These have not been geologically tested or determined. This category consists of the majority of the quartzites, chert, the mylonites, and anorthosite.

(3) Based on geological analysis. For some raw materials with known sources, geological analyses have been performed of both sources and artefacts, and the results indicate close resemblance. These are diabase, rhyolite, fine, green quartzite 1, sandstone and slate.

The geological tests of the diabase, rhyolite, sandstone and slates should also be seen as a test of the reliability of the visual classification of these raw materials and ultimately of the reference-system as such. The geological analyses show that few of the artefact samples did not match samples from the presumed source. It can, therefore, be concluded that the visual classification can be seen as a fairly reliable method for most of these raw materials.

These results are important for the below interpretations of boundaries and connections, primarily because they make up the empirical basis for a discussion of the character and direction of the regional flow of artefacts.

Chapter 9

Interpreting the boundaries

Introduction

In the preceding chapters I analysed the spatial distribution of a number of early Neolithic cultural traits and practices. On the basis of this regional analysis of raw materials, types, and techniques I tentatively concluded that, on the coast, five districts could be distinguished: (1) northern and southern Sunnmøre, (2) Nordfjord, (3) Sunnfjord and Sogn, (4) Nordhordland, and (5) Midthordland and Sunnhordland. There were major differences between most of the mountain sites and the coastal sites in general. Below, the boundaries between these districts will be investigated further in order to decide whether they were constructed on the basis of ethnic differences. As I suggested in Chapter 3, four archaeological criteria should be met if ethnicity was to be regarded as a relevant interpretation: (1) co-variation of cultural traits, (2) sudden fall-off in cultural traits or raw material use, (3) crossing, and (4) historical continuity.

The different boundaries will be evaluated on the basis of these criteria. I will explore in detail alternative interpretations of the differences between the districts. In the case of historical continuity, I will mainly draw on the results of Guro Skjelstad (2003), who has analysed the middle and late Mesolithic site assemblages in the coastal region applying the same methodological tools as in the present work. In the final section, I will discuss sedentism as a factor in the development of ethnicity in the coastal region.

Coast - mountain

There are significant typological differences between coastal and mountain sites. First of all, the adze/axe types are different. The mountain sites only have thin-butted flint axes, while the Vespestad adze dominates heavily on the coast. The struck projectile-points are also different. On the coast, blade tanged points dominate almost completely. Only a few transverse points are found. The mountain assemblages, except Styggvasshelleren and Blånut IV, on the other hand, consist of a variety of types: blade points, flake points, single-edged points and transverse points. With regard to the slate points, there are some similarities: early Neolithic types are found at both mountain sites and coastal sites. Points of the Pyhensilta type are, however, only found at the mountain sites.

Technologically, the situation at the mountain sites is also very different from the coastal sites. Cylindrical cores are only present at one site. Together with the fairly low number of small blades and blade points, this indicates that cylindrical core technique was hardly practised in the mountains. Bipolar technique has a relatively low frequency except at Styggvasshelleren. The category other cores is well represented at several sites. This should probably be interpreted as indicating that flake technique was the most important way of lithic reduction at the mountain sites. Slate was reduced at several of the mountain sites.

With regard to raw materials, the differences between the coastal sites and the mountain sites are very distinct. A few raw materials from the coast: rhyolite, anorthosite, and red slate occur at some of the mountain sites. These raw materials are, however, only minimally represented. Fine green quartzite 1, while very common at the some of the mountain sites, has only been found as a few flakes on the coast. Some more common categories are present at both mountain and coastal sites: flint, quartz crystal, fine white quartz, medium and coarse quartz, green slate and grey slate. Slate may, in some cases, be traced back to the west coast (see Chapter 8). Flint may also originate in western Norway, but may just as likely have been acquired through eastern networks. Quartz crystal and quartz cannot be related to specific sources.

There were also differences during the late Mesolithic. Svein Indrelid has pointed out that contrary to the west coast, flint and fine quartzites dominate at the mountain, and transverse points are only found at the mountain sites. At a more general level between eastern lowland and western coast, the Nøstvet-adze is only found in the east, while the point-butted adzes with round cross-sections dominates in the west (Indrelid 1994:279 ff). There are also significant differences in the rock-art traditions (Fig. 120).

On the basis of this evidence it may therefore be concluded that the four criteria: co variation, sudden fall-off, crossing and historical continuity are all satisfied.

Below, I will discuss the social significance of this boundary. First, however, it is necessary to decide where the mountain-people came from.

Who used the mountain sites?

It is generally held that the mountains between eastern and western Norway were not inhabited on an annual basis during the Neolithic, i.e. that these sites – contrary to the large coastal habitations, were used on a short term basis. Still, there is disagreement on the question of origins; whether they came from western Norway, from the eastern river valleys or from the east coast. Regarding the sites at Hardangervidda, Johs. Bøe (1942:82) argued that they had been used by western populations, but this was countered by Gutorm Gjessing (1945:188), who suggested that they had originated at the eastern lowlands. In line with this, Anders Hagen and Irmelin Martens have partly argued for a southern origin, partly that they were interior hunter-gatherers (Martens & Hagen 1961). Svein Indrelid (1973:95; 1994, Indrelid & Moe 1983) has suggested that Hardangervidda was mainly used from the west, but he also argued for utilisation from the west coast, which would imply a degree of social interaction at Hardangervidda between eastern and western groups (Indrelid 1994:302). A theory of a mainly eastern connection is also supported by Trond Klungseth Lødøen (1995). The sites at Lærdal were, according to Arne B. Johansen, mainly used from the west (A. B. Johansen 1978:283). A western origin is also argued by Kjersti Randers (1986:95) for the sites in Breheimen. Similarly, A. B. Olsen (1992:248 ff) has argued that sites at Hardangervidda, the mountains of Årdal and Lærdal were mainly used by groups from the west coast.

In their discussions of the mountain sites, A. B. Johansen (1969:98) and Indrelid (1973:96) have pointed out that functional differences between sites may result in significant differences in the lithic industries at sites left by the same group of people. In line with this, one might argue that the differences between the mountain sites and the coastal sites could be explained as a result of different activity patterns by the same coastal populations. However, such an interpretation is not supported here. I do not find it likely that task groups from hunter-fisher settlements on the coast would have left all of their raw materials, types and technologies behind when they travelled to their field camps in the highlands. A certain degree of technological, typological and raw material similarity is to be expected if sites were occupied by the same groups.

A significant amount of coastal traits are only found at Styggvasshelleren, at which several raw materials originate on the west coast: anorthosite, rhyolite, fine black quartzite 1, and green slate. Typologically, it also resembles the coastal sites by the presence of blade tanged points and absence of single edged points and transverse points. Technologically, the site is dominated by bipolar technique, which is common at the northern part of the coast. In the previous chapter I also showed that rhyolite, slate and probably anorthosite at this site originated on the west coast. In addition, it is also important to note that a Mesolithic adze with round cross-section of diabase from Stakaneset was found in an older stratigraphical layer at Styggvasshelleren. A Vestland adze of basaltic rock was found at the nearby site Hella. Furthermore, bones of marine fish as well as, pumice and blue mylonite were clearly carried from the west coast (Randers 1986:87 ff). Thus, I agree with Kjersti Randers who concludes that Breheimen was probably used by western groups. With regard to the problem of site types, the composition of tools at the site does not distinguish Styggvasshelleren significantly from the coastal sites in this analysis. This might perhaps be expected, considering the field camp character of the site, which is indicated by the location. A special purpose character is only identified archaeologically by the large number of slate points. The relative lack of blanks and almost total absence of slate flakes strongly indicate that these points were prepared beforehand and carried to Styggvasshelleren as finished products.

As for the remaining mountain sites, it is less likely that they were used by people from the west coast. If coastal task groups had occupied them, I find it odd that they only carried with them flint and left out all of the raw materials that were, in fact, much more common than flint on the west coast. Secondly, it does not make sense that they did not carry Vespestad adzes from the coast. One might argue that they did not need adzes during these forays. However, the fairly large number of ground flint fragments (presumably of thin-butted axes) at the sites demonstrates that such tools were, in fact, common at the mountain sites. Thirdly, I also find it strange that the composition of projectile points deviates so much from the west coast. It is possible that reindeer hunting required different points than was needed in the western lowlands, but one would still expect a much greater typological overlap between the mountain and the coast if it was the same people who used the sites. It should also be remembered that at Styggvasshelleren, which was interpreted as a site used by western groups, had exactly the same repertoire of points as the sites at the west coast. Finally, it is highly surprising that the technologies that were common on the coast were not applied. The general raw material surplus at the mountain sites may explain why bipolar technique was not applied to any extent, but it does not explain the absence of cylindrical core technique. To my judgement, several of the fine quartzites that dominate at the mountain sites could successfully have been reduced by cylindrical core technique.

On the basis of this discussion, I find it unlikely that the mountain sites other than Styggvasshelleren were used by groups from the west coast. A much more likely alternative is that they were used by groups from eastern Norway. Typologically, such a theory is supported by the fragments of thin-butted flint axes, a type that is frequently found in eastern Norway. The composition of projectile points also supports this theory: tanged points, single-edged points as well as transverse points occur at several of the late phase 4/early Neolithic sites in this region (e.g. Glørstad 1998). In this respect, it is important that a large portion of the tanged points at the mountain sites were made from flakes, not blades. As noted in Chapter 7, phase 4/early Neolithic tanged points found in eastern Norway are also mainly made from flakes. This indicates that the projectiles at the mountain sites were largely made as eastern types, using eastern techniques. The flake technique, which was common at the mountain sites, also dominated in eastern Norway in this period. Thus, in line with Indrelid (1994:278), I conclude here that these mountain sites were part of an exploitation of the highlands from eastern or south-eastern Norway.

Having established the main cultural influences of the mountain hunters, I will now explore the possible ethnic relations of these groups with the coastal populations in western Norway. I will also discuss the character of the connections between the mountains and the eastern coast and interior.

East-west relations

All of the sites at Hardangervidda in this analysis are located in the middle of or in the eastern part of the plateau. A mainly eastern connection is therefore not surprising. The western part, which is closer to the fjord systems, has been surveyed, but few sites have been found and none of these are early Neolithic (Indrelid 1994:259). As Indrelid points out, one possible explanation for the lack of sites here may be that the reindeer mainly inhabited the eastern part of the plateau (Indrelid 1994:258-260). In any case, there is no certain evidence that the southern highlands were exploited from the west in this period. Still, west coast populations were not unfamiliar with the idea of mountain hunting. The site at Styggvasshelleren shows that the groups from western Norway exploited the mountains and the resources there. A few reindeer boned have been identified at the coastal sites Skipshelleren and Flatøy XI (H. Olsen 1976:41, Bjørgo 1981:105). Furthermore, knowledge of this animal in the west is illustrated by images of a possible reindeer at Vangdal, Hardanger (Bakka 1966:85) and at Vingen in Nordfjord (Bøe 1932: tafel 2, no.18). These carvings probably indicate that reindeer had an elevated position in the cosmogony of these people. Thus, I find it likely that the southern highlands were attractive hunting grounds for the western as well as the eastern populations.

What interests us here, however, is whether eastern and western forays were contexts also for social interaction between these different groups. As was pointed out above, coastal and mountain sites do not share many elements. The only indications of direct contact at the sites in this analysis are a few flakes and points of rhyolite at the Hardangervidda-sites, slate of western origin at Mørkedøla I, and flakes of fine, green quartzite 1, probably from the Lærdal Mountains, at some of the coastal sites. In his analysis of the Hardangervidda sites, Indrelid counted altogether 27 artefacts of rhyolite from Bømlo: three A-points, one retouched blade, two retouched flakes, one cylindrical core and 18 flakes (Indrelid 1994:302). At the site 512 Nordmannslågen, which is the site at Hardangervidda with the largest quantity of rhyolite, this raw material constitutes less than 0, 5% of the total. At the contemporary sites on the west coast, rhyolite dominates heavily. As far north as Nordhordland, which is situated at an almost equal distance from the quarry at Bømlo as Hardangervidda, rhyolite represents between 45 and 60% at the sites. These low shares of rhyolite at Hardangervidda are indeed weak indications of direct contact between east and west. It is even possible to imagine that some of the artefacts, at least the ones that are found in the southern and eastern part of the mountains, came through eastern coastal networks. The rhyolite may, for example, have been transported from Bømlo to the east along the southern coast and taken to Hardangervidda by eastern groups. Such a theory may be supported by the fact that that one of the two tanged points of rhyolite at 760 Finnsbergvatn is made from a flake, which was the eastern way of producing these points. The same coastal network is also most likely responsible for the spread of the few sherds of TRB pottery and the 15 thin-butted axes that have been found in western Norway.

In a context of ethnicity, this lack of contact across the plateaus is interesting. As will be discussed in the succeeding paragraphs on the coastal situation, the marked differences are always combined with extensive contacts across the boundaries. This evidence for contacts and relations makes arguing for ethnicity as a factor possible. The east-west situation is different. Above, I have shown that there were profound cultural differences. I have also shown that there was little or no contact, at least not directly, between the coast and the plateau. This indicates that they were mutually exclusive, culturally different populations. Instead of being an ethnic relation, it is quite possible that the differences between east and west are rather a result of different populations being integrated in separate contact networks (e.g. Sackett 1985). There were few or no contexts for the exchange of information about technologies, types and other cultural values. The mountain plateau and the mountain ridges between east and west may thus be seen as insurmountable barriers and a major cause for the isolation.

In a recent contribution, Trond Klungseth Lødøen has presented arguments for such a view. Lødøen test-pit surveyed intensely along the Sognefjord, and the meagre results lead him to conclude that Mesolithic and early/middle Neolithic habitation sites were generally lacking in the middle and inner parts of the fjord (Lødøen 1995:60-63). The same lack of habitation sites can also be observed for the inner parts of the other large fjord systems in this region: Nordfjord and Hardangerfjord. The distance from the main areas of habitation in the coastal and mountain districts is considerable. A number of stone adzes had, however, been delivered to the museum from the middle and inner parts of the Sognefjord. Lødøen found that several of them had been found at strange and highly visible locations; elevated from the shoreline and under or beside large stones or boulders. Some had also been found in brooks and screes. Lødøen argued, on the basis of the odd locations and on the general lack of habitation sites, that a number of these adzes probably were votive offerings (Lødøen 1995:87-105). Furthermore, he suggested that the fjord systems have been regarded as "ritual landscapes" which were accessible only for chosen groups and individuals from the coast during rites de passage and "vision quests". The fjords, let alone the plateaus, may consequently have been peripheral in the consciousness of the majority of the population (Lødøen 1995:106-117). In this perspective, the fjord systems are obstacles instead of channels of communication. Only in rare cases, when particularly daring shamans or neophytes found their way to the plateaus, would east-west relations be established. But the stories of such infrequent encounters would probably have little or no impact on the construction of ethnic identities and out-group categories on the coast.

Another explanation for the lack of contact may be hostile relations. As argued above, the mountain resources – reindeer in particular – were probably regarded as valuable for the hunters at either side of the plateau, and it is likely that the western groups would have hunted there if they could. It is quite possible that the highland territories were dominated and controlled by the eastern populations, who have denied other groups access at Hardangervidda and other alpine regions between east and west. If this theory is correct, one might ask why Breheimen was not included in such an eastern domain. As is pointed out by several scholars, the reindeer populations were fairly small on the western side compared to the east (e.g. Indrelid 1977, Gustafson 1980, Randers 1986:91). Breheimen is nevertheless likely to have been out of reach by the eastern hunters because of the long distances across the intermediate Jotunheimen, which is an area without reindeer, and this may have left it open for exploitation from the west. A hostile relation is, after all, a relation. This implies that the populations knew about each other and related to each other as groups. In order to maintain the dominance and control over a territory, and also to justify a lack of access, a great deal of categorisation of the other – often in negative terms – was probably needed. Therefore, if the situation was one of competition and perhaps hostility between the east and west, the relation may perhaps best be termed as ethnic.

In this perspective, the fjords cannot solely be interpreted as ritual landscapes. They should also be regarded as practical landscapes for travelling and hunting. This does not mean that Lødøen's theory of the fjords as accessible for only selected persons is wrong and that the fjord-adzes were not votive offerings. It is perfectly possible to imagine task group travels to Breheimen and Lærdal-Hemsedal through the Sognefjord both as hunting expeditions in a utilitarian logistical sense and at the same time as a means – through different types of rites de passage – of initiating young boys or girls into manhood or womanhood.[1] Such initiation rites might, for example, have consisted of an element of reindeer hunting. Another important element would be to approach strangers in the correct way. During such forays, it would be essential to leave certain marks on the land in order to create a social landscape, to pay tributes to the paths that were travelled by the ancestors, but also, as Lødøen himself has discussed, to symbolise the territorial domains of the coastal populations. Based on this discussion, I suggest that ethnicity played a role in the relations between the eastern and western groups.

Relations between the mountain and the eastern lowland

It was concluded above that the majority of the mountain sites were used by groups from the eastern lowlands, first of all because of the significant technological and typological overlap with the eastern assemblages in general. With regard to raw material use, however, the mountain sites differ greatly from each other and also from most other early Neolithic sites in lowland and coastal eastern Norway. Thus, the specific origin of these people is difficult to decide.

The large amount of flint at Sumtangen was taken by Bøe to indicate that the earliest hunters used the mountains during seasonal migrations from the coast (Bøe 1942:82). Some years later, however, the excavations at Gyrinosvatn uncovered sites with only small amounts of flint. Martens

1 See Bergsvik (2002b) for a development of this theory in a context of social inequality.

and Hagen argued that the sites had been occupied by groups that stayed in the interior on an annual basis, but that they had contact with farming populations in the south (Martens & Hagen 1961:46). Some years later, Indrelid held that Hardangervidda was most likely used by migratory herding groups from winter bases in the lowland around the Oslo fjord. The main arguments for this theory is the early evidence for ribwort (Plantago lanceolata), nettle (Urtica), and mugwort (Artemisia), the dominance of flint, the presence of TRB pottery at two sites, and the flakes of thin-butted flint axes at several sites at Hardangervidda. Several thin-butted axes had been found in river systems leading up to Hardangervidda from the Oslo fjord (Indrelid & Moe 1983). Recent ethnological data on long-distance movements of cattle in eastern Norway also shows that it is possible to carry out such a mobility pattern in practice (Hougen 1947:86-88, Indrelid 1994:294:266). Alternatively, according to Indrelid, Hardangervidda had been exploited by populations in the interior who had adopted a pastoral economy from the agricultural groups on the eastern coast (Indrelid 1994:294, 295). In his thesis on the Stone Age of Telemark, Mikkelsen also argued that mobile hunters/herders alternated between upper and lower parts of Telemark during the early Neolithic. This theory is based mainly on pollen from ribwort and on TRB flint axes (Mikkelsen 1989:169).[2] Both Indrelid and Mikkelsen argue that during the preceding late Mesolithic period, upper Telemark and Hardangervidda were primarily used by residentially mobile hunters who had their winter settlements on the east coast or by fjords close to the coast.

During the last few years, large excavation projects have been carried out at Svevollen, Hedmark (Fuglestvedt 1995), Rødsmoen, Hedmark (Boaz 1997) and Dokka, Oppland (Boaz 1998) (Fig. 1). These projects in the interior lowland of eastern Norway have produced results which are important for the evaluation of the eastern relations. Fuglestvedt and Boaz have suggested, in line with Martens and Hagen, that the sites at Svevollen, Rødsmoen and Dokka in the late Mesolithic and the early Neolithic were most likely the results of inland hunter-gatherers who mainly hunted elk and beaver. For Svevollen and Rødsmoen, this theory is supported by the presence of numerous house depressions, accompanied by large amounts of fire-cracked stones (for boiling) which are interpreted as winter dwellings. The house-sites are part of an inland phenomenon which stretches far into present day Sweden (e.g. Lundberg 1997). The sites at Dokka lack house-pits, but heaps of fire-cracked stones are found. A mainly interior attachment is also indicated by a marked dominance of locally procured lithics. Pollen analyses from Dokka and Rødsmoen has not produced any plants that are favoured by herded animals. Therefore, these sites at these locations are interpreted as hunter-gatherer settlements (Boaz 1997:143). Significant amounts of flint are present at both Dokka and Rødsmoen. There are also typological similarities with the coast, particularly with regard to projectile points. In addition, fragments of thin-butted flint axes are identified at Dokka as well as Rødsmoen.

If the evidence from the different interior mountain, lowland and coastal assemblages are seen together, the situation is complex and diversified. Different, often locally procured, lithic raw materials are used in different districts, and there is also a variety of food-preparing techniques and architectural solutions. The economical basis varies considerably, between elk/beaver hunting, pastoralism and combinations of hunting, pastoralism and cultivation. At the same time, similar types of projectile points are used throughout the area, thin-butted axes are found everywhere, and flint from the coast has a major share of the lithics in most districts. Joel Boaz has drawn three important conclusions from his investigations of the interior: (1) that the coast and the interior in eastern Norway were inhabited by distinct populations already during the late Mesolithic, (2) that different, indigenous groups exploited different parts of the interior, and (3) that these different interior groups had close relations with coastal populations (Boaz 1998:348).

These new results and interpretations are important for the evaluation of the eastern connections of the mountain sites. On the basis of the above, I believe that Indrelid's alternative mobility pattern of farmers/herders between the coast and Hardangervidda is unlikely to have happened. I admit that the proportions of flint, TRB pottery, thin butted axes and the possible herded animals may support such a theory. My objections are merely circumstantial. First, considering that there were probably much better possibilities for grazing in the lowlands than in the mountains, it makes no sense that farmers would take domestic animals through dense primeval forests back and forth between the Oslo fjord and the plateau. Second, the recent results from eastern Norway referred to above indicate that different areas were inhabited by populations that were culturally different. If populations of farmers inhabited the inner parts of the Oslo fjord, it may be expected from the above

2 Recently, several botanists and archaeologists have questioned the use of pollen data, particularly the presence of Plantago lanceolata, as a secure indicator of a pastoralism. It is argued that, although coinciding with a type of landscape that was produced by agriculture and grazing animals, they may just as likely be present as a result of near-site factors such as firewood collection and trampling, and they may have been spread by hunter-gatherers who had been in contact with agricultural groups (e.g. Behre 1981, Welinder 1985, Rowley-Conwy 1995, Prescott 1996). This critique is important, but no one has yet aimed it at Indrelid's and Mikkelsen's work in particular. Until a critical evaluation of their interpretation of the pollen data from Telemark and Hardangervidda is performed, I regard the data as possible evidence for early Neolithic pastoralism in these areas.

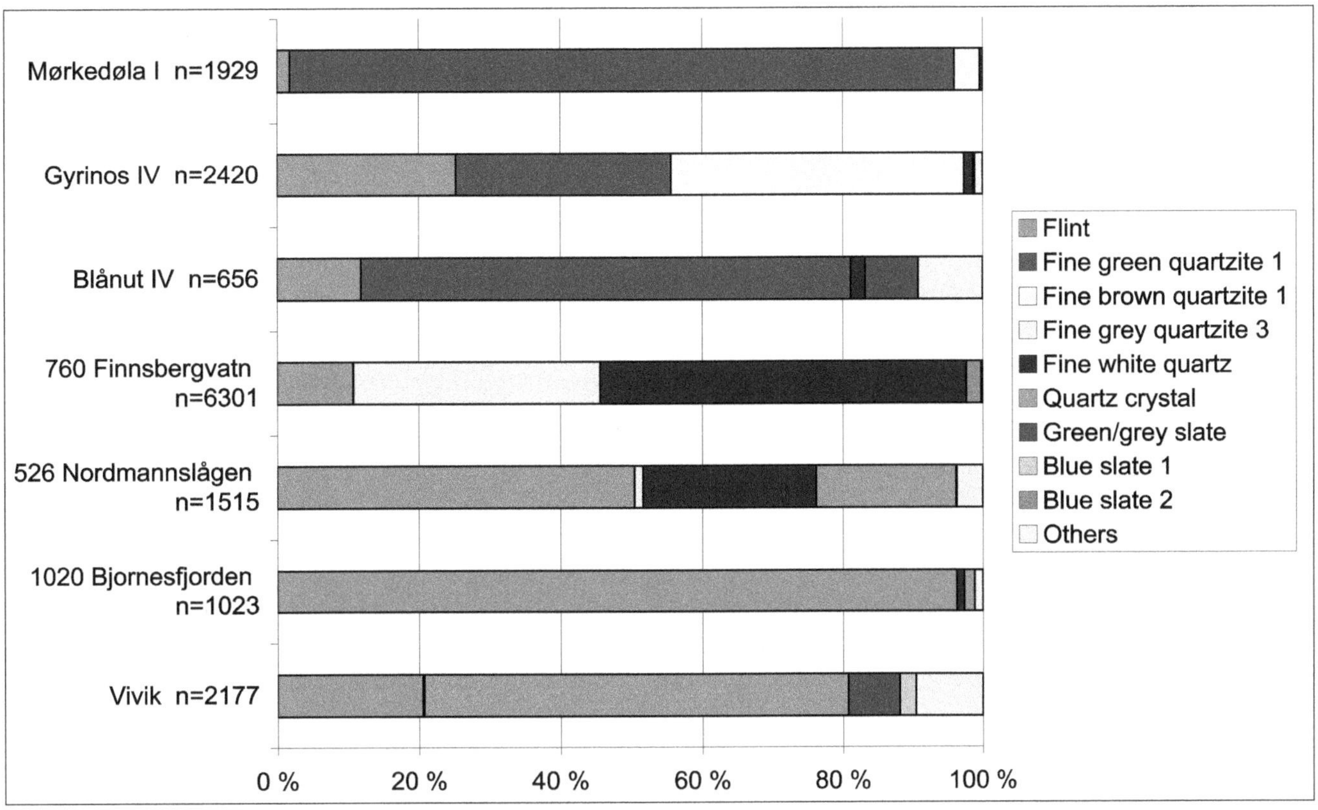

Figure 118. *Percentages of different raw materials at the mountain sites*

that these may have been ethnically different from the foraging populations in the north and west. If this was the case, I find it unlikely that these agricultural groups had direct access to inland resources and that they had safe conduct to areas that were situated far from home, deep into the domains of the interior hunter-gatherers. Instead, I find that Indrelid's and Mikkelsen's alternative of the plateau being used by hunter-gatherers or hunters-herders from the nearby lowland eastern interior valleys a more plausible interpretation of the mountain sites.

Interestingly, the sites in the mountains are internally quite different, particularly with regard to the use of raw materials (Fig. 118). Already in 1961, Hagen remarked on a marked tendency towards an increased use of quartz, quartzite and slate at the northern sites compared to the southern sites, where flint is more commonly found (Martens & Hagen 1961:46). A. B. Johansen has also pointed out that the different sites at Lærdalsfjella may have been utilised by groups that had different raw material preferences (A. B. Johansen 1978:297). If Boaz's assertion of a territorial subdivision of the interior is correct, these different populations would have had access to different parts of the highland; local groups in Telemark and Numedal, who had adopted live-stock keeping, were perhaps the ones who exploited Haukelifjell and Hardangervidda. The sites further to the north were exploited by groups who remained solely hunters-gatherers.

Åsa Lundberg (1997) has convincingly argued that the interior river valleys of Norrland, Sweden were occupied by separate "regional bands" during the late Mesolithic and Neolithic periods. As will be pointed out below, the west coast was subdivided by several ethnic boundaries. On the basis of the available data from eastern Norway, I find it likely that such boundaries existed in this region as well, perhaps between groups who lived in the major river valleys, and that these groups controlled different parts of the highlands.

These suggestions of different eastern areas of origin for the mountain sites are, of course, tentative, and should only be seen as a working hypothesis for future research in eastern Norway. In order to have more reliable data on the character of these relations, the lithic material from the lowland interior sites should be compared to the mountain assemblages as well as to the sites at the eastern and southern coast of Norway.

Coastal similarities and differences

Within the coastal region, the cultural differences are not as marked as between the coast and the mountain. A number of similarities are apparent. In Chapter 4 I concluded that there were no significant differences in site locations neither at the micro level nor at the macro level. This was interpreted to indicate that the landscape perceptions were similar and could be related to similar subsistence-settlement patterns. Previous research on

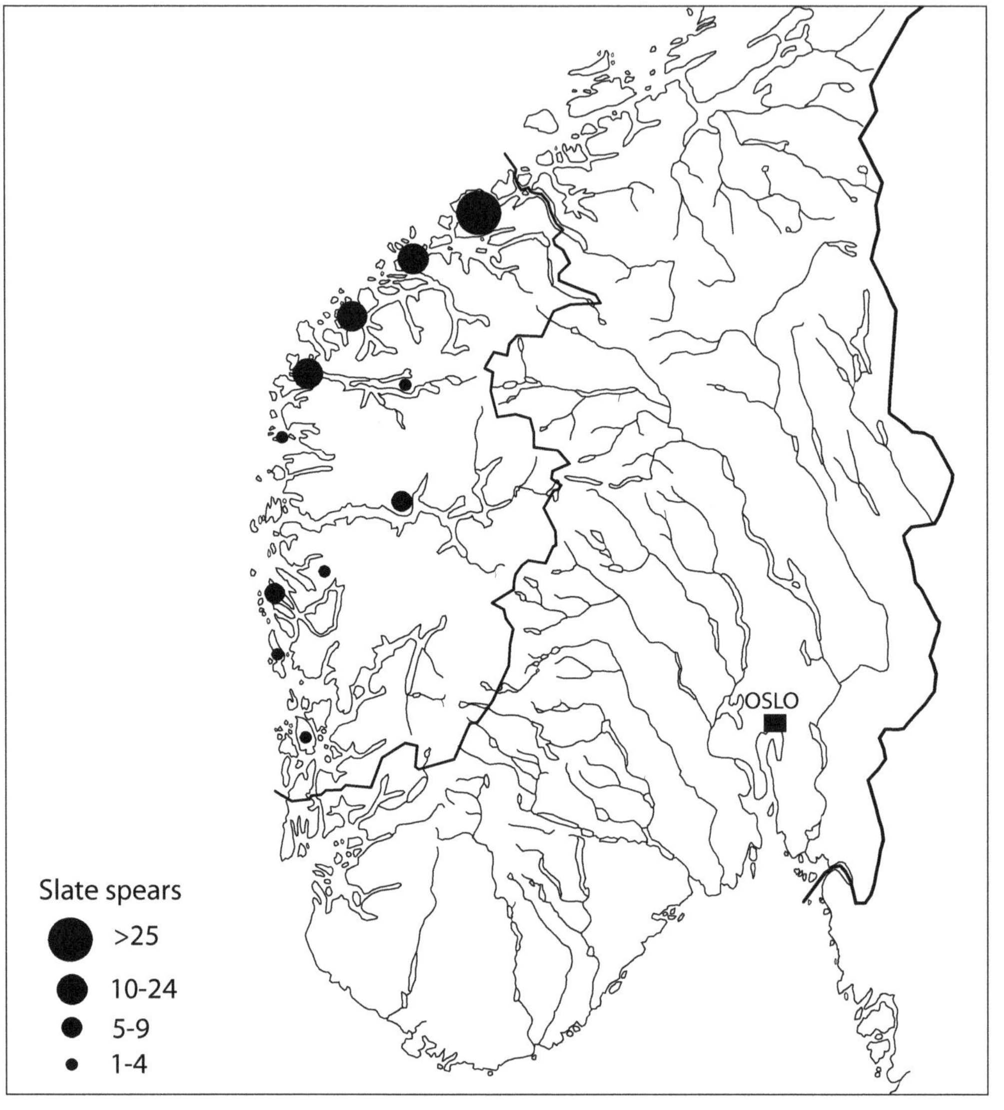

Figure 119. *Distribution of slate spears in the counties Møre og Romsdal (not Nordmøre), Sogn og Fjordane and Hordaland. Data from Romsdal is exstracted from Ramstad (1999). Data from the remaining districts is produced for this work.*

subsistence-settlement patterns also strongly indicates that the early Neolithic populations in this region were mainly sedentary hunter-fishers who stayed at the coast most of the time and relied on a large variety of marine and terrestrial resources (e.g. Hufthammer 1992, A. B. Olsen 1992, Bergsvik 2001b). Widespread similarities are also present in the lithic data analysed here; many raw materials, types, and techniques are spread throughout the coastal region. On the other hand, I have also presented a number of cultural traits that were not shared along the coast. There were major distribution boundaries in the use of raw materials, types and techniques. In the following paragraphs, these boundaries will be examined.

Sunnmøre-Nordfjord

Technologically, sites in the two districts are very different: dominance of the bipolar technique has a southern boundary in southern Sunnmøre, while cylindrical cores have a northern distribution boundary in the same area. The distribution limit of some raw materials is clearly coincidental with this technological boundary. Medium and coarse grey mylonite, anorthosite, and coarse quartzite have a northern distribution boundary at Sunnmøre. In addition, several frequently used raw materials have a marked fall-off: flint clearly dominates at the Sunnmøre sites, and it drops to about 10-20% south of Stad. In the other direction, chert, fine, grey quartzite 1, red, green and brown slate, which are frequently used at the Sunnfjord and Nordfjord sites, have a marked fall-off at the boundary to Sunnmøre. Slate knives, which are dated broadly to the Neolithic period, also have a southern distribution boundary at Stad (Fig. 109) (Søborg 1988:226).

Guro Skjelstad has shown that several late Mesolithic traits also coincide with this boundary. The late Mesolithic residential sites in Sunnmøre are dominated by flint (> 70%) while contemporary sites in Nordfjord and Sunnfjord have 35% or less of this raw material. In addition, fine white/grey quartzes, which are common in Nordfjord and southwards, have a marked fall-off in

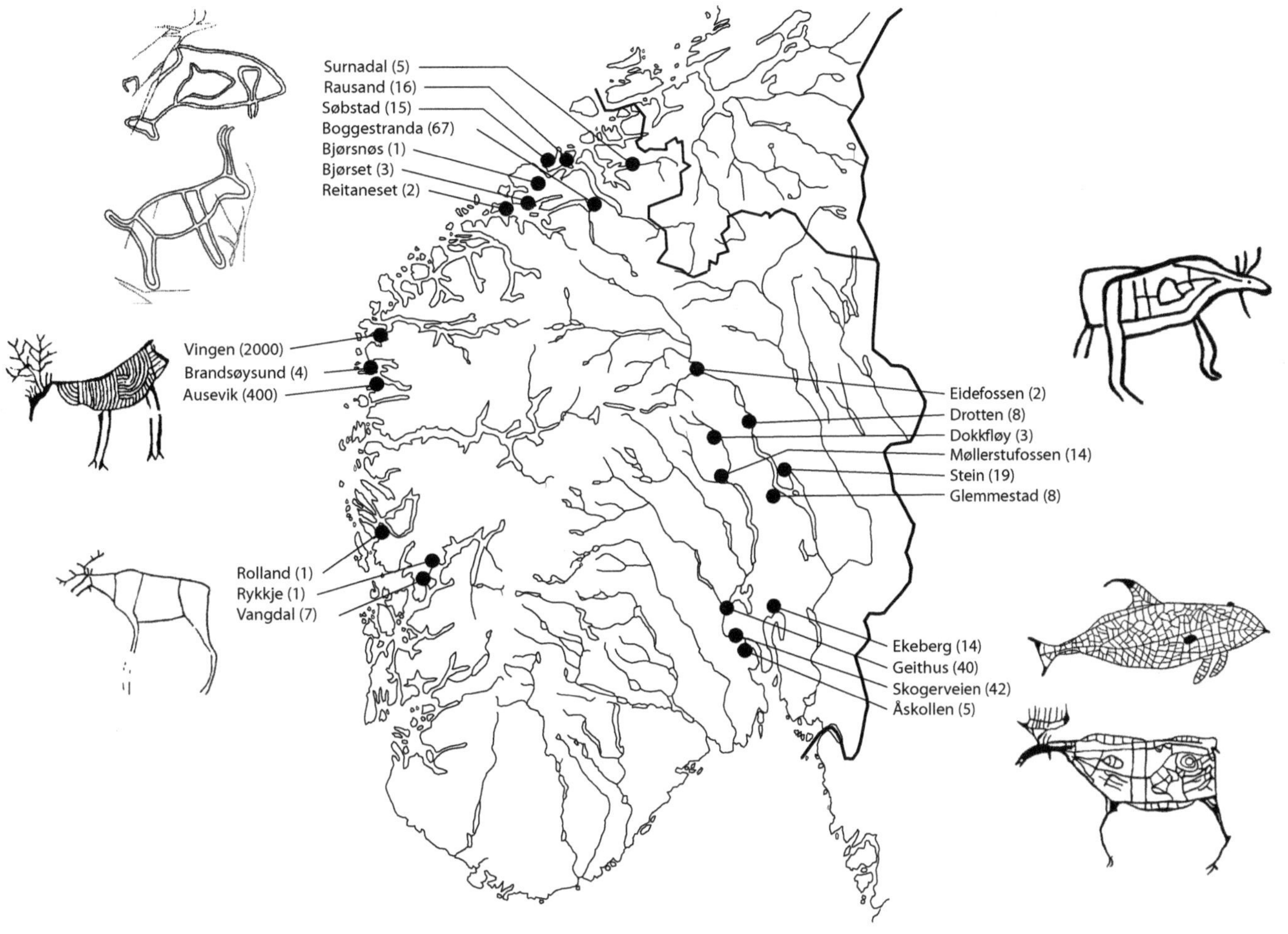

Figure 120. *Hunters rock-art sites (petroglyphs) in Norway to the south of Trøndelag. Data and animal figures reproduced from several authors (Bakka 1966, Hagen 1977, Mikkelsen 1977, Walderhaug 1998, Paasche 2000, Ramstad 2000).*

Sunnmøre. The same is true for sandstone. Blue and yellow mylonites, while dominating in Nordfjord and Sunnfjord, are absent in Sunnmøre. Soapstone net-sinkers are common in Nordfjord and southwards and rare in Sunnmøre (Skjelstad 2003:64-97).

The boundary also continues into the succeeding Middle Neolithic period· Slate spearheads[3], which have been classified for this project (see list in appendix 9), are also mainly found to the north of Nordfjord (Fig. 119).

Finally, there are major differences in the hunter's rock-art traditions (Fig. 120). In Sunnmøre, no petroglyphs have yet been found, but the Romsdal carvings consist of a variety of marine and terrestrial mammals; whales and deer. These are situated at several different sites which have only a few carvings each. In Nordfjord and Sunnfjord, on the other hand, the rock art mainly concentrate at two locations, Vingen and Ausevik, and both have large amounts of carvings, primarily images of deer. Only one small site is found at Brandsøysund (Bøe 1932, Walderhaug 1998, Ramstad 2000b).

The marked co-variation of distributions and raw material fall-off around Stad, and the historical continuity of these patterns should probably be interpreted in social terms. However, before a conclusion about the social significance of this boundary is drawn too firmly, it is necessary to consider (1) the dependency between raw materials and techniques, (2) techniques and social and ideological restrictions, (3) the southern boundary of slate knives, and (4) whether there was a continuous social landscape.

The dependency between raw materials and techniques

Flint and bipolar technique are the two variables that most clearly distinguish sites in Sunnmøre from sites in the districts further south, where platform technique and other raw materials dominate. It is, therefore, important to investigate whether these patterns are due to differences in raw material availability in these districts and whether the different raw materials required different technologies.

3 The only safely dated excavated units in western Norway with slate spearheads are middle Neolithic (Bergsvik 2002a:294). In eastern Norway, however, they also appear in early Neolithic contexts (Boaz 1997:94).

With regard to the availability problem, Kristian Pettersen (1986) has argued that a large number of icebergs carrying flint are likely to have drifted ashore on the long stretches of beaches at the islands in the northern part of western Norway. Flint deposits may have been more numerous in this area compared to further south, where the beaches were smaller and less frequent. Another important point is that, although outcrops of usable quartz were probably available, Sunnmøre largely consists of gneiss, which is often very coarse and not very fit for working stone tools (Sigmond et al. 1984). In contrast, numerous large outcrops of high quality slate, chert, mylonite, and chert were available in Nordfjord and Sunnfjord. The combination of high frequency of flint and scarcity of usable lithic bedrock sources may at least partly explain the dominance of flint at the Mesolithic and Neolithic sites in Sunnmøre, while scarcity of flint and abundance of flint-alternatives probably explain the situation to the south of Stad.

On the problem of the relationship between technique and raw materials, it is evident that flint as such may be treated with a number of techniques. Still, factors like the quality and size of nodules may restrict the possibilities of the knapper. While the flint deposits in Sunnmøre during the early postglacial times may have been more numerous than compared to other districts, it was still a limited resource, because of the exhaustible nature of the (ice-dropped) beach deposits. Consequently, it is therefore to be expected, as Anathon Bjørn (1921b:44) noted a long time ago, that during the latter part of the Stone Age, flint collectors in this area would have to manage with nodules that were both inferior and smaller than those of their predecessors. Such a theory is supported by data from the Sunnmøre sites. At some of these sites, particularly at 13 and 26 Valderøya, a high number of small flint nodules were documented. It should also be noted that very few flint flakes larger than 4 cm were identified at the Sunnmøre sites. If there was an abundance of large-sized, good quality flint nodules in this area, it is hard to understand why these small specimens were reduced to tools. A relevant explanation may be that the flint nodules left in Sunnmøre during the early Neolithic were generally small, and that they were mainly suitable for reduction using bipolar technique. The large number of bipolar cores and the low number of platform cores in Sunnmøre may be a result of the flint nodules generally being too small for platform reduction when the cylindrical core technique was introduced during the transition to the early Neolithic. In other words, although the populations in Sunnmøre may have known about the platform technique, and may have wanted to use it, the quality of their raw materials generally did not allow for it to be applied. Instead, they continued to use the bipolar technique.

Techniques and social and ideological restrictions

In the districts to the south of Sunnmøre, the bipolar technique also dominated the lithic industries during the late Mesolithic, but in contrast to the north, where it continued into the Neolithic, the technique almost disappeared during the transition to the Neolithic. It was still applied on flint and on some of the quartzes and quartzites to a minor degree in the south, but the raw materials that dominate at the sites: chert, blue mylonite, grey fine quartzite 2, and rhyolite have almost only been reduced by platform techniques. Very few bipolar cores of these raw materials have been found. Considering that the bipolar technique was a traditional technique in the area, with deep roots into the Mesolithic period (e.g. Skjelstad 2003:86), and that it is a very effective means for a maximum utilisation of raw materials, it is difficult to understand why it was practised so little to the south of Stad.

One possible explanation may be that there was an abundance of raw materials in these districts, which made it unnecessary to economise and to reduce everything by bipolar technique. However, if this was the case, one would expect to find numerous large flakes and unreduced blocks of raw material at the sites. Blocks and large flakes are seldom found. Instead, the flake material at all of the sites in Nordfjord is dominated by specimens that are smaller than 4 cm, which is largely the same pattern as for flint in Sunnmøre. Although a certain amount of these small flakes are the results of platform preparation of the cylindrical cores, this does not explain the lack of large blocks and flakes (see the mountain site Mørkedøla I and Gyrinos IV for good examples of raw material waste). A more likely interpretation may be that the availability was limited and that people economised. However, this was not done by using the bipolar technique.

Another possibility is that the early Neolithic raw materials in Nordfjord were not practically suitable for bipolar reduction. But considering that chert, blue mylonite, grey fine quartzite 2, and rhyolite all have cleavage properties that are largely comparable to flint, I also find this alternative less probable.

A third possibility is that there were social or ideological technological restrictions attached to these particular raw materials; they may have been reserved for the domain of cylindrical core technique, and as such they could not be subjected to bipolar reduction at a later stage.

In order to explore this alternative, I investigated the products of the different techniques in more detail. This was done because I assumed that it may not have been only the techniques in themselves that were socially or ideologically distinctive, but also the tools that were produced and the tasks that they were produced for.

The primary end product of the cylindrical core technique was blades (mainly small blades), and these were either manufactured further into A-points or were used directly for scraping or cutting. For all of the relevant raw materials at all of the sites, however, there is a marked dominance of unused blades compared to A-points and retouched blades. This pattern is puzzling, and indicates that the blades were actually discarded before they came into use as tools. The question is: which tools? They were probably not discarded knives or scrapers, because the blades still have intact sharp edges. A more likely option is that they are discarded blade points (A-points). The early Neolithic blade points are almost entirely made of blades where the distal end is a natural, unretouched point. One might argue that if this theory is correct, the percentage of A-points found at the sites would be much larger (they represent only about 2% of the total amount of blades and points): after all, the Neolithic knappers would not miss in 49 out of 50 attempts to make a pointed blade. Most likely they did not miss that much at all. A possible explanation for the low percentage of A-points is that the successful specimens were shafted and carried away: they would only accidentally be found at the settlement sites.

There are exceptions to this pattern: in Nordfjord flint is reduced both by cylindrical core technique and bipolar technique, and was used for A-points as well as scrapers. The same is to a certain extent true for blue mylonite. There is also an A-point of fine, white quartz (17 Havnen) which is a raw material that is mainly reduced using the bipolar technique. Nevertheless, the trend is quite clear: the raw materials chert, blue mylonite, grey fine quartzite 2, and rhyolite have almost only been reduced by cylindrical core technique and were mainly shaped into blades and blade points. A likely interpretation of this pattern is that these raw materials were mainly restricted to the domain of the cylindrical core technique and production of blade points. There may, in other words, have been a close relationship between the cylindrical core technique and hunting. Nevertheless, one should be careful in relating ethnicity to this pattern; even if the material culture in this case probably has had symbolic significance, it was not necessarily related to ethnic identity.

The southern boundary of slate knives

Regarding the southern distribution limit of slate knives, William Fitzhugh (1974:54) has put forward that the knives could be associated with the production of skin boats, which were highly dependent upon "soft" preparation of the hides. A possible reason why the knives are not found may be that such boats were not produced in the south. Søborg has argued that three factors may have been important. (1) In line with Fitzhugh, he suggested that the slate knives probably had been applied as flensing knives (similar to the Eskimo ulo), for marine butchering marine mammals. Seals and whales, he argued, may have been more frequently hunted in Sunnmøre and Romsdal than to the south of Stad. (2) The long distance from the northern slate sources (see Chapter 8) and (3) The boundary marked a social difference between southern social groups (Søborg 1988).

With regard to the problem of boat technology, this cannot be evaluated, because no remains of boats have yet been identified from Stone Age southern Norway, and the rock art in this area includes no images of boats.

Søborg supports his first argument by referring to the fact that seals are more frequent to the north than to the south of Stad today. He also adds that in hunter's rock art to the north of Stad, marine mammals occupy a more significant part than at Vingen and Ausevik (Fig. 120). Nevertheless, in historical times, all types of sea mammals were extensively hunted in the entire coastal region (e.g. Brøgger 1979). The faunal data from a number of excavated Neolithic sites also shows that seals represented a significant share to the south of Stad (e.g. Hufthammer 1992). Finally, the site location analysis (Chapter 4) shows that the Neolithic sites in coastal western Norway were mainly oriented towards the marine environment. Although sea mammals may have been more important to the north of Stad, these differences are probably only a matter of degrees. More fundamentally I also find that the explicit link that is often assumed between sea mammal hunting and slate knives (e.g. Fitzhugh 1974) is weakly founded, at least for areas such as coastal Norway, where terrestrial game was quite as important as marine species; I would presume that slate knives are effective also for flensing and butchering deer or wild boar. Terrestrial taboos might, of course, be attached to the knives, but until now, I have not seen convincing arguments for this being the case in a Norwegian Neolithic context (but see Eliasson & Joelsson 1989 for north-eastern Sweden).

The second argument of the distance from the sources is also difficult to evaluate directly, considering that none of the slate knives have yet been investigated geologically. It is probably correct that the knives with green bands originate at the margin of the Caledonian mountain range further north and east. Considering that few slate knives have been found close to the Devonian sandstone/siltstone sequences in Sunnfjord, I agree with Søborg that the slate knives found in Sunnmøre do not originate here. They probably came from the north. On the other hand, I disagree with Søborg concerning the Devonian sequences as irrelevant because of inferior quality. To my judgement, the slates in Sunnfjord are perfectly suitable for production of knives. When this was not done, it has other causes than lack of suitable raw material.

In conclusion, I find that Søborg's suggestions of distance to lithic sources and differences in the economic base as causes for the southern distribution boundary are problematic. Below, his third alternative of socially constructed boundary will be explored further.

A continuous social landscape?

Based on the above discussion, a suggestion can be made that the marked boundary between Sunnmøre and Nordfjord is a result of differences in the availability of raw materials in the two districts and by different properties or ideological/social restrictions on different raw materials. One relevant interpretation is that Nordfjord and Sunnmøre were not separated by an ethnic boundary, but that it was a continuous social landscape inhabited by closely related local groups which used different raw materials and different techniques when operating in the two different districts. However, if this was the case, one would expect to find significant evidence for large scale mobility and contact between Sunnmøre and Nordfjord.

Indirect evidence for this type of contact is provided by the types that were analysed here. These show that there was a typological consensus to the north and south of Stad: Vespestad adzes, slate points, A-points and retouched flakes and blades were found in Sunnmøre as well as in Fjordane, and no particular fall-off could be documented. These crosscutting typological distributions indicate that communication took place. If the mobility between these districts happened frequently, however, one would expect that people carried southern raw materials with them to Sunnmøre and flint in the other direction, because one could not always trust that nodules were easy to find or that quarries were readily available. A case of frequent mobility would thus result in a situation of raw material overlap between the two districts.

In theory, much of the flint at the Nordfjord sites may have been transported from the north, considering the relative abundance in Sunnmøre compared to the situation to the south of Stad. However, if this was the case, one would expect that the flint at the Nordfjord sites was reduced mainly using the bipolar technique, which was typical for the Sunnmøre sites. This is not confirmed by the core material. At the Nordfjord sites, flint bipolar cores are only found at 17 Havnen, and represents less than 2% of the total amount of flint artefacts. In contrast, bipolar cores of flint make up between 10 and 14% of the total in Sunnmøre. This indicates strongly that flint was not carried from Sunnmøre to Nordfjord. Instead, the inhabitants of Nordfjord probably collected their flint at local beaches, and reduced them mainly with the technologies which dominated in that district.

The raw materials pumice, fine white quartz, quartz crystal, medium and coarse quartz, and medium grey quartzite are found in both districts. Still, as was argued in Chapter 8, these are difficult to relate to specific sources, because they occur in small outcrops throughout the region and they may have been procured locally and independently in both districts. Slate, sandstone, blue mylonite, chert, rhyolite, greenstone and diabase, can all be geologically related to sources to the south of Stad and they were positively transported from Nordfjord to Sunnmøre. With the exception of grey slate at 9 Bustadvika, these southern raw materials make up very small shares (2% or below) of the total amount of raw materials at the Sunnmøre sites.

One might argue that the two sites Holvikshaugen and 9 Bustadvika in southern Sunnmøre, which have smaller shares of flint than the sites further north, could represent a overlapping zone between Sunnmøre and Nordfjord. At these sites, flint was replaced by large amounts of quartz/quartz crystal, which could also be quarried or collected locally. The composition of raw materials at these sites thus cannot be used to support a theory of an overlap-zone to Nordfjord.

Taken together, the raw material evidence clearly shows very little overlap. This indicates that the mobility between Nordfjord and Sunnmøre was not happening on a large scale. Instead of frequent mobility across, the data may more likely be interpreted as activities that were organised on a smaller scale, perhaps as barter expeditions or as gift exchanges in the context of task group mobility (see Chapter 10). On the basis of the above discussion, I suggest that the differences between Nordfjord and Sunnmøre should be interpreted as a significant rupture in communication between two different populations. The question is: what was the reason for this breakage?

The development of an ethnic boundary

One possibility is that people were restricted by natural hindrances. Sunnmøre and Nordfjord are divided by the peninsula Stad, which stretches far into the North Sea (Fig. 25). Traditionally as well as today, Stad is feared by crews of small boats because of the often unpredictable weather conditions, and because of the dangerous steep cliffs and the few harbours of refuge at the westernmost coast. During the Medieval period, Selje just south of Stad served as a secure port for northbound trade vessels waiting for the weather to change at Stad. The place-name Drage (dra = pull), Medieval written sources, and archaeological remains of wooden logs in bogs also tell stories of iron age traffic of boats across land at the lowest and narrowest part of Stad (e.g. Djupedal 1966). The Neolithic shorelines and weather conditions were not significantly different from the medieval situation. It is to be expected that Stad was a major obstacle for naval communication along the coast during the Stone Age as well.

The observed differences between sites in Sunnmøre and Fjordane may be explained by differences in the availability of raw materials as well as by the fact that Stad effectively stopped communication along the coast. This breakage may have resulted in different traditions for raw material use and technological solutions developing independently in the different districts. As was suggested for the coast-mountain relations, the differences may, in other words, largely be a matter of isocrestic variation. As Sackett (1985) points out in his critique of Wiessner, ethnicity may have been irrelevant or unimportant, because people on different sides of this boundary were very little in touch.

However, when I hesitate to abandon the theory of ethnicity in this case, it has two main reasons: the historical background for the establishment of the boundary and the active role that material culture may have played for the development of ethnic identities

In the above paragraphs, I have pointed out that there were cultural differences between Sunnmøre and Nordfjord already during the late Mesolithic, but these were not as marked as in the early Neolithic. For example, bipolar and microblade techniques were applied in both Sunnmøre and Nordfjord, and adzes made of diabase (from Sunnfjord) were distributed throughout the entire area. Thus, Guro Skjelstad, who has analysed the Mesolithic data in both districts, downplays the social significance of the boundary, even if she does not exclude the possibility of ethnicity as a factor (Skjelstad 2003:118).

During the transition to the Neolithic, the bipolar technique continued in the north, as well as the intensive use of flint. Slate knives were imported from further north. In the south, however, the majority of the late Mesolithic raw materials were replaced by new ones, and the slate industry (for production of projectile points) and the cylindrical core technique were introduced. The bipolar technique almost vanished. The differences were much more marked during the Neolithic than during the Mesolithic; the marked rupture in communication that can be seen in the Neolithic was not present in the preceding period. This means that the early Neolithic differences did not gradually evolve in Sackett's sense during several hundred years as a result of less and less contact. Instead, the differences appear to have developed fast, and they were most likely the results of conscious actions; people to the south of Stad broke the traditions.

I believe that during the transition to the Neolithic in this area, the material culture became meaningful in the social relations; it served as resources for in-group identification and out-group categorisation between populations to the north and south of Stad. A possible explanation for the fast changes may, as Morten Ramstad (1998) has pointed out for the middle Neolithic, be that they were the results of deliberate, historically determined choices with negative reference to the others. These choices would have developed as a result of experiences at the boundaries. Below, I suggest a possible scenario for such a development.

A number of new cultural traits, such as Vespestad adzes, cylindrical core technique, blade points, slate points, and major utilisation of rhyolite from Siggjo were invented and distributed among local groups in the area between Nordfjord and Rogaland during the transition to the early Neolithic. A significant flow of types, raw materials and techniques within this area indicates strongly that the social networks between these groups were quite advanced. People in Nordfjord were part of the networks; already during the late Mesolithic they received many of their important raw materials, such as diabase and yellow mylonite, from the south. It is, therefore, not surprising that they followed closely, and surely also actively contributed to the development of these new cultural traits during the period of transition.

It is quite likely that the populations in Sunnmøre knew that production of slate points and the cylindrical core technique were introduced in the south. However, few of their own raw materials could be reduced by these techniques. Therefore, if they converted, they would be dependent on a large-scale import of suitable lithics from the south. Instead, they chose to avoid these techniques and continued the old bipolar flint tradition. Initially, ethnicity may have been irrelevant for this choice. However, when meeting people from the south, the situation may have changed. In such contexts of interaction with out-groups, all that was experienced as different was made explicit and had a potential for being received as symbols of ethnicity. In such situations, the significant differences in technologies may have taken on new layers of meaning; the Nordfjord people categorised the people from Sunnmøre, perhaps in negative terms, by their extensive flint-use, their "old fashioned" bipolar technique and their slate knives. In contrast, their own identities were symbolised by applying a variety of raw materials, slate points and cylindrical core technique. At the other side of Stad, the situation was the other way around. If this is correct, it is likely that the major differences that I have observed are partly due to ethnicity "working back" on the choices that were made: people from Sunnmøre avoided flint alternatives and cylindrical core technique because they were associated with the out-groups in the south, while the groups in Nordfjord did not use the bipolar technique and slate knives for similar reasons. Central for this interpretation is that the boundary was socially constructed. It was a result of deliberate actions, even if no one beforehand predicted or wanted such a development to take place.

Nordfjord – Sunnfjord

Technologically and typologically, the two districts are identical. In addition, there are similarities with regard to the distributions of chert and rhyolite (both reduced by the cylindrical core technique) and red and green slate, which are all well represented at several sites in both districts. The main portion of raw material at the sites, however, diverges from this uniform pattern. Blue mylonite, medium grey quartzite 1, and medium and coarse grey mylonite, all reduced from cylindrical cores, which make up a significant portion of the lithics in Sunnfjord, have a marked fall-off on the boundary to Nordfjord. The reverse is the case for fine grey quartzite 2 (cylindrical), grey slate and brown slate. During the Mesolithic, there were also few differences, except that quartz crystal and yellow mylonite are more frequent in Nordfjord than in Sunnfjord. The reverse is the case for blue mylonite (Skjelstad 2003:69-73).

These differences may be interpreted as a result of the same households or local groups alternated between sites in Sunnfjord and Nordfjord because several of the quartzites, the mylonites, chert, and slate, were available locally in these districts (see Chapter 8), and were most likely quarried by the households or local groups that consumed them at these sites. The raw materials that dominate at the Sunnfjord sites may thus have been quarried when living in the south, and visa versa. Nevertheless, a quite as likely interpretation of the differences and similarities is that the two districts were occupied by two populations which had much in common and much contact, but which were separated by a territorial ethnic boundary. In order to distinguish between these two alternatives, it is necessary to confer with other types of data that is relevant for this problem. Below, I will discuss previous interpretations of rock art and the distribution of diabase adzes in Sunnfjord and Nordfjord.

Vingen and Stakaneset

In 1984, A. B. Olsen and S. Alsaker pointed out that during the Mesolithic and Neolithic periods there was a stability in the use of diabase for adzes in the districts between Sogn and Sunnmøre. This stability was interpreted as indicating a contact network in this area which was maintained by collective ceremonies at three ritual centres: the quarry at Stakaneset, Sunnfjord, and the rock-art sites Ausevik, Sunnfjord and Vingen, Nordfjord. They argued that all groups in the distribution area from Sogn to Sunnmøre had direct access to the quarry at Stakaneset. The most important argument for this was that 60-70% of all adzes in this area were made of diabase. If these had been distributed "down-the-line" (cf. Renfrew 1977), one would have expected a gradual fall-off in the distribution pattern with increasing distance to the quarry. Another argument was that the local area around Stakaneset is characterised by numerous small workshop-sites, which were taken to indicate that several small groups had been involved in quarrying as well as working of diabase (A. B. Olsen & S. Alsaker 1984, see also Mandt 1991:476).

A. B. Olsen's and S. Alsaker's theory was later discussed by Eva Walderhaug, who argued that the idea of "ceremonial centres" at Stakaneset and Vingen was problematic, considering the geographical distance between these sites. Instead, she suggested that they may have represented "reservoirs of symbolical-ideological capital" of vital importance for the regional populations, and that they were controlled by different social groups. Vingen was controlled by groups in Nordfjord/Sunnmøre, while Stakaneset was controlled by people from Sunnfjord and Sogn. The balance between these two groups was maintained by letting groups from Sogn and Sunnfjord have access to Vingen and people from Sunnmøre and Nordfjord could quarry at Stakaneset. Instead of being ritual centres for a larger region, Vingen and Stakaneset may have been situated on either side of a socially constructed boundary between Nordfjord and Sunnfjord. This boundary was, according to Walderhaug, strengthened at the end of the middle Neolithic period. At this point, the southern groups developed strong relations with agricultural groups further to the south. They established the rock-art site Ausevik, which took over the function that Vingen had previously had for these groups. At the same time she argued that these southern groups took increasingly control of the quarry at Stakaneset (Walderhaug 1994:101 ff).

Before these interpretations are discussed further, two important issues need to be clarified here. One of these is the dating of Vingen and Ausevik, the other is the change in the distribution of diabase during the Mesolithic-Neolithic transition.

The dating of the rock-art

Both A. B. Olsen and S. Alsaker and Walderhaug accepted Egil Bakka's (1979) dating of Vingen and Ausevik. On the basis of shoreline-curves, Bakka estimated a Tapes maximum in Vingen at about 10 m above sea level. Thereafter, he calculated that the lowest petroglyphs (8,25 m above sea level) in Vingen (at Elva) could not have been made earlier than the end of the middle Neolithic period. These were assumed to be the latest in a typological-chronological framework that Bakka developed for Vingen. The majority of the carvings were dated to the Neolithic, while some of the oldest were made during the late Mesolithic. Bakka found that the rock-art in Ausevik was very similar to the carvings at Elva, and accordingly suggested a late middle Neolithic date for Ausevik as well.

Today, Bakka's shoreline dating is less secure. New calculations of Tapes maximum at Skatestraumen, a few kilometres to the west of Vingen (Bergsvik 2002a:301), indicate that the Tapes-maximum at Vingen is as low as 7,5–8 m above sea level, which is 2–3 m lower than Bakka's estimate from 1979. This implies that all of the carvings in Vingen may have been made before or during Tapes maximum. These new results imply that Bakka's shoreline dating of the carvings at Elva to the late middle Neolithic no longer can be accepted. Considering that Ausevik is also situated above the maximum Tapes level in Flora, the same argument applies. In reality, this means that both Vingen and Ausevik lack a properly dated fix. One might also question Bakka's premise that the rock art were shore bound and his relative typological-chronological framework. If this is done, both sites may have been used during the entire postglacial period (Dommasnes 1997:175). However, if one still accepts that the carvings at Elva are typologically youngest and that they also were shore bound, the fix is still there, except that it is situated much earlier. The typologically youngest carvings at Elva and Ausevik may instead have been produced during the latter part of the late Mesolithic. Such an age determination is accordance with results from recent excavations at Vingen (Lødøen 2001), which have exclusively late Mesolithic dates. Lødøen convincingly relates the Mesolithic activities in Vingen to the production of rock art.

Changes in the distribution of diabase

During the transition to the Neolithic, there are significant changes in the distribution patterns of diabase. The share of diabase, which had as much as 60-70% of the "adze market" from Sogn to Sunnmøre during the Mesolithic, is reduced to about 20-30% in the Neolithic. The diabase is first of all replaced by greenstone (Fig. 62. See also A. B. Olsen 1981: Fig. 154). This important change has not been sufficiently considered in the interpretations presented by the above authors. The change indicates that, although the quarry at Stakaneset was still used as a lithic source during the Neolithic, and may have served a social or ceremonial function in that period, its regional importance in this respect must have been reduced considerably compared to the previous period.

Discussion

The reviews of diabase distribution and rock-art dating are important, because they suggest that the final rock-art production of Vingen, the establishment of Ausevik, and the reduction in the distribution of diabase adzes most likely happened already during the transition to the early Neolithic. If Walderhaug's theory of a marked social boundary between Sunnfjord and Nordfjord is correct, it should be related to social processes and events that took place already about 5200 BP (4000 BC), not as late as 4000 BP (2600 BC).

In order to have a better understanding of these processes, it is necessary, in this case, to refer to the late Mesolithic data from recent excavations in the district (Bergsvik 2002a, Skjelstad 2003). The raw materials chert, slate, sandstone and soapstone found at the late Mesolithic sites at Skatestraumen, Nordfjord, were mainly quarried in the district. However, significant portions also come from Sunnfjord: all adzes are made of diabase from Stakaneset, and yellow mylonite, which represents significant amounts, was also most likely quarried in Sunnfjord. As pointed out above, however, the new raw materials that dominate during the Neolithic are almost all from sources in Nordfjord. Only blue mylonite, which is relatively rare at these sites, can be related to sources in Sunnfjord. Diabase was used during the early phases of the early Neolithic (see 1 Haukedal, phase 2), but quickly disappears from the sites at Skatestraumen; only one of the 11 Neolithic adzes found at 17 Havnen were made of diabase. In contrast, 22 of 24 Mesolithic adzes were of diabase (Bergsvik 2002:284). The termination in the use of raw materials from Sunnfjord may partly be explained by the new cylindrical technique, which demanded other properties of the raw materials than yellow mylonite could offer. People had to find new sources, and found them closer to the sites. However, the abolishment of diabase cannot be explained in this fashion. Diabase should be regarded as an excellent raw material for Neolithic as well as Mesolithic stone adzes. Interestingly, Neolithic Vespestad adzes in Sunnmøre make up as much as 20-40% of the total. The change from diabase to greenstone at Skatestraumen should not be explained as a result of the introduction of new technologies or adze types; instead, it may have been a result of a change in the social relations towards the south during the transition to the Neolithic.

On the basis of the above, it can be argued that the Mesolithic local groups from Skatestraumen had direct access to the diabase-quarry at Stakaneset. In combination with these raw material expeditions, which probably also had social and ideological aspects, yellow mylonite was quarried in the nearby sources. As is suggested by Walderhaug, such access by the Nordfjord groups to important reservoirs of symbolical-ideological capital may have been balanced by admitting the populations of Sunnfjord to Vingen. Consequently, if there was a socially constructed boundary between Sunnfjord and Nordfjord in this period, it was open for large scale trespassing.

During the transition to the Neolithic, however, this situation changed. The local groups at Skatestraumen ceased to use adzes from Stakaneset. Parallel to this, the populations in Sunnfjord in Ausevik established an alternative rock-art site to Vingen. This means that two important arenas for the maintenance of broad alliances between Sunnfjord and Nordfjord were unwound, and the mobility across cooled off. This indicates that the boundary between these two districts was more clearly defined than before; a social distance towards the Flora-group was symbolised by not using diabase from Stakaneset. In order to obtain raw materials for their adzes, they instead applied the quarry 4 greenstone (from the north?). The groups at Sunnmøre, however, could freely apply the diabase from Sunnfjord, because for them, this raw material did not carry any negative out-group associations.

As is indicated by the flow of "exotic" raw materials such as rhyolite and diabase, communications still took place, but their character had changed compared to the previous period (see Chapter 10). The question is: should these differences between Nordfjord and Sunnfjord be related to ethnicity? If the situation is compared to the relations between Nordfjord and Sunnmøre, which was characterised by deep traditions of major differences in technique as well as in raw material use, the differences towards Sunnfjord appear to be much less significant. Culturally, they were not very different. Still, this does not necessarily imply that the relations were "less" ethnic. As was pointed out in Chapter 3, ethnic relations with close neighbouring groups are often tense, because they compete for the same resources. Competition for the same resources may have been less relevant towards groups from Sunnmøre because of the natural boundary Stad. Towards the south, however, Vingen and Stakaneset may quite likely have been potential sources of conflict. It is also possible to imagine that other resources on the intersection between the territories of these groups may have caused problems. Local disputes may have arisen over the rights to exploit attractive fishing sites and hunting grounds. The chert sources at Frøya, which is situated at equal distances from Flora and Skatestraumen, would also have considerable conflict potential. Although not yet localised, the slate and sandstone quarries may also have been problematic.

Thus, it could be argued that, despite that differences between Sunnfjord and Nordfjord did not have a deep history, and that they had fairly similar material cultures, the relations between these two districts may still have had an ethnic component, which was based on marked political conflicts over the rights to exploit common resources of vital importance.

Sunnfjord and Sogn

The difference between Sunnfjord and Sogn is hard to evaluate, because of the general lack of excavated sites in Sogn. The only excavated site from Sogn, Masnesberget, is different from the Sunnfjord assemblages in a number of respects; some traits are much more frequent and others are not present at all. But as pointed out in Chapter 7, it is likely that most of the "extreme" values at Masnesberget are the results of the site being collected by amateurs. However, one diverging trait, the dominance of fine black quartzite 1, can probably not be explained by different excavation techniques. Even if its relative value compared to other categories is somewhat adjusted, it would still make up a large share of the raw materials that are found at the site. If we take into account that this category is frequently found at other, test-pit surveyed Neolithic sites in Sogn (Lødøen 1995), and that it is only minimally present in the adjacent districts, one may conclude that fine black quartzite 1 was a significant part of the cultural repertoire in Sogn in contrast to these other districts.

Guro Skjelstad has analysed two good late Mesolithic sites in Sogn. She found that these did not diverge significantly from the sites in Sunnfjord and Nordhordland, and argued that these three districts are likely to have been inhabited by the same social groups (Skjelstad 2003:122). As was the case with the Nordfjord-Sunnfjord situation, a possible Neolithic boundary between Sunnfjord and Sogn would not have its roots in cultural differences the preceding period.

It cannot be excluded that a political process comparable to the one outlined in the above paragraph on Nordfjord and Sunnfjord took place between populations in Sunnfjord and Sogn. Perhaps future excavations of early Neolithic sites in Sogn will provide data that may support such a theory. However, at the moment, I find it hard to argue for such a development solely on the basis of one diverging cultural trait. Below, Sunnfjord and Sogn will, therefore, be presented together in contrast to Nordhordland.

Sunnfjord/Sogn, Nordhordland, and Midthordland

Technologically and typologically, the three districts are quite similar, but there are three differences that may be significant: slate points were not produced at the sites to the south of Nordhordland, there are much more A-points at the sites in Nordhordland than to the north of that district, and there are more retouched flakes and blades at sites in Nordhordland and southwards than in the north.

Between Sunnfjord/Sogn and Nordhordland, the raw materials fine blue quartzite 7, medium grey quartzite 1, and coarse, grey mylonite have southern distribution limits. Chert and red and grey slate also have a marked southern fall-off here, while fine black quartzite has a marked fall-off to the south of Sogn. In the other direction, fine blue quartzite has a northern distribution limit to the north of Nordhordland. Rhyolite from Bømlo, which makes up between 40 and 60% of the total at all sites in Nordhordland, Midthordland and Sunnhordland has a marked fall-off, and represents no more than 5-10% at the sites in Sogn and Sunnfjord.

Between Nordhordland and Midthordland, chert, red slate, and brown slate have southern distribution limits to the south of Nordhordland. Sandstone, anorthosite, green slate, and blue mylonite also have a marked southern fall-off. In the other direction, medium grey quartzite 10 and 11 have a northern distribution limit to the north of Midthordland. As noted in Chapter 7, Snekkevik 1 has a raw material composition very similar to sites further to the south, and might "neutralise" the differences between Nordhordland and Midthordland. However, considering that the sample-size from this site is small, I choose to attach more importance to the tendencies at the site Kotedalen, which has large sample-sizes.

With regard to basaltic rocks (adze material), the results from the sites indicate that greenstone is generally much used throughout the region; no fall-off to the north of Nordhordland was found. However, it should be remembered that possible differences between quarry 4 greenstone and greenstone from Bømlo are masked by a grouping of these categories (see Chapter 8). Diabase at the sites, on the other hand, shows a significant fall-off to the south of Sogn. The distribution patterns of greenstone Vespestad adzes of largely confirm this pattern at the sites. Diabase, on the other hand, diverges from the pattern at the sites by a gradual fall-off from the quarry at Stakaneset to Midthordland.

The deep historical roots of these boundaries on either side of Nordhordland are also supported by the Mesolithic data. The distribution of Mesolithic adzes of diabase and greenstone overlapped in the district of Nordhordland, and there was a significant fall-off of these raw materials to the south as well as to the north of this district (A. B. Olsen & S. Alsaker 1984).[4] The finishing process in the manufacturing process of these adzes are different north and south of Nordhordland (A. B. Olsen 1981, S. Alsaker 1987), and the Mesolithic adzes with round cross-sections are also characterised by different shapes (Gjerland 1984). Furthermore, the distribution of Mesolithic cross-shaped and star-shaped clubs also coincides with this boundary (Gräslund 1962, Solberg 1989a). Finally, Guro Skjelstad has found that tabular sandstone knives have a southern distribution limit at the boundary to Midthordland, there is also a southern fall-off in the use of grinding slabs and soapstone net-sinkers here. In addition, sandstone, yellow mylonite and blue mylonite from Sunnfjord have a distribution boundary between Nordhordland and Midthordland, while the use of flint increases to the south of Nordhordland (Skjelstad 2003:108).

In sum, the data indicates that there are coincidental distribution boundaries and marked fall-off of types, technologies and raw materials between Sunnfjord/Sogn and Nordhordland as well as between Nordhordland and Midthordland. These boundaries can also be documented by the Mesolithic data. Importantly, the early Neolithic boundaries can be observed without considering the distribution patterns of greenstone.

Discussion

It should be kept in mind that the district of Nordhordland is not particularly rich in bedrock lithic resources. Only anorthosite (2,5%) could positively be identified as coming from this area. Contrary to their contemporaries in Sunnmøre, the local populations lacked rich flint beach deposits. The scattered sources of quartz and quartzite may not have satisfied their quality demands. As a result, they were dependent upon large scale supplies from the outside. The raw materials diabase, chert, slate, sandstone and blue mylonite were all taken out in Sunnfjord or Nordfjord, and rhyolite and (some of) the greenstone was quarried in Sunnhordland. Nordhordland is positioned half-way between these source-areas. Thus, it might be argued that such a distribution pattern is to be expected and is normal because of the almost equal distances to the sources. However, this does not explain the marked fall-off of the distributions just here. There are no natural barriers in this area, such as mountain plateaus or long stretches of water that would hinder communication and prevent further spread of artefacts. As pointed out above, the Neolithic subsistence-settlement patterns were similar along the entire coastline, so that the discontinuity can not be explained by different subsistence practices in the north and south. Still, a discontinuity can be observed, and a major rupture in communication has apparently taken place; although certain raw materials passed the district of Nordhordland, the majority did not go any further. I believe that such a rupture cannot be interpreted without considering the element of ethnicity.

4 One might argue that, similar to the Neolithic Vespestad adzes, the raw material composition of the Mesolithic adzes should have been reconsidered before it was used to support such a claim. Such a re-analysis was not possible to carry out for the present work. However, I find it likely that A. B. Olsen's and S. Alsaker's raw material determinations for the Mesolithic adzes are more reliable than the Neolithic (greenstone) data. The main reason for this is that, in contrast to the Neolithic, where bluish greenstone (quarry 4 greenstone) dominated in all of the districts between Nordhordland and Romsdal, Mesolithic adzes found in these districts were largely identified as diabase (60-70%). Quarry 4 greenstone represented a relatively small share, and was probably generally infrequent as adze material during this period. This implies that the raw material of the Mesolithic adzes classified as greenstone most likely came from the quarries at Bømlo

The situation differs from the boundaries further north in that there is a fall-off towards both sides of one district – Nordhordland – in which most of the raw materials, types and technologies are present. In a context of ethnicity it is possible that the ethnic boundary was changing back and forth within the district through time. However, if this was the case, one would expect that the sites in Nordhordland were characterised either by elements mainly from the north or from the south, and perhaps also that this composition was reversed through time. The site Kotedalen at Fosnstraumen, which has provided significant amounts of data from two early Neolithic phases, may be used to investigate this problem. The oldest phase 12 at Kotedalen is dominated by rhyolite from Bømlo (66%). However, as much as 17,5% of the raw materials are identified as blue mylonite from Flora. It should also be noted that slate and chert from the north were reduced at the site. Such a composition of raw materials is particularly clear at the succeeding phase 13 at the site, which also has significant shares of rhyolite from the south (40%). In this phase, however, the northern elements also make up large shares: slate and blue mylonite represent at total of 20 and 11,5% respectively. Although southern relations seem stronger during this phase as well, the northern connections are particularly clear through the extensive slate industry at this site. A similar composition of raw materials is seen in even greater scale at the nearby site Ramsvikneset (Bakka 1993), which was occupied at the same time as Kotedalen.

It is important to note that this situation also characterises the preceding period. On the basis of the Mesolithic adze-distributions, A. B. Olsen and S. Alsaker pointed out that Nordhordland is an overlap-zone: southern as well as northern cultural traits are found in this district, and these were interpreted as a "contact zone" between northern and southern "social territories" (A. B. Olsen & S. Alsaker 1984:98). The idea of Nordhordland as a Mesolithic "contact-zone" was supported and further developed by Bruen Olsen and myself in a recent paper. We concluded, based mainly on the fact that there were many greenstone adzes and few diabase adzes in Nordhordland, that the groups in Nordhordland probably identified stronger with groups in the south than in the north. The northern relations were primarily organised as exchange networks in a context of ethnicity (Bergsvik & A. B. Olsen 2003:391). This interpretation has, however, been countered by Skjelstad, who has argued that her site data from Nordhordland indicates rather that the populations who lived there probably identified with groups to the north and that the ethnic relations instead went southwards (Skjelstad 2003:127).

The disagreement concerning the direction of the Mesolithic affiliations is based on the assumption that the populations of Nordhordland should either have northern or southern identities. This assumption may, however, be incorrect the in Mesolithic as well as the Neolithic. As is indicated by the Neolithic site data from Kotedalen discussed above, the populations that inhabited the site during the early Neolithic cannot clearly be related to a southern or northern side of a social boundary. The situation of the Nordhordland sites is instead characterised as a "mixture" of elements of northern and southern origin. If it had not been for the marked raw material fall-off at both sides of Nordhordland, this mixture would not have drawn so much attention, however, the abrupt halt in the spread of certain raw materials and technologies indicates that the groups in Nordhordland had given them a symbolic significance, and/or were collectively categorised by them; the very blend of northern and southern elements may itself have been important resources for the ethnic identity of local groups from this district: they were different from the populations to the north by their extensive use of rhyolite. This raw material dominated heavily in the south, and is likely to have been related to their identity as "southerners" in opposition to groups to the north of the boundary. The early use of pottery, which until now has not been recorded further north than Nordhordland, may also have been related to such an identity. Extensive use of slate, anorthosite and blue mylonite may have had the same function as ethnic idioms in relations with groups in Midthordland.

It is still unclear what particular historical circumstances could have led to a construction of social boundaries to the south and north of Nordhordland, but the Mesolithic information indicates that the process started early. The Mesolithic data, which almost entirely consists of types and raw materials that were not in use during the later periods, emphasises the historical depth of these boundaries, and strongly indicate that their social significance persisted despite radical shifts in the cultural content. It appears, therefore, that they were maintained and renewed for several thousand years among the local populations. Thus, the development in Nordhordland is a good example of the changeable nature of cultural content (cf. Barth 1969), but it is also an example of ethnicity being strong and very resistant to change.

Midthordland and Sunnhordland

There is a tendency that thin-butted axes concentrate in Sunnhordland. Except from this, the two districts are typologically and technologically similar. There are also great similarities in raw material use: rhyolite clearly dominates at almost all of the sites. Fine blue quartzite 8, which similar to rhyolite was reduced using the cylindrical core technique, also appears in fairly large quantities at sites in both Midthordland and Sunnhordland. The only differences between the districts are a slight increase in the use of flint at the southernmost sites and presence of medium grey quartzite 10 and 11 at the sites Nilsvik 4 and 17 and at Austvik III. A large degree of similarity is also observed in the Mesolithic data from Sunnhordland and Midthordland (Skjelstad 2003:108).

The differences that are found between the two districts indicate that the flow of goods did not include everything. However, I find it unlikely that a slightly larger percentage of flint in the south was symbolically meaningful in contexts of ethnicity. It is also difficult to imagine medium grey quartzite 10 and 11 as symbolically important, considering that it only makes up an average of 3-4 % at the relevant sites and that almost no recognisable tools were made from them. These particular raw materials were probably procured at local lithic sources. A likely interpretation of the above data is that there was a large-scale communication of cultural values and flow of raw materials across the two districts. If there were socially constructed boundaries within the area, these cannot be detected by the information available here.

The increasing percentage of flint in the south, and the slight concentration of thin-butted axes close to the boundary between the counties Hordaland and Rogaland are interesting, and indicates, as Gutorm Gjessing (1945:330) already has pointed out, that the early Neolithic assemblages in Rogaland are quite different from those further to the north. In the future, a study of raw materials, types and techniques should be performed in Rogaland as well.

Sedentism and ethnicity

Above, I have argued that there were several ethnic boundaries in western Norway during the early Neolithic. The attention has mainly been directed towards identifying such boundaries, but I have also discussed different reasons why ethnicity may have arisen between particular districts. In the following, I will focus more on possible causes. As I pointed out in Chapter 2, sedentary populations are particularly predisposed for developing ethnic relations with neighbouring groups. Below, I will develop this idea further by investigating why sedentism would be important for the development of ethnic relations between the coastal populations.

Sedentary hunter-fishers

During the last few years, several scholars have suggested that the coast of western Norway was occupied by sedentary groups during the Mesolithic and Neolithic periods (Bergsvik 1991; 1995, A. B. Olsen 1992, Warren 1994, Nærøy 2000). Sedentism is defined here as "a situation in which a group spends over six months a year continuously at one locus, even if at other times during the year the group leaves, returning to the community after short, often seasonal absences" (Kent 1989:2). On the basis of this definition, a residential site from a sedentary local group may be recognised by three archaeological attributes: (1) a wide resource base, (2) continuous use of the site for at least 6 months and (3) frequent reoccupation of the same site. If the site meets these criteria, it may be distinguished from a site left by a residentially mobile local group as well as from an aggregation camp.

The term "local group" is defined here as a politically and territorially autonomous social unit consisting of a collection of households (Drucker 1983). It is assumed that residential sites are associated with this unit or households within this unit.

Based mainly on data from the site Kotedalen at the tidal current channel Fosnstraumen, the problem of sedentism is thoroughly discussed elsewhere on the basis of archaeological, botanical, and faunal data (Bergsvik 2001b). Here, only the conclusions of that paper will be referred to:

(1) The groups that lived at Kotedalen during the Mesolithic and Neolithic exploited a wide range of resources, (2) all of the occupational phases were the results of more than six months of continuous occupation, and (3) although the site periodically may have been left by entire households, these returned to and reoccupied the site after short absences. Thus, the late Mesolithic as well as Neolithic phases had clearly been occupied by sedentary households. Although sedentism characterised both periods, it was also clear that the Neolithic sites had a markedly broader subsistence base, the households stayed for longer periods at the time, and there were more frequent reoccupations of the same sites than during the late Mesolithic.

Kotedalen was probably not unique, neither in a local nor in a regional perspective. At Fosnstraumen, close to Kotedalen, two other early Neolithic sites which share many of the characteristics of Kotedalen are excavated: Ramsvikneset (Bakka 1964; 1993 Nærøy 1993b), and Snekkevik 1 (Bergsvik 1988). Such sites are also investigated in other local areas that are similar to Fosnstraumen both with regard to communications and resource concentrations. At Skatestraumen, Nordfjord, the sites 1 and 6 Haukedal, 17 Havnen, 23 and 28 Gloføyk are of similar type (Bergsvik 2002a). At Brandsøysund, Flora, Vikja I (A. B. Olsen 1981) belongs to this category, and also 4 and 17 Nilsvik at Bjorøy (K. Kristoffersen 1995), possibly also Sokkamyro at southern Bømlo (Shetelig 1922) and Nordøy grendahus in northern Sunnmøre (Søborg 1994). None of the excavations at the above sites have produced as much faunal and botanical data as Kotedalen. This is lack of palynological evidence is, however, due to less detailed excavation techniques and/or inferior preservation conditions. It is probably not a result of differences in site type. Similar to Kotedalen, these sites should probably all be characterised as results of long, continuous occupations by households. Corresponding functional interpretations also probably apply for several of the only surveyed long-term sites at Fosnstraumen, Vatlestraumen, and Skatestraumen (see Chapter 4).

The sedentary sites are mainly found close to tidal current channels. Surveys and excavations areas outside of these channels, on the other hand, have generally produced small short-term sites (e.g. Bøe 1923, Bjerck & Ringstad 1985, S. Kristoffersen 1990, Nærøy 1994, Bergsvik 2001c, K. Kristoffersen & Warren 2001). These small sites are generally interpreted as results of seasonal task group expeditions from the sedentary sites in the channels (Bjørgo 1981:161, Nærøy 2000:207).

On the basis of the above, it is possible to conclude that a series of tidal current channels during 5000 years localised settlement concentrations of sedentary households along the west coast of Norway. These results are of utmost importance for the understanding of the ethnic relations that were discussed above, because they may provide some of the "flesh to the bones" of ethnicity. Stated more precisely, they may indicate what economical and cosmological factors have been important for the forming of in-group identities as well as out-group categories.

Economical, political, and cosmological aspects of sedentism

The channels offer denser and more stable concentrations of fish, marine mammals and marine birds than any other type of biotope on the coast (Aksnes 1988). For most of these locations, the communications are also very good into the fjord systems in the east and to the north and south along the coast. These factors may largely explain the situation of long-term habitations and frequent reoccupations of the same channels: Sedentism combined with seasonal task group mobility was probably the most convenient way of exploiting the coastal resources.

There are, however, other factors that may have been quite as important for the local populations. Although the earliest site occupants may have been primarily concerned with the rich resources, one may expect that their predecessors were equally driven by strong attachments to these places. As Ericka Engelstad (1990) and Kenneth Ames (1991) point out, sedentism may be understood not only as a way of organising around the resources, but also a way of constructing social landscapes, where residential patterns and associated social and cultural patterns become spatially fixed and maintained over some period of time. The important question is which social and cultural patterns that is relevant, or to phrase it more bluntly: What made the groups capable of returning to the same sites and maintain the same settlement patterns over a long period of time? What both Ames and Engelstad point to is probably the important link between sedentism and territoriality. According to Maurice Godelier, the term "territory":

> *"...is used to designate a portion of nature and space that is claimed by a given society, the society guaranteeing all, or only some, of its members stable rights of access to control and use of all or part of the resources found therein, and which it (the society) is capable of exploiting" (Godelier 1979:138).*

Obviously, politics is an important an element of this definition, which implies that the character of the territoriality is clearly influenced by the social strategies of individual agents.

Several other ethnographic examples from northern areas similar to western Norway illustrate the issue of territoriality (e.g. Balikci 1970, Watanabe 1972, Burch 1980, Townshend 1980). The Indians of the north-west coast of America, who were both sedentary and strongly territorial, are a case in point. In this case, the rights to exploit the resources within an area are vested in the local groups. These resources imply specific locales, such as village sites, fishing streams, berry grounds and hunting grounds, beaches, firewood, fresh water, sacred sites and particular trade routes. The locales are usually defended while terrain or water between these places might be open to anyone. In this area, the political leadership is ascribed, and there are usually three classes: nobles, commoners, and slaves (e.g. Drucker 1951, Richardson 1982, de Laguna 1983, Blackman 1990).

Others have discussed local group territoriality for sedentary coastal hunter-fishers elsewhere in Scandinavia (e.g. Rowley-Conwy 1983, Schanche 1995). I find that territoriality may also be a relevant interpretation of the current data, because the households and local groups in western Norway would probably not have returned to the same sites and locations if they were not attached to them, had the rights to exploit the resources in them, and were prepared to defend these rights. The large degree of site reoccupation at the tidal current channels therefore indicates that the populations in these areas were territorial already from the late Mesolithic onwards. Even if several local groups, from time to time may have congregated at the channels for social purposes (Bostwick Bjerck 1987:148), one may assume that only one local group in each area managed these locales and resources. I find it reasonable to expect that the strong continuity of occupation demonstrated here is the result of inherited ownership to locales and resources within such local groups; descendants of the earliest occupants of the channels are likely to have maintained the knowledge of the best procurement sites and transferred these to the next generation.

Yet, such territorial relationships should not be seen as exclusive formalised ownership characteristic of modern capitalist states. While there undoubtedly is an element of self-interest in the land-rights, there are surely other aspects that may be quite as important. Many examples from hunter-gatherer territoriality show that people belong to the land and consequently have specific obligations towards the land, which is closely related to the web of ancestral stories and creation myths that are attached to the particular landscape. Places and resources have to be tended regularly to please the ancestral beings. For the Aborigines of Australia, Howard Morphy points out that people demonstrate attachment to land by displaying knowledge of the creation myths of the topography. This again is related to knowledge of the topography itself, a knowledge that can only be acquired by spending time with the older generation in that particular landscape (Morphy 1995). Such territorial notions are also found in North America (Riches 1982:107 ff). Thus, hunter-gatherer territoriality can often be seen as a sum of socio-economic and mythological attachments to the land (e.g. Stanner 1965, Peterson & Long 1986) which again makes it much stronger than if only one of these elements were present.

The territorial attachments would be relevant for the tidal current channels. But one would also expect that short-term sites and other locales such as rock art sites and lithic raw material quarries outside of the channels were included in the same social landscape as well, and that only segments of the local groups had the rights to exploit the resources.

If such complex territorial notions were relevant here, it would be difficult for competing groups to establish themselves at particular sites, even when the local groups were temporarily absent. When arriving at the channels, competing groups would immediately see the empty houses and other marks on the land. At Kotedalen, they would for example have seen a large forest clearance behind the residential site (Kaland 1992:87). Although made for domestic activities, these marks undoubtedly also served symbolic ends. The landscapes were loaded with countless numbers of stories, origin myths, and memories. It is easy to imagine that the bonds and attachments to these places were strong, and that they played important parts in the forming of the identities of the people who lived there.

In a context of ethnicity, one might expect that these territorial rights and attachments, in which senses of belonging were quite as important as economic maximising, were important cultural resources for ethnicity. This is so, because it would imply that in-groups would be the ones that shared the rights to exploit the same resources. They would also share the origin myths of the same landscape, and they would share the knowledge of the paths and places within the territory. In contrast, neighbouring and more distant local groups, who did not have these rights and attachments, would be regarded as out-groups.

Ethnic boundaries between local groups or "cultures"?

A consequence of the above reasoning is that the locus of ethnicity is found primarily between local groups, which might be thought of as a relatively low level of social organisation. This is not unreasonable, because the ethnographic data referred to in Chapter 3 indicate strongly that "local band" (Drucker 1983) and "regional band" (Helm 1968) is the highest political and territorial level among hunter-gatherers in the recent past, and is also the most inclusive level, to which most in-group references are made. The boundaries that were identified above on the basis of lithic data does not contradict such a theory at all: some of the marked raw material fall-offs and distribution limits of technologies and styles may quite well indicate the boundaries between neighbouring local groups which had their sedentary sites and core settlement area in places such as Skatestraumen, Brandsøysund, Fosnstraumen, and Bjorøy.

Nevertheless, an objection against this interpretation might be that by relating it to the local group, I have reduced ethnicity to be a matter of small-scale politics of identity; I have really only been talking about is hunter-gatherer territoriality, not ethnicity: ethnicity is found on higher levels. The question is, however: did higher levels of social organisation exist in early Neolithic Norway? Were several local groups united into "tribes"? Did the ordinary man and woman identify with such "tribes"?

These questions are difficult to answer with certainty. I cannot, on the basis of the data presented here, exclude the possibility of such higher levels of social organisation as loci for ethnic identities. It should, for example, be pointed out that there were significant variation within the area of analysis as regards the number of common cultural traits and the degree of communication across the boundaries; the Sunnmøre-Nordfjord boundary and the coast-mountain boundaries were much more marked than the others. Much more raw materials, types and technologies had distribution limits and marked fall-offs at these two boundaries than anywhere else. Another

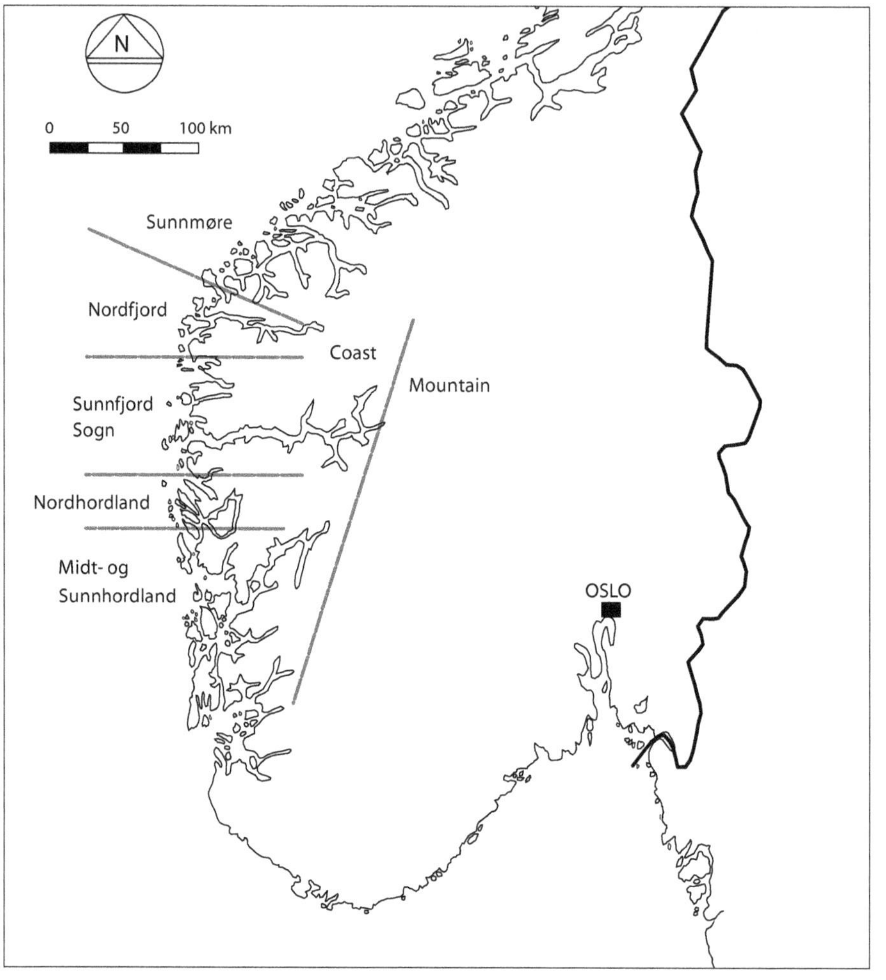

Figure 121. *Suggested location of ethnic boundaries in early Neolithic southern Norway.*

point is that there were also significant differences in the production of rock-art between eastern and western Norway, and between sites to the north and south of Stad. This indicates that there may have been significant differences also in the world-view of the populations already during the late Mesolithic. Furthermore, only for these two boundaries was the topic of isolation seriously brought up. Finally, one should remember that while the other boundaries are relatively new discoveries, these two have been known by arhaeologists for a long time. The boundaries between east and west and between Nordfjord and Sunnmøre may, therefore, have been much stronger than the ones at the coast of western Norway. This means that that the different boundaries probably had different significance and force.

Consequently, it is perhaps likely that coastal populations from Nordfjord thought differently about the people to the north of Stad than the ones in Sunnfjord, and that the relations were more frequent and perhaps more familiar towards the south than towards the north. Much less frequent communication certainly also characterised the coast-mountain relations compared to the different districts at the coast. From the early Neolithic and onwards, the local groups at Sunnmøre and Nordfjord probably directed their most important external relations northwards and southwards respectively. The contacts over Stad may have cooled off. As Ramstad (1998) has pointed out, this would imply that people one each side of this boundary participated in different, relatively isolated fields of interaction for exchanges of values and norms. This may have been related to a symbolic communication of conformity with groups in the north and the south (e.g. B. Olsen 1988). Together, these factors led to a strengthening of the differences that already existed.

This probably meant that if a man from Nordfjord travelled to, say, Southern Bømlo in the early Neolithic, he would have to pass three of four boundaries which may be characterised as ethnic. Still, he would have experienced that people lived more or less in the same manner as him, and used the same techniques and tool-types. He would have recognised a large number of cultural traits, and would probably not have had problems in communicating about values and traditions. A travel from Nordfjord to Sunnmøre or to Hardangervidda would have been much more challenging.

In a sense, these more marked boundaries are coincidental with the previously proposed boundaries between the "slate culture", "TRB culture", and the "dwelling site culture" which were isolated by A. W. Brøgger and G.

Gjessing many years ago. However, contrary to Brøgger and Gjessing, who assumed that the ordinary man and woman identified with these "cultures" and were aware that that they lived in the centre of in the periphery of them, I do not believe that these "cultures" were social and political entities. Instead of being coincidental with the distribution of ethnic groups, I believe that the old "cultures" should simply be seen as distribution areas for certain styles and traditions. There may be several reasons why these traditions and styles developed and were maintained (e.g. Odner 2000), but these causes have little to do with the organisation of large social and political units; these large scale cultural differences were not socially organised. I suggest that the lack of dynamics in previous research on the problem of Neolithic cultural variation is due to a focus on social entities that were too large. By concentrating on large regions and "cultures", one has missed what was going on "on the ground". Most likely, few people – if any – were conscious about these large-scale differences during the Stone Age. Therefore, to say that some boundaries were more marked than other is not the same as saying that they delineated different levels of the social organisation.

To conclude, I believe that the most important identities of the average members of the society were connected to the social entity that mattered to them in daily life. If it is correct that these populations were sedentary and territorial, this would be the local group. Such groups may have counted up to 50–60 households together, and these households had their residential sites on a handful of central locations in the relevant district. The most important out-groups of these local groups would be the ones that lived in the neighbouring districts, with whom they would have exchange networks as well as on-going conflicts and negotiations over economic as well as ritual resources. People outside of these spaces would be less important to them, simply because they were less in touch.

Conclusions

Above, I have discussed whether ethnicity was a factor that may help understanding some of the marked cultural differences that were present between sites in districts in western Norway and at the mountains plateaus between eastern and western Norway.

Mainly on the basis of a discussion of marked cultural differences and breakage in communication, I have found that ethnic boundaries can be delineated between (1) Sunnmøre and Nordfjord, (2) Nordfjord and Sunnfjord, (3) Sunnfjord/Sogn and Nordhordland, (4) Nordhordland and Midthordland/Sunnhordland, and (5) between the coastal and the mountain sites (Fig. 121). Several of these boundaries were established already during the preceding Mesolithic period, but it appears that, although the social organisation of the boundaries themselves persisted, much of the cultural content that constituted these old boundaries disappeared and was replaced by different raw materials, types and techniques during the transition to the Neolithic. The maintenance and establishment of these boundaries were unique processes, which all required local explanations. One important cause for this variety of explanations may be that the local groups were not unified politically. The region of western Norway probably consisted of a series of politically autonomous local groups who, because of their marked sedentary life style, were culturally different. This resulted in a series of different boundary processes.

In the above discussions I have mainly concentrated on identifying the boundaries by referring to the historical depth, the cultural content, and the rupture in communication. Less attention was given to ethnicity as an organisational feature and to the boundaries as contexts of ethnic relations. In the next chapter I will explore the character of the inter-group relations in more detail.

Chapter 10

Exploring the connections

Introduction

In the preceding chapters I showed that there are marked discontinuities in several areas in the distribution of raw materials, types, and techniques. These areas of discontinuity were interpreted as ethnic boundaries between sedentary hunter-fisher populations.

However, even if there are marked discontinuities, the boundaries at the coast are not absolute. Quite a few artefacts have been distributed far beyond these proposed ethnic boundaries (Fig. 122). One might argue that such crosscutting distributions provide arguments against the existence of an ethnic boundary. However, an untidy and unsystematic distribution of cultural traits is not an argument in itself against the existence of social boundaries (Shennan 1989). It is rather a necessary condition, considering that ethnicity is a product of social interaction rather than isolation from other groups. Opposing ethnic groups should therefore be expected to share a number of cultural features.

Nevertheless, a sharing of cultural features such as techniques or styles does not say anything about the maintenance of contact networks, or about the character of the exchange relations, considering that they are the ideas of how to make things rather than the things themselves. Transmission of knowledge about techniques or styles may also happen during sporadic contacts. Thus, the character of the connections is best studied in the distributions of raw materials.

Below, the issue of cross-boundary relations is exclusively dealt with on the basis of the distribution of raw materials with known sources. First, I will refer to different types of inter-group flow of artefacts in small-scale societies. It will be distinguished between gift-exchange/commodity-barter on the one hand and direct access on the other. Secondly, by means of the chaîne opératoire terminology I will investigate whether the cross-boundary flow of individual raw materials in western Norway can be related to either of these two forms of acquisition. Thirdly, by means of the concept of task group mobility, I will discuss the socio-political aspects of the connections in the context of ethnicity.

Direct access, gift-exchange, and commodity-barter

There are three main fashions in which raw artefacts may be distributed in small-scale societies: direct access, gift-exchange and commodity-barter. Direct access refers to a situation when the (local) groups who have used and discarded the artefacts also have procured (quarried or collected) and manufactured them. Gift-exchange and commodity-barter both refer to goods being acquired through middlemen. In the case of gift-exchange, a received gift cannot immediately be re-paid by a return-gift. This means that a future debt and therefore a social obligation is created on the part of the receiver (e.g. Sahlins 1972, Mauss 1990 [1925]). Commodity-barter, on the other hand, takes place simultaneously between two parties who both have (equally valued) goods that the other desires. After the goods have been bartered, social relationships may exist between the parties (which may ease future transactions of the same kind), but there are no significant obligations. Thus, the main difference between the two types of acquisition is that while gift-exchange creates an asymmetrical social relation (debt), commodity-barter is characterised by a relative freedom and balance (Humphrey & Hugh-Jones 1992:18).

Archaeologically, it is difficult to distinguish between gift-exchange and commodity-barter. In both cases, one may expect that the goods were of high value. For the time being I find no convincing arguments from which I may develop different archaeological expectations in order to distinguish between barter and exchange in a Stone Age context. Below, these two phenomena will therefore be termed "exchange/barter". It is perhaps less problematic to distinguish between direct access and exchange/barter. It is assumed here that exchanged/bartered raw materials have higher values than goods that are procured by direct access to the raw material sources. Based on this assumption, it may be possible to relate different phases of the chaîne opératoire to these two different types of acquisition.

Chaîne opératoire

The chaîne opératoire (the operative process) approach has first of all been developed within French archaeology

Figure 122. Cross-boundary flow of raw materials with known sources in early Neolithic southern Norway. Sources are indicated approximately by letters and arrows indicate approximate distribution limits.
SL: *Slates*
SA: *Sandstone*
D: *Diabase*
R: *Rhyolite*
CH: *Chert*
MY: *Mylonites*
A: *Anorthosite*
GQ: *Fine green quartzite 1*
TRB: *TRB pottery and thin-butted axes*
FL: *Flint*

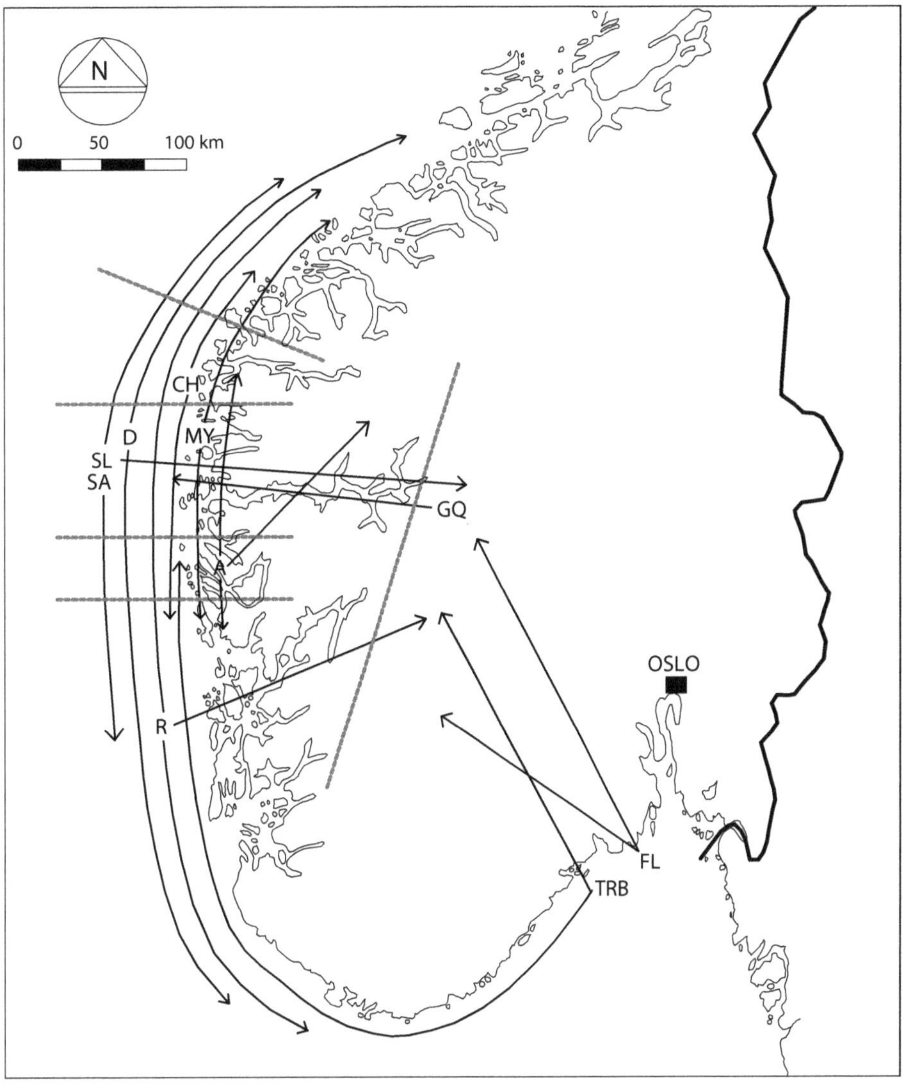

Operational stage	Activity	Product	Discard
Quarrying	Quarrying and testing	Blocks	Discarded blocks, debris of different sizes
Preparation	Production of cores or blanks	Cores, blanks	Large amounts of flakes of different sizes, discarded blanks
	Adzes: grinding of blanks	Finished adzes	Discared adzes
Basic production	Slate points:grinding of blanks	Finished slate points	Discarded slate points
	Grinding slabs: preparation of grinding surface	Finished grinding slabs	Used-up or broken slabs
	A-points: production of blades and suitable flakes	Blades, flakes	Middle-sized and small-sized flakes, used up cores and discarded blades
Modification	A-points: retouch of tang	A-points	Small flakes, discarded A-points
Utilisation	Shafting and use, resharpening, re-sharpening, re-use etc.	Re-used products	Small flakes, discarded products
Discard	Discard or deposition	All stages	All stages

Table 5. *Model of chaîne opératoire of the lithic artefact-types analysed. Modified from Eriksen (2000:81).*

and ethnology since the 1960's (e.g. Leroi-Gourhan 1964, Leroi-Gourhan 1965, Lemonnier 1976, Pelegrin et al. 1988). The important contribution of the approach is that it takes the whole cognitive operative process of tool production – not only the end products – into consideration, and that it enables for a study of how technology is involved in the social system. Aside from studying the tools themselves, one also analyses the way the tools have been produced and tries to understand the basic technological knowledge (Eriksen 2000:77, Barndon 2002). In this contribution, I will not perform a complete chaîne opératoire analysis for any of the lithic raw materials. However, I find that the terminology developed within the approach is suitable for my purpose, which is to distinguish between socially meaningful technological stages in the lives of the lithic tools. The terms applied here are the ones related to the entire operative process from acquisition to discard: quarrying/collecting, preparation, basic production, modification, application, and discard. In Table 5 I present a model which describes the different activities that can be related to these six terms. The expected products and the discard from the operative processes of making adzes, slate points, and A-points are also described.

At the outset I assume that there is a close relationship between the number of operational stages performed at a site and the type of acquisition. If the artefacts have been bartered or received as gifts before arriving at the site, I expect the preparatory operational stages (which is likely to have left much more debris than later stages) to already have been performed. As a result, the number of discarded blanks, cores, and finished products would be quite large compared to the amount of debris from the preparatory stages of tool production. I assume that blanks/cores/finished products would be more valuable as gifts or objects of barter than would unprepared blocks. Reverse, if the artefact has been quarried or collected by the site occupants themselves, one would expect to find the preparatory stages of the tool production as well.

In order to distinguish between these two interpretative options, I will compare the relative percentages of debris[1] (large (>4 cm), medium (4><1 cm) and small (<1 cm) flakes) with the percentages of products (cores, blades, blanks, and finished products) at all of the sites for each of the raw materials. Not all of the raw materials will be compared here. I concentrate on the ones that are spread across the ethnic boundaries and for which the quarries or source areas are assumed to be known.

Flint

As pointed out in Chapter 7, flint could probably be found at beaches throughout the coastal area and is a likely candidate for direct access in all of the districts. Due to the distance between the coast and the mountains, and also because of the possible ethnic boundaries between the coast and the inland, one may expect that they were bartered/exchanged by the groups who occupied the mountain sites. As a result, if the above assumptions are correct, preparatory stages as well as basic production, modification, and utilisation should be found at all of the coastal sites, while only basic production and later operational stages would be found in the mountains.

1 A number of the artefacts classified as flakes have most likely also been used as cutting tools, but in order to distinguish these, a detailed inspection of the stone implement is needed. Such an inspection could not be performed for the present work. Here, only flakes with scraping retouch are recorded as "retouched flakes".

For the coastal sites, the results of the analysis (Figs. 123, 63) largely match these expectations. The relative distribution between flake sizes is quite similar. The number of products compared to flakes is also quite constant. Based on these data, it can be argued that flint arrived at the sites as unprepared blocks or nodules. Such a theory is supported in that several nodules (particularly at Sunnmøre) and a number of pieces with cortex were identified. If the above assumptions are correct, one may, therefore, conclude that flint was mainly acquired by direct access to the beach deposits throughout the coastal area.

Contrary to the expectations, the data from the mountain sites shows almost the same pattern as the coastal sites: people who occupied these sites did not bring with them only prepared cores or tools. The large amounts of flakes of all sizes indicate that entire blocks or nodules were also carried to the highlands and reduced there (see Coulson 1986 for an example of this practice). One interpretation of this pattern is that groups that exploited the mountains also had direct access to the flint deposits at the coast. An alternative explanation, which is more in accordance with the theory of an ethnic boundary between inland and coastal groups, is that the flint nodules were bartered or exchanged from coastal groups. This alternative would, however, imply that the initial assumption of a relationship between direct access and presence of blocks/preparatory stage is incorrect. Instead, blocks/nodules should also be regarded as valued goods, and as such, just as attractive as objects of exchange/barter as blanks or finished products.

The problem of direct access vs. barter/exchange across ethnic boundaries is of crucial importance for the problem at hand, and will be discussed at length below. However, before such a discussion can take place, it is necessary to investigate whether the raw materials with known sources also display the same kind of tendencies as flint.

Rhyolite

Rhyolite was quarried at Siggjo, Bømlo. It was spread almost throughout the area of analysis, but the northern boundary of the main distribution area is situated in Nordhordland. To the north of this district, rhyolite makes up less than 5% of the total amount of raw materials (Figs. 124, 84). Rhyolite is also rare at the mountain sites.

There is a fairly marked difference between the number of operational stages performed on rhyolite within the main distribution area and districts outside of this area (Figs. 124, 84). As far north as Nordhordland, the number of rhyolite flakes clearly exceeds the number of blades, cores and tools. It is also important to note that large rhyolite flakes are almost exclusively found within the main distribution area. This indicates that blocks taken directly from the quarry were reduced at the sites themselves as far north as Nordhordland. To the north of this district, and at the mountain sites, a significant amount of flakes were found. However, tools and blades constitute larger percentages at these sites than within the main distribution area. This implies that outside of this area, rhyolite was generally not acquired as unreduced blocks. Instead they came as already reduced blocks, prepared cores, or as finished products.

This result indicates that, within the main distribution area, rhyolite was primarily acquired by direct access to the quarry at Siggjo. Groups living to the north and east, however, first of all have secured it through exchange relations with the former populations.

Blue mylonite

Blue mylonite, which was probably quarried in Sunnfjord, has been identified at sites from Midthordland in the south to northern Sunnmøre in the north, but the main distribution area is Nordfjord, Sunnfjord, and Nordhordland (Figs. 125, 82). The share of flakes compared to blades/cores and tools are somewhat higher in Sunnfjord. However, a high number of flakes of all sizes found at sites in Nordfjord and Nordhordland still indicate that the raw material also arrived there mainly as unprepared blocks, not as finished tools or prepared cores. The operational sequences performed on blue mylonite were therefore mainly the same at the Sunnfjord sites as in the more peripheral districts Nordfjord and Nordhordland. In Sunnmøre, the raw material is mainly represented by tools, while only a few flakes have been found in Midthordland. This result indicates that groups from Nordfjord as well as Nordhordland had direct access to the mylonite sources in Sunnfjord.

Coarse, and medium grey mylonite

Grey mylonite was probably quarried in Sunnfjord. It is mainly found in this district, although some items have been identified in Nordfjord and Nordhordland (Figs. 126, 81). The number of flakes compared to tools and cores/blades indicate that it was brought to all of the sites as blocks. This may mean that it was procured by means of direct access across the boundaries.

Chert

Chert was most likely quarried at the island of Frøya in Bremanger, Nordfjord. The largest quantities of chert are found at sites in Sunnfjord and Nordfjord, but it has been identified at most of the sites from northern Sunnmøre to Midthordland (Figs. 127, 83). Large flakes are also exclusively found in Nordfjord and Sunnfjord, and here flakes dominate tools and cores/blades. At these sites, therefore, chert arrived as blocks. At sites further away from the source (Nordhordland), flakes also make up the largest shares, while at the Sunnmøre sites, finished products dominate. Chert was distributed across the boundaries to distant sites as blocks well as prepared artefacts.

Fine, green quartzite 1

This raw material was most likely taken out at the quarry Kjølskarvet II in the Lærdal Mountains, and has been identified in large quantities at mountain sites quite close to the source. Only a few possible specimens of fine green quartzite have been found at the coast (Figs. 128, 64). At the mountain sites, the large number of flakes clearly indicates that it was carried to the sites as blocks. At the coastal sites, the data indicates that it first of all came as finished products, cores or large flakes.

Red, green and grey slate

Green slate is represented in all of the coastal and mountain districts, while red slate is only found between northern Sunnmøre and Nordhordland (Figs. 129-131, 85-87). At the sites in Nordfjord, Sunnfjord and Nordhordland, green slate is generally represented by a fairly large amount of flakes compared to blanks and points, indicating that that blocks were reduced into blanks and projectile points. To the north and south of these districts, points and blanks have much larger shares. Interestingly, while red slate displays the same general pattern at sites in Nordfjord and Sunnfjord, it has only been found as blanks or points in Nordhordland and in Sunnmøre.

Grey slate only appears in significant amounts in Nordfjord and in Sunnmøre. At all of the sites, flakes dominate and it was probably distributed as blocks.

The variation in operational sequences is great between the mountain sites as well. At Styggvasshelleren, slate projectile points clearly dominated. Only a very few flakes were identified. Such a pattern is also found at Mørkedøla I, Blånut IV, and Gyrinos IV. This indicates that slate was taken to these sites as blanks or finished products, not as blocks. At 760 Finnsbergvatn and Vivik, the pattern is very different. At these sites, the flakes dominate the finished products, indicating that slate came as unreduced blocks.

As noted in Chapter 8, much of the slate at the mountain sites may have other sources than the Devonian sand and siltstone basins in Sunnfjord. Before the problem of provenance is solved, the slate-data should be treated with caution. For the slate in the coastal assemblages, it is argued that most of it probably originates in the Devonian Basins. Admittedly, the distribution patterns are somewhat confusing with regard to the problem at hand: the different colour variations of slate are not distributed in the same manner, and the diversity is significant with regard to the problem of boundaries and operational process.

Anorthosite

Anorthosite probably originates in Nordhordland. At the sites in this district, a few points and a large number of flakes were found. Outside of Nordhordland, mainly points and blanks have been identified (Figs. 132, 91).

Greenstone, diabase and sandstone

Diabase from Stakaneset, Flora was used in the entire coastal region, but has a main distribution area between Sunnmøre and Nordhordland (Figs. 59, 62). Greenstone is more difficult to evaluate (see Chapter 8), but both quarry 4 greenstone and greenstone from Bømlo appear to have been spread widely (Figs. 63, 62). Significant amounts of debris from adze production are only found at sites close to the quarries at Stakaneset (Kleiva, Vikja I, Neset II) and Hespriholmen (Sokkamyro). At all other sites, these raw materials appear first of all as blanks, finished adzes or as flakes of ground adzes. As has been already pointed out by Bruen Olsen and S. Alsaker (1984:82), this indicates that they were distributed from these sites as blanks or finished products.

Sandstone from Sunnfjord has been spread throughout the coastal area of analysis, but the main distribution area is between Nordfjord and Nordhordland (Fig. 92). It only appears as grinding slabs or fragments or such. There are no indications that preparation of the slabs has been performed at any of the excavated sites. Similar to the adzes, they were probably prepared close to the quarries and distributed as finished products from these production sites.

According to the above, adzes/blanks as well as grindstone slabs were primarily bartered/exchanged from the groups that lived close to the sources. If this is correct, however, the idea of direct access only taking place within boundaries and exchange/barter only across, will have to be modified.

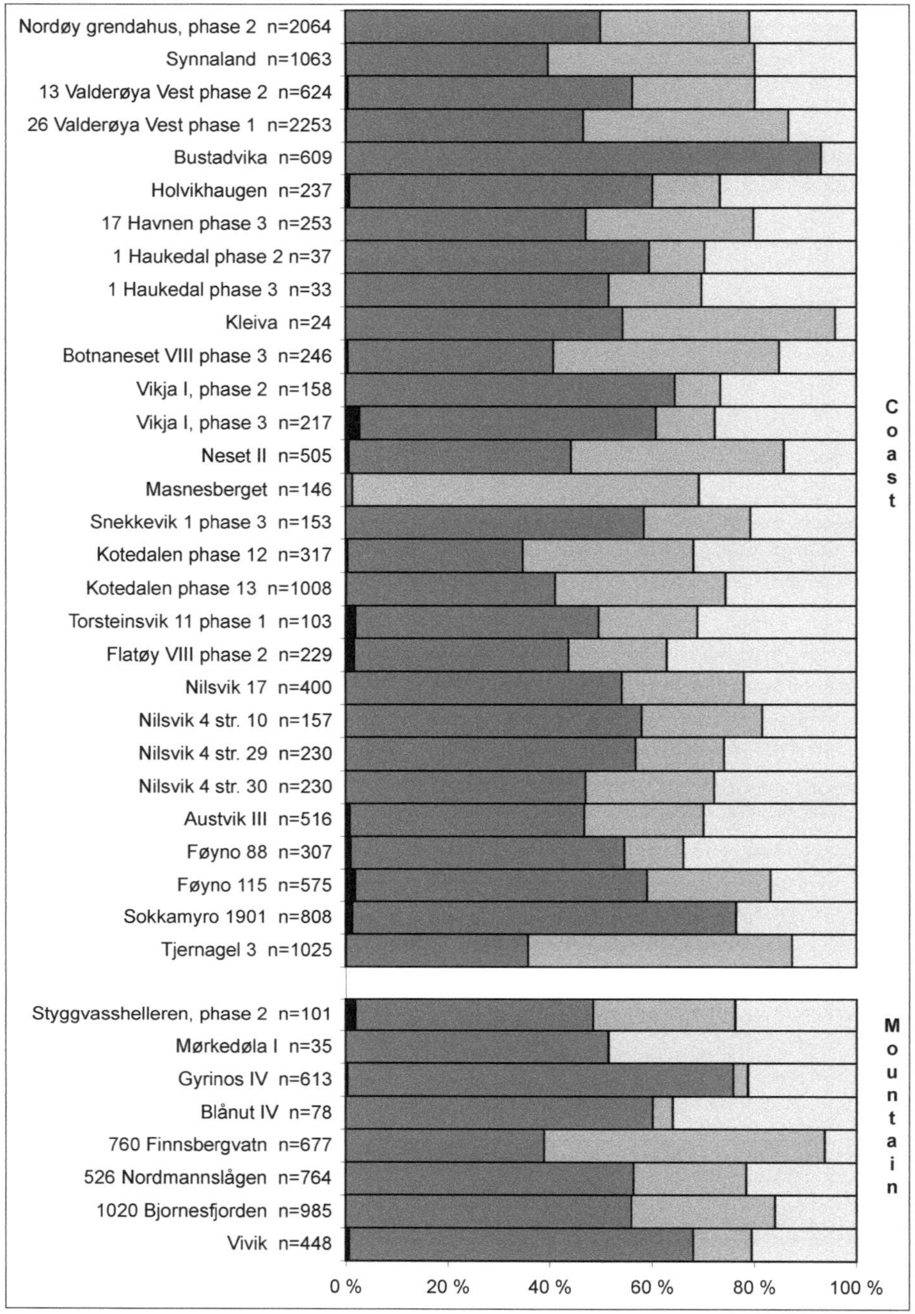

Figure 123. *Percentages of flint flakes and points/blanks/blades/cores at excavated sites.*
Black: large flakes, dark grey: medium sized flakes, grey: small flakes, light grey: points/blanks/blades/cores.

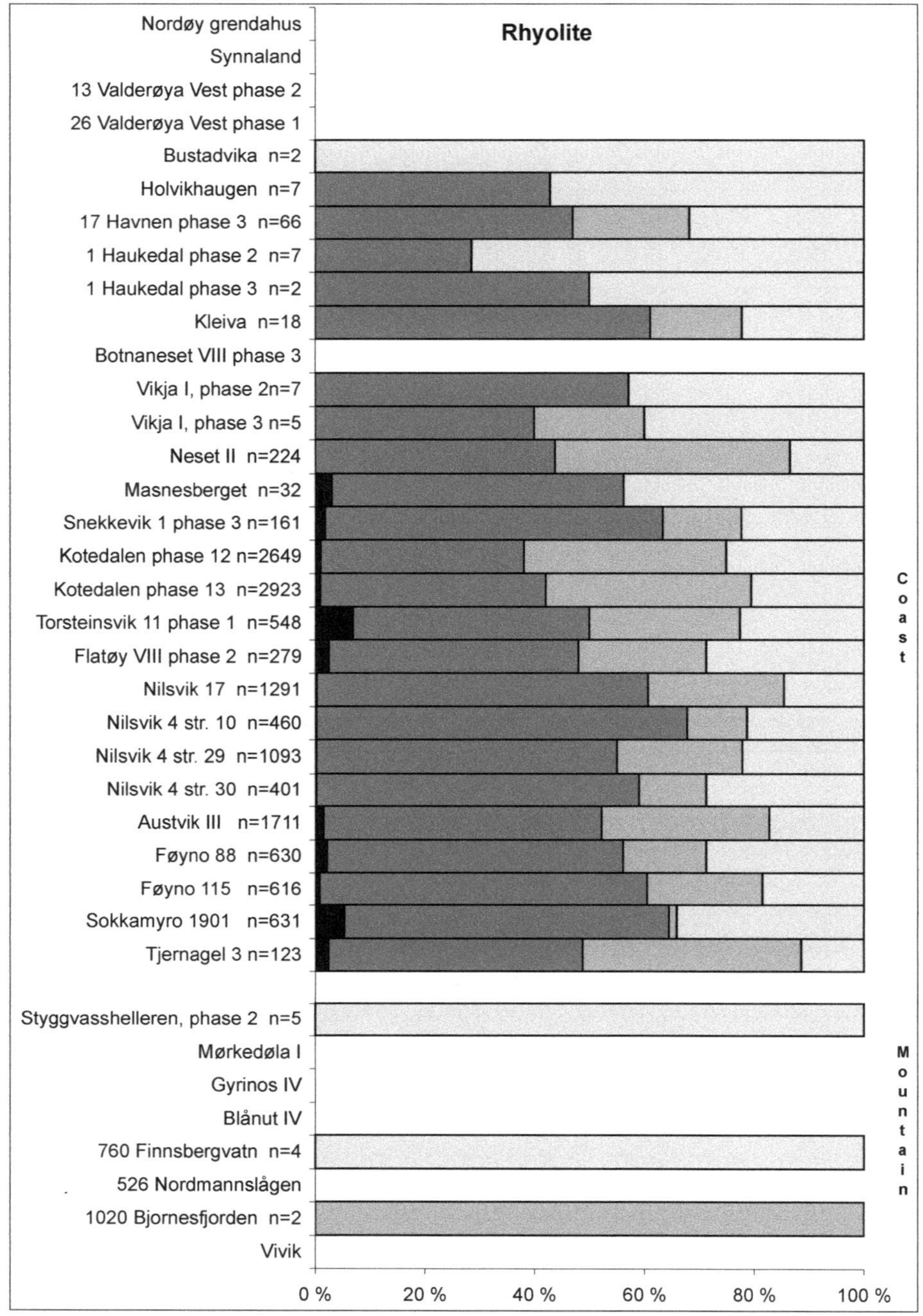

Figure 124. *Percentages of rhyolite flakes and points/blanks/blades/cores at excavated sites.*
Black: large flakes, dark grey: medium sized flakes, grey: small flakes, light grey: points/blanks/blades/cores.

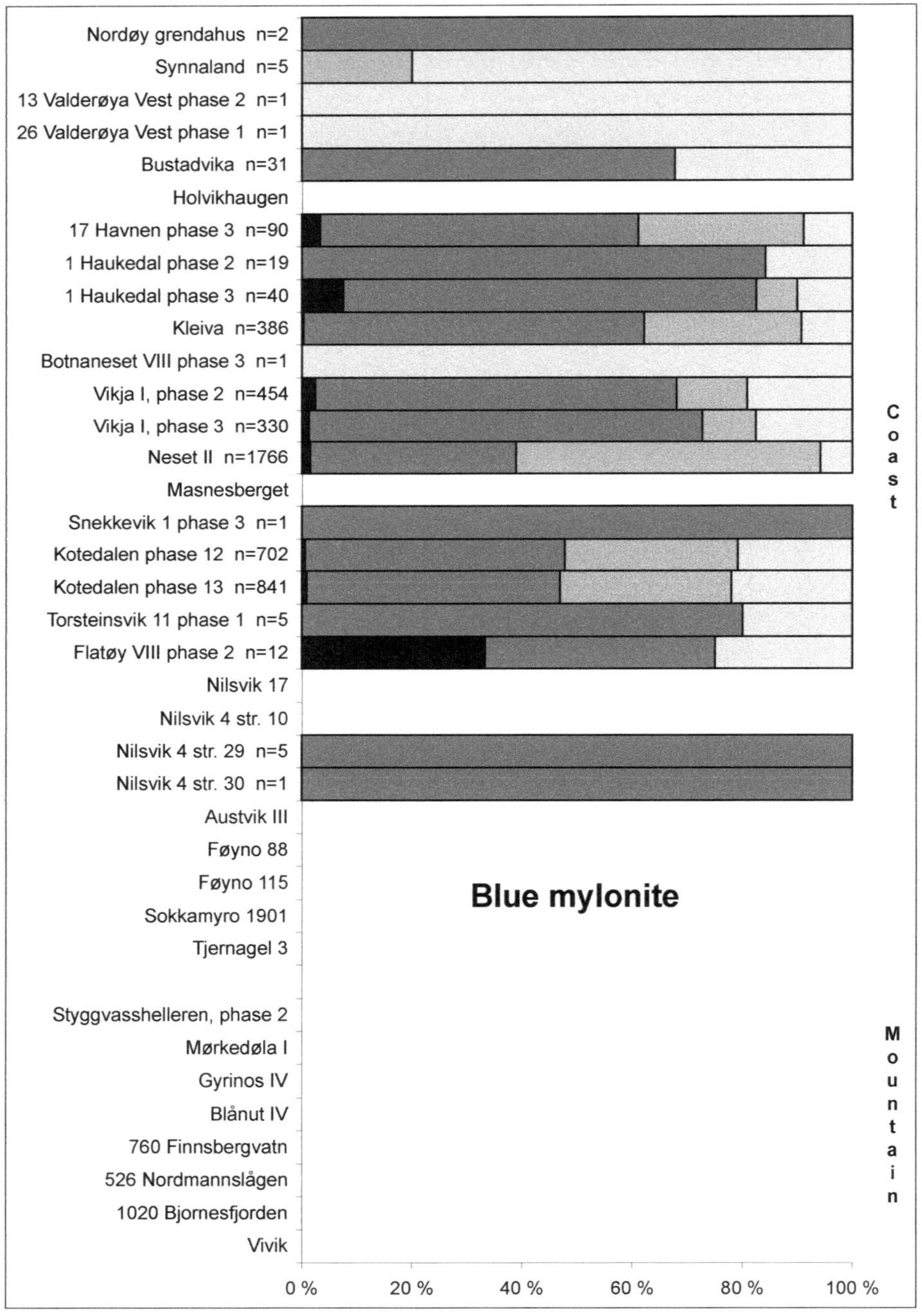

Figure 125. *Percentages of blue mylonite flakes and points/blanks/blades/cores at excavated sites. Black: large flakes, dark grey: medium sized flakes, grey: small flakes, light grey: points/blanks/blades/cores.*

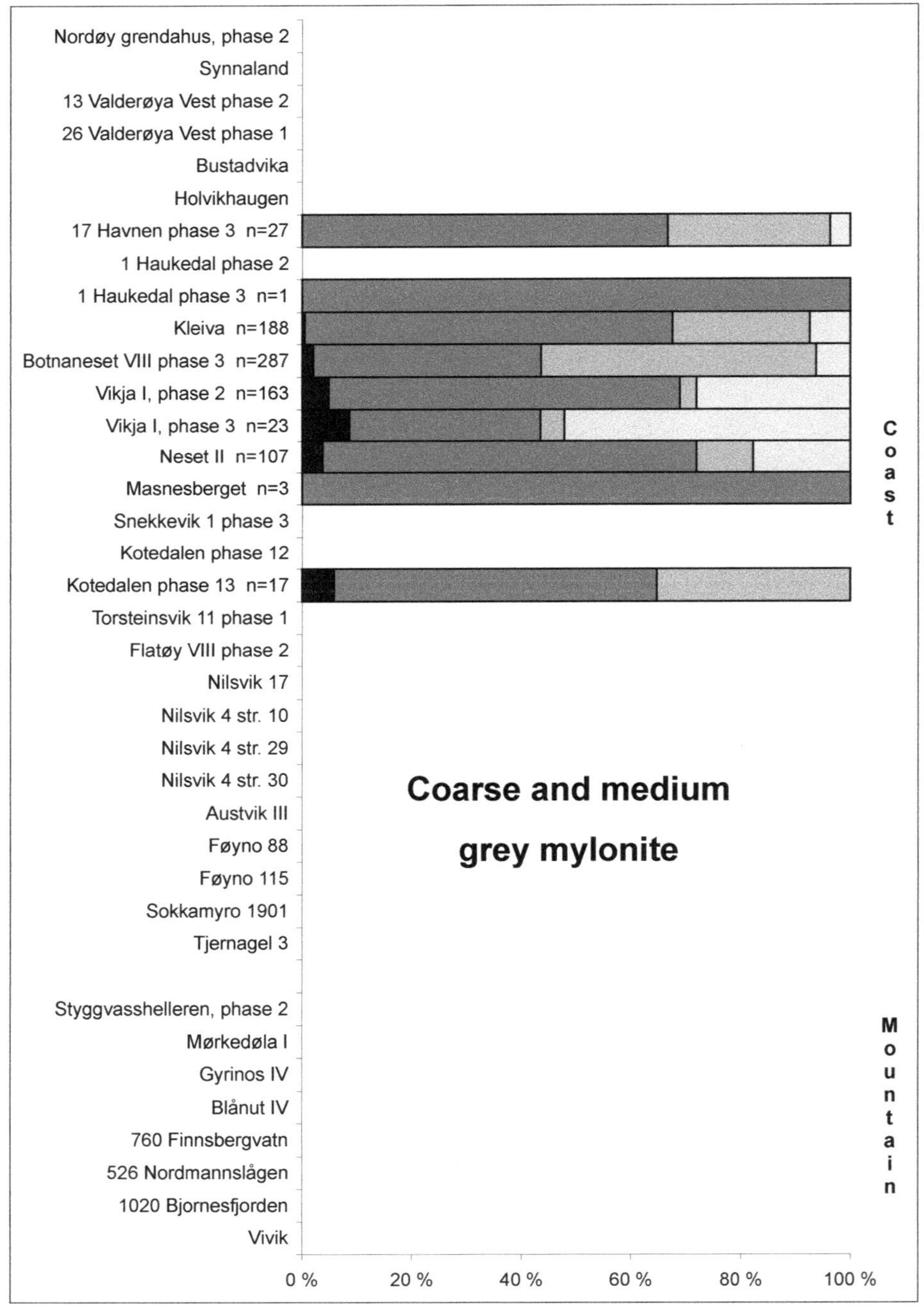

Figure 126. *Percentages of coarse and medium grey mylonite flakes and points/blanks/blades/cores at excavated sites. Black: large flakes, dark grey: medium sized flakes, grey: small flakes, light grey: points/blanks/blades/cores.*

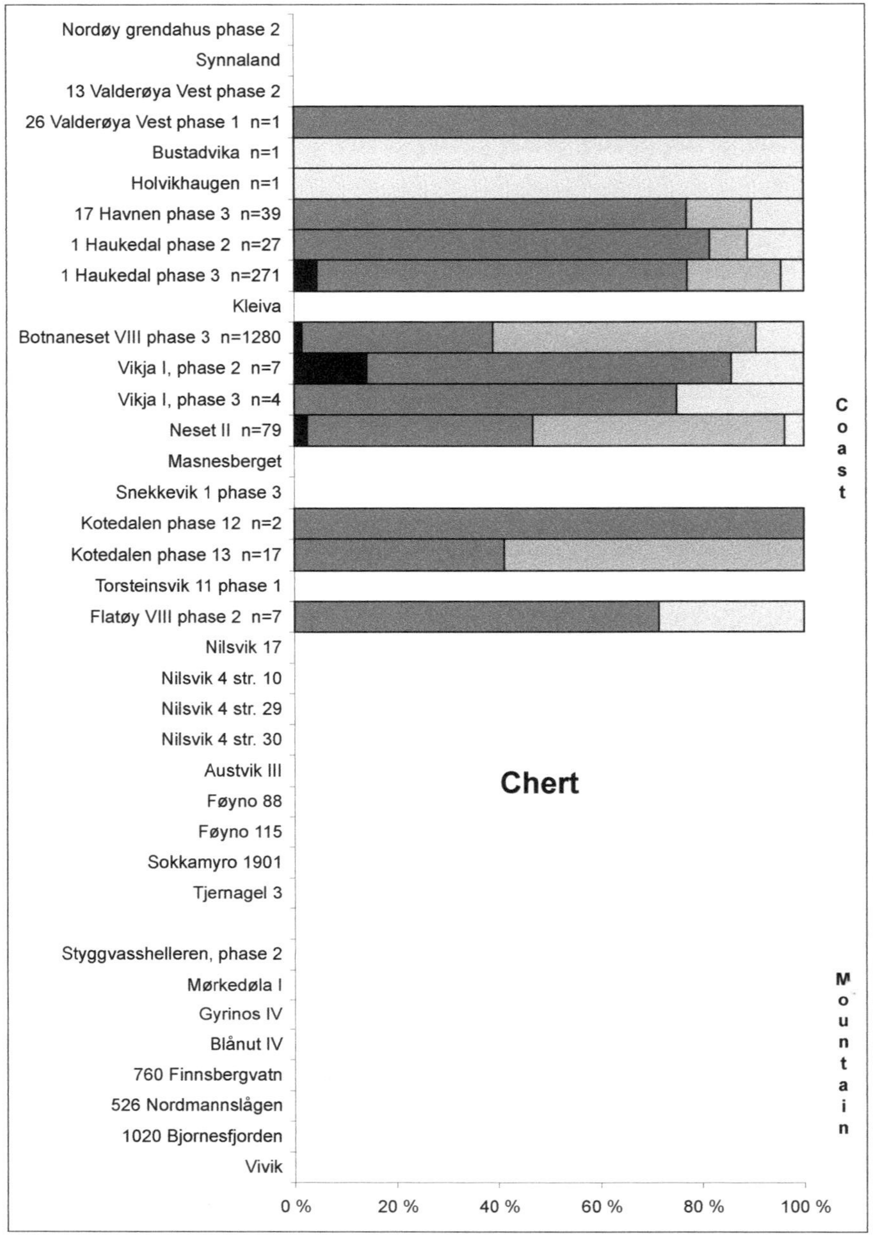

Figure 127. *Percentages of chert flakes and points/blanks/blades/cores at excavated sites. Black: large flakes, dark grey: medium sized flakes, grey: small flakes, light grey: points/blanks/blades/cores. Figure 128. Percentages of fine green quartzite 1 flakes and points/blanks/blades/cores at excavated sites. Black: large flakes, dark grey: medium sized flakes, light grey: small flakes, light grey: points/ blanks/blades/cores.*

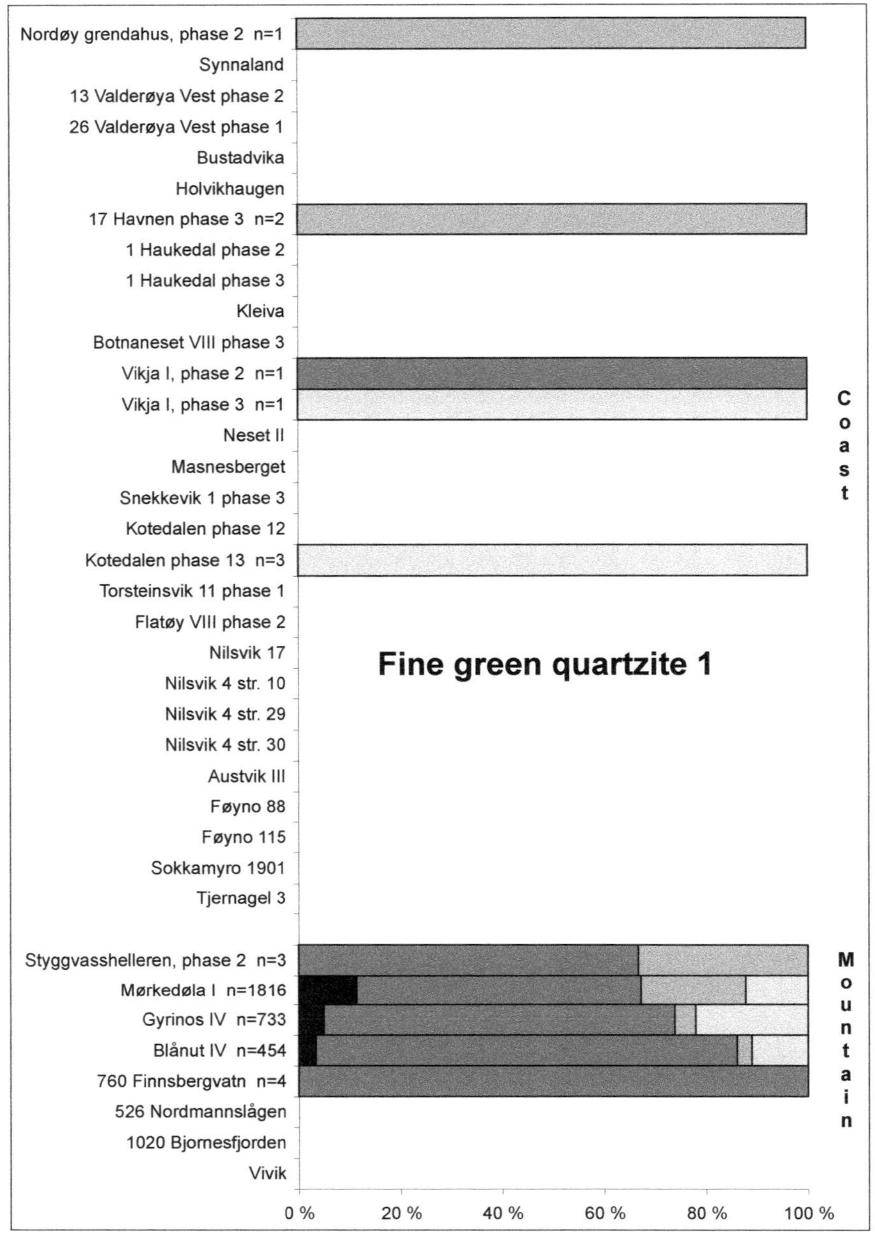

Figure 128. *Percentages of fine green quartzite 1 flakes and points/blanks/blades/cores at excavated sites. Black: large flakes, dark grey: medium sized flakes, light grey: small flakes, horizontal ribs: points/ blanks/blades/cores.*

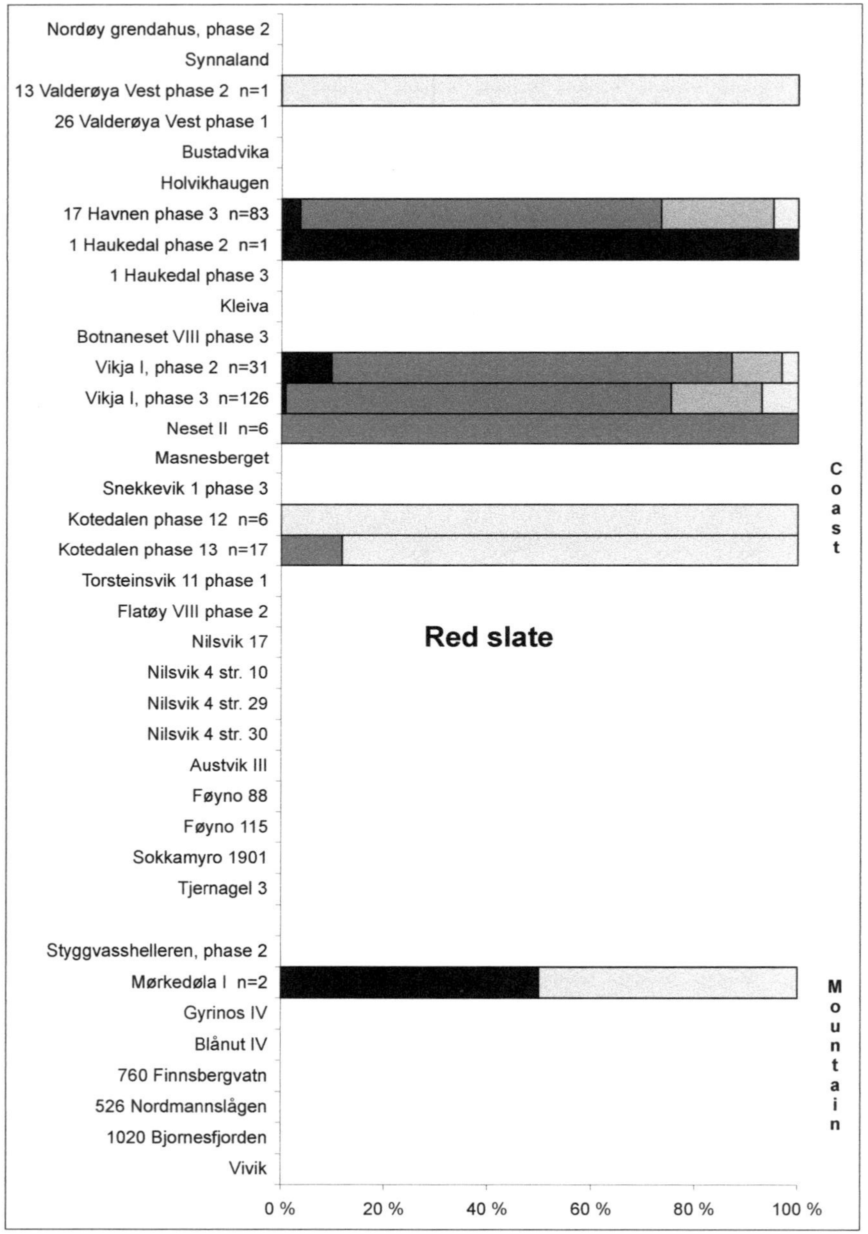

Figure 129. *Percentages of red slate flakes and points/blanks at excavated sites. Black: large flakes, dark grey: medium sized flakes, grey: small flakes, light grey: points/blanks/blades/cores.*

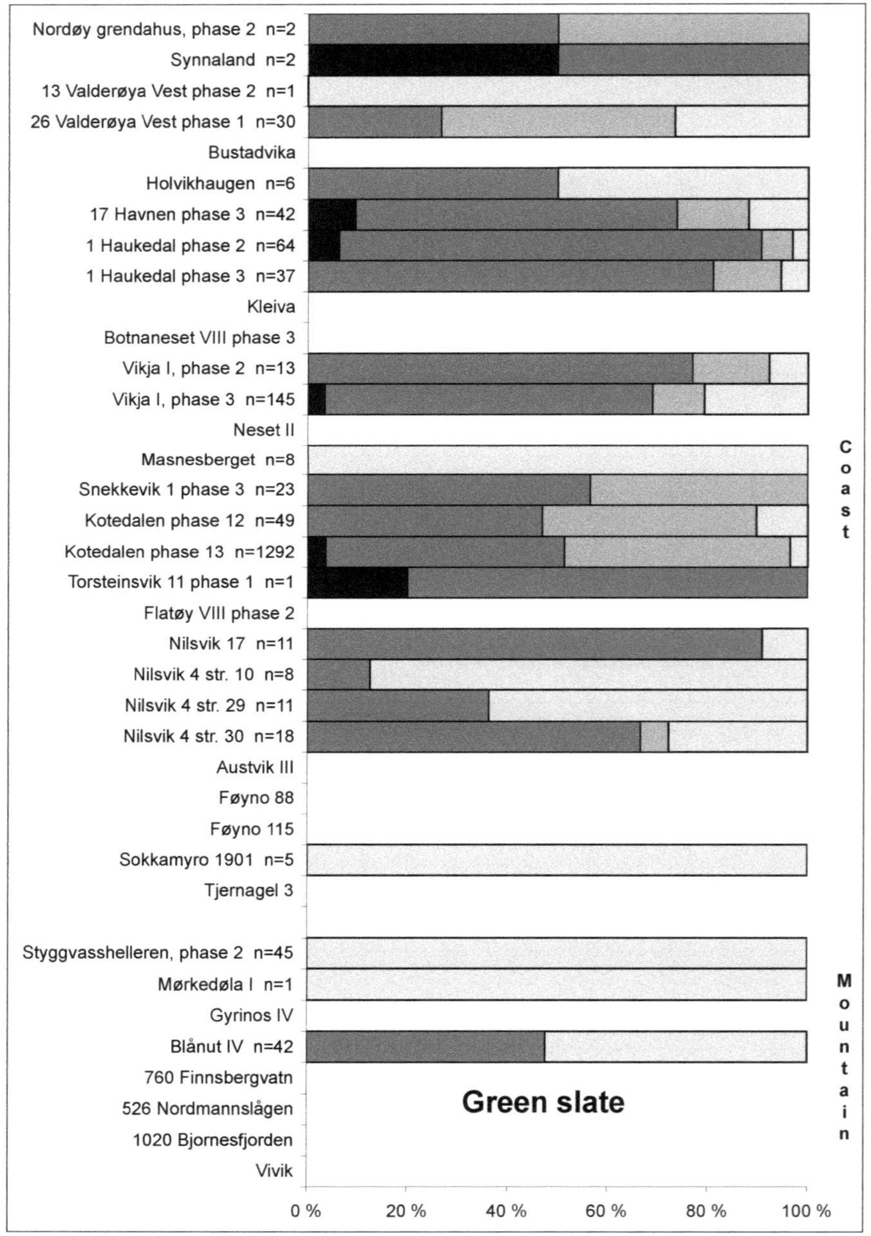

Figure 130. *Percentages of green slate flakes and points/blanks at excavated sites. Black: large flakes, dark grey: medium sized flakes, grey: small flakes, light grey: points/blanks/blades/cores. Figure 131. Percentages of grey slate Black: large flakes, dark grey: medium sized flakes, grey: small flakes, light grey: points/blanks/blades/cores.*

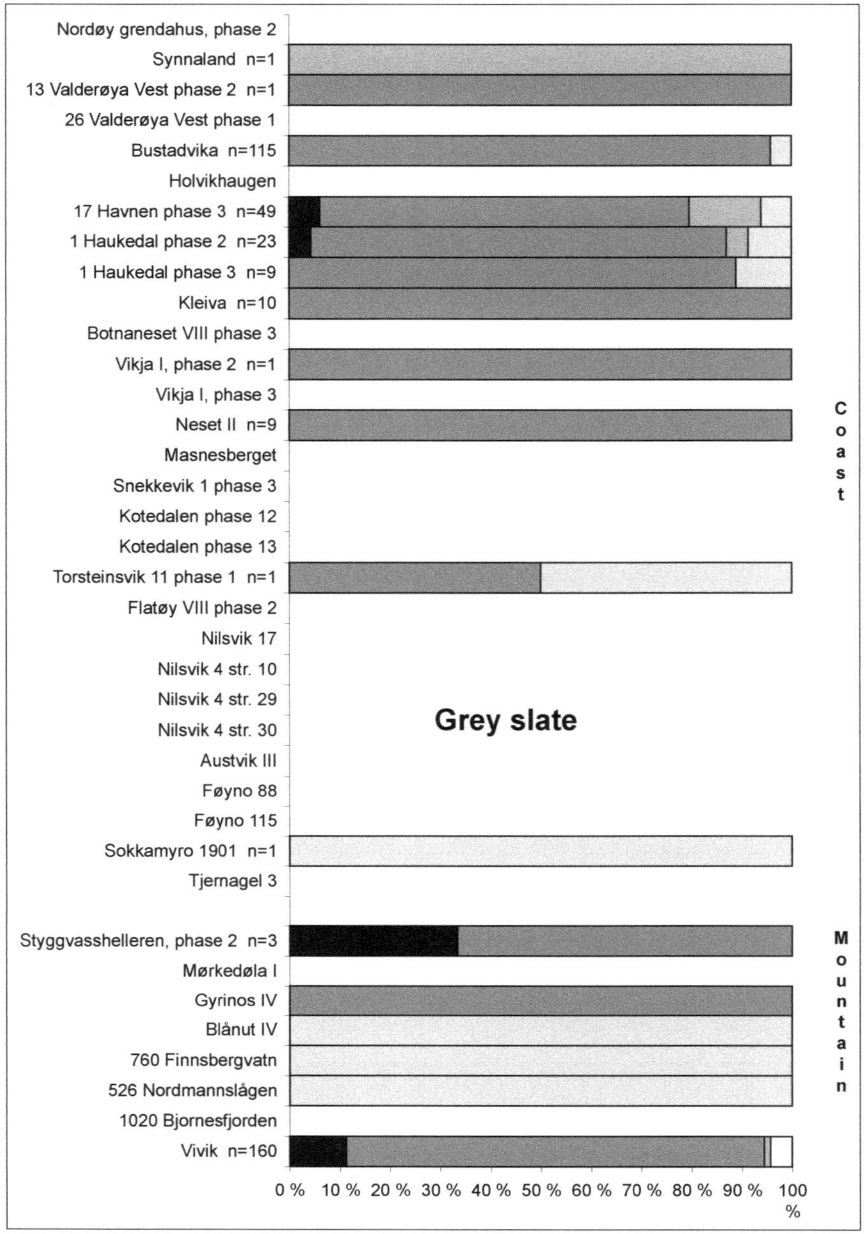

Figure 131. *Percentages of grey slate flakes and points/blanks at excavated sites. Black: large flakes, dark grey: medium sized flakes, light grey: small flakes, horizontal ribs: points/blanks.*

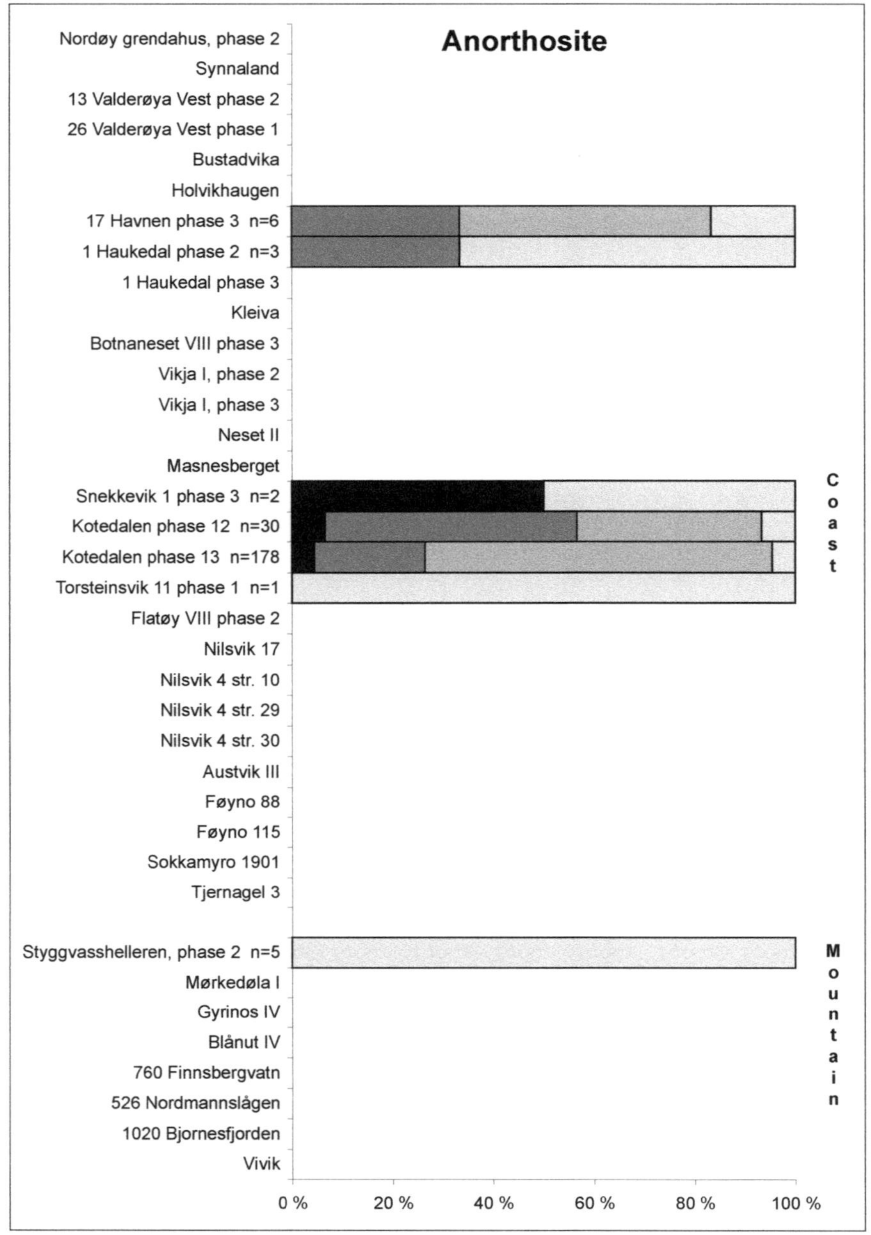

Figure 132. *Percentages of anorthosite flakes and points/blanks at excavated sites. Black: large flakes, dark grey: medium sized flakes, grey: small flakes, light grey: points/blanks/blades/cores.*

Discussion

In summary, the above analysis shows that there is no simple one-to-one relationship between the location of ethnic boundaries and the documented steps in the chaîne opératoire of the individual raw materials. The distribution patterns of some of the raw materials largely confirm the expectations of a change in operational stages across the proposed boundaries. This is the case with rhyolite to the north of Nordhordland and fine green quartzite 1 between the mountain sites and the coastal sites. It is also the case with red and green slate between Nordfjord and Sunnmøre and red slate between Sunnfjord and Sogn/Nordhordland. Treatment of anorthosite also changes markedly towards the south and north of Nordhordland.

On the other hand, several of the raw materials do not match the expected pattern. Flint passed as nodules between the coast and the mountains, blue and grey mylonite passed as blocks from Sunnfjord across the boundaries to Nordfjord as well as Nordhordland. The same applies to green slate and chert.

Greenstone, diabase and sandstone have been distributed as blanks and finished products throughout the region.

An illustrative example of the complexity of the situation is the raw material composition during phase 13 at Kotedalen in Nordhordland (Fig. 5 and 6). At this site, rhyolite from Bømlo represents more than 40 % of the total amount of lithics in phase 13, while green slate and blue, medium mylonite from Flora/Bremanger constitute 20 % and 10 % respectively. These three raw materials were reduced in a fairly similar fashion at the site (Fig. 133). Judging from the amount of debris of medium and large sizes, and the amount of cores/blanks/blades, it is apparent that core preparation, core reduction as well as finishing of products of rhyolite, mylonite, and slate tools were performed at the site itself. This means that these particular raw materials were mainly carried to the site as blocks, not as finished products. Greenstone from Bømlo and diabase/red slate from Sunnfjord, on the other hand, were treated in a strikingly different manner at the site. The amount of finished products made by these raw materials clearly exceeds the amount of debris. This indicates that red slate, greenstone, and diabase mainly were brought to the site as finished products or as blanks ready for grinding, not as unprepared blocks.

According to the initial assumptions, therefore, the data from Kotedalen indicates that the group that lived at this site had direct access to quarries in Flora in the north as well as Bømlo in the south. Ethnic boundaries had to be crossed in both cases. At the same time, greenstone, diabase and red slate (+ chert, grey mylonite) were bartered/exchanged from these neighbouring groups.

Seen together, this result is contradictory, and calls for a discussion of unreduced blocks being attractive as objects of exchange/barter and of direct access as a possible mode of acquisition across ethnic boundaries

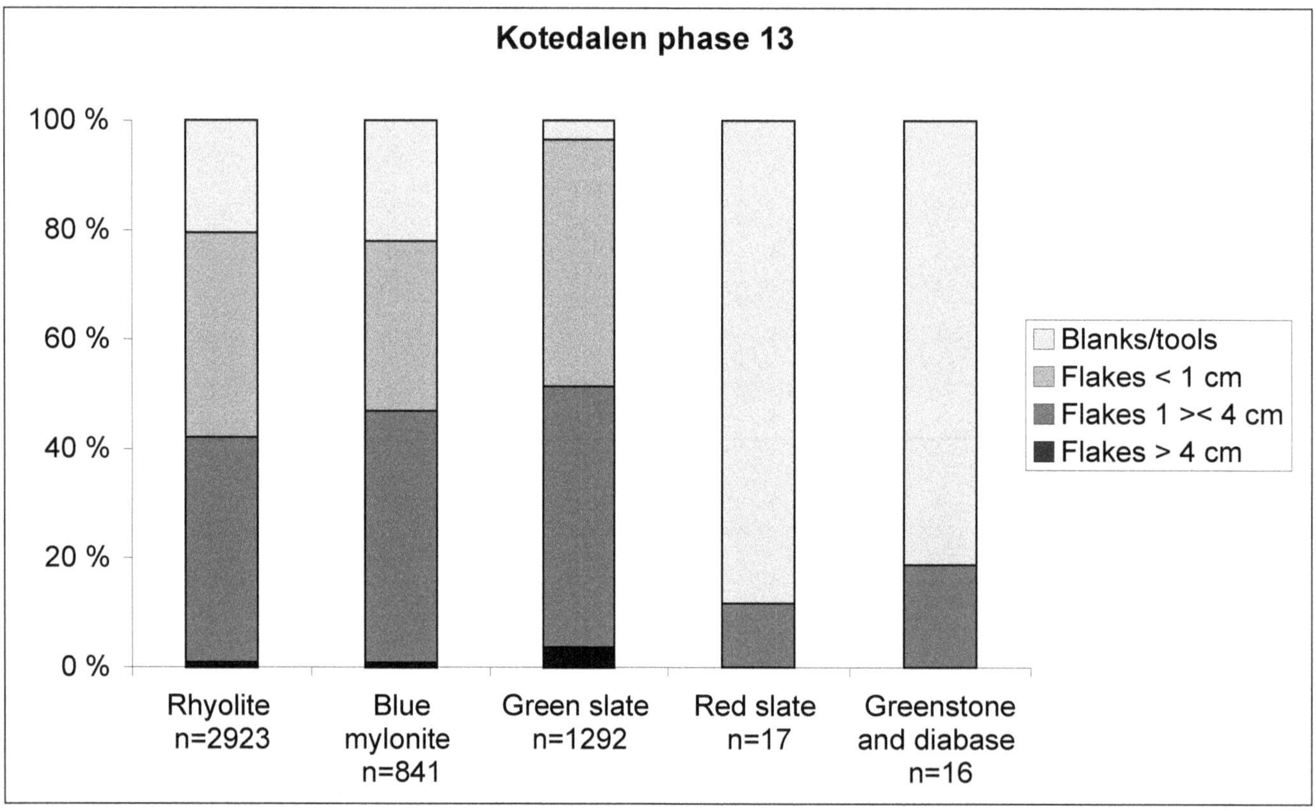

Figure 133. *Percentages of rhyolite, blue mylonite, green slate, red slate and greenstone/diabase flakes and points/blanks/blades/cores at Kotedalen phase 13, Nordhordland. Black: large flakes, dark grey: medium sized flakes, light grey: small flakes, horizontal ribs: points/blanks/blades/cores.*

Exchange/barter of blocks

One alternative interpretation of the above result may be that the initial assumption was wrong; unreduced blocks may have been just as attractive as objects of barter/exchange as blanks and finished tools. When establishing the assumption of a value-distinction between blocks and blanks/finished products in the first place, this was done partly because of the additional work that had been put into the objects, work that may have required special skills that may not have been present among the receivers. Another cause was that the finished object was thought of as more valuable because of its new shape and (humanly made) form, in contrast to (almost natural) coarse blocks. Thirdly, in contrast to blanks and finished tools, blocks have a problem with dead weight.

I believe that this distinction of value is still relevant for grindstone slabs, adzes/adze blanks vs. blocks of adze material, because the sandstone and basaltic rocks would probably be unsuitable for other types of stone tools due to their coarse nature; most of the flakes from the process of preparation are nothing more than waste material. For all blocks or nodules that were intended for all other types of tools, however, the distinction may be more problematic, first of all because blocks of slate or rhyolite or flint would be much more economical than blocks of basalt. In principle, a skilled knapper – and the regional distribution of the different reduction techniques shows that these skills were spread across ethnic boundaries – would be able to effectively utilise the blocks almost anywhere. Important is also the fact that flake material from the production of blanks and cores (slate probably excepted) would be more than dead weight; they were also potential tools.

On the basis of these arguments one may question the assumption of a distinction of value between blocks and blanks/finished tools of slate/struck materials; it is likely that blocks of rhyolite, mylonite, and quartzite were quite as attractive as blanks and finished products.

A relevant interpretation of the cross-boundary distribution of the blocks may therefore be that they were not necessarily acquired by direct access to the quarries. Instead, they may have been given as gifts or were bartered across the ethnic boundaries. If this is true, all goods that passed the boundaries may have been bartered or exchanged. In fact, it may imply that only the populations that lived close to the quarries had direct access to them and that everyone else would have to acquire the raw materials through these groups. In order to explore the latter alternative, I will discuss the social contexts of the raw material extraction.

Quarries and workshops

The extraction of greenstone, diabase and rhyolite is well known as a result of the extensive work of Haakon Shetelig (1922), Asle Bruen Olsen (1981), and Sigmund Alsaker (1987, Olsen & Alsaker 1984). With regard to the greenstone and diabase, these scholars have convincingly argued that the main reduction sequences took place on workshop-sites in the production area not more than 10 km from the quarries during the Mesolithic as well as during the Neolithic (Fig. 134). Outside of the production area, only blanks and finished adzes have been found. The production was organised differently in the two periods. A. B. Olsen and S. Alsaker have argued that the Mesolithic populations mainly secured their demands for greenstone and diabase by direct access to the sources. The main argument for this interpretation was the high number of Mesolithic small workshop sites close to the quarries. These sites were associated with small, independent task groups who were assumed to be responsible for quarrying and manufacturing as well as dispersal of the stone adzes (A. B. Olsen & S. Alsaker 1984, Bergsvik & A. B. Olsen 2003). The Neolithic data on workshop sites is very different. For Flora-diabase, the data has not yet been analysed for this purpose. At Bømlo, however, the reduction sequences of greenstone in the Neolithic mainly took place at one site, Sokkamyro, Southern Bømlo (Fig. 132), which was also a residential site (see Chapter 6). Shetelig, who excavated at Sokkamyro during several periods between 1901 and 1939, found debris from adze production and domestic activities in as much as 9000 m² at the site (Shetelig 1922, S. Alsaker 1987, A. B. Olsen 2002). The immense concentration of Neolithic greenstone debris at Sokkamyro indicates that the organisation of the production of adzes had changed from being a spatially dispersed activity during the Mesolithic to a spatially concentrated activity during the Neolithic.

The location of the quarries of slate and mylonite in Bremanger and Flora are not yet known, but the rhyolite-quarry at the top of Siggjo, Bømlo was surveyed and investigated by Sigmund Alsaker in the late 1970s (S. Alsaker 1987). At the quarry itself, numerous heaps of large, struck pieces of rhyolite were found. No other convincing rhyolite workshops have been found elsewhere at Bømlo[2]. Thus, it can be argued that all of

2 At the quarry itself, S. Alsaker found numerous heaps of large, struck pieces of rhyolite. These heaps he interpreted as results of the first stage of tool production (preparation of blocks). He assumed that the blocks produced at Siggjo were carried down to shore bound workshops for further reduction (preparation of cores) and distribution. A problem is that, despite of extensive surveys, no sites of this type have been yet been found at Bømlo. S. Alsaker argued that one site to the north of Siggjo, Rubbestadneset II, could be characterised as a rhyolite-workshop (S. Alsaker 1987:51-54). S. Alsaker's interpretation of the site may, however, be questioned. In connection to the present work I investigated the lithic inventories of Rubbestadneset II, and found that it cannot be distinguished from other residential sites, such as Kotedalen, in terms of the reduction sequences of rhyolite. I believe therefore, contrary to S. Alsaker, that all of the preparatory stages of rhyolite took place at the quarry itself.

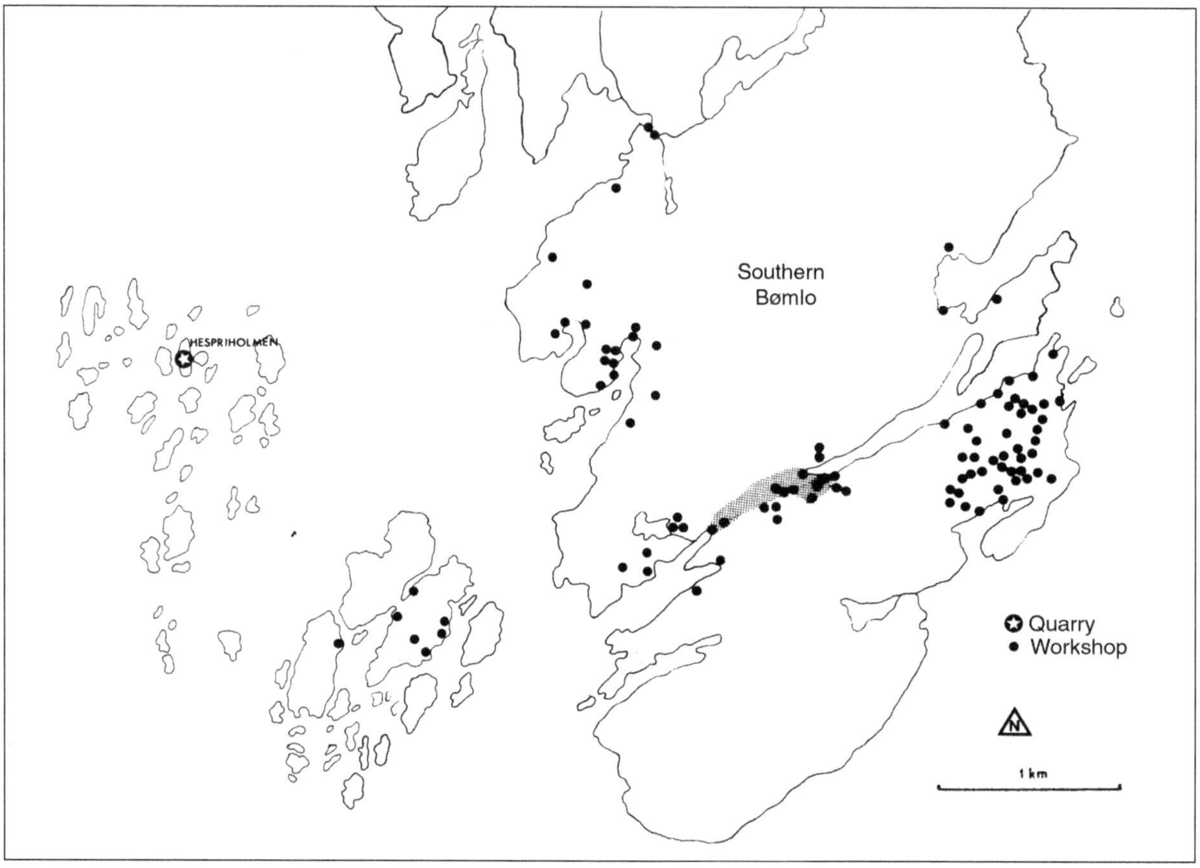

Figure 134. *The greenstone quarry at Hespriholmen is situated on a small, naked islet (marked by a star) far off the main island of Bømlo. Heavy winds and waves make Hespriholmen hard to access by small boats. Normally, safe journeys can only be arranged during the summer months. The small workshop sites are mainly Mesolithic, while the hatched area shows the extension of the Neolithic site Sokkamyro, which was both a workshop site and residential site. Modified from Alsaker (1987, Figure 30).*

the preparatory operational stages of rhyolite took place at the quarry itself; when blocks or prepared cores were carried down from Siggjo, they were ready for distribution to the residential sites. If this is so, it appears that, similar to the Neolithic greenstone workshop at Sokkamyro, the Neolithic rhyolite workshop at the top of the mountain Siggjo was spatially restricted to a fairly small area.

These observations and interpretations from the raw material sources are interesting, because they may indicate that greenstone and rhyolite production at Bømlo during the Neolithic was a tightly organised, restricted, and controlled activity. If the production was not organised and controlled, one would expect a much more dispersed pattern, such as the Mesolithic distribution of workshops (see S. Kristoffersen 1995b for a similar interpretation on the basis of data from Brandasund at northern Bømlo). But if this conclusion is accepted, what does it imply for the problem at hand?

One relevant interpretation is that sedentary local groups in the Bømlo district, perhaps based at the residential site at Sokkamyro, were the only ones with access to the quarries in this district, and that they performed all preparatory stages of the tool manufacture at Sokkamyro and Siggjo themselves. They may also have organised the further distribution of blocks as well as blanks/tools from these sites. If this was the case, populations from districts such as Nordhordland would have to depend on barter relations or exchange networks if they wanted blocks and blanks from Bømlo and Flora.

A conclusion that most or all of the inter-ethnic bonds were arranged in the contexts of exchange and barter may be seen as supportive for the theory of ethnic boundaries and consequently as sufficient for the problem at hand. However, by advocating such a theory, I have not really moved much beyond the previous contributions on the subject in early Neolithic Norway, where terms like "social networks" and "social channels" have been applied in order to characterise the transference of goods and ideas across space (e.g. Bostwick Bjerck 1988, A. B. Olsen 1992, Prescott 1996). The main problem with these terms is that they are not very specific with regard to what really happened "on the ground": how were the networks organised, and who did the travelling? This lack of purposeful agents, has, in effect, made it difficult to develop theories of why ethnic boundaries developed in the first place.

In the next section, I will investigate these problems by applying the concept of task group mobility, which is a well-known phenomenon among mobile hunter-gatherers and sedentary hunter-fishers. I argue that, in Neolithic Norway, task groups were significant agents in the inter-ethnic relations, and that the development of these relations was closely connected to the emergence of social inequality in this area.

Task group mobility across ethnic boundaries

The anthropologist June Helm (1965; 1968) is the first to define task group mobility as an explicit element in hunter-gatherer social organisation. Based mainly on her fieldwork among the Arctic Dene and Dogrib, she defines task groups as groups of persons mobilised for specific work tasks. The expeditions can last for one night up to several weeks, depending of the nature of the tasks, varying from fishing or moose hunting to trading parties or war parties. The size and the composition of the task groups vary accordingly. With regard to composition, she distinguishes between two different types of task groups. One type comprises one or several conjugal pairs and dependants (households). Another type consists of able-bodied men (crews) drawn from one or several households. Both households and crews may be organised for trapping, fishing, or moose hunting, but only crews take care of trading and warfare (Helm 1968:120, see also Bergsvik 2002b for a discussion of the concept).

In my view, the flexible and universal institution of task group mobility as defined by Helm is useful for understanding the development and character of ethnic relations, not only because it helps understanding how people adapt to a given environment, but because it provides a link between the local groups and the rest of the world. In other words, task groups may, through their direct relations with neighbouring groups and distant powers, be pivotal in the processes of change. The development of social inequality is particularly interesting (e.g. Bender 1978 and contributions in Price & Brown 1985, Gregg 1991, and Price & Feinman 1995). I take social inequality to exist "...when socially distinct entities have differential access to strategic resources, and this differentiation gives those with access the ability to control the actions of others" (Paynter 1989:369-370). In the context of social inequality, Robert Kelly expresses the importance of inter-group relations the following way:

"Mobile and sedentary peoples have different types of inter-group contacts. When groups are residentially mobile, many individuals maintain their own relations with other groups through individual trade or acts of sharing. In the sedentary case, one or two persons maintain a social link with another group, while everyone else maintains his or her own relationship through these particular individuals" (Kelly 1991:143-144, italics mine).

Kelly's point is worth noting, because he focuses on the messengers. In his example, the messengers are the only persons with access to distant knowledge. As will be argued below, this type of knowledge is the ultimate attribute of leadership. In order to understand the development of social inequality among hunter-gatherers, we have to consider the character of distant knowledge; how it is acquired, and how it is manipulated. Mary Helms (1988) does this most coherently. Helms' main postulate is that space and distance are not neutral concepts, but instead deeply associated with sociology, politics, and especially ideology (Helms 1988:4). Through her extensive ethnographic survey, Helms demonstrates that traditional societies (which include sedentary chiefdoms as well as more mobile and egalitarian hunter-gatherers) have strikingly similar attitudes towards distant realms. Distant places and beings are often considered as sacred, mythical, and dangerous, compared to the safe homely spaces. These distant realms are often also seen as places where space and time melt together and where ancestral beings can be encountered. In effect, long distance travels are of utmost importance in the cosmogony of the local populations. Individuals or groups that make such journeys are consequently seen as particularly wise and brave on their return. Furthermore, the things they bring home are considered more powerful than items of more local origin. One might, according to Helms, expect long distance associations to have greater visibility in societies with centralised polities and hierarchically focused political-religious statuses. Alfred Gell, for instance, have argued that long distance inter-tribal commodity-barter among foraging groups in old Melanesia is not necessarily a basis for intra-group political power (Gell 1992:149). Yet, Helms maintains that hunter-gatherer societies may also be relevant in this context:

> *"...long-distance associations can benefit individual political-religious statuses as much as ranked positions. Shamans or lineage headmen or band leaders of more egalitarian societies, also being specialists in political-religious affairs and therefore also separated from society as ritual and knowledge experts, may be expected to pursue long-distance geographical contacts, too, and for essentially the same reasons" (Helms 1988:170).*

It could, of course, be argued that I am mixing two different institutions here: task groups should be kept analytically separate from Helms's individual agents; they have completely different agendas. To a certain extent, this is true; most likely there was a major difference between picking of berries in the next valley and visiting a great shaman 200 km away. Still, it might be argued that the participants in all types of task groups may gain some degree of self-realisation, escape for a while from the limitations of society and acquire fame (Helms 1988:170 ff.). This freedom may of course be utilised in different ways, but the main point here is that task group mobility as an institution, even in its more modest forms, and can be a pathway to power for socially ambitious individuals.

One such pathway may go through control of the task groups themselves. Again, such control is dependent on the distribution of knowledge. The leader of a hunt or a quarry-party knows the techniques and the rituals that have to be performed. He knows how to hunt, how to quarry, and how to approach strangers in the correct way. He also knows the paths and the travel routes. The expeditions may thus serve as contexts for initiations into manhood, where such knowledge is distributed in portions, and where some are favoured (e.g. Barth 1975). Through such uneven distributions of knowledge, the expeditions may serve as a tools for controlling other members of the task groups, in general as well as during the expeditions themselves. A classic ethnographic example of such a combination of task group leadership and leadership in general is the umialik (leader of the whale hunt) among the north-western Alaskan Eskimo. The umialik not only leads the whale hunt, he also initiates feasts, wars and trade expeditions: He is a politically powerful individual (Sheenan 1985). But this was not always so. Glenn Sheenan argues that prior to the adoption of whaling, the umialik was a successful hunter, but otherwise not differentiated from his fellows. Thus, the umialik transformed his role over time.

In the following, I will present data from Nordhordland, western Norway (Nærøy 1994, Bergsvik 1995) which indicates that task group mobility increased during the transition to the early Neolithic.

Task group mobility in Mesolithic and Neolithic western Norway

In Chapter 9, I suggested that the late Mesolithic populations of western Norway were probably sedentary at several tidal current channels along the coast, and that resources outside of these channels had to be acquired by means of task group mobility. During the transition to the early Neolithic, the local groups became even more attached to the channels (longer stays and more frequent reoccupations (see also Bergsvik 2001b).

Earlier interpretations of data from Nordhordland have taken up the problem of task group mobility as related to sedentism for the Mesolithic as well as the Neolithic periods. This was mainly done from data provided by a systematic test-pit survey of the area around Fosnstraumen (Fig. 5). The survey identified altogether 18 Mesolithic and 45 Neolithic sites (Fig. 135). These sites were functionally classified as long-term camps and short-term camps on the basis of archaeological differences (see Chapter 4 for archaeological criteria for distinguishing between these types of sites).

A site location analysis demonstrated a marked difference in spatial distribution between long-term and short-term sites in both periods. The long-term camps were all concentrated around the narrowest point in Fosnstraumen. Here, the harbour conditions, view and drainage conditions were good. Due to their long-term use and generally favourable locations, these sites were held to be the results of long occupations by one or several households. In contrast, the short-term camps showed no particular attachment to the narrowest point of the channel. They also had moderate view of sea, varying slope, and the drainage was often bad. Their relations to the contemporary shoreline were not as close as for the long-term camps. This indicates that most of them were not chosen to meet general, but rather specific aims. It is therefore argued that the short-term camps have been used by task groups (Bergsvik 1995:113). At least some of the many short-term sites at Fosnstraumen may be a matter of task groups working to ensure firewood or vegetable food or other raw materials for the residential group in Kotedalen. Others may have been the results of the households temporarily moving out of their winter dwellings.

Admittedly, test-pits and site location analysis do not provide substantial information about the character of such short-term occupation, such as types of activities and group composition. Recently, however, Arne Johan Nærøy has analysed several Neolithic site at Kollnes in Øygarden (Fig. 8), which is situated about 20 km to the south-west of Fosnstraumen. After the initial surveys, these sites could all be classified as "short-term camps" by the criteria used at Fosnstraumen. The sites were excavated, and from his investigations of activity patterns, Nærøy concluded that some of the sites were the results of specialised activities with only males present, while more generalised (but still relatively short-term) activities had taken place at others, indicating female as well as male presence (Nærøy 2000:197 ff). Although other interpretations of Nærøy's data are possible, his result is nevertheless indicative of a complex mobility pattern on the task group level during the early Neolithic period – sometimes involving crews, sometimes entire households. A similar variability can probably also be found among the short-term camps at Fosnstraumen. Thus, the data from Fosnstraumen and Kollsnes strongly indicates that task group mobility was an important element both in the Mesolithic and the Neolithic settlement patterns.

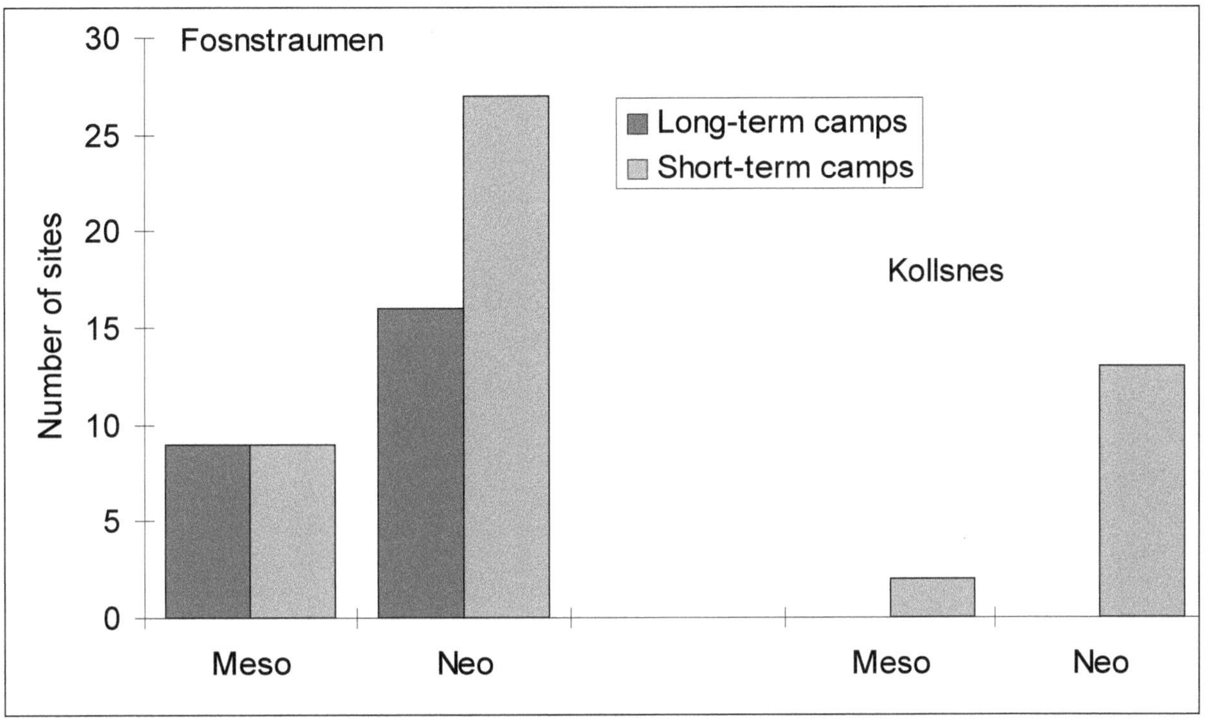

Figure 135. *Test-pit surveyed sites at Fosnstraumen and Kollsnes. The data from Kollsnes is extracted from Nærøy (1994).*

Important here is that there is a significant higher amount of short-term camps in the Neolithic as opposed to the Mesolithic period at Fosnstraumen as well as at Kollsnes (Fig. 135). This increase may partly be due to demographic changes, but considering that the number of short-term camps is three times as high (the number of long-term camps only double), it is reasonable to suggest that task group mobility was more important in the Neolithic than in the Mesolithic period. I believe that the change in the frequency in task-group mobility during the transition to the Neolithic period is strongly related to the general political development in the region, in which increased sedentism, growing social inequality, and more marked ethnic relations should be understood as different aspects of the same process. Below, I will attempt to outline this development.

Sedentism, social inequality and ethnicity

Although the late Mesolithic populations in western Norway were basically sedentary at tidal current channels, there was still frequent mobility at the local group level; households moved temporarily out of the channels in order to hunt, fish, or procure raw materials. As pointed out above, the Mesolithic groups from Skatestraumen probably had direct access to the quarries in Sunnfjord and Hespriholmen at Bømlo was accessible for people from a larger region. The Mesolithic period, therefore, appears as socially open; in the sense that most people had face-to-face relations with individuals from other local groups and that most of them had the possibility to establish external alliances. Such broad-scale alliances and networks may, for example, have been created and maintained during inter-regional gatherings at the quarry at Stakaneset or Hespriholmen, or at the rock-art site in Vingen. These alliances may have been important for creation of peaceful relations towards neighbouring local groups and within the region at large.

During the transition to the Neolithic, several changes took place. The local groups in the entire region became more sedentary than before, which first of all meant that they stayed in the channels for longer periods of the time. However, the needs for external supplies had not decreased. This implied that the mobility related to hunting, fishing and acquisition of raw materials was transferred to small households and crews, which belonged to lower levels of the social organisation. Being smaller, these task groups may also have been faster and more flexible travellers. With regard to the character of the external relations, this change in the mobility pattern was fundamental, because it probably meant that, contrary to the Mesolithic, when all members of the local groups in principle had the possibility to create external alliances, these alliances would now instead first of all be taken care of by mobile crews and small households. The task groups had, in other words, become key-personnel in the inter-group relations, while the majority of the population was generally prevented from face-to-face contacts with outsiders.

This development should probably be associated with the emergence of social inequality. Bryan Hayden points out that social inequality is likely to develop sooner or later among hunter-fishers in areas with an abundance of marine recourses. In such a context, competition between individuals will occur if there exists a way to transform abundant resources into highly desired, scarcer goods and services (Hayden 1994:239). In line with this, I would suggest that the increased sedentism during the transition to the Neolithic in western Norway left several already existing institutions for providing goods and services open for control by socially ambitious individuals. Task group mobility was such an institution, which had deep roots in the preceding Mesolithic period. If a prosperous individual from a particular household during the early Neolithic managed to secure the control of the most attractive task group moves, his or her household would be a step closer to their social goals.

An apparent source-critical problem is that house sites and graves, which might have revealed status differences, are very few or have not been found at all. Other symbols of elevated status, like heavy, crafted goods have not been documented (Nygaard 1988). Theoretically, the use and deposition of the diabase and greenstone adzes, and the rhyolite and slate projectile points might indicate social differences, but these artefacts seem to have been distributed everywhere and were also discarded in a similar fashion as the raw materials that were procured in the Nordhordland district. Still, I would hesitate to label these populations as egalitarian on such grounds. A well-known strategy for climbing in the social ladder is to establish debts and dependencies through gift giving (e.g. Sahlins 1963, Mauss 1990 [1925]). Hayden (1990) has elaborated on this theory by introducing the concept of competitive feasting. Hayden's main point is that big-men build up and maintain elevated social positions by acquiring new and exotic foods and beverages, but also artefacts and distant knowledge. The items and knowledge are kept for a while, but are later displayed and given away at feasts. In this model, leadership has quite weak foundations and will result in a spread instead of accumulation of valuable items. Kristina Jennbert (1984) has suggested that such a process may be responsible for the introduction of agriculture in southern Scandinavia, and the idea of social inequality has been discussed by other scholars for the late Mesolithic as well as early and middle Neolithic populations of Norway and Sweden (e.g. Taffinder 1998, Nordquist 2001, Glørstad 2002b).

The theory of feasting may fit well with the data from western Norway, where a wide spread of artefacts is documented. Previously it has been suggested that the southern Scandinavian TRB items may have been introduced through such "social channels" among the hunter-fisher populations in western Norway (A. B. Olsen 1992, Prescott 1996, Trones 1998). It has also been suggested that that adzes from Hespriholmen and Stakaneset were distributed through "social networks" (Bostwick Bjerck 1988). Similar interpretations would of course also be relevant for the widespread sandstone grinding slabs, slate, blue mylonite, Siggjo-rhyolite, and of course also the agricultural products that were consumed at Kotedalen during the middle Neolithic (Hjelle 1992).

A problem that is neither developed by Hayden in his general model nor in the regionally specific interpretations, is how such "social channels" or "social networks" actually and practically were organised across space between the sedentary local groups. My contention here is that task groups took care of this mobility in all of its different forms: Task groups attended feasts or social and ritual gatherings from a wider region, each representing their local group. But before the feast could take place at all, locally derived task groups would already have bartered, quarried, hunted and gathered the valuable items which were to be consumed and distributed at these occasions.

I find that social inequality as related to task group mobility is important for understanding the maintenance of old as well as the emergence of new ethnic boundaries. As pointed out above, what makes an item attractive and exotic is that it comes from a strange and foreign country. It is therefore necessary to create and maintain the idea of these distant realms. This may have been done by establishing ethnic boundaries. Such boundaries would probably be easier to strengthen if only a few individuals were responsible for the external social interaction. In a situation where the personal bonds of the ordinary members of the local groups did not include other people than co-residents or the members own population, it may have been of vital importance for leaders, on the basis of repeatedly expressing negative categorisations of the other, to be able to mobilise own local groups against neighbouring groups. For these leaders, a construction of conflicts would clearly serve their own ends.

However, these leaders could not solely focus on the establishment of conflicts and boundaries, because this would, aside from leading to a continuously violent and stressful life, probably prevent themselves from access to the distant realms. Another agenda would consequently be needed; they would be dependant on relations with leaders of other local groups in order to secure their own acquisition of non-local goods and attractive raw materials, because these goods would be necessary for maintaining their own statuses locally, and also because it would prevent conflicts with these neighbours. It is thus quite possible that other identities, for example class or clan identities that referred to common sacred sites or ancestral houses far away from where people actually lived, and which were maintained by means of exchanges of marriage partners, were quite as important as ethnicity to these leaders or big-men. Such conflicting identities may, as Max Gluckman and Ernest Burch have emphasised in two very different contexts (Gluckman 1963:1-26, Burch & Correll 1972), be exactly what would create some kind of stability in the social system.

Conclusion

The main intention of this chapter was to investigate the character of the cross-boundary relations. On the basis of a chaîne opératoire analysis of the widely spread raw materials with known sources it was concluded that the flow of items across the proposed boundaries (Fig. 122) first of all happened in a context of gift exchange or commodity barter. It was also argued that the social interaction between the local groups probably had been organised by big-men and that task groups were instrumental in these relations. Finally, it was suggested that for these leaders, which were dependent upon access to exotic items in order to maintain their own positions locally, ethnicity may have been important in order to prevent other people from access to these distant goods.

Chapter 11

Conclusions

Research on the early Neolithic of southern Norway has been carried out for more than 130 years. During this period, one has gradually realised that that the area was inhabited by populations that were different in many respects. Agricultural groups lived around the Oslo fjord. In the forests further north there were elk hunters. Hunter-herders occupied the valleys and the mountain plateaus between east and west. The entire west coast of Norway was inhabited by sedentary hunter-fishers. Previous interpretations have mainly applied the culture-concept in order to characterise this diversity: the agriculturists belonged to the "TRB culture". Groups further north have been related to the "slate culture", while the western hunter-fishers have been labelled "dwelling site culture".

One important problem with the archaeological culture-concept is that while it is clearly an etic and descriptive term, it is nevertheless thought of as emic, in the sense that one assumes that people identified with it and that they were aware that they lived in core-areas or in the peripheries of such "cultures". Another problem is that one has concentrated much on the circumscription of large "cultural" areas; much attention has been paid to isolation and less to interaction. This focus on large and isolated social entities has made it difficult to identify the social dynamics within the "cultures" as well as between them. The causes for change have thus often been sought in outside factors, such as immigrations or climatic changes.

When commencing on the present work, I wanted to approach the diversity of early Neolithic Norway from a different angle. By applying the concept of ethnicity, I hoped to achieve a better understanding of the regional differences; why they were established and how they were maintained. The idea was that the differences emerged in contexts of ethnicity in territorial boundary areas, and that different regional cultural practices constituted the resources for such a process to take place. In order to operate this approach, the concepts "ethnicity" and "culture" were thought to be of different orders. Ethnicity was, in line with recent anthropological thinking, seen as an organisational feature, in principle empty of content. Culture was, in contrast to both the old culture-concept referred to above and ethnicity, seen as learnt practices and traditions of for example language, technique, religion, morality or symbolism, which may be shared across ethnic boundaries, but which also, when they are not shared with opposing groups, provide the necessary fuel for the emergence, maintenance and renewal of ethnicity. On the basis of these definitions, I argued that the following criteria should be satisfied if a boundary was to be interpreted as ethnic: (1) there should be a co-variation of several cultural traits, (2) a sudden fall-off in the distribution of lithic raw materials would have to be observed, (3) there should be evidence for crossing, and (4) the boundary should have a deep history.

Western Norway and the mountain plateaus between eastern and western Norway were chosen as the area of analysis, and the following cultural practices were investigated: site locations, lithic raw materials, types, and technologies. 37 excavated early Neolithic sites were re-analysed for this thesis on the basis of a reference system for raw materials. In addition, the museum collections of Vespestad adzes were re-classified. Geological isotope analyses and mineralogical analyses were performed for a number of the raw materials in order to relate them to specific bedrock sources.

The site location analysis, in which six test-pit surveyed local areas were compared at the coast of western Norway, showed that there were significant differences in the choices of site locations in the Neolithic. These differences could, however, largely be explained as results of differences in site types, physiography and background environment. There were no indications that the variation was related to cultural differences within the region of analysis.

The analysis of the lithic data, however, showed that the distribution boundaries of certain types and techniques coincided with the distributions of certain raw materials. Furthermore, these raw material distributions were marked by abrupt fall-offs. The situation was therefore one of isolation and communication at the same time, which, according to current theories in the social sciences is what characterises ethnic relations. The boundaries between the different ethnic groups seem to have been maintained and renewed through hundreds of years.

Several boundaries were tentatively delineated. On the basis of a discussion of the differences, it was concluded that ethnicity probably was an important factor for the formation of boundaries between (1) Sunnmøre and Nordfjord (2) Nordfjord and Sunnfjord, (3) Sunnfjord/Sogn and Nordhordland (4) Nordhordland and Midthordland/Sunnhordland and (5) coastal and mountain sites. Most likely, ethnicity played a role among these populations in already during the Mesolithic, but it became much more important in the early Neolithic. Material culture would be central in this process, because it constituted the resources for ethnicity to develop and was in turn actively used as symbols of these differences.

The fact that people were sedentary was important for the development of ethnicity. Local groups throughout the coastal region were hunter-fishers, and they were, most likely, organised in relatively small social units, such as local groups. Most people would not travel far from their place of birth during their lifetime; for example, the ones that were born in Nordhordland may never have gone to Sogn or Sunnhordland. This would mean that they had profound knowledge about and strong attachments to particular places and landscapes in their districts. These kinds of attachments were important for the development of ethnic identities among the local groups, and they were effective means of defining out-groups; the ones that lacked the attachments could be categorised as strangers.

The cross-boundary relations were explored by studying the widely spread raw materials with known sources. It appeared that blocks as well as blanks and finished products had been distributed across the boundaries. This traffic of goods was interpreted as mainly a result of gift exchange/commodity barter. It was suggested that big-men or similar types of leaders organised the inter-group relations, and that task groups were instrumental in this interaction. Furthermore, it was held that, for these leaders, ethnicity would have a purpose; a double agenda of contact and conflict with out-groups would have served their social ambitions.

Conclusively, the thesis raises doubts about the validity of the large "cultures" which were thought by earlier scholars to be the locus of ethnic identity. Today, these "cultures" may instead be seen as archaeological constructions which have little to do with the social and political boundaries that existed during the Stone Age. Instead, it appears that ethnic identities in this period were related to relatively small social units; ethnicity is argued primarily to be a way of organising the social relations between territorial local groups.

References

Aksnes, D. L. 1988. Biologiske forutsetninger for ekstensivt oppdrett av torsk i fjorder - en modellstudie av den biologiske bæreevnen Masfjorden. *Sogn og Fjordane Distriktshøgskule Skrifter* 1988(2), 88-107.

Alsaker, E. 2002. Geologisk beskrivelse av referansesamlingen for råstoffer fra Skatestraumprosjektet. In Bergsvik, K. A. (ed.), *Arkeologiske undersøkelser ved Skatestraumen. Bind 1*, 349-352. Bergen University Museum, University of Bergen, Bergen.

Alsaker, S. 1982. Bømlo. Steinalderens råstoffsentrum på Sørvestlandet. Unpublished Mag. art. thesis, University of Bergen.

Alsaker, S. 1987. *Bømlo - steinalderens råstoffsentrum på Sørvestlandet.* Arkeologiske avhandlinger 4. Historisk museum, University of Bergen, Bergen.

Ames, K. M. 1991. Sedentism: a temporal shift or a transitional change in hunter-gatherer mobility patterns? In Gregg, S. A. (ed.), *Between Bands and States*, 108-134. Southern Illinois University Press, Carbondale.

Amundsen, Ø. M. 2000. Neolittkum i Agder og Telemark. En komparativ analyse av keramikk og flintøkser. Unpublished Cand. philol. thesis, University of Oslo.

Andresen, L. 1979. *Monthly and annual frequencies of current wind forces and wind directions in western Norway and Trøndelag.* Climatological Summaries for Norway. University of Oslo, Oslo.

Anfinset, N. 1994. Trekantsambandet. Funnrapport B-15288, Nautøysundet lok. 115, Føyno, gnr. 51, bnr. 2, Stord k., Hordaland. Unpublished report in the topographical archive, Bergen University Museum, University of Bergen.

Anfinset, N. 1998. Samfunnsstruktur, identitet og spesialister. Et eksempel fra Nepal. In Welinder, S. (ed.), *Etnicitet eller kultur*, 11-20. Mitthögskulan, Östersund.

Baglo, C. 2001. Vitenskapelige stereotypier. Om konstruksjonen av samene som kulturhistorisk enhet i tida fram mot 1910. Unpublished Cand. philol. thesis, University of Tromsø.

Bakka, E. 1964. Steinaldergranskingar i Nordhordland 1960-63. *Frå Fjon til Fusa. Årbok for Nord- og Midthordland Sogelag* 1964, 9-42.

Bakka, E. 1966. To helleristningar frå steinalderen i Hardanger. *Viking* XXX, 77-95.

Bakka, E. 1973. Omkring problemet om kulturdualisme i Sør-Noreg. In Simonsen, P. & Munch, G. S. (eds.), *Bonde-veidemann, bofast-ikke bofast i nordens forshistorie*, 109-128. Universitetsforlaget, Tromsø.

Bakka, E. 1976. Comments on typological and chronological problems. *Norwegian Archaeological Review* 9(1), 16-25.

Bakka, E. 1979. On shoreline dating of Arctic rock carvings in Vingen, Western Norway. *Norwegian Archaeological Review* 12(2), 115-122.

Bakka, E. 1993. Ramsvikneset - a sub neolithic dwelling place in Western Norway. In Solberg, B. (ed.), *Minneskrift Egil Bakka. Arkeologiske skrifter 7*, 21-69. Historisk museum, University of Bergen, Bergen.

Bakka, E. & Kaland, P. E. 1971. Early farming in Hordaland, Western Norway. *Norwegian Archaeological Review* 4(2), 1-35.

Balikci, A. 1970. *The Netsilik Eskimo.* Natural History Press, New York.

Ballin, T. B. & Jensen, O. L. 1995. *Farsundprosjektet - steinalderbopladser på Lista.* Varia 29. Universitetets Oldsaksamling, Oslo.

Bang-Andersen, S. 1981. *En fangstboplass på Eigerøy - boplassbruk og miljøtilpasning i sørvestnorsk yngre steinalder.* AmS-Skrifter 6. Arkeologisk museum i Stavanger, Stavanger.

Barlindhaug, S. 1996. Hvor skal vi bygge og hvor skal vi bo? En analyse av lokaliseringsfaktorer i tidlig eldre steinalder i Troms. Unpublished Cand. philol. thesis. University of Tromsø.

Barndon, R. 2002. The chaîne opératiore, social change and modifications or the technological milieu. In Mandt, G. & Dommasnes, L. H. (eds.), *Arkeologiske skrifter fra Universitetet i Bergen 11*, 5-22. Department of Archaeology, University of Bergen, Bergen.

Barrett, J. C. 1994. *Fragments From Antiquity.* Blackwell, Oxford.

Barth, F. 1966. *Models of Social Organization.* Royal Anthropological Institute Occational paper no. 23. The University Press, Glasgow.

Barth, F. 1969. Introduction. In Barth, F. (ed.), *Ethnic Groups and Boundaries. The Social Organization of Culture Difference*, 9-38. Universitetsforlaget, Oslo.

Barth, F. 1975. *Ritual and Knowledge among the Baktaman of New Guinea.* Universitetsforlaget, Oslo.

Becker, C. J. 1951. Den grubekeramiske kultur i Danmark. *KUML. Aarbøker for Nordisk Oldkyndighed og Historie* 1950, 153-274.

Behre, K. E. 1981. The interpretation of anthropogenic indicators in pollen diagrams. *Pollen et Spores* 13(2), 225-245.

Bender, B. 1978. Gatherer-hunter to farmer: a social perspective. *World Archaeology* 10(2), 204-222.

Berg, E. 1986. De neolittiske flintøksene i Vest-Norge. Typologisk-kronologisk inndeling. Bakgrunnen for flintøksenes oppkomst og forholdet til de lokale tradisjonene. Unpublished Cand. philol. thesis, University of Bergen.

Berg, E. 1995. *Dobbeltspor/E6 prosjektet. Steinalderlokaliteter fra senmesolittisk tid i Vestby, Akershus.* Varia 32. Universitetets Oldsaksamling, Oslo.

Berg-Hansen, I. M. 1999. The availability of flint at Lista and Jæren, Southwestern Norway. In Boaz, J. (ed.), *The Mesolithic of Middle Scandinavia*, 255-266. Universitetets Oldsaksamling, Oslo.

Berg-Hansen, I. M. 2001. Registrering som erfaring. En undersøkelse av metoden for steinalderregistrering i Norge med eksempel fra Lista i Vest-Agder. Unpublished Mag. art. thesis, University of Oslo.
Bergsvik, K. A. 1988. Innberetning fra utgravningene av lok. 1 Snekkevik, Austrheim kommune, Hordaland. Unpublished report in the topographical archive, Bergen University Museum, University of Bergen, Bergen.
Bergsvik, K. A. 1991. Ervervs- og bosetningsmønstre langs kysten av Nordhordland i steinalder belyst ved funn fra Fosnstraumen. En arkeologisk og geografisk analyse. Unpublished Cand. philol. thesis, University of Bergen.
Bergsvik, K. A. 1995. Bosetningsmønstre på kysten av Nordhordland i steinalder. En geografisk analyse. In Bergsvik, K. A., Nygård, S. & Nærøy, A.J. (eds.), *Steinalderkonferansen i Bergen 1993. Arkeologiske skrifter 8*, 111-130. Arkeologisk institutt, University of Bergen, Bergen.
Bergsvik, K. A. 1998. Referansesystem for råstoffer ved Skatestraumprosjektet. Unpublished report in the topographical archive, Bergen University Museum, University of Bergen.
Bergsvik, K. A. 1999. A new reference-system for classification of lithic raw materials. A case-study from Skatestraumen, Western Norway. In Boaz, J. (ed.), *The Mesolithic of Middle Scandinavia*, 283-298. Universitetets Oldsaksamling, University of Oslo, Oslo.
Bergsvik, K. A. 2001a. Pragmatisme og arkeologi. *Primitive Tider* 4, 73-90.
Bergsvik, K. A. 2001b. Sedentary and mobile hunter-fishers in stone age western Norway. *Arctic Anthropology* 38(1), 2-26.
Bergsvik, K. A. 2001c. Strømmer og steder i vestnorsk steinalder. *Viking* LXIV, 11-34.
Bergsvik, K. A. 2002a. *Arkeologiske undersøkelser ved Skatestraumen. Bind 1.* Arkeologiske avhandlinger og rapporter fra Universitetet i Bergen 7. Bergen University Museum, University of Bergen, Bergen.
Bergsvik, K. A. 2002b. Task groups and social inequality in early Neolithic western Norway. *Norwegian Archaeological Review* 35(1), 1-25.
Bergsvik, K. A. 2003. Mesolithic ethnicity - too hard to handle? In Larsson, L., Kindgren, H., Knutsson, K., Loeffler, D. & Åkerlund, A. (ed.), *Mesolithic on the Move*, 290-301. Oxbow Books, Oxford.
Bergsvik, K. A. & Olsen, A. B. 2003. Traffic in stone adzes in Mesolithic western Norway. In Larsson, L., Kindgren, H., Knutsson, K., Loeffler, D. & Åkerlund, A. (ed.), *Mesolithic on the Move*, 395-404. Oxbow Books, Oxford.
Berndt, R. M. 1959. The concept of "tribe" in the Western desert of Australia. *Oceania* 30, 81-107.
Berndt, R. M. & Berndt, C. H. 1964. *The World of the First Australians: an Introduction to the Traditional Life of the First Australians.* Angus and Robertson, London.
Binford, L. R. 1965. Archaeological systematics and the study of culture process. *American Antiquity* 31, 203-210.
Binford, L. R. 1980. Willow smoke and dog's tails: hunter gatherer settlement systems and archaeological site formation. *American Antiquity* 45, 4-20.
Binford, L. R. 1982. The archaeology of place. *Journal of Anthropological Archaeology* 1(1), 5-31.
Birdsell, J. 1953. Some environmental and cultural factors influencing the structure of Australian aboriginal populations. *American Naturalist* 87(834), 171-207.
Bjerck, H. B. 1983. Kronologisk og geografisk fordeling av mesolitiske element i Vest- og Midt-Norge. Unpublished Mag. art. thesis, University of Bergen.
Bjerck, H. B. 1984. Undersøkelsen av Torsteinsvik, lok. 11, Torsteinsvik, Øygarden k., Hordaland. Unpublished report in the topographical archive, Bergen University Museum, University of Bergen.
Bjerck, H. B. 1986. The Fosna-Nøstvedt problem. A consideration of archaeological units and chronozones in the south norwegian Mesolithic period. *Norwegian Archaeological Review* 19(2), 103-121.
Bjerck, H. B. 1989. *Forskningsstyrt kulturminneforvaltning på Vega, Nordland. En studie av steinaldermenneskenes boplassmønstre og arkeologiske letemetoder.* Gunneria 61. Universitetet i Trondheim. Vitenskapsmuseet, Trondheim.
Bjerck, H. B. 1990. Mesolithic site types and settlement patterns at Vega, Northern Norway. *Acta Archaeologica* 60, 1-32.
Bjerck, H.B. & Ringstad, B. 1985. *De kulturhistoriske undersøkelsene på Tjernagel, Sveio.* Arkeologiske rapporter 9. Historisk museum, University of Bergen, Bergen.
Bjørgo, T. 1981. Flatøy. Et eksempel på steinalderens kronologi og livbergingsmåte i Nordhordland. Unpublished Mag. art. thesis, University of Bergen.
Bjørn, A. 1921a. *Stenalderboplassene i Allanhenget i Kristiansund.* De Kongelige Norske Vitenskapers Selskaps Skrifter 1920. Nr. 27. Aktietrykkeriet i Trondhjem, Trondhjem.
Bjørn, A. 1921b. *Træk av Søndmørs stenalder.* Bergens Museums Årbok 1919-20. Historisk-antikvarisk række nr. 4. Bergens Museum, Bergen.
Bjørn, A. 1924. *Stenalderstudier.* Videnskabs-Selskabets Skrifter. II. Math.-Naturv. Klasse. 1924. No. 5. Jacob Dybwad, Kristiania.
Bjørn, A. 1931. Hovedlinjer i den norske nasjons tilblivelseshistorie. *Naturen* 55, 224-245.
Blackman, M.B. 1990. Haida: traditional culture. In Suttles, W. (ed.), *Volume 7: Northwest Coast*, 240-260. Smithsonian Institution, Washington.
Boaz, J. 1997. *Steinalderundersøkelsene på Rødsmoen.* Varia 41. Universitetets Oldsaksamling, Oslo.
Boaz, J. 1998. *Hunter-Gatherer site variability: Changing patterns of site utilization in the interior of eastern Norway, between 8000 and 2500 B.P.* Universitetets Oldsaksamlings Skrifter. Ny rekke nr. 20. Universitetets Oldsaksamling, Oslo.
Bostwick Bjerck, L. G. 1987. In-context pollen diagrams from 3 archaeological sites in western Norway: Towards a unified model of land use in the late Mesolithic and Neolithic periods, Unpublished Mag. art. thesis, University of Bergen.
Bostwick Bjerck, L. G. 1988. Remodeling the Neolithic in southern Norway: another attack on a traditional problem. *Norwegian Archaeological Review* 21(1), 21-33.
Bourdieu, P. 1977. *Outline of a theory of practice.* Translated by R. Nice. Cambridge University Press, Cambridge.
Bronk Ramsey, C. 2000. OxCal version 3.8. *http://www.rlaha.ox.ac.uk/orau/index.htm* .

Bryhni, I. & Lyse, K. 1985. The kalvåg Mélange Norwegian Caledonites. In Gee, D. G. & Sturt, B. A. (eds.), *The Caledonite Orogen - Scandinavia and Related Areas*, 417-427. John Wiley & Sons Ltd, New York.
Brøgger, A. W. 1906. *Studier over Norges stenalder. I. Øxer uten skafthul fra Yngre Stenalder i det Sydøstlige Norge*. Videnskabs-Selskabets Skrifter. I. Math.-Naturv. Klasse. 1906. No. 2. Jacob Dybwad, Christiania.
Brøgger, A. W. 1907. *Norges Vestlands stenalder. Typologiske studier*. Bergens Museums Aarbok 1907. No. 1. Bergens Museum, Bergen.
Brøgger, A. W. 1909. *Den Arktiske stenalder i Norge*. Videnskabs-Selskabets Skrifter. II. Hist.-Filos. Klasse. 1909. No. 1. Jacob Dybwad, Christiania.
Brøgger, A. W. 1925. *Det norske folk i oldtiden*. Instituttet for Sammenliknende Kulturforskning, Oslo.
Brøgger, A. W. 1979. Vårsel og vintersel. *Universitetets Oldsaksamlings Årbok* 1979, 21-38.
Burch, E. S., Jr. 1980. Traditional Eskimo societies in Northwest Alaska. *Senri Ethnological Studies* 4, 253-304.
Burch, E. S., Jr. 1988. War and trade. In Fitzhugh, W.W. & Crowell, A. (eds.), *Crossroads of Contintets: Cultures of Siberia and Alaska*, 227-240. Smithsonian Institution Press, Washington D.C.
Burch, E. S., Jr & Correll, T. C. 1972. Alliance and conflict: inter-regional relations in North Alaska. In Guemple, D. L. (ed.), *Alliance in Eskimo society. Proceedings of the American Ethnological Society 1971. Supplement*, 17-39. Washington University Press, Seattle.
Bøe, J. 1923. *Steinaldersboplassene paa Nappen i Søndhordland*. Bergens Museums Aarbok 1921-22. Hist.-antikv. række nr. 3. Bergens Museum, Bergen.
Bøe, J. 1932. *Felszeichnungen im Westlichen Norwegen I. Die Zeichnungsgebiete in Vingen und Hennøya. Bergens Museums Skrifter nr. 15*. A.S. John Griegs boktrykkeri, Bergen.
Bøe, J. 1942. *Til høgfjellets forhistorie. Boplassen på Sumtangen ved Finsevatn på Hardangervidda*. Bergens Museums Skrifter nr. 21. A.S. John Griegs Boktrykkeri, Bergen.
Bågenholm, G. 1996. *Etnicitet som problem i arkeologisk forskning*. Gotarc Serie C. Arkeologiska Skrifter No. 11. Göteborgs Universitet, Göteborg.
Callahan, E. 1987. *An evaluation of the Lithic Technology in Middle Sweden During the Mesolithic and Neolithic*. Aun. 8. Societas Archaeologica Upsaliensis, Uppsala.
Childe, V. G. 1925. *The Dawn of European Civilization*. Keagan Paul, Trench, Trubner & Co, London.
Childe, V. G. 1929. *The Danube in Prehistory*. Clarendon, Oxford
Clark, G. A. 1994. Migration as an explanatory concept in Paleolithic archaeology. *Journal of Archaeological Method and Theory* 1(4), 305-343.
Clark, G. 1975. The Earlier Stone Age Settlement of Scandinavia. Cambridge University Press, Cambridge.
Clarke, D. L. 1968. *Analytical Archaeology*. Cambridge University Press, Cambridge.
Cohen, A. 1974. Introduction: The lesson of ethnicity. In Cohen, A. (ed.), *Urban Ethnicity*, ix-xxiv. Tavistock Publications, London.
Cohen, A. 1978. Ethnicity: Problem and focus in anthropology. *Annual Review of Anthroplogy* 7, 379-403.
Cohen, A. P. 1985. *The Symbolic Construction of Community*. Ellis Harwood/Tavistock, London.
Cohmap Members 1988. Climatic changes over the last 18.000 years: observations and model simulations. *Science* 241(8), 1043-1052.
Comaroff, J. & Comaroff, J. 1992. *Ethnography and the Historical Imagination*. Westview Press, Boulder.
Conkey, M. 1990. Experimenting with style in Archaeoogy: Some historical and theoretical issues. In Conkey, M. & Hastorf, C. (eds.), *The Uses of Style in Archaeology*. Cambridge University Press, Cambridge.
Cornell, S. 1996. The variable ties that bind: content and circumstance in ethnic processes. *Ethnic and Racial Studies* 19(2), 265-289.
Coulson, S. 1986. Refitted flint nodules form Songa, Telemark. *Universitetets Oldsaksamlings Årbok* 1984/1985, 17-22.
Damas, D. 1963. *Igluligmiut Kinship and Local Groupings: a Structural Approach*. Anthreopological Series 65, Bulletin 196. National Museum of Canada, Ottawa.
David, B. & Cole, N. 1990. Rock art and inter-regional interaction in northeastern Australian prehistory. *Antiquity* 64, 788-806.
de Laguna, F. 1972. *Under Mount Saint Elias: The History and Culture of the Yakutat Tlingit*. Smithsonian Contributions to Anthropology, Vol. 7, Pt. I. Smithsonian University Press, Washington D.C.
de Laguna, F. 1983. Aboriginal Tlingit sociopolitical organization. In Tooker, E. (ed.), *The Development of Political Organization in Native North America*, 71-85. American Ethnological Society, Washington.
Djupedal, R. 1966. *Selja i tusen år*. Eget forlag, Selje.
Dommasnes, L. H. 1997. *Tradisjon og handling i førkristen vestnorsk gravskikk. 1. Undersøkelser på et gravfelt på Vereide i Gloppen, Sogn og Fjordane*. Arkeologiske rapporter 21. Arkeologisk institutt, Bergen University Museum, University of Bergen, Bergen.
Drucker, P. 1951. *The Northern and Central Nootkan tribes*. Bureau of American Ethnology Bulletin 144. Bureau of American Ethnology, Washington.
Drucker, P. 1983. Ecology and political organization on the Northwest Coast of America. In Tooker, E. (ed.), *The Development of Political Organization in Native North America*, 86-98. American Ethnological Society, Washington.
Dyrkorn, P. & Rich, D. M. 1973. Dagbok fra utgravning av steinalderboplassen lok. 526 Nordmannslågen, Hardangervidda, Ullensvang k. Hordaland. Unpublished report in the topographical archive, Bergen University Museum, University of Bergen.
Earle, T. K. & Ericcson, J. E. 1977. Exchange systems in archaeological perspective. In Earle, T. K. & Ericcson, J. E. (ed.s), *Exchange Systems in Prehistory*, 3-12. Academic Press, New York.

Ebbesen, K. 1984. Traktbægerkulturens grønstensøkser. *KUML Aarbøker for Nordisk Oldkyndighed og Historie* 1984, 113-153.
Edgren, T. & Törnblom, L. 1992. *Finlands historia. Bind 1.* Schildts Förlag AB, Esbo.
Eliasson, L. & Joelsson, J. 1989. Knivar och spetsar. En jämförande studie av skifferföremål från Västerbottens kust- och inland. Unpublished C-uppsats, University of Umeå.
Engelstad, E. 1990. The meaning of sedentism and mobility in an archaeological and historic context. *Acta Borealia* 7(2), 21-35.
Eriksen, B. V. 1991. *Change and Continuity in a Prehistoric Hunter-Gatherer Society: a Study of Cultural Adaptation in Late Glacial - Early Postglacial Southwestern Germany.* Archaeologia Venatoria, Tübingen.
Eriksen, B. V. 2000. "Chaîne opératoire" - den operative proces og kunsten at tænke som en flinthugger. In Eriksen, B.V. (ed.), *Flintstudier. En håndbog i systematiske analyser av flintinventarer,* 75-100, Arhus Universitetsforlag, Århus.
Eriksen, T. H. 1992. *Us and Them in Modern Societies: Ethnicity and Nationalism in Mauritius, Trinidad and Beyond.* Scandinavian University Press, Oslo.
Espedal, O. 1965. Innberetning om undersøkelsen av steinalderboplassen Mørkedøla I, Eldrevann, Borgund s., Lærdal pgd., Sogn og Fjordane sommaren 1965. Unpublished report in the topographical archive, Bergen University Museum, University of Bergen.
Fitzhugh, W. 1974. Ground slates in the Scandinavian Younger Stone Age with reference to Circumpolar maritime adaptations. *Proceedings of the Prehistoric Society* 1974(80), 45-58.
Forsberg, L. 1985. *Site Variability and Settlement Patterns.* Archaeology and Environment 5. Department of Archaeology, University of Umeå, Umeå.
Fried, M. M. 1968. On the concepts of "tribe" and "tribal society". In Helm, J. (ed.), *Essays on the Problem of Tribe. Proceedings of the annual 1967 spring meeting of the American Ethnological Society,* 3-20. University of Washington Press, Seattle.
Fuglestvedt, I. 1995. Svevollen - spor av senmeolittisk bosetning i lavlandets indre skoggsone. In Bergsvik, K. A., Nygård, S. & Nærøy, A. J. (eds.), *Steinalderkonferansen i Bergen 1993. Arkeologiske skrifter 8,* 95-110. Arkeologisk institutt, University of Bergen, Bergen.
Fulsås, N. 1999. *Historie og nasjon. Ernst Sars og striden om norsk kultur.* Universitetsforlaget, Oslo.
Furnes, H., Roberts, D., Sturt, B. A., Thon, A. & Gale, G. H. 1980. Ophiolite fragments in the Scandinavian Caledonites. In Panatyitou, A. (eds.), *Ophiolites,* 107-114, Geological Surveys Department, Cyprys.
Førland, E.J. 1984. Lokalklima på vestlandskysten. *Klima* 6, 24-36.
Gell, A. 1992. Inter-tribal commodity barter and reproductive gift-exchange in old Melanesia. In Humphrey, C. & Hugh-Jones, S. (eds.), *Barter, Exchange and Value. An Anthropological Approach,* 142-168. Cambridge University Press, Cambridge.
Gendel, P. A. 1984. *Mesolithic Social Territories in Northwestern Europe.* B.A.R. International series 218, Oxford.
Giddens, A. 1984. *The Constitution of Society: Outline of the Theory of Structuration.* Polity Press, Cambridge.
Gjerland, B. 1984. Bergartsøkser i Vest-Noreg. Distribusjon sett i forhold til praktisk funksjon, økonomisk tilpasning of tradisjon i steinalderen. Unpublished Mag. art. thesis, University of Bergen.
Gjessing, G. 1935. *Fra steinalder til jernalder i Finnmark. Etnologiske problemer.* Instituttet for sammenliknende kulturforskning. Serie CIII - 3. Aschehoug, Oslo.
Gjessing, G. 1942. *Yngre steinalder i Nord-Norge.* Instituttet for Sammenliknende Kulturforskning. Serie B: Skrifter XXXIX. Instituttet for Sammenliknende Kulturforskning, Oslo.
Gjessing, G. 1945. *Norges stenalder.* Norsk Arkeologisk Selskap, Oslo.
Gjessing, H. 1920. *Rogalands stenalder.* Stavanger Museum, Stavanger.
Gjærevoll, O. 1992. *Plantegeografi.* Tapir Forlag, Trondheim.
Gluckman, M. 1963. *Custom and Conflict in Africa.* Basil Blackwell, Oxford.
Glørstad, H. 1996. *Neolittiske smuler: Små teoretiske og praktiske bidrag til debatten om neolittisk keramikk og kronologi i Sør-Norge.* Varia 33. Universitetets Oldsaksamling, Oslo.
Glørstad, H. 1998. Senmesolitikum i Østfold - et kronologisk perspektiv. In Østmo, E. (ed.), *Fra Østfolds oldtid,* 69-82. Universitetets Oldsaksamling, Oslo.
Glørstad, H. 1999. Lokaliteten Botne II - Et nøkkelhull til det sosiale livet i mesolittikum i Sør-Norge. *Viking* LXII, 31-68.
Glørstad, H. 2002a. *Neolittisk renessanse. Hypoarkeologiske tekster om neolitikum i Sør-Norge.* Acta Humaniora nr. 143. Det historisk-filosofiske fakultet, Oslo.
Glørstad, H. 2002b. Østnorske skafthullhakker fra mesolittikum. Arkeologisk og forhistorisk betydning - illustrert med et eksempelstudium fra vestsiden av Oslofjorden. *Viking* LXV, 7-47.
Godelier, M. 1979. Territory and property in primitive society. In von Cranach, M., Foppa, K., Lepenies, W. & Ploog, D. (eds.), *Human Ethology,* 133-155. Cambridge University Press, Cambridge.
Goodby, R. G. 1998. Technological patterning and social boundaries: ceramic variability in southern New England. In Stark, M. T. (ed.), *The Archaeology of Social Boundaries,* 161-182. Smithsonian Institution Press, Washington.
Gould, R.A. 1980. *Living Archaeology.* Cambridge University Press, Cambridge.
Graburn, N. H. H. & Strong, B. S. 1973. *Circumpolar Peoples: an Athroplogical Perspective.* Goodyear regional anthropology series. Goodyear, Pacific Palisades, California.
Greenachre, M. J. 1993. *Correspondence Analysis in Practice.* Academic Press, London.
Gregg, S. A. 1991 (ed). *Between Bands and States.* Southern Illinois University Press, Carbondale.
Gräslund, B. 1962. Skafthålsförsedda spetsredskap av sten. *Tor* VIII, 105-150.
Gustafson, G. 1899. *En stenalders boplads paa Jæren.* Bergens Museums Aarbok 1899. No. 1. Bergens Museum, Bergen.
Gustafson, G. 1906. *Norges oldtid. Mindesmærker og oldsager.* Norsk Folkemuseum, Kristiania.
Gustafson, L. 1980. Om "vestgrensa" i høyfjellet. *Arkeo* 1980, 6-10.
Gustafson, L. 1983. *Arkeologiske registreringer i Vossovassdraget.* Arkeologiske rapporter 6. Historisk museum, University of Bergen, Bergen.

Hagen, A. 1946. Frå innlandets steinalder. Hedmark fylke. *Viking* X, 1-93.
Hagen, A. 1963. Mesolittiske jegergrupper i norske høyfjell. Synsmåter om Fosnakulturens innvandring til Vest-Norge. *Universitetets Oldsaksamlings Årbok* 1960-1961, 109-142.
Hagen, A. 1967. *Norges Oldtid,* first edition, Cappelen, Oslo.
Hagen, A. 1970. Refleksjoner over noen arkeologiske problemstillinger og tolkningsmuligheter. *Viking* XXXIV, 111-162.
Hagen, A. 1977. *Bergkunst: Jegerfolkets helleristninger of malinger i norsk steinalder.* Cappelen, Oslo.
Hagen, A. 1983. *Norges Oldtid,* third edition, Cappelen, Oslo.
Hahn, J. 1979. Essai sur l'écologie du Magdalénien dans le Jura suabe. In de Sonneville-Bordes, D. (ed.), *La fin de temps glacieres en Europe,* 203-211. Centre Nacional de Recherche Scientifique, Paris.
Hallgren, F. 1998. Etnicitet under stenåldern i Mellansverige under stenåldern. In Welinder, S. (ed.), *Etnicitet eller kultur,* 61-77. Mitthögskulan, Östersund.
Hamilton, A. 1980. Dual social systems: technology, labour and women's secret rites in the western desert of Australia. *Oceania* 51, 4-19.
Handelman, D. 1977. The organization of ethnicity. *Ethnic Groups* 1, 187-200.
Hansen, A. M. 1904. *Landnåm i Norge. En Udsigt over Bosætningens Historie.* W.C. Fabritius & Sønner A/S, Kristiania.
Hauken, Å. D. 1983. Utgravning av stenålderslokalitet 3 Tjernagel, Sveio k., Hordaland, Unpublished report in the topographical archive. Bergen University Museum, University of Bergen.
Hayden, B. 1990. Nimrods, piscators, pluckers and planters: The emergence of food production. *Journal of Anthropological Archaeology* 9(1), 31-69.
Hayden, B. 1994. Competition, labor, and complex hunter-gatherers. In Burch, J.E.S. & Ellanna, L.J. (eds.), *Key Issues in Hunter-Gatherer Research,* 223-239. Berg, Oxford.
Hedeager, L. 1997. *Skygger af en anden virkelighed: Oldnordiske myter.* Samleren, København.
Hegmon, M. 1998. Technology, style and social practices: archeological approaches. In Stark, M. T. (ed.), *The Archaeology of Social Boundaries,* 264-280. Smithsonian Institution Press, Washington.
Helliksen, W. 1996. *Evolusjonisme i norsk arkeologi. Diskutert med utgangspunkt i A. W. Brøggers hovedverk fra 1909-25.* Varia 37. Universitetets Oldsaksamling, Oslo.
Helm, J. 1965. Bilaterality in the socio-territorial organization of the Arctic Drainage Dene. *Ethnology* 4, 361-385.
Helm, J. 1968. The nature of Dogrib socioterritorial groups. In Lee, R. B. & DeVore, I. (eds.), *Man the Hunter,* 118-125. Aldine, Chicago.
Helms, M. W. 1988. *Ulysses' Sail: an Ethnographic Odyssey of Power, Knowledge, and Geographical Distance.* Princeton University Press, Princeton, New Yersey.
Helskog, K. 1980. The chronology of the younger stone age in Varanger, North Norway. Revisited. *Norwegian Archaeological Review* 13(1), 47-54.
Helskog, K., Indrelid, S. & Mikkelsen, E. 1976. Morfologisk klassifisering av slåtte steinartefakter. *Universitetets Oldsaksamlings Årbok* 1972-1974, 9-40.
Hesjedal, A. 2000. Samisk forhistorie i norsk arkeologi 1900 – 2000. Unpublished Dr. art. thesis, University of Tromsø.
Hildebrand, H. 1866. *Svenska folket under hednatiden.* Stockholm.
Hinsch, E. 1948. Boplass-kulturen på Mørekysten. *Viking* XII, 89-132.
Hinsch, E. 1955. Traktbegerkultur-Megalitkultur. En studie av Øst-Norges eldste, neolitiske gruppe. *Universitetets Oldsaksamlings Årbok* 1951-1953, 10-177.
Hjelle, K. L. 1992. Pollenanalytiske undersøkelser innenfor boplassen på Kotedalen. In Hjelle, K. L., Hufthammer, A. K., Kaland, P. E., Olsen, A. B. & Soltvedt, E. C. (eds.), *Kotedalen - en boplass gjennom 5000 år. Bind 2. Naturvitenskapelige undersøkelser,* 91-122. Historisk museum, University of Bergen, Bergen.
Hjelle, K. L., Hufthammer, A. K., Kaland, P. E., Olsen, A. B. & Soltvedt, E. C. 1992 (eds.). *Kotedalen - en boplass gjennom 5000 år. Bind 2: Naturvitenskapelige undersøkelser,* Historisk museum, University of Bergen, Bergen.
Hjellebrekke, J. 1999. *Innføring i korrespondanseanalyse.* Fagbokforlaget, Bergen.
Hodder, I. 1982. *Symbols in Action. Ethnoarchaeological Studies of Material Culture.* Cambridge University Press, Cambridge.
Hougen, B. 1947. *Fra seter til gård. Studier i norsk bosetningshistorie.* Norsk Arkeologisk Selskap, Oslo.
Hufthammer, A. K. 1992. De osteologiske undersøkelsene fra Kotedalen. In Hjelle, K. L., Hufthammer, A. K., Kaland, P. E., Olsen, A. B. & Soltvedt, E. C. (eds.), *Kotedalen - en boplass gjennom 5000 år. Bind 2. Naturvitenskapelige undersøkelser,* 9-64. Historisk museum, University of Bergen, Bergen.
Humphrey, C. & Hugh-Jones, S. 1992. Introduction: Barter, exchange and value. In Humphrey, C. & Hugh-Jones, S. (eds.), *Barter, Exchange and Value. An Anthropological Approach,* 1-20. Cambridge University Press, Cambridge.
Hvoslef, S. 1967. Innberetning om undersøkelsen av steinalderboplassen Mørkedøla I, Eldrevann, Borgund s., Lærdal pgd., Sogn og Fjordane. Ettergravninger sommeren 1967. Unpublished report in the topographical archive, Bergen University Museum, University of Bergen.
Härke, H. 1995. 'The hun is a methodical chap': reflections on the German pre- and protohistory. In Ucko, P. J. (ed.), *Theory and Archaeology: a World Perspective,* 187-222. Routledge, London.
Høeg, H. I. 1996. *Pollenanalytiske undersøkelser i "Østerdalsområdet" med hovedvekt på Rødsmoen, Åmot i Hedmark.* Varia 39. Universitetets Oldsaksamling, Oslo.
Håland, R. 1977. Archaeological classification and ethnic groups. *Norwegian Archaeological Review* 10(1), 1-31.
Indrelid, S. 1973. *Hein 33 - en steinalderboplass på Hardangervidda: Forsøk på kronologisk og kulturell analyse.* Årbok for Universitetet i Bergen 1972. Humanistisk serie nr. 1., Bergen.
Indrelid, S. 1977. Eldre steinalder i sørnorske fjell. *Viking* XL, 129-148.

Indrelid, S. 1994. *Fangstfolk og bønder i fjellet. Bidrag til Hardangerviddas førhistorie 8500-2500 før nåtid.* Universitetets Oldsaksamlings Skrifter. Ny rekke nr. 17. Universitetets Oldsaksamling, Oslo.
Indrelid, S. & Moe, D. 1983. Februk på Hardangervidda i yngre steinalder. *Viking* XLVI, 36-71.
Ingold, T. 1986. *Evolution and Social Life.* Cambridge University Press, Cambridge.
Ingstad, A. S. 1965. Sluppan. II. Oldsaksmaterialet. *Viking* XXIX, 83-107.
Ingstad, A. S. 1970. Steinalderboplassen Rognlien i Eidanger. Et bidrag til belysning av yngre steinalder i Telemark. *Universitetets Oldsaksamlings Årbok* 1967-1968, 19-139.
Isaksson, S. 1996. Chemical analysis of organic residues on pottery. In Østmo, E. (ed.), *Auve bind 2. Tekniske og naturvitenskapelige undersøkelser. Norske Oldfunn XVII*, 27-42. Universitetets Oldsaksamling, Oslo.
Isaksson, S. 1999. Appendix 2. Analys av organiska belägningar på kermaik från Västra Norge. In Åstveit, L. I. (ed.), Keramikk i vitenskapelig kontekst: En studie over neolittisk keramikkmateriale fra Radøy, Hordaland. Unpublished Cand. philol. thesis, University of Bergen.
Jenkins, R. 1996. *Social Identity.* Routledge, London.
Jenkins, R. 1997. *Rethinking Etnicity: Arguments and Explorations.* London. Sage Publications.
Jennbert, K. 1984. *Den produktiva gåvan: Tradisjon och innovation i Sydskandinavien för omkring 5300 år sedan.* Acta Archaeologica Lundensia. Series in 4°. N° 16. CWK Gleerup, Lund.
Jewitt, J. R. 1987. *The Adventures and Sufferings of John R. Jewitt, Captive of Maquinna.* Jewitt's narrative annotated and illustrated by Hilary Stewart. Original narrative published in 1815. Douglas & McIntyre, Vancouver.
Johannessen, L. 2002. Arkeologiske undersøkelser på steinalderboplassen Synnaland I, Unpublished report in the topographical archive. Bergen University Museum, University of Bergen, Bergen.
Johansen, A. B. 1969. *Høyfjellsfunn ved Lærdalsvassdraget. I. Den teoretiske bakgrunnen og det første analyseforsøk.* Universitetet i Bergens Årbok 1969. Humanistisk Serie nr. 4. University of Bergen, Bergen.
Johansen, A. B. 1976. Comments on typological and chronological problems. *Norwegian Archaeological Review* 9(1), 28-31.
Johansen, A. B. 1978. *Høyfjellsfunn i Lærdalsfjella II.* Universitetsforlaget, Oslo.
Johansen, E. 1957. Norsk strandflint - er den hentet i Danmark-Skåne eller i Norske strande? *Meddelelser fra Dansk Geologisk Forening* 13(4), 257-258.
Johansen, K. B. 2000. Byttenetverk, allianser og aggresjon. Mellomneolittiske boplasser rundt Oslofjorden og Bohuslän. Unpublished Cand. philol. thesis, University of Bergen.
Jones, S. 1997. *The Archaeology of Ethnicity. Constructing Identities in the Past and Present.* Routledge, London.
Kaland, P. E. 1984. Holocene shore displacement and and shorelines in Hordaland, Western Norway. *Boreas* 13, 203-242.
Kaland, P. E. 1992. Pollenanalytiske undersøkelser utenfor boplassen i Kotedalen. In Hjelle, K. L., Hufthammer, A. K., Kaland, P. E., Olsen, A. B. & Soltvedt, E. C. (eds.), *Kotedalen - en boplass gjennom 5000 år. Bind 2. Naturvitenskapelige undersøkelser*, 65-90. Historisk museum, University of Bergen, Bergen.
Kaldhol, H. 1947. Et gammelt steinbrot på kalsedon på Ytrehorn i Horninndal. *Naturen* 1947, 62-63.
Kelly, R. L. 1991. Sedentism, sociopolitical inequality, and resource fluctuations. In Gregg, S. A. (ed.), *Between Bands and States*, 135-158. Southern Illinois University Press, Carbondale.
Kelly, R. L. 1995. *The Foraging Spectrum. Diversity in Hunter-Gatherer Lifeways.* Smithsonian Institution Press, Washington.
Kent, S. 1989. Cross-cultural perceptions of farmers as hunters and the value of meat. In Kent, S. (ed.), *Farmers as Hunters: the Implications of Sedentism*, 1-17. Cambridge University Press, Cambridge.
Keyser, R. 1868 [1839]. *Om Nordmændenes Herkomst og Folkeslægtskab. Samlede Afhandlinger, 263-462.* B. T. Walling, Christiania.
Kildal, E. S. 1970. *Geologisk kart over Norge. Berggrunnskart. Måløy, 1:250 000, norsk utgave.* Norges Geologiske Undersøkelse, Oslo.
Kleppe, E. J. 1977. Archaeological material and ethnic identification. A study of Lappish material from Varanger. *Norwegian Archaeological Review* 10(1), 32-45.
Knudsen, A. M. 1981. Rapport over utgravning av steinalderlokalitet på Synnaland, gnr. 28, bnr. 82, Brattvåg, Haram k., Sogn og Fjordane. Unpublished report in the topographical archive, Bergen University Museum, University of Bergen.
Koch, E. 1998. *Neolithic Bog Pots: From Zealand, Møn, Lolland and Falster.* Nordiske fortidsminder. Serie B ; vol. 16. Det Kgl. Nordiske Oldskriftselskab, København.
Kohl, P. L. 1993. Limits to a post-processual archaeology (or, dangers of a new scolasticism). In Yoffee, N. & Sherratt, A. (ed.), *Archaeological Theory: Who Sets the Agenda?*, 13-19. Cambridge University Press, Cambridge.
Kolderup, N. H. 1925. Petrologische Untersuchungen über das Material für Steinzeit-werkzeuge im Westlichen Norwegen. *Tschermaks mineralogische und petrografische Mitteilungen* 38:165.
Kossina, G. 1902. Die indogermanische Frage, archäologisch beantwortet. *Zeitschrift für Ethnologie* 1902, 161-222.
Kristoffersen, K. 1995. *De arkeologiske undersøkelsene på Bjorøy 1992-1995.* Arkeologiske rapporter 20. Arkeologisk institutt, University of Bergen, Bergen.
Kristoffersen, K. K. & Warren, E. J. 2001. *Kulturminner i trekant-traséen. De arkeologiske undersøkelsene i forbindelse med utbygging av Trekantsambandet i kommunene Bømlo, Sveio og Stord i Sunnhordland.* Arkeologiske avhandlinger og rapporter fra Universitetet i Bergen 6. Arkeologisk institutt, University of Bergen, Bergen.
Kristoffersen, K. K. 2001a. Lokaliseringsanalyse. In Kristoffersen, K. K. & Warren, E. J. (eds.), *Kulturminner i trekant-traséen. De arkeologiske undersøkelsene i forbindelse med utbygging av Trekantsambandet i kommunene Bømlo, Sveio og Stord i Sunnhordland*, 217-232. Arkeologisk institutt, University of Bergen, Bergen.
Kristoffersen, K. K. 2001b. Neolittiske lokaliteter og lokaliteter fra sen steinbrukende tid. In Kristoffersen, K. K. & Warren, E. J. (eds.), *Kulturminner i trekant-traséen. De arkeologiske undersøkelsene i forbindelse med utbygging av Trekantsambandet i kommunene Bømlo, Sveio og Stord i Sunnhordland. Arkeologiske avhandlinger og rapporter fra Universitetet i Bergen 6*, 64-173.

Arkeologisk institutt, University of Bergen, Bergen.
Kristoffersen, S. 1990. *FV 018 Austvik - Brandasund 1988-1990.* Arkeologiske rapporter 13. Historisk museum, University of Bergen, Bergen.
Kristoffersen, S. 1995. Neolithic hunting stations on the coast of Hordaland, Western Norway. Site organization. In Bergsvik, K. A., Nygård, S. & Nærøy, A. J. (eds.), *Steinalderkonferansen i Bergen 1993. Arkeologiske skrifter 8*, 151-173. Arkeologisk institutt, University of Bergen, Bergen.
Krzywinski, K. & Stabell, B. 1984. Late Weichselian sea level changes at Sotra, Hordaland, western Norway. *Boreas* 13(2), 243-258.
Kuper, A. 1988. *The Invention of Primitive Society.* Routledge, London.
Kvamme, K. & Jochim, M. 1989. The environmental basis of Mesolithic settlement. In Bonsall, C. (eds.), *The Mesolithic in Europe: Papers Presented at the Third International Symposium, Edinburg 1985*, 1-12. John Donald, Edinburg.
Landmark, T. & Stylegar, F. -A. 1998. Et nyoppdaget diabasbrudd ved Mandal og den sørlandske diabasprovins i eldre steinalder. *Primitive Tider* 1, 77-89.
Larsson, M. 1985. *The early Neolithic Funnel-Beaker Culture in South-west Scania, Sweden. Social and Economic Change 3000-2500 BC. BAR 264.* BAR International Series, Oxford.
Larsson, M. & Olsson, E. 1997 (eds.). *Regionalt och interregionalt. Stenalderundersökningar i Syd- och Mellansverige.* Arkeologiske undersøkningar. Skrifter nr 23, Riksantikvarieämbetet, Stocholm.
Lechtman, H. 1977. Style in technology: some early thoughts. In Lechtman, H. & Merrill, R. S. (eds.), *Material Culture: Style, Organization and Dynamics of Technology*, 3-20. West Publishing, New York and St. Paul, Minnesota.
Lemonnier, P. 1976. La Description des Chaînes Opératoires: Contribution à l'Analyse des Systèmes Techniques. *Techniques et Culture* 1, 100-151.
Lemonnier, P. 1986. The study of material culture today: Toward an anthropology of technical systems. *Journal of Anthropological Archaeology* 5, 147-186.
Leroi-Gourhan, A. 1964. *La Geste et la Parole I. Technique et Langage.* Albin Michel, Paris.
Leroi-Gourhan, A. 1965. *La Geste et la Parole II. Memoire et les Rythmes.* Albin Michel, Paris.
Lindblom, I. 1980. *Slipte spisser av bergart i Rogaland. Funnspredning og mulige årsaker til denne.* AmS - Varia 9. Arkeologisk museum i Stavanger, Stavanger.
Lindblom, I. 1984. Former for økologisk tilpasning i Mesolitikum, Østfold. *Universitetets Oldsaksamlings Årbok* 1982/1983, 43-86.
Lourandos, H. 1998. *Continent of Hunter-Gatherers: New Perspectives in Australian Prehistory.* Cambridge University Press, Cambridge.
Lundberg, Å. 1997. *Vinterbyar - et bandsamhälles territorier i Norrlands inland, 4500-2500 f.Kr.* Studia Archaeologica Universitas Umensis 8. Umeå universitet, Arkeologiska institutionen, Umeå.
Lødøen, T. K. 1995. Landskapet som rituell sfære i steinalder. En kontekstuell studie av bergartsøkser fra Sogn. Unpublished Cand. philol. thesis, University of Bergen.
Lødøen, T. K. 2001. Contextualizing rock art in order to investigate Stone Age ideology. Results from an ongoing project. In Helskog, K. (ed.), *Theoretical Perspectives in Rock Art Research*, 211-223, Novus forlag, Oslo.
MacEachern, S. 1998. Scale, style and cultural variation: technological traditions in the Northern Mandera mountains. In Stark, M. T. (ed.), *The Archaeology of Social Boundaries*, 107-131. Smithsonian Institution Press, Washington.
Madsen, T. & Petersen, J. E. 1984. Tidligneolitiske anlæg ved Mosegården. Regionale og kronologiske forskelle i tidligneolitikum. *KUML Aarbøker for Nordisk Oldkyndighed og Historie* 1982-1983, 61-120.
Mandt, G. 1991. Vestnorske ristninger i tid og rom. Kronologiske, korologiske og kontekstuelle studier. Bind 1 og 2. Unpublished Dr. philos. thesis, University of Bergen.
Martens, I. & Hagen, A. 1961. *Arkeologiske undersøkelser langs elver og vann. Gyrinosvatn, Hallingdal og Tokke-Vinje vassdraget, Telemark.* Norske Oldfunn X. Universitetets Oldsaksamling, Oslo.
Mauss, M. 1990 [1925]. *The Gift: The Form and Reason for Exchange in Archaic Societies.* Translated by W.D. Halls. Routledge, London.
McCarthy, F. D. 1939. "Trade" in Aboriginal Australia and "trade" relationships with Torres strait, New Guinea and Malaya (part 1). *Oceania* 9(4), 405-438.
McCarthy, F. D. 1940. "Trade" in Aboriginal Australia and "trade" relationships with Torres strait, New Guinea and Malaya (part 2 and 3). *Oceania* 10(1 and 2), 80-105, 171-195.
McKennan, R. A. 1969. Athapaskan groupings and social organization in central Alaska. In Damas, D. (ed.), *Contributions to Anthropology: Band Societies*, 93-114. National Museums of Canada, Ottawa.
Meinander, C. F. 1939. Pyheensilta stenåldersboplads. *Finskt Museum* LIV-LV, 14-25.
Meinander, C. F. 1961. De subneolitiska kulturgrupperna i norra Europa. *Societas Scientarium Fennica. Årsbok* XXXIX B(4), 3-23.
Meschke, C. 1977. *Early Norrland Sites on the Umeälven: a Cultural-Historical Survey.* Early Norrland 3. Kungl. Vitterhets Historie och Antikvitets Akademien, Uppsala.
Mikkelsen, E. 1975. Mesolithic in South-eastern Norway. *Norwegian Archaeological Review* 8(1), 19-35.
Mikkelsen, E. 1976. Comments on typological and chronological problems. *Norwegian Archaeological Review* 9(1), 32-36.
Mikkelsen, E. 1977. Østnorske veideristninger - Kronologi og økokulturelt miljø. *Viking* XL, 147-201.
Mikkelsen, E. 1982. Introduksjonen av jordbruk i Øst-Norge. In Sjøvold, T. (ed.), *Introduksjonen av jordbruk i Norden*, 129-141. Universitetsforlaget, Oslo.
Mikkelsen, E. 1984. Neolitiseringen i Øst-Norge. *Universitetets Oldsaksamlings Årbok* 1982-1983, 87-128.

Mikkelsen, E. 1989. *Fra jeger til bonde. Utviklingen av jordbrukssamfunn i Telemark i steinalder og bronsealder*. Universitetets Oldsaksamlings Skrifter. Ny rekke nr. 11. Universitetets Oldsaksamling, Oslo.

Milnes, A. G., Dietler, T. N. & Koestler, A. G. 1988. The Sognefjord north shore log - a 25 km depth section through Caledonizised basement in western Norway. *Norges Geologiske Undersøkelser. Special Publications* 3, 114-121.

Montelius, O. 1872-1874. *Sveriges forntid: försök till framställning af den svenska fornforskningens resultat*. Norstedt, Stockholm.

Montelius, O. 1919. *Vår Forntid*. Norstedt, Stockholm.

Morphy, H. 1995. Landscape and the reproduction of the ancestral past. In Hirsch, E. & O'Hanlon, M. (eds.), *The Anthropology of Landscape: Perspectives on Space and Place*, 184-209. Oxford University Press, Oxford.

Mortensen, M. 1991. Kosesteiner; spill, samlemani eller bare tilfeldighet. *Arkeo* 1991(1), 29-30.

Munch, P. A. 1852. *Det Norske Folks Historie. Bind I*. Chr. Tønsberg, Christiania.

Myhre, B. 1963. Innberetning om fortsatt graving på steinalderboplassen foran Knut Haukelids gamle hytte i Vivik, Holmevatn, Grungedal s., Vinje pgd., Telemark. Unpublished report in the topographical archive, Oldsaksamlingen, University of Oslo.

Myhre, L. N. 1994. *Arkeologi og politikk. En arkeo-politisk analyse av faghistoria i tida 1900-1960*. Varia 26. Universitetets Oldsaksamling, Oslo.

Narmo, L. E. 1993. Steinalder på Romsdalskysten. *Årbok for Romsdalsmuseet* 1993, 9-34.

Nesje, A. & Kvamme, M. 1991. Holocene glacier and climate variations in western Norway: Evidence for early Holocene glacier demise and multiple Neolglacial events. *Geology* 19, 610-512.

Nielsen, P. O. 1978. Die Flintbeile der frühen Trichterbecherkultur in Dänemark. *Acta Archaeologica* 48, 61-138.

Nielsen, Y. 1891. Lappernes Framrykning mot Syd i Trondhjems Stift og Hedemarkens Amt. *Det Norske Geografiske Selskabs Årbog* 1889-1890, 19-39.

Nilsson, S. 1938. *Skandinaviska Nordens urinnvånare, ett försök i comparativa ethnographien. Hefte 1 och 2*. Schmidt, Christianstad.

Noe-Nygaard, A. 1979. *Vulkaner*. Gyldendal Nordisk Forlag, København.

Nordquist, P. 2001. *Hierarkiseringsprocesser : om konstruktion av social ojämlikhet i Skåne, 5500-1100 f. Kr*. Studia Archaeologica Universitatis Umensis 13. Arkeologiska Institutionen, Umeå universitet, Umeå.

Nordås, J. 1987. Appendix A. Generell oversikt over Siggjo-området, midtre del av Bømlo. In Alsaker, S. (ed.), *Bømlo - steinalderens råstoffsentrum på Sørvestlandet*, 106-108. Historisk museum, University of Bergen, Bergen.

Nummedal, A. & Bjørn, A. 1930. Boplassfund fra yngre stenalder i Aust-Agder. *Universitetets Oldsaksamlings Årbok* 1929, 21-109.

Nygaard, S. 1974. Håvikboplassene på Karmøy. *Stavanger Museums Årbok* 1973, 5-36.

Nygaard, S. 1987. Socio-economic developments along the southwestern coast of Norway between 10.000 and 4.000 bc. In Rowley-Conwy, P., Zvelebil, M. & Blankholm, H. P. (eds.), *Mesolithic Northwest Europe:Rrecent Trends*, 147-154. The Department of Archaeology and Prehistory, University of Sheffield, Sheffield.

Nygaard, S. 1988. Too many people, ritual porridge eating, or was Hinsch right after all? In Indrelid, S., Kaland, S. & Solberg, B. (eds.), *Festskrift til Anders Hagen. Arkeologiske skrifter 4*, 147-155. Historisk museum, University of Bergen, Bergen.

Nærøy, A. J. 1987. Redskapstradisjon i Nordhordland fra 5500 til 4000 før nåtid - en lokalkronologisk studie. Unpublished Cand. philol. thesis, University of Bergen.

Nærøy, A. J. 1988. Teknologiske endringer ved overgangen fra eldre til yngre steinalder på Vestlandet. I Indrelid, S., Kaland, S. & Solberg, B. (eds.), *Festskrift til Anders Hagen. Arkeologiske skrifter 4*, 205-213. Historisk museum, University of Bergen, Bergen.

Nærøy, A. J. 1993a. Chronological and technological changes in Western Norway 6000-3800 BP. *Acta Archaeologica* 63-1992, 76-95.

Nærøy, A. J. 1993b. The site Ramsvikneset, Straume, Hordaland, Western Norway. An introduction to Egil Bakka's article. In Solberg, B. (ed.), *Minneskrift Egil Bakka. Arkeologiske skrifter 7*, 17-20. Historisk museum, University of Bergen, Bergen.

Nærøy, A. J. 1994. *Troll-prosjektet. Arkeologiske undersøkelser på Kollsnes, Øygarden k., Hordaland, 1989-1992*. Arkeologiske rapporter 19. Historisk museum, University of Bergen, Bergen.

Nærøy, A. J. 2000. *Stone Age Living Spaces in Western Norway. BAR 857*. BAR International Series, Oxford.

Odner, K. 1965. Vivik ved Holmevatn på Haukelifjell. Arkeologiske undersøkelser i Røldal-Suldal I. *Viking* XXIX, 201-242.

Odner, K. 1968. Vivik near Holmevatn on Haukelifjell. A study in South Norwegian Stone Age chronology. *Norwegian Archaeological Review* 1, 80-83.

Odner, K. 1983. *Finner og Terfinner. Etniske prosesser i det nordlige Fenno-Skandinavia*. Occational papers in Social Anthropology no. 9. University of Oslo, Oslo.

Odner, K. 1985. Saamis (Lapps), Finns and Scandinavians in history and prehistory. Ethnic origins and ethnic processes in Fenno-Scandiavia. *Norwegian Archaeological Review* 18(1), 1-12.

Odner. K. 2000. *Tradition and Transmission. Bantu, Indo-European, and Circumpolar Great Traditions*. Bergen Studies in Social Anthropology 54. Norse Publications. Bergen.

Olsen, A. B. 1976a. Innberetning fra utgravning av steinalderlokalitet på Berknes, Volda k., Møre og Romsdal. Unpublished report in the topographical archive, Bergen University Museum, University of Bergen, Bergen.

Olsen, A. B. 1976b. Innberetning fra utgravning av steinalderlokaliteten Vikja I, Brandsøy, Flora k. Sogn og Fjordane. Unpublished report in the topographical archive, Bergen University Museum, University of Bergen, Bergen.

Olsen, A. B. 1977. Grønnsteinsindustri i Sunnfjord i steinalderen. Funn av verkstedsboplasser i Florøområdet. *Arkeo* 1977, 11-14.

Olsen, A. B. 1981. Bruk av diabas i vestnorsk steinalder. Unpublished Mag. art. thesis, University of Bergen.

Olsen, A. B. 1983. Arkeologi. In Bostwick Bjerck, L. & Olsen, A. B. (eds.), *Kulturhistoriske undersøkelser på Botnaneset, Flora 1981-82. Arkeologiske rapporter 5*, 7-132. Historisk museum, University of Bergen, Bergen.
Olsen, A. B. 1992. *Kotedalen - en boplass gjennom 5000 år. Bind 1: Fangstbosetning og tidlig jordbruk i vestnorsk steinalder: Nye funn og nye perspektiver.* Historisk museum, University of Bergen, Bergen.
Olsen, A. B. 2002. Omregulering av deler av reguleringsplan for Hovland, Vold, Vespestad gnr. 65, bnr. 1., Bømlo k., Hordaland. Unpublished report in the topographical archive, Bergen University Museum, University of Bergen.
Olsen, A. B. & Alsaker, S. 1984. Greenstone and diabase utilization in the stone age of Westen Norway: Technological and socio-cultural aspects of axe and adze production and distribution. *Norwegian Archaeological Review* 17(2), 71-103.
Olsen, B. 1985a. Arkeologi og etnisitet. Et teoretisk of empirisk bidrag. In Næss, J. R. (ed.), *Arkeologi og etnisitet*, 25-31. Arkeologisk museum i Stavanger, Stavanger.
Olsen, B. 1985b. Comments on Saamis, Finns and Scandinavians in history and prehistory. *Norwegian Archaeological Review* 18(1), 13-18.
Olsen, B. 1988. Interaction between hunter-gatherers and farmers: ethnographical and archaeological perspectives. *Archeologia Polski* XXXIII(2), 425-434.
Olsen, B. 1991. Kjelmøyfunnenes (virknings) historie og arkeologi. *Viking* LIV, 65-87.
Olsen, B. & Kobylínski, Z. 1991. Ethnicity and anthropological and archaeological research: a norwegian-polish perspective. *Archeologia Polona* 29, 5-27.
Olsen, H. 1976. Skipshelleren, osteologisk materiale, Unpublished thesis, University of Bergen, Bergen.
Paynter, R. 1989. The archaeology of equality and inequality. *Annual Review of Anthropology* 18, 369-399.
Pedersen, R. B. & Dunning, G. R. 1997. Evolution of arc crust and relations between contrasting sources: U-Pb (age), Nd and Sr isotope systematics of the ophiolitic terrain of SW Norway. *Contrib. Mineral. Petrol* 128, 1-15.
Pelegrin, J., Karlin, C. & Bodu, P. 1988. "Chaînes Opératoires": un outil pour le préhistorien. *Technologie Préhistoriques. Notes et Monographies Techniques* 25, 55-62.
Persson, P. 1998. Gropkeramikfenomenet på vestkusten. *In Situ. Vestsvensk arkeologisk tidskrift* 1998, 63-84.
Persson, P. 1999. *Nelitikums början. Undersökningar kring jurdbrukets introduktion i Nordeuropa*. Kust til kust. GOTARC serie B nr. 11. Dpts. of archaeology and ancient History, University of Uppsala and Göteborg, Göteborg.
Peterson, N. & Long, J. 1986. *Australian Territorial Organization*. Oceania Monograph 30. University of Sydney, Sydney.
Pettersen, K. 1986. Den fantastiske flinten. *Spor* 1, 12-16.
Prescott, C. 1996. Was there really a Neolithic in Norway? *Antiquity* 70, 77-86.
Price, T. D. 1980. Regional approaches to human adaptation in the Mesolithic in the North European plain. In Gramsch, B. (ed.), *Mesoliticum in Europa*, 217-234. Veröffentlichungen des Museums für Ur- und Frühgeshichte Potsdam, Band 14-15, Berlin.
Price, T. D. & Brown, J. A. 1985 (eds.). *Prehistoric Hunter-Gatherers: The Emergence of Cultural Complexity*. Studies in Archaeology, Academic Press Inc, New York.
Price, T. D. & Feinman, G. M. 1995 (eds.). *Foundations of Social Inequality. Fundamental Issues in Archaeology*. Plenum Press, New York and London.
Paasche, K. 2000. Nye helleristninger på Geithus. Dokumentasjon, motiv og kunst. *Universitetets Oldsaksamlings Årbok* 1999, 25-38.
Ramstad, M. 1998. Common group identity with or without ethnicity? The Norwegian west coast during the late stone age. In Anderson, A. -C., Gillberg, Å., Jensen, O. W., Karlsson, H. & Rolöf, M. V. (eds.), *The Kaleidoscopic past: Proceedings from the 5th Nordic TAG Conference Göteborg, 2-5 April 1997*, 355-368. Göteborg University, Department of Archaeology, Göteborg.
Ramstad, M. 1999. Brytninga mellom nord og sør: En faghistorisk og lokalkronologisk studie over Møre i yngre steinalder. Unpublished Cand. philol. thesis, University of Bergen.
Ramstad, M. 2000a. Brytninga mellom nord og sør - ei faghistorisk analyse med utgangspunkt i Møre i yngre steinalder. *Primitive Tider* 3, 54-79.
Ramstad, M. 2000b. Veideristningene på Møre. Teori, kronologi og dateringsmetoder. *Viking* LXIII, 51-86.
Ramstad, M. & Olsen, A. B. 2002. Innberetning fra utgravningen av steinalderboplassen Kleiva, Flora, Sogn og Fjordane. Unpublished report in the topographical archive, Bergen University Museum, University of Bergen.
Randers, K. 1986. *Breheimenundersøkelsene 1982-1984 I:Høyfjellet*. Arkeologiske rapporter 10. Historisk museum, University of Bergen, Bergen.
Randers, K. & Höglin, S. 1988. *Valderøya Vest 1986-1987*. Arkeologiske rapporter 12. Historisk museum, University of Bergen, Bergen.
Ravnås, R. & Furnes, H. 1995. The use of geochemical data in determining the provenance and tectonic setting of ancient sedimentary successions: the Kalvåg Melange, western Norwegian Caledonites. *Spec. Publs int. Ass. Sediment.* 22, 237-264.
Renfrew, C. 1977. Alternative models for exchange and spatial distribution. In Earle, T. & Ericcson, J. E. (eds.), *Exchange Systems in Prehistory*, 72-91. Academic Press, New York.
Richardson, A. 1982. The control of productive resources on the Nortwest Coast of North America. In Williams, N. M. & Hunn, E. S. (eds.), *Resource Managers: North American and Australian Hunter-Gatherers*, 93-112. American Association for the Advancement of Science. Selected Symposium 67, Boulder, Colorado.
Riches, D. 1982. *Northern Nomadic Hunter-Gatherers. A Humanistic Approach*. Academic Press, New York.
Roland, H. 1994. Trekantsambandet Sveio-Stord-Bømlo. Rapport fra utgravning av lok. 88, Arahaugen, Føyno, Stord k., Hordaland. Unpublished report in the topographical archive, Bergen University Museum, University of Bergen.
Rowley-Conwy, P. 1983. Sedentary hunters: the Ertebølle example. In Bailey, G. (ed.), *Hunter-Gatherer Economy in Prehistory*, 111-126. Cambridge University Press, Cambridge.

Rowley-Conwy, P. 1995. Making the first farmers younger: the West European evidence. *Current Anthropology* 36(2), 346-353.
Rygh, K. 1897. *Trøndelagen i forhistorisk tid. En archæologisk oversikt.* Aktietrykkeriet i Trondhjem, Trondhjem.
Rygh, K. 1903. *Spidser og knive af skifer.* Det Kgl. Norske Vitenskapers Selskabs Skrifter. 1902. No. 3. Aktietrykkeriet i Trondhjem, Trondhjem.
Rygh, O. 1966. De i siste Halvaar 1866 til Universitetets Samling indkomne Sager, ældre end Reformationen. *Foreningen til Norske Fortidsmindesmærkers Bevaring. Aarsberetning for* 1966, 85-105.
Rygh, O. 1872. Om Affaldsdyngen ved Stenkjær. *Foreningen til Norske Fortidsminnesmerkers Bevaring. Aarsberetning* 1871, 100-117.
Rygh, O. 1876a. Bidrag til oversigt over den skandinaviske stenalder i Norge. *Christiania Videnskabs Selskabs Forhandlinger. No* 9, 1-12.
Rygh, O. 1876b. Sur le Groupe Arctique de l'Age de la Pierre Polie en Norvège. *Congrès International d'Anthtropologie & d'Archaéologie Préhistoriques: compte rendu de la 7e session Stockholm 1874,* 177-187.
Sackett, J. R. 1985. Style and ethnicity in the Kalahari: A reply to Wiessner. *American Antiquity* 50(1), 154-159.
Sackett, J. R. 1986. Style, function, and assemblage variability - a reply to Binford. *American Antiquity* 51(3), 628-34.
Sahlins, M. D. 1963. Poor man, rich man, big man, chief: Political types in Polynesia and Melanesia. *Comparative studies in Society and History* 5, 285-303.
Sahlins, M. D. 1972. *Stone Age Economics.* Tavistock, London.
Schaller, E. 1984. Organisasjonsmønstre i steinalderen i sørnorske fjellstrøk. Unpublished Mag. art. thesis, University of Oslo.
Schaller-Åhrberg, E. 1990. Refitting as a method to separate mixed sites: a test with unexpected results. In Cziesla, E., Eickhoff, S., Arts, N. & Winter, D. (eds.), *The Big Puzzle (International Symposium on Refitting Stone Artefacts),* 611-622. Holos Verlag, Bonn.
Schanche, K. 1995. Var det landsbyer i Finnmark i yngre steinalder? In Bergsvik, K. A., Nygård, S. & Nærøy, A. J. (eds.), *Steinalderkonferansen i Bergen 1993. Arkeologiske skrifter 8,* 174-186. Arkeologisk institutt, University of Bergen, Bergen.
Schetelig, H. 1901. *Et bosted fra stenalderen paa Bømmeløen.* Bergens Museums Aarbok 1901. No. 5. Bergens Museum, Bergen.
Service, E. R. 1971. *Primitive Social Organization: an Evolutionary Perspective.* Studies in Anthropology. Random House, New York.
Sheenan, G. W. 1985. Whaling as and Organizing Focus in Northwestern Alaskan Eskimo Societies. In Price, T. D. & Brown, J. A. (eds.), *Prehistoric Hunter-Gatherers: The Emergence of Cultural Complexity,* 123-154. Academic Press Inc, New York.
Shennan, S. 1989. Introduction: Archaeological approaches to cultural identity. In Shennan, S. (ed.), *Archaeological Approaches to Cultural identity,* 1-32. Unwyn & Hyman, London.
Shennan, S. 1997. *Quantifying Archeology.* Edinburgh University Press, Edinburgh.
Shetelig, H. 1922. *Primitive tider i Norge.* John Griegs Forlag, Bergen.
Shetelig, H. 1930. *Fra oldtiden til omkring 1000 e.Kr.* Det Norske Folks Liv og Historie gjennom Tidene. Aschehoug, Oslo.
Shetelig, H. not dated. Plandrawings from Sokkamyro on file from the farm Vespestad, Bømlo. Topographical archive.Bergen University Museum, University of Bergen.
Sigmond, E. M. O., Gustavson, M. & Roberts, D. 1984. *Berggrunnskart over Norge - M. 1:1 million.* Norges Geologiske Undersøkelse, Oslo.
Simonsen, P. 1959. Fundforhold og dateringer. In Torgersen, J., Getz, B. & Simonsen, P. (eds.), *Varanger-funnene I,* 5-19 Tromsø Museum, Tromsø.
Simpson, D. N. 1992. *Archaeological investigations at Krossnes, Flatøy 1988-1991.* Arkeologiske rapporter 18. Historisk museum, University of Bergen, Bergen.
Simpson, D. N. 1996. Aspects of weathering of rhyolite and typological and technological considerations of this material based on results of refitting. *Norwegian Archaeological Review* 29(2), 79-88.
Simpson, D. N. 1999. Lithic raw material frequencies and the construction of site chronology. In Boaz, J. (ed.), *The Mesolithic of Central Scandinavia,* 299-316. Universitetets Oldsaksamling, Oslo.
Simpson, D. N. 2003. Arkeologiske undersøkelser på Mjølstadneset, Herøy k., Møre og Romsdal – 1999. Unpublished report in the topographical archive, Bergen University Museum, University of Bergen.
Singer, M. 1968. The concept of culture. In Sills, D. L. (ed.), *International Encyclopedia of the Social Sciences,* 527-43. Macmillan & Free Press, London.
Skandfer, M. 2002. Keramikk og politikk. *Viking* LXV, 137-157.
Skjelstad, G. 2003. Regionalitet i vestnorsk mesolitikum. Råstoffbruk og sosiale grenser på Vestlandskysten i mellom- og senmesolitikum. Unpublished Cand. philol. thesis, University of Bergen, Bergen.
Skjerlie, F. & Olsen, A. B.. 1981. Appendix A1. Mineralogisk beskrivelse av norske diabasforekomster. In Olsen, A. B. (ed.), *Bruk av diabas i vestnorsk steinalder.* Unpublished Mag. art. thesis. University of i Bergen, Bergen.
Skjølsvold, A. 1977. *Slettabøboplassen.* Arkeologisk museum i Stavanger skrifter 2. Arkeologisk museum i Stavanger, Stavanger.
Skjølsvold, A. 1980a. *Boplassen på Holeheia i Klepp.* AmS-Varia 7. Arkeologisk museum i Stavanger, Stavanger.
Skjølsvold, A. 1980b. *Boplassen på Nygård i Ogna.* AmS-Varia 6. Arkeologisk museum i Stavanger, Stavanger.
Solberg, B. 1989a. Køller, klubber og hakker av stein. Lite påaktede gjenstandsgrupper i vestnorsk yngre steinalder. *Universitetets Oldsaksamlings Årbok* 1986-88, 81-102.
Solberg, B. 1989b. The Neolithic transition in Southern Scandinavia: internal development or migration? *Oxford Journal of Archaeology* 8(3), 261-296.
Stanner, W.E.H. 1965. Aboriginal territorial organization: estate, range, domain and regime. *Oceania* 36(1), 1-26.
Stark, M. T. 1998. Technical choices and social boundaries in material culture patterning. In Stark, M. T. (ed.), *The archaeology of Social Boundaries,* 1-11. Smithsonian Institution Press, Washington.

Statens Kartverk 1997. *Tidevannstabeller for den norske kyst med Svalbard samt Dover, England 1997.* Statens Karverk, Sjøkartverket, Stavanger.

Steel, R., Siedlecka, A. & Roberts, D. 1985. The old red sandstone basins of Norway and their deformation: a review. In Gee, D. G. & Sturt, B. A. (eds.), *The Caledonite Orogen - Scandinavia and Related Areas*, 293-315. John Wiley & Sons, Chisester.

Storli, I. 1993. Fra "kultur" til "natur". Om konstitueringa av den "arktiske" steinalderen. *Viking* LVI, 7-22.

Stuiver, M., Reimer, P. J., Bard, E., Beck, J. W., Burr, G. S., Hughen, K. A., Kromer, B., McCormac, G., van der Plicht, J. & Spurk, M. 1998. INTCAL98 Radiocarbon Age Calibration, 24000-0 cal BP. *Radiocarbon* 40(3), 1041-1083.

Suttles, W. 1987. Cultural diversity within the Coast Salish continuum. In Auger, R., Glass, M. F., MacEachern, S. & McCartney, P. H. (eds.), *Ethnicity and Culture*, 243-249. Archaeological association, University of Calgary, Calgary, Alta.

Svendsen, J. I. & Warren, E. 2001. Isavsmeltning- og strandforskyvningskurve. In Kristoffersen, K. K. & Warren, E. J. (eds.), *Kulturminner i trekant-traséen. De arkeologiske undersøkelsene i forbindelse med utbygging av Trekantsambandet i kommunene Bømlo, Sveio og Stord i Sunnhordland*, 297-309. Arkeologisk institutt, University of Bergen, Bergen.

Søborg, H. C. 1986. Skiferkniver sør for Polarsirkelen i Norge. En analyse av attributter, typer og geografisk fordeling med bakgrunn i det fennoskandinaviske skiferkompleks. Unpublished Mag. art. thesis, University of Bergen.

Søborg, H. C. 1988. Knivskarpe grenser for skiferbruk i steinalderen. In Indrelid, S., Kaland, S. & Solberg, B. (eds.), *Festskrift til Anders Hagen. Arkeologiske skrifter 4*, 225-241. Historisk museum, University of Bergen, Bergen.

Søborg, H. C. 1994. Innberetning fra utgravningen av lokaliteten Nordøy grendahus, Longva, Flemsøy, Haram k. Møre og Romsdal. Unpublished report in the topographical archive, Bergen University Museum, University of Bergen.

Taffinder, J. 1998. *The Allure of the Exotic. The Social Use of Non-Local Raw Materials During the Stone Age in Sweden.* Aun 25. Department of Archaeology and Ancient history, University of Uppsala, Uppsala.

Tindale, N. B. 1974. *Aboriginal Tribes of Australia.* University of California Press, Berkeley.

Townshend, J. B. 1979. Indian or Eskimo? Interaction and identity in Southern Alaska. *Arctic Anthropology* 16, 160-182.

Townshend, J. B. 1980. Ranked societies of the north Pacific rim. *Senri Ethnological Studies* 4, 123-156.

Trigger, B. G. 1989. *A History of Archaeological Thought.* Cambridge University Press, Cambridge.

Trigger, B. G. 1995. Romanticism, nationalism, and archaeology. In Kohl, P. & Fawcett, C. (eds.), *Nationalism, Politics and the Practice of Archaeology*, 263-279. Routledge, London.

Trones, J. I. 1996. Bjorøy i yngre steinalder. Et tolkningsforslag om sosial organisering. Unpublished Cand. philol. thesis, University of Bergen.

Trones, J. I. 1998. Om ulike kvinner og menn i yngre steinalder. Tolkninger av boplassorganisering på Vestlandet. *Viking* LXI, 37-70.

Tørhaug, V. 2002. Berget 2 - En boplass fra senmesolitikum-tidligneolitikum med traktbegerkeramikk. In Glørstad, H. (ed.), *Svinesundprosjektet. Bind I. Utgravninger avsluttet i 2001*, 73-108. Universitetets Kulturhistoriske Museer. Oldsaksamlingen, Oslo.

von Düben, G. 1873. *Om Lappland och lapparna: företrädesvis de svenske: ethnografiska studier.* Norstedt & söner, Stockholm.

Walderhaug, E. M. 1994. Ansiktet er av stein: Ausevik i Flora - en analyse av bergkunst og Kontekst. Unpublished Cand. philol. thesis, University of Bergen.

Walderhaug, E. M. 1998. Changing art in a changing society: the hunter's rock-art of western Norway. In Chippindale, C. & Taçon, P. S. C. (eds.), *The Archaeology of Rock-Art*, 285-301. Cambridge University Press, Cambridge.

Wallén, C. C. 1968. Climate. In Sømme, A. (ed.), *A geography of Norden*, 48-61. Heinemann, London.

Warren, J. E. 1994. Coastal sedentism during the Atlantic period in Nordhordland, Western Norway? The middle and late Mesolithic components at Kotedalen. Unpublished Upublisert MA thesis, Memorial University of Newfoundland, St. John's.

Watanabe, H. 1972. The Ainu. In Bicchieri, M. G. (ed.), *Hunters and Gatherers Today*, 451-484. Holt, Rinehart and Winston, New York.

Welinder, S. 1985. Comments on early agriculture in Scandinavia. *Norwegian Archaeological Review* 18(1-2), 94-96.

Welsch, R. L. & Terrell, J. E. 1998. Material culture, social fields, and social boundaries on the Sepic Coast of new Guinea. In Stark, M. T. (ed.), *The Archaeology of Social Boundaries*, 50-77. Smithsonian Institution Press, Washington.

White, R. 1985. *Upper Paleolithic Land Use in the Périgord.* BAR International Series 253, Oxford.

Wiessner, P. 1983. Style and ethnicity in the Kalahari San projectile point. *American Antiquity* 48, 254-76.

Wiessner, P. 1985. Style or isocrestic variation? A reply to Sackett. *American Antiquity* 50(1), 160-166.

Winther, T. 1878. Om den såkaldte "Arktiske gruppe" af Stensager, med specielt Hensyn til de i Tromsø Museum Oppbevarede. *Foreningen til Norske Fortidsminnesmerkers Bevaring. Aarsberetning* 1877, 105-146.

Wobst, M. 1977. Stylistic behaviour and information exhange. In Cleland, C. E. (ed.), *For the Director: Research Essays in Honor of James B. Griffin*, 317-342. University of Michigan Press, Ann Arbor.

Wobst, M. 1978. The archaeo-ethnology of hunter-gatherers or the tyranny of the ethnographic record in archaeology. *American Antiquity* 43(2), 147-178.

Wolf, E. 1982. *Europe and the People without History.* University of California Press, Berkeley.

Zvelebil, M. 1998. Agricultural frontiers, Neolithic origins, and the transition to farming in the Baltic basin. In Zvelebil, M., Dennell, R. & Domanska, L. (eds.), *Harvesting the Sea, Farming the Forest*, 9-27. Sheffield Academic Press, Sheffield.

Østmo, E. 1976. Torsrød. En senmesolittisk kystboplass i Vestfold. *Universitetets Oldsaksamlings Årbok* 1972-1974, 41-52.

Østmo, E. 1984. Auve. Noen inntrykk fra en mellomneolittisk kystboplass i Vestfold. *Viking* XLVII - 1983, 42-65.

Østmo, E. 1985. En dysse på Holtenes i Hurum. Nytt lys på Østnorsk traktbegerkultur. *Viking* XLVIII-1984, 70-82.

Østmo, E. 1988. *Etableringen av jordbrukskultur i Østfold i steinalderen.* Universitetets Oldsaksamlings Skrifter. Ny rekke nr. 10. Universitetets Oldsaksamling, Oslo.

Østmo, E. 1989. Begrepet arkeologisk kultur og dets anvendelighet ved studiet av bronsealder og keltertid. *Universitetets Oldsaksamlings Årbok* 1986-1988, 7-17.
Østmo, E. 1993. Auve i Sandefjord-sanddynen, snorstempelkeramikken og C-14 dateringen. *Viking* LVI, 70-82.
Østmo, E. 2000. Elleve trøndske økser. Traktbegerkulturen nordafjells. *Primitive Tider* 3, 80-101.
Åstveit, L. I. 1999. Keramikk i vitenskapelig kontekst: En studie over neolittisk keramikkmateriale fra Radøy, Hordaland. Unpublished Cand. philol. thesis, University of Bergen.

APPENDIX 1

SCORES
SITE LOCATION ANALYSES

Skatestraumen

Fosnstraumen

Soltveit

Kollsnes

Bjorøy

Stord-Bømlo

SKATESTRAUMEN								
Site	**Harbour**	**Slope**	**Shelter**	**Accessible site area**	**View**	**Distance to fresh water**	**M.A.S.L.**	**Site type**
3	1	2	2	1	2	2	1	Short-term camp
4	1	2	1	1	3	3	2	Short-term camp
7	2	2	2	2	2	2	2	Short-term camp
8	2	3	2	3	2	2	2	Short-term camp
10	2	3	2	3	2	2	2	Short-term camp
15	3	3	2	3	2	2	1	Short-term camp
16	3	3	1	3	2	3	1	Short-term camp
31	2	3	2	3	2	3	1	Short-term camp
38	2	3	1	1	2	3	2	Short-term camp
42	2	3	1	3	2	2	2	Short-term camp
46	2	3	1	3	2	2	2	Short-term camp
54	3	3	1	3	2	3	2	Short-term camp
61	3	2	1	1	2	2	2	Short-term camp
63	2	2	1	1	2	2	2	Short-term camp
68	1	2	1	3	2	2	3	Short-term camp
70	1	2	2	3	2	2	2	Short-term camp
72	2	2	1	3	2	2	2	Short-term camp
73	2	2	1	3	2	2	2	Short-term camp
78	1	3	1	3	2	3	3	Short-term camp
79	2	3	1	3	2	3	2	Short-term camp
81	3	3	3	2	2	2	2	Short-term camp
96	3	3	3	3	1	1	3	Short-term camp
111	3	3	2	3	2	2	2	Short-term camp
112	3	3	2	3	3	1	2	Short-term camp
113	1	2	1	2	3	1	2	Short-term camp
115	1	2	1	2	3	1	2	Short-term camp
125	3	2	2	2	2	2	2	Short-term camp
149	3	2	3	2	1	3	2	Short-term camp
150	3	3	2	3	2	2	2	Short-term camp
151	2	3	2	3	2	3	2	Short-term camp
1	2	3	2	2	2	2	2	Long-term camp
6	2	3	1	2	3	3	2	Long-term camp
12	2	2	2	3	2	2	2	Long-term camp
14	3	3	2	3	1	3	2	Long-term camp
17	3	2	2	2	2	3	2	Long-term camp
19	3	3	2	3	2	1	3	Long-term camp
20	3	3	2	3	2	2	1	Long-term camp
23	3	3	3	2	2	2	2	Long-term camp
26	3	2	2	2	2	2	3	Long-term camp
27	1	2	1	3	2	2	1	Long-term camp
28	1	3	2	2	2	2	1	Long-term camp
29	2	2	2	2	2	2	2	Long-term camp
30	2	3	2	3	2	3	1	Long-term camp
33	1	3	1	3	2	3	2	Long-term camp
34	1	3	1	3	2	3	2	Long-term camp
35	2	3	1	3	2	3	2	Long-term camp
40	2	3	2	3	2	2	2	Long-term camp
47	2	3	1	3	2	2	1	Long-term camp
50	1	3	1	3	2	2	1	Long-term camp
55	3	3	1	3	2	3	2	Long-term camp

SKATESTRAUMEN (continued)								
67	2	3	1	3	2	3	3	Long-term camp
71	2	2	1	3	2	2	2	Long-term camp
76	2	2	1	3	2	2	2	Long-term camp
83	3	3	3	2	2	2	2	Long-term camp
84	3	3	3	2	2	2	2	Long-term camp
85	3	3	2	3	2	2	2	Long-term camp
88	2	3	2	3	3	2	2	Long-term camp
89	2	3	2	3	2	1	2	Long-term camp
90	3	2	3	2	1	3	2	Long-term camp
98	3	2	3	2	2	2	2	Long-term camp
102	3	3	3	3	2	2	3	Long-term camp
106	2	3	2	2	3	1	2	Long-term camp
108	3	3	3	3	2	2	2	Long-term camp
109	3	3	2	3	2	2	2	Long-term camp
110	3	2	2	3	2	2	2	Long-term camp
118	3	2	3	3	1	2	2	Long-term camp
130	3	2	2	2	2	2	2	Long-term camp
132	3	3	2	3	2	2	2	Long-term camp
133	3	3	2	3	2	2	2	Long-term camp
135	2	3	1	3	3	1	2	Long-term camp
136	2	3	2	3	3	3	2	Long-term camp

FOSNSTRAUMEN								
Site	**Harbour**	**Slope**	**Shelter**	**Accessible site area**	**View**	**Distance to fresh water**	**M.A.S.L.**	**Site type**
B1	2	3	1	3	2	2	3	Short-term camp
H2	2	3	1	3	1	1	2	Short-term camp
H3	2	2	1	3	2	1	2	Short-term camp
R1	2	3	2	2	2	1	3	Short-term camp
R6	2	3	2	3	2	1	3	Short-term camp
R12	3	3	2	3	1	1	2	Short-term camp
R15	3	2	2	3	1	1	2	Short-term camp
R18	3	3	1	3	1	1	1	Short-term camp
R19	3	2	2	3	2	1	2	Short-term camp
R22	3	3	2	3	2	1	1	Short-term camp
R26	1	2	3	2	2	1	3	Short-term camp
R29	3	2	2	3	1	1	2	Short-term camp
R32	2	2	2	3	2	1	2	Short-term camp
R35	2	2	2	2	2	1	3	Short-term camp
R36	1	3	2	3	2	1	2	Short-term camp
R37	1	3	1	3	1	1	2	Short-term camp
S2	3	3	1	2	2	1	2	Short-term camp
S5	3	3	2	3	1	3	1	Short-term camp
S8	3	3	1	3	1	1	2	Short-term camp
S12	3	3	2	3	1	2	2	Short-term camp
S14	2	2	1	2	3	1	2	Short-term camp
S17	2	2	2	2	1	2	3	Short-term camp
S22	3	2	2	3	2	1	3	Short-term camp
S24	3	3	2	3	1	1	3	Short-term camp
S26	3	2	3	3	3	1	3	Short-term camp
R4	2	3	2	3	1	1	3	Short-term camp
R23	3	3	2	3	1	1	2	Short-term camp

FOSNSTRAUMEN (continued)								
B2	2	3	1	3	2	2	3	Long-term camp
I1	3	3	3	3	2	1	3	Long-term camp
S1	3	3	3	3	2	1	2	Long-term camp
R2	2	3	3	3	2	1	2	Long-term camp
R3	2	3	3	3	2	1	2	Long-term camp
R9	2	3	2	3	2	1	2	Long-term camp
R25	3	2	2	3	2	1	2	Long-term camp
R30	3	3	2	2	2	1	2	Long-term camp
R38	3	3	2	3	1	2	2	Long-term camp
RA1	3	3	3	3	2	1	3	Long-term camp
S1	3	3	2	3	2	1	2	Long-term camp
S4	3	3	3	3	2	1	2	Long-term camp
S6	3	3	2	3	2	1	2	Long-term camp
S13	2	3	1	3	3	2	2	Long-term camp
S15	2	2	1	3	2	0	2	Long-term camp
S25	3	3	2	3	2	0	2	Long-term camp

SOLTVEIT								
Site	**Harbour**	**Slope**	**Shelter**	**Accessible site area**	**View**	**Distance to fresh water**	**M.A.S.L.**	**Site type**
3 Nord	3	3	2	2	2	1	2	Short-term camp

KOLLSNES								
Site	**Harbour**	**Slope**	**Shelter**	**Accessible site area**	**View**	**Distance to fresh water**	**M.A.S.L.**	**Site type**
1S	3	2	3	3	1	3	2	Short-term camp
5R	2	3	2	3	1	3	2	Short-term camp
6R	2	2	3	2	2	1	2	Short-term camp
33B	3	2	2	3	1	3	2	Short-term camp
3S	3	3	3	3	1	2	3	Short-term camp
4R	3	3	3	3	2	3	3	Short-term camp
12S	3	2	2	2	1	1	2	Short-term camp
16B	3	2	3	2	1	3	3	Short-term camp
18H	1	2	3	2	3	2	1	Short-term camp
30B	3	3	3	2	1	3	3	Short-term camp
31B	3	3	3	2	1	3	3	Short-term camp
17B	3	3	3	3	1	3	2	Long-term camp

BJORØY								
Site	**Harbour**	**Slope**	**Shelter**	**Accessible site area**	**View**	**Distance to fresh water**	**M.A.S.L.**	**Site type**
25	2	3	3	1	3	1	2	Short-term camp
21	3	2	1	2	2	2	3	Long-term camp
4	3	3	3	3	2	1	3	Long-term camp
5	3	2	2	2	2	1	3	Long-term camp
6	3	2	2	2	2	1	3	Long-term camp
7	3	3	3	2	1	1	3	Long-term camp
17	3	3	2	3	2	1	3	Long-term camp

STORD-BØMLO								
Site	**Harbour**	**Slope**	**Shelter**	**Accessible site area**	**View**	**Distance to fresh water**	**M.A.S.L.**	**Site type**
N39	3	3	3	3	3	3	3	Short-term camp
F49	3	3	2	3	1	3	2	Short-term camp
F55	3	3	3	3	2	1	2	Short-term camp

STORD-BØMLO (continued)								
B59	3	2	3	3	2	0	3	Short-term camp
B63	3	3	2	3	3	3	1	Short-term camp
F88	3	3	2	3	3	1	2	Short-term camp
S103	3	3	2	3	3	3	2	Short-term camp
F115	3	3	3	3	3	2	3	Short-term camp
N138	3	2	3	3	1	2	3	Short-term camp
B13	3	2	3	3	2	3	2	Short-term camp
S32	3	3	3	3	2	1	3	Short-term camp
B60	3	3	2	3	3	3	1	Short-term camp
S102	3	2	3	3	3	1	3	Short-term camp
F113	3	2	2	3	2	2	3	Short-term camp

APPENDIX 2

REFERENCE COLLECTION

INDIVIDUAL AND GROUPED RAW MATERIALS

Individual and grouped raw materials classified by the reference system	
Raw materials	**Groupings**
Basaltic rock	
Greenstone	From Hespriholmen, Stegahaugen, and greenstone quarry 4
Diabase	
Flint	Fine, medium and coarse
Fine flint	
Medium flint	
Coarse flint	
Fine green quartzite 1	
Fine brown quartzite 1	
Fine black quartzite 1	
Other fine black quartzites	Fine black quartzite 2, Fine black quartzite 3
Fine blue quartzite 1	
Fine blue quartzite 7	
Fine blue quartzite 8	
Fine grey quartzite 2	
Fine grey quartzite 3	
Fine grey quartzite 11	
Other fine grey quartzites	Fine grey quartzite 10, Fine grey quartzite 15, Fine grey quartzite17, Fine grey quartzite19
Fine white quartzite 16	
Medium green quartzite	Medium green quartzite 2, Medium green quartzite 4, Medium green quartzite 11, Medium green quartzite 12, Medium green quartzite 13
Medium grey quartzite 1	
Medium grey quartzite 8	
Medium grey quartzite 9	
Medium grey quartzite 10	
Medium grey quartzite 11	
Other medium quartzites	Medium grey quartzite 3, Medium, grey quartzite 7, Medium blue quartzite 7, Medium white quartzite 5, Medium brown quartzite 1
Coarse quartzites	
Quartz crystal	
Fine white quartz	
Other fine quartzes	Fine blue quartz 1, Fine grey quartz 2, Fine yellow quartz 1, Fine, yellow quartz 2, Fine yellow yartz 3, Fine red quartz 3
Medium quartz	
Coarse quartz	
Medium grey mylonite 1	
Medium grey mylonite 3	Medium grey mylonite 3, Fine grey mylonite 1
Coarse grey mylonite	Coarse grey mylonite, Coarse grey mylonite 1
Blue mylonite	Fine blue mylonite 1, Fine blue mylonite 4, Medium blue mylonite 1, Medium blue mylonite 2, Medium blue mylonite 3, Medium blue quartzite 1
Green mylonite	Fine, green mylonite 1 Medium, green mylonite 1
Other mylonites	Medium green mylonite 2, Medium green mylonite 3, Medium green mylonite 4, Medium blue mylonite 3, Medium blue mylonite 5
Rhyolite	
Chert	
Jasper	
Anorthosite	
Red slate	
Green slate	
Grey slate	
Blue slate 1	
Blue slate 2	
Slate other	
Soapstone	
Soapstone 8	
Sandstone	
Pumice	
Amber	
Undetermined rock	

APPENDIX 3

RAW-DATA
FROM THE EXCAVATED SITES

Nordøy grendahus phase 2																					
Types	Basaltic rock	Fine flint	Medium flint	Coarse flint	Fine green quartzite 1	Fine blue quartzite 7	Other fine grey quartzites	Medium green quartzite	Coarse quartzite	Quartz crystal	Fine white quartz	Coarse quartz	Green mylonite	Blue mylonite	Green slate	Sandstone	Soapstone	Pumice	Undetermined	Pottery	Sum
Flakes > 4cm	1																				1
Flakes 1 >< 4cm	1	124	786	108		1	2	1		12	16	68		2	1						1122
Flakes < 1cm		140	440	16	1					13	15	101			1						727
Small blades 8 >< 12mm		1	15					3	1												20
Microblades <8mm		9	13																		22
Microblades <8mm (microblade technique)			3																		3
Bipolar cores		25	214	13				1		6	13	15									287
Other cores		2	23	6																	31
Flakes struck from ground stone artefacts	1																				1
Struck chisels/blanks (not ground)	1																				1
Tabular sandstone knives																1					1
Transverse points			1																		1
Tanged points (A-type)		1																			1
Tanged points (B-type)		1																			1
Slate points with diamond cross section - undetermined shoulders																	1				1
Retouched flake		8	82	8									1								99
Net sinkers with a groove																			1		1
Grinding slabs																2			4		6
Pumice without grooves																		8			8
Pumice with grooves																		29			29
Flint nodules			7																		7
Pottery with no/undetermined decoration																				3	3
Sum	4	311	1584	151	1	1	2	5	1	31	44	184	1	2	2	3	1	37	5	3	2373
Percentage	0,2	13,1	66,8	6,4	0,0	0,0	0,1	0,2	0,0	1,3	1,9	7,8	0,0	0,1	0,1	0,1	0,0	1,6	0,2	0,1	100,0

Synnaland															
Type	Basaltic rock	Fine flint	Medium flint	Coarse flint	Other fine grey quartzites	Quartz crystal	Fine white quartz	Coarse quartz	Green mylonite	Blue mylonite	Blue slate	Green slate	Grey slate	Undetermined	Sum
Flakes > 4cm			1									1			2
Flakes 1 >< 4cm		27	331	60	1	3	2	62	1			1			488
Flakes < 1cm		38	333	58		4		12		1			1		447
Small blades 8 >< 12mm			3				1								4
Microblades <8mm			1	1						1					3
Microblades <8mm (microblade technique)			2							1					3
Cylindrical cores				1											1
Bipolar cores		15	128	15		1		18		2					179
Other cores				1				2							3
Struck chisels/blanks (not ground)	1														1
Ground adzes - rectangular cross-section (Vestland/Vespestad adzes)	1														1
Slate points with diamond cross section - contracting shoulders											1				1
Retouched flake		1	34	4				1							40
Ordinary blades with retouch			4	1											5
Hammerstones														1	1
Flint nodules		1	3												4
Sum	2	82	840	141	1	8	3	95	1	5	1	2	1	1	1183
Percentage	0,17	6,93	71,01	11,92	0,08	0,68	0,25	8,03	0,08	0,42	0,08	0,17	0,08	0,08	100,00

Valderøya Vest 26 phase 2

Type	Basaltic rock	Fine flint	Medium flint	Coarse flint	Other fine black quartzites	Medium grey quartzite 9	Coarse quartzite	Quartz crystal	Fine white quartz	Coarse quartz	Blue mylonite	Chert	Green slate	Grey slate	Sandstone	Pumice	Amber	Sum
Flakes > 4cm			1															1
Flakes 1 >< 4cm	4	99	817	114			2	4	5	29		1	8	1				1084
Flakes < 1cm		116	716	58				5	6	28			14		1		7	951
Ordinary blades > 12mm			1															1
Small blades 8 >< 12mm			11	1		2												14
Microblades <8mm			4															4
Microblades <8mm (microblade technique)			1															1
Crested blades			1															1
One-sided cores with two platforms			1															1
Bipolar cores		18	184	29				3	5	6								245
Other cores				2														2
Flakes struck from ground stone artefacts	2																	2
Ground adzes - rectangular cross-section (Vestland/Vespestad adzes)	1																	1
Tanged points (A-type)					1						1							2
Slate points with diamond cross section - square shoulders													1					1
Slate points with diamond cross section - undetermined shoulders													1					1
Slate points with planoconvex cross section - hanging shoulders													1					1
Slate points with planoconvex cross section - contracting shoulders													1					1
Slate points - chipped blanks													1					1
Slate points - other/undeterminded cross sections													1					1
Slate points - preforms													2					2
Retouched flake		5	36															41
Grinding slabs															2			2
Pumice with grooves																1		1
Flint nodules			38															38
Sum	7	238	1811	204	1	2	2	12	16	63	1	1	30	1	3	1	7	2400
Percentage	0,3	9,9	75,5	8,5	0,0	0,1	0,1	0,5	0,7	2,6	0,0	0,0	1,3	0,0	0,1	0,0	0,3	100,0

Valderøya Vest 13 phase 1																	
Type	Basaltic rock	Fine flint	Medium flint	Coarse flint	Other fine black quartzites	Fine white quartzite 16	Medium grey quartzite 1	Quartz crystal	Fine white quartz	Medium quartz	Coarse quartz	Blue mylonite	Green slate	Red slate	Sandstone	Soapstone	Sum
Flakes > 4cm			1	2							1						4
Flakes 1 >< 4cm	2	41	239	59		2	1	2	1	1	1						349
Flakes < 1cm		16	124	6		1		2			1						150
Small blades 8 >< 12mm			8		1												9
Small blades 8 >< 12mm (microblade technique)		1	1														2
Microblades <8mm		2	3									1					6
Microblades <8mm (microblade technique)			2														2
Bipolar cores		2	63	16				1		1							83
Other cores with one platform				2													2
Other cores with two platforms			1														1
Other cores			2														2
Flakes struck from ground stone artefacts	1																1
Ground adzes - rectangular cross-section (Vestland/Vespestad adzes)	3																3
Other four-sided stone chisels	1																1
Tanged points (A-points)			1														1
Slate points with diamond cross section - contracting shoulders																1	1
Slate points with diamond cross section - undetermined shoulders													1	1		1	3
Slate points - ground blanks																1	1
Retouched flakes		2	13	2													17
Grinding slabs															3		3
Flint nodules			15														15
Sum	7	64	473	87	1	3	1	5	1	2	3	1	1	1	3	3	656
Percentage	1,1	9,8	72,1	13,3	0,2	0,5	0,2	0,8	0,2	0,3	0,5	0,2	0,2	0,2	0,5	0,5	100,0

9 Bustadvika

Types	Basaltic rock	Greenstone	Flint	Other medium quartzites	Quartz crystal	Fine white quartz	Coarse quartz	Blue mylonite	Rhyolite	Chert	Grey slate	Slate other	Sandstone	Pumice	Undetermined	Sum
Flakes	5	2	566	12	414	340	25	21			115	4	12		6	1522
Ordinary blades > 12mm								2								2
Small blades 8 >< 12mm			5					3	1						1	10
Microblades <8mm			7													7
Other core fragments			1					2	1							4
One sided cores with one platform			1					1								2
Other cores with one platform			1													1
One-sided cores with two platforms			1													1
Bipolar cores			12	1	26	3										42
Other cores			1													1
Ground adzes - rectangular cross-section (Vestland/Vespestad adzes)	1															1
Flakes struck from ground stone artefacts	6	1									1					8
Vespestad adzes	1	1														2
Struck blanks - long and slim	1															1
Other bi-convex chisels	1															1
Tabular sandstone knives with transversely ground side													1			1
Tanged points (A-type)			1					1								2
Flake drills			2													2
Slate points with diamond cross section - undetermined shoulders											1					1
Slate points - ground blanks											2					2
Slate points - preforms											1					1
Retouched flakes			9		1											10
Ordinary blades with other retouch										1						1
Small blades with retouch					1			1								2
Microblades with retouch			1													1
Grinding slabs													4			4
Anvil stones															2	2
Small round smooth stones															1	1
Pumice without grooves														209		209
Pumice with grooves														1		1
Other artefacts															1	1
Flint nodules			1													1
Sum	15	4	609	13	442	343	25	31	2	1	120	4	17	210	11	1847
Percentage	0,8	0,2	33,0	0,7	23,9	18,6	1,4	1,7	0,1	0,1	6,5	0,2	0,9	11,4	0,6	100,0

Holsvikhaugen Types	Basaltic rock	Greenstone	Diabase	Fine flint	Midium flint	Coarse flint	Fine blue quartzite 7	Fine white quatzite 16	Medium green quartzite	Mediium grey quartzite 9	Quartz crystal	Fine white quartz	Coarse quartz	Chert	Rhyolite	Green slate	Sandstone	Sum
Flakes > 4cm				1	1					1								3
Flakes 1 >< 4cm	1			5	125	10		2	2		15	17	105		3	3		288
Flakes < 1cm				3	27	1		1	1		9	10	38					90
Small blades 8 >< 12mm					6		1					1		1	3			12
Microblades <8mm					1													1
Microblades <8mm (microblade technique)					2													2
Other core fragments					2													2
One sided cores with one platform					1													1
Bipolar cores				1	23	3					5	8	27					67
Other cores				1	2	1					1	1						6
Flakes struck from ground stone artefacts		1																1
Struck chisels/blanks (not ground)	1																	1
Vespestad adzes	1		1															2
Tabular sandstone knives with crescent shape and straight edge																	1	1
Tanged points (A-type)															1			1
Slate points with diamond cross section - square shoulders																1		1
Slate points with diamond cross section - contracting shoulders																1		1
Slate points - ground blanks																1		1
Retouched flake				1	19				1			1						22
Grinding slabs																	2	2
Small round smooth stones	1																	1
Flint nodules					1													1
Sum	4	1	1	12	210	15	1	3	4	1	30	38	170	1	7	6	3	507
Percentage	0,8	0,2	0,2	2,4	41,4	3,0	0,2	0,6	0,8	0,2	5,9	7,5	33,5	0,2	1,4	1,2	0,6	100,0

17 Havnen phase 3

Types	Basaltic rock	Greenstone	Fine flint	Medium flint	Coarse flint	Fine green quartzite 1	Fine blue quartzite 7	Fine grey quartzite 2	Other fine grey quartzites	Fine white quartzite 16	Fine black quartzite 1	Medium green quartzite	Medium grey quartzite 9	Coarse quartzite	Quartz crystal	Other fine quartzes	Fine white quartz	Medium quartz	Coarse quartz	Green mylonite	Medium grey mylonite 3	Blue mylonite	Coarse grey mylonite	Chert	Rhyolite	Brown slate	Green slate	Grey slate	Red slate	Slate other	Anorthosite	Sandstone	Soapstone	Pumice	Sum
Flakes > 4cm								8						1					9			3				5	4	3	3				1		37
Flakes 1 >< 4cm	1	1	14	89	16		5	238	10	2		7	3	11	12	11	94	38	112	17	14	52	4	30	31	18	27	36	58	4	3		10	4	972
Flakes < 1cm			19	53	11	2		64	8			1	1	2	5	4	65	13	39	19	5	27	3	5	14	5	6	7	18		2		7		405
Ordinary blades > 12mm								5																	3										8
Small blades 8 >< 12mm			2	6	2		1	36				1					1					4		1	16										70
Microblades <8mm			1																																1
Crested blades																									1										1
Other core fragments			7	5	2				1							1	1	1		1															19
Cylindrical cores								1			1																								2
Other cores with one platform								7																											7
Bipolar cores			2	2	1				1							1	3	1	2																13
Other cores				1				2								1				1		1		1											7
Flakes struck from ground stone artefacts	1	2																									1		4			2			10
Ground adzes - rectangular cross-section (Vestland/Vespestad adzes)		1																																	1
Vestland chisles		1																																	1
Tabular sandstone knives																														2					2
Tanged points (A-type)								1					1				1					1		1											5
Slate points with bi-convex cross section - contracting shoulders																										1									1
Slate points with diamond cross section - undetermined shoulders																												2							2
Slate points with triangular cross section - contracting shoulders																												1							1
Slate points - ground blanks																											1								1
Slate points - chipped blanks																											2				1				3
Slate points - other/undeterminded cross sections																											1								1
Retouched flake			5	11				6	1								1		1	1			1	1	1										29
Ordinary blades with other retouch				1	1																	1													3
Small blades with other retouch					2			2																											4
Grinding slabs																																15			15
Hammerstones																			1													1			2
Small round smooth stones																																1			1
Pumice with grooves																																		2	2
Sum	2	5	50	168	35	2	6	370	21	2	1	9	5	14	17	18	166	53	164	39	19	89	8	39	66	29	42	49	83	6	6	19	18	6	1626
Percentage	0,1	0,3	3,1	10,3	2,2	0,1	0,4	22,8	1,3	0,1	0,1	0,6	0,3	0,9	1,1	1,1	10,2	3,3	10,1	2,4	1,2	5,5	0,5	2,4	4,1	1,8	2,6	3,0	5,1	0,4	0,4	1,2	1,1	0,4	100,0

1 Haukedal phase 2																																	
Type	Basaltic rock	Greenstone	Diabase	Fine flint	Medium flint	Coarse flint	Fine grey quartzite 2	Fine black quartzite 1	Other fine black quartzites	Other medium quartzites	Medium green quartzite	Medium grey quartzite 9	Other medium quartzites	Coarse quartzite	Quartz crystal	Other fine quartzes	Fine white quartz	Medium quartz	Coarse quartz	Green mylonite	Blue mylonite	Chert	Rhyolite	Brown slate	Green slate	Grey slate	Red slate	Slate other	Anorthosite	Sandstone	Soapstone	Undetermined	Sum
Flakes > 4cm														1				2	8						4	1	1	2		1	1		21
Flakes 1 >< 4cm	8		3	3	11	8	2		3	5	4	4	2		3	2	21	28	122	3	16	22	2	6	54	19		11	1	6	1	4	370
Flakes < 1cm					3	1								1	1		6	2	13	2		2			4	1							36
Small blades 8 >< 12mm				1																		2	3										6
Microblades <8mm								1									1																2
Crested blades																					1	1											2
Other core fragments					4							1					1				1												7
Bipolar cores																				1													1
Other cores										1										3	1												5
Flakes struck from ground stone artefacts		1	3																							1							5
Other four-sided stone chisels	1																																1
Struck blanks - short and broad			1																														1
Transverse points					1																												1
Tanged points (A-type)																							2										2
Slate points with bi-convex cross section - square shoulders																									1								1
Slate points with bi-convex cross section - undetermined shoulders																													1				1
Slate points with diamond cross section - square shoulders																													1				1
Slate points with diamond cross section - undetermined shoulders																									1								1
Slate points - preforms																										1							1
Retouched flake				2	2															2													6
Ordinary blades with other retouch					1																												1
Grinding slabs																														11			11
Sum	9	1	7	6	22	9	2	1	3	6	4	5	2	2	4	2	29	32	143	11	19	27	7	6	64	23	1	13	3	18	2	4	483
Percentage	1,9	0,2	1,5	1,2	4,6	1,9	0,4	0,2	0,6	1,2	0,8	1,0	0,4	0,4	0,8	0,4	6,0	6,6	29,6	2,3	3,9	5,6	1,5	1,2	13,3	4,8	0,2	2,7	0,6	3,7	0,4	0,8	100,0

1 Haukedal phase 3																												
Types	Basaltic rock	Greenstone	Diabase	Fine flint	Medium flint	Coarse flint	Fine grey quartzite 2	Fine black quartzite 1	Medium green quartzite	Other medium quartzites	Coarse grey quartzite	Quartz crystal	Fine white quartz	Medium white quartz	Coarse quartz	Blue mylonite	Green mylonite	Coarse grey mylonite	Chert	Rhyolite	Other slate	Brown slate	Green slate	Grey slate	Sandstone	Soapstone	Pumice	Other	Sum
Flakes > 4cm	1		1						1		2			2	9	3			12		1							1	33
Flakes 1 >< 4cm	12	1	3	2	8	7	1	4	14	1	4	1	13	13	82	30	2	1	197	1	7	1	30	8		1		6	450
Flakes < 1cm	3			1	3	2			2			1	1	1	22	3	1		50				5						95
Small blades 8 >< 12mm						1													5	1									7
Microblades <8mm				1												1													2
Other core fragments				4	3										1				3										11
Cylindrical cores																			1										1
Bipolar cores															1	1			1										3
Other cores									1							1													2
Other platform cores																			2										2
Flakes struck from ground stone artefacts	1	1																						1				1	4
Ground adzes - rectangular cross-section (Vestland/ Vespestad adzes)		1																											1
Slate points with diamond cross section - undetermined shoulders																							1						1
Slate points - preforms																							1						1
Retouched flake					1		1																						2
Grinding slabs																									5				5
Hammerstones											1																		1
Pumice without grooves																											7		7
Sum	17	3	4	8	15	10	2	4	18	1	7	2	14	16	115	39	3	1	271	2	8	1	37	9	5	1	7	8	628
Percentage	2,7	0,5	0,6	1,3	2,4	1,6	0,3	0,6	2,9	0,2	1,1	0,3	2,2	2,5	18,3	6,2	0,5	0,2	43,2	0,3	1,3	0,2	5,9	1,4	0,8	0,2	1,1	1,3	100,0

Kleiva																										
Types	Diabase production	Basaltic rock	Greenstone 2	Diabase	Fine flint	Medium flint	Coarse flint	Fine blue quartzite 7	Other fine grey quartzites	Medium green quartzite	Medium grey quartzite 9	Other medium quartzites	Coarse quartzite	Quartz crystal	Fine white quartz	Medium quartz	Coarse quartz	Blue mylonite	Medium grey mylonite 3	Coarse grey mylonite	Rhyolite	Grey slate	Slate other	Sandstone	Soapstone	Pumice	Sum
Flakes > 4cm	3								1									1	1								3
Flakes 1 >< 4cm	37					9	4		8	3	7	1	15	4	5	32	6	239	48	78	11	10	1				481
Flakes < 1cm					1	9			5		3		1		1	13		110	12	35	3						193
Ordinary blades > 12mm																		4	2	3	2						11
Small blades 8 >< 12mm																		6									6
Microblades <8mm						1										1				1							3
Other core fragments																		3		1							4
Bipolar cores																		5									5
Other cores								1			1				1			5	2	1	2						13
Flakes struck from ground stone artefacts			1																								1
Struck chisels/blanks (not ground)				9																							9
Ground adzes - rectangular cross-section (Vestland/Vespestad adzes)				1																							1
Other four-sided adzes			1																								1
Tabular sandstone knives																								2			2
Tanged points (A-type)																		1	1	1							3
Retouched flakes										1			1		1			10	1	1							15
Ordinary blades with other retouch																		1									1
Small blades with other retouch																		1									1
Grinding slabs																								3			3
Anvil stones		2											1														3
Pumice without grooves																										1	1
Soapstone vessels																									1		1
Sum	40	2	2	10	1	19	4	1	14	4	11	1	18	4	8	46	6	386	67	121	18	10	1	5	1	1	761
Percentage		0,3	0,3	1,3	0,1	2,5	0,5	0,1	1,8	0,5	1,4	0,1	2,4	0,5	1,1	6,1	0,8	50,7	8,8	15,9	2,4	1,3	0,1	0,7	0,1	0,1	100,0

Botnaneset VIII phase 3

Types	Diabase	Flint	Medium grey quartzite 1	Coarse grey quartzite	Coarse quartz	Blue mylonite	Medium grey mylonite 1	Chert	Sandstone	Sum
Flakes > 4cm	5	1	8	1			6	19		40
Flakes 1 >< 4cm		99	88	10	2		119	480		798
Flakes < 1cm		109	114				144	660		1027
Ordinary blades > 12mm			2				3	7		12
Small blades 8 >< 12mm		1	3				4	32		40
Microblades <8mm		14	6				6	51		77
Microblades <8mm (microblade technique)		5								5
Keeled cores		1								1
Other cores with one platform		3	5				3	20		31
Cylindrical cores							2			2
Bipolar cores		3		1						4
Other cores		6	3							9
Struck chisels/blanks (not ground)	1									1
Tanged points (A-type)								2		2
Retouched flake		4	1			1		6		12
Ordinary blades with retouch								1		1
Small blades with retouch			1					2		3
Grinding slabs									1	1
Hammerstones									2	2
Sum	6	246	231	12	2	1	287	1280	3	2068
Percentage	0,29	11,90	11,17	0,58	0,01	0,05	13,88	61,90	0,15	100,00

Neset II																						
Types	Diabase production	Fine flint	Medium flint	Fine blue quartzite 7	Fine white quartzite 16	Other medium quartzites	Medium grey quartzite 1	Quartz crystal	Other fine quartzes	Fine white quartz	Coarse quartz	Blue mylonite	Green mylonite	Medium grey mylonite 1	Coarse grey mylonite	Chert	Rhyolite	Grey slate	Red slate	Sandstone	Soapstone	Pumice	Sum
Flakes > 4cm	460		3	5		1	41				1	26	2	4		2							85
Flakes 1 >< 4cm	340	26	193	4		8	771	32		101	114	661	11	73		35	98	9	6		1		2143
Flakes < 1cm		33	177			6	485	36	2	117	78	976	15	11		39	96						2071
Ordinary blades > 12mm			1				4					5											10
Small blades 8 >< 12mm			16				31			6		33		7		3	23						119
Microblades <8mm			20		5		32			6		33		1			5						102
Microblades <8mm (microblade technique)		4	1																				5
Conical cores			2							1		1											4
Cylindrical cores							1																1
Other cores with two platforms												1											1
Bipolar cores			2					1		3	2												8
Other cores			3		1	1	8			4	11	20		1	4		1						54
Struck chisels/blanks (not ground)	10																						0
Tanged points (A-type)			2				2					1		2			1						8
Retouched flake		3	17				1		1	2	1	9		4									38
Grinding slabs																				7			7
Hammerstones											5		1							34			40
Small round smooth stones											1												1
Pumice without grooves																						1	1
Pumice with grooves																						4	4
Quartz crystal, entire crystal								3															3
Sum	810	66	437	9	6	16	1376	72	3	240	213	1766	29	103	4	79	224	9	6	41	1	5	4705
Percentage		1,4	9,3	0,2	0,1	0,3	29,2	1,5	0,1	5,1	4,5	37,5	0,6	2,2	0,1	1,7	4,8	0,2	0,1	0,9	0,0	0,1	100,0

Vikja I phase 2																														
Types	Diabase production	Basaltic rock	Diabase	Flint	Fine green quartzite 1	Fine blue quartzite 7	Other fine grey quartzites	Fine white quartzite 16	Medium, grey quartzite 1	Coarse quartzite	Quartz crystal	Fine white quartz	Medium white quartz	Coarse quartz	Green mylonite	Blue mylonite	Other mylonites	Medium grey mylonite 1	Medium grey mylonite 3	Coarse grey mylonite	Chert	Rhyolite	Slate other	Green slate	Grey slate	Red slate	Sandstone	Soapstone	Pumice	Modern	Sum
Flakes > 4cm	99								24				3	4		11	1		8		1					3	1	1			57
Flakes 1 >< 4cm	138	1		102	1	1	1		200		25	91	9	131	12	298	2	7	98		5	4		10	1	24					1023
Flakes < 1cm				14					18		1	15	3	24	2	58			5				1	2		3					146
Ordinary blades > 12mm				2					3							6		1	2		1										15
Small blades 8 >< 12mm				4					13			2				41	2	3	26			1									92
Microblades <8mm				9					1			2				5	1		2												20
Conical cores				1																											1
Cylindrical cores									2			1		1		3		3				1									11
Other cores with two platforms				2					2			2		1		10		2													19
Bipolar cores				2					2		3	10			1	3															21
Other cores				6					3			2	1	3	1	8		5		1											30
Flakes struck from ground stone artefacts			2																												2
Struck chisels/blanks (not ground)	6																														0
Struck chisels/blanks ground at the edge			2																												2
Vestland adzes		1																													1
Adzes with bi-convex, oval cross section			1																												1
Tanged points (A-type)																2						1									3
Slate points - ground blanks																										1					1
Slate points - chipped blanks																								1							1
Retouched flake			1	16				1	5		4	7		1	1	8			1												45
Net sinkers with a groove																												1			1
Grinding slabs																											14				14
Hammerstones										4																	10				14
Pumice with grooves																													1		1
Historical artefacts																														5	5
Sum	243	2	6	158	1	1	1	1	273	4	33	132	16	165	17	453	6	21	142	1	7	7	1	13	1	31	25	2	1	5	1526
Percentage		0,1	0,4	10,4	0,1	0,1	0,1	0,1	17,9	0,3	2,2	8,7	1,1	10,8	1,1	29,7	0,4	1,4	9,3	0,1	0,5	0,5	0,1	0,9	0,1	2,0	1,6	0,1	0,1	0,3	100,0

Vikja I phase 3

Types	Basaltic rock	Diabase	Flint	Fine green quartzite 1	Fine white quartzite 16	Medium grey quartzite 1	Coarse quartzite	Quartz crystal	Fine white quartz	Coarse quartz	Green mylonite	Blue mylonite	Medium grey mylonite 3	Coarse grey mylonite	Jasper	Chert	Rhyolite	Slate other	Green slate	Red slate	Sandstone	Other	Pumice	Modern	Sum
Flakes > 4cm		3	6			145	1		1	5		4	1	1					5	1					173
Flakes 1 >< 4cm		12	126		10	459	4	1	105	61	4	236	7	1	1	3	2	5	95	94		42			1268
Flakes < 1cm			25			37			20	13	2	32	1				1		15	22		2			170
Ordinary blades > 12mm		1	3			8						1					1				1				15
Small blades 8 >< 12mm		2	14			4		1	2			16	8			1	1								49
Microblades <8mm			3			1			2			2													8
Other cores with one platform			6									7	1												14
Cylindrical cores												1													1
Bipolar cores			8			2		1	8		1	3	1												24
Other cores			3			12				3		15													33
Flakes struck from ground stone artefacts	2																								2
Ground adzes - rectangular cross-section (Vestland/Vespestad adzes)	8	3																							11
Transverse points			1																						1
Tanged points (A-type)			3																						3
Slate points with diamond cross section - hanging shoulders																			3						3
Slate points with diamond cross section - square shoulders																			1						1
Slate points with diamond cross section - undetermined shoulders																			9						9
Slate points with planoconvex cross section - contracting shoulders																			1						1
Slate points with planoconvex cross section - contracting shoulders																			3						3
Slate points - ground blanks																			4	6					10
Slate points - chipped blanks																			8	3					11
Slate points - other/undetermined cross sections																			1						1
Retouched flake			16		1	10			2			11	2												42
Ordinary blades with retouch			3	1		1			1			2													8
Grinding slabs																					8				8
Hammerstones																					10				10
Small round smooth stones																					1				1
Pumice without grooves																							1		1
Pumice with grooves																							10		10
Historical artefacts																								1	1
Sum	10	21	217	1	11	679	5	3	141	82	7	330	21	2	1	4	5	5	145	126	20	44	11	1	1892
Percentage	0,5	1,1	11,5	0,1	0,6	35,9	0,3	0,2	7,5	4,3	0,4	17,4	1,1	0,1	0,1	0,2	0,3	0,3	7,7	6,7	1,1	2,3	0,6	0,1	100,0

Masnesberget

Types	Basaltic rock	Greenstone	Diabase	Flint	Other fine grey quartzites	Fine black quartzite 1	Other medium quartzites	Medium grey quartzite 1	Coarse quartzite	Quartz crystal	Green mylonite	Medium grey mylonite 3	Rhyolite	Green slate	Sandstone	Sum
Flakes > 4cm				2		14		2					1			19
Flakes 1 >< 4cm				99	2	110	1	11				3	17			243
Ordinary blades > 12mm				1		9										10
Small blades 8 >< 12mm				7		11		2					8			28
Microblades <8mm				4		1										5
Crested blades				1												1
Conical cores				2												2
Other cores with one platform						2										2
Cylindrical cores						1							3			4
One-sided cores with two platforms				1		4										5
Other cores with two platforms													2			2
Bipolar cores				2				1		1						4
Other cores				4		3		2	1	1	3					14
Other struck adzes	6															6
Other ground adzes	3															3
Struck chisels/blanks (not ground)	6	1														7
Ground adzes - round cross-section	1		11													12
Ground adzes - rectangular cross-section (Vestland/Vespestad adzes)	1	1														2
Vestland adzes		2														2
Vespestad adzes		3														3
Chisels with bi-convex, oval cross sections	1		2													3
Slate points with diamond cross section - hanging shoulders														1		1
Slate points with diamond cross section - square shoulders														1		1
Slate points with diamond cross section - undetermined shoulders														1		1
Slate points - chipped blanks														5		5
Retouched flake				23				1					1			25
Grinding slabs															21	21
Hammerstones															2	2
Small round smooth stones															1	1
Sum	18	7	13	146	2	155	1	19	1	2	3	3	32	8	24	434
Percentage	4,1	1,6	3,0	33,6	0,5	35,7	0,2	4,4	0,2	0,5	0,7	0,7	7,4	1,8	5,5	100,0

Snekkevik 1 phase 3														
Types	Fine flint	Medium flint	Coarse flint	Other fine grey quartzites	Fine white quartzite 16	Quartz crystal	Fine white quartz	Other medium quartzites	Coarse quartz	Blue mylonite	Rhyolite	Green slate	Anorthosite	Sum
Flakes > 4cm											3			3
Flakes 1 >< 4cm	13	90	3		2		3	1	3	1	99	13	1	229
Flakes < 1cm	3	32				1	9		3		23	10		81
Ordinary blades > 12mm		4				1					1			6
Small blades 8 >< 12mm		4									16			20
Microblades <8mm		4									1			5
Other core fragments											8			8
Cylindrical cores											3			3
Bipolar cores		2							1		1			4
Other cores		2									1			3
Tanged points (A-type)											3			3
Slate points with diamond cross section - undetermined shoulders													1	1
Retouched flake	1	13						2			2			18
Ordinary blades with retouch		1												1
Small blades with retouch		2												2
Sum	17	154	3	0	2	2	12	3	7	1	161	23	2	387
Percentage	4,4	39,8	0,8	0,0	0,5	0,5	3,1	0,8	1,8	0,3	41,6	5,9	0,5	100,0

Kotedalen phase 12																													
Types	Basaltic rocks	Greenstone	Diabase	Fine flint	Medium flint	Fine blue quartzite 1	Fine blue quartzite 7	Fine grey quartzite 2	Other fine grey quartzites	Fine white quartzite 16	Fine black quartzite 1	Medium green quartzite	Medium grey quartzite 9	Coarse quartzite	Quartz crystal	Other fine quartzes	Fine white quartz	Medium quartz	Coarse quartz	Blue mylonite	Chert	Rhyolite	Brown slate	Green slate	Red slate	Anorthosite	Sandstone	Pumice	Sum
Flakes > 4cm			1		1							1		2			1		7	4		27				2			46
Flakes 1 >< 4cm	2	1		38	71		5		2		3	1	1	3	5		17	2	50	331	2	982		23		15			1554
Flakes < 1cm				21	85		3		2				3		12	2	9	4	69	221		976	3	21		11			1442
Ordinary blades > 12mm				2	8	2											1			23		139							175
Small blades 8 >< 12mm				6	9												1	1	1	36		243							297
Microblades <8mm				6	15					1							1			30		98							151
Crested blades					1															2		6							9
Other core fragments				1	3												1			15		50							70
Keeled cores																						1							1
One sided cores with one platform					1																	1							2
Other cores with one platform																				1		5							6
Cylindrical cores					1															1		12							14
One-sided cores with two platforms																				1		6							7
Other cores with two platforms					1															1		7							9
Bipolar cores				1	4												1			2		3							11
Other cores					2	1														2		3							8
Flakes struck from ground artefacts	15	5																						1	3				24
Ground adzes - round cross-section	1																												1
Tanged points (A-type)				1	2			1												8		31							43
Slate points with bi-convex cross section - contracting shoulders																								1					1
Slate points with bi-convex cross section - undetermined shoulders																								1	1	1			3
Slate points with diamond cross section - undetermined shoulders																										1			1
Slate points - chipped blanks																								1					1
Slate points - other/undeterminded cross sections																									1				1
Retouched flake				4	15												1			16		29							65
Ordinary blades with retouch					6															5		17							28
Small blades with retouch				1	10															2		10							23
Microblades with retouch					1															1		1							3
Retouched crested blades																						2							2
Grinding slabs	1																							1	1		3		6
Pumice without grooves																												8	8
Pumice with grooves																												2	2
Sum	19	6	1	81	236	3	8	1	4	1	3	2	4	5	17	2	33	7	127	702	2	2649	3	49	6	30	3	10	4014
Percentage	0,5	0,2	0,0	2,0	5,9	0,1	0,2	0,0	0,1	0,0	0,1	0,1	0,1	0,1	0,4	0,1	0,8	0,2	3,2	17,5	0,1	66,0	0,1	1,2	0,2	0,7	0,1	0,3	100,0

Kotedalen phase 13																																					
Types	Basaltic rock	Greenstone	Diabase	Fine flint	Medium flint	Fine green quartzite 1	Fine blue quartzite 1	Fine blue quartzite 7	Fine grey quartzite 2	Other fine grey quartzites	Fine white quartzite 16	Fine black quartzite 1	Other medium quartzites	Medium green quartzite	Medium grey quartzite 9	Coarse quartzite	Quartz crystal	Other fine quartzes	Fine white quartz	Medium quartz	Coarse quartz	Green mylonite	Blue mylonite	Other mylonites	Medium grey mylnite 1	Chert	Rhyolite	Brown slate	Green slate	Red slate	Slate other	Anorthosite	Sandstone	Soapstone	Pumice	Pottery	Sum
Flakes > 4cm	1				1			1			1	1				1			1		23		7		1		27		48			8					121
Flakes 1 >< 4cm	17	3		55	358		8	4	2	3	8	4		7	11	5	8	2	66	9	185	1	387	5	10	7	1202	111	616	2		39	1	5			3141
Flakes < 1cm				51	285		4	2	2	1	6	1	1	5	2	1	3		39	3	191	1	262		6	10	1095	69	583			123					2746
Ordinary blades > 12mm				9	7		1				1					1							27				88										134
Small blades 8 >< 12mm				1	47	1					1								1				56				212										319
Microblades <8mm				11	26		2				1					1			3		1		29				82										156
Crested blades					1																						8										9
Other core fragments					9		1														1		7				32										50
Keeled cores																			1																		1
One sided cores with one platform					1																																1
Other cores with one platform					1																		2				4										7
Cylindrical cores																							1				4										5
One-sided cores with two platforms					1																						3										4
Other cores with two platforms																							2				6										8
Bipolar cores					4														2		2		6				12										26
Other cores																							1				2										3
Flakes struck from ground artefacts	32	6	4																										6	5			1				54
Ground adzes - rectangular cross-section (Vestland/Vespestad adzes)		1	2																																		3
Tanged points (A-type)				4	4																		10				38										56
Blade drills				1																																	1
Slate points with bi-convex cross section - hanging shoulders																														1							1
Slate points with bi-convex cross section - square shoulders																													1	1							2
Slate points with bi-convex cross section - contracting shoulders																													1			1					2
Slate points with bi-convex cross section - undetermined shoulders																													5	1	1	4					11
Slate points with diamond cross section - hanging shoulders																													1			1					2
Slate points with diamond cross section - square shoulders																													1								1
Slate points with diamond cross section - undetermined shoulders																													2	1		2					5
Slate points with triangular cross section - contracting shoulders																													1	1							2
Slate points with planoconvex cross section - contracting shoulders																													2								2
Slate points - ground blanks																													1								1
Slate points - chipped blanks																													19	1							20
Slate points - preforms																													2	1							3
Slate points - other/undeterminded cross sections																													2	1							3
Retouched flake				3	66	2	1														1		28				57										158
Ordinary blades with retouch				10	25																		9				23										67
Small blades with retouch				5	14			1							1					1			3				23										48
Microblades with retouch				2	3																		2				4										11
Retouched crested blades				2	1																						1										4
Retouched flakes from ground objects																													1								1
Grinding slabs																														2			8				10
Hammerstones	1																																				1
Small round smooth stones	1																																				1
Pumice without grooves																																			27		27
Pumice with grooves																																			16		16
Pottery with undetermined cord stamp impressions (locally produced)																																				1	1
Pottery with impressions from real cord (TRB type)																																				1	1
Pottery with no/undetermined decoration (TRB type)																																				20	20
Sum	52	10	6	154	854	3	17	8	4	4	18	6	1	12	14	9	11	2	113	13	404	2	839	5	17	17	2923	180	1292	17	1	178	10	5	43	22	7266
Percentage	0,7	0,1	0,1	2,1	11,8	0,0	0,2	0,1	0,1	0,1	0,2	0,1	0,0	0,2	0,2	0,1	0,2	0,0	1,6	0,2	5,6	0,0	11,5	0,1	0,2	0,2	40,2	2,5	17,8	0,2	0,0	2,5	0,1	0,1	0,6	0,3	100,0

Flatøy VIII phase 2																				
Types	Basaltic rock	Greenstone	Fine flint	Medium flint	Fine blue quartzite 1	Fine blue quartzite 7	Other fine grey quartzites	Fine white quartzite 16	Medium green quartzite 2	Medium grey quartzite 9	Coarse quartzite	Quartz crystal	Fine white quartz	Coarse quartz	Blue mylonite	Other mylonites	Chert	Rhyolite	Pumice	Sum
Flakes > 4cm			1	3	1				1	1	1		3	1	4			7		23
Flakes 1 >< 4cm	2	1	19	77	2	1	1	9	6	2		2	18	12	5	1	5	127		290
Flakes < 1cm		1	17	27				1	4			1	10	1				65		127
Ordinary blades > 12mm			1																	1
Small blades 8 >< 12mm			11	14					2				2		2		2	63		96
Microblades <8mm			6	11			1											2		20
Crested blades															1			1		2
Other core fragments			1	2		1														4
Conical cores				3														1		4
Cylindrical cores																		5		5
Bipolar cores			2	4																6
Other cores			2															1		3
Flakes struck from ground stone artefacts	1																			1
Ground adzes - rectangular cross-section (Vestland/Vespestad adzes)	1																			1
Tanged points (A-type)			1	2					1									4		8
Retouched flake			3	21														2		26
Ordinary blades with retouch				1														1		2
Grinding slabs	2																			2
Pumice with grooves																			1	1
Sum	6	2	64	165	3	2	2	10	14	3	1	3	33	14	12	1	7	279	1	622
Percentage	1,0	0,3	10,3	26,5	0,5	0,3	0,3	1,6	2,3	0,5	0,2	0,5	5,3	2,3	1,9	0,2	1,1	44,9	0,2	100,0

Torsteinsvik 11 phase 1															
Types	Basaltic rock	Fine flint	Medium flint	Other medium quartzites	Coarse quartzite	Fine white quartz	Coarse quartz	Blue mylonite	Rhyolite	Green slate	Grey slate	Anorthosite	Sandstone	Pumice	Sum
Flakes > 4cm			2		1				38	1					42
Flakes 1 >< 4cm		5	44	1	1	1	3	4	236	4	1				300
Flakes < 1cm		1	19	1			1		150				1		173
Ordinary blades > 12mm			1						1						2
Small blades 8 >< 12mm			1					1	63						65
Microblades <8mm									14						14
Cylindrical cores									3						3
Bipolar cores		4	5						10						19
Other cores							1		1						2
Flakes struck from ground stone artefacts	2														2
Ground adzes - rectangular cross-section (Vestland/Vespestad adzes)	1														1
Tanged points (A-type)			1						4						5
Slate points with diamond cross section - undetermined shoulders											1				1
Slate points - chipped blanks												1			1
Retouched flake		7	11						18						36
Ordinary blades with retouch		2							10						12
Pumice with grooves														1	1
Sum	3	19	84	2	2	1	5	5	548	5	2	1	1	1	679
Percentage	0,4	2,8	12,4	0,3	0,3	0,1	0,7	0,7	80,7	0,7	0,3	0,1	0,1	0,1	100,0

Nilsvik 17															
Types	Basaltic rock	Greenstone	Flint	Fine blue quartzite 8	Medium grey quartzite 10	Medium grey quartzite 9	Quartz crystal	Fine white quartz	Coarse quartz	Jasper	Rhyolite	Green slate	Sandstone	Pumice	Sum
Flakes > 4cm	1			7	2				5		4				19
Flakes 1 >< 4cm	3	1	216	116	48	1	1	46	208	1	780	10	3		1434
Flakes < 1cm	2		96	58	2		1	24	80		319		1		583
Ordinary blades > 12mm				8							9				17
Small blades 8 >< 12mm			13	10	1						51				75
Microblades <8mm			3	7							15				25
Crested blades											1				1
Other core fragments			7	2	1				3		12				25
Conical cores											1				1
Cylindrical cores											2				2
Bipolar cores	1		6					1	5		8				21
Other cores											6				6
Flakes struck from ground artefacts	3											1	2		6
Flakes struck from ground flint artefacts			1												1
Ground adzes - rectangular cross-section (Vestland/Vespestad adzes)	1	2													3
Vestland chisles		1													1
Tanged points (A-type)			6	1			1				21				29
Retouched flake			36					1	1		51				89
Ordinary blades with retouch			8					1			3				12
Small blades with retouch			8	1							6				15
Microblades with retouch											2				2
Grinding slabs									1				1		2
Hammerstones	1														1
Small round smooth stones	2														2
Pumice without grooves														30	30
Pumice with grooves														5	5
Sum	14	4	400	210	54	1	3	73	303	1	1291	11	7	35	2407
Percentage	0,6	0,2	16,6	8,7	2,2	0,0	0,1	3,0	12,6	0,0	53,6	0,5	0,3	1,5	100,0

Nilsvik 4 structure 10

Types	Greenstone	Flint	Medium grey quartzite 10	Quartz crystal	Fine white quartz	Coarse quartz	Other mylonites	Rhyolite	Green slate	Sandstone	Pumice	Undetermined	Sum
Flakes > 4cm			3					1					4
Flakes 1 >< 4cm		91	21	1	23	96	1	311	1				545
Flakes < 1cm		37	2	1	11	13		50					114
Ordinary blades > 12mm						1							1
Small blades 8 >< 12mm		1			1	1		24					27
Microblades <8mm		2				2		11					15
Crested blades		1						5					6
Other core fragments		8						27					35
Cylindrical cores								2					2
Other cores					1								1
Flakes struck from ground artefacts	1												1
Struck chisels/blanks ground at the edge	1												1
Vestland chisels	1												1
Tanged points (A-type)								6					6
Slate points with diamond cross section - square shoulders									1				1
Slate points with diamond cross section - contracting shoulders									1				1
Slate points - ground blanks									2				2
Slate points - chipped blanks									2				2
Slate points - other/undeterminded cross sections									1				1
Retouched flakes		11			2	1		14					28
Ordinary blades with retouch		2						5					7
Small blades with retouch		1						3					4
Microblades with retouch		2						1					3
Retouched crested blades		1											1
Grinding slabs						1				1		5	7
Pumice with grooves											1		1
Sum	3	157	26	2	38	115	1	460	8	1	1	5	817
Percentage	0,4	19,2	3,2	0,2	4,7	14,1	0,1	56,3	1,0	0,1	0,1	0,6	100,0

Nilsvik 4 structure 29																		
Types	Greenstone	Flint	Fine blue quartzite 1	Fine black quartzite 1	Medium green quartzite	Medium grey quartzite 10	Medium grey quartzite 11	Quartz crystal	Fine white quartz	Medium quartz	Coarse quartz	Blue mylonite	Jasper	Rhyolite	Green slate	Sandstone	Pumice	Undetermined	Sum
Flakes > 4cm											2			3					5
Flakes 1 >< 4cm	2	125	1	1	1	16	9		12	8	110	1	1	599	4	2			892
Flakes < 1cm		38				1	1	1	6	4	23			249		1			324
Ordinary blades > 12mm		2												28					30
Small blades 8 >< 12mm		4												35					39
Microblades <8mm		6												17					23
Crested blades		2												5					7
Other core fragments		11							1		1			69					82
Cylindrical cores														5					5
Bipolar cores														2					2
Other cores											1			1					2
Flakes struck from ground artefacts	4														1				5
Tanged points (A-type)		2												12					14
Slate points with diamond cross section															1				1
Slate points with planoconvex cross section															2				2
Slate points with planoconvex cross section - square shoulders															2				2
Slate points - ground blanks															1				1
Retouched flake		25				3			1		3			43					75
Ordinary blades with retouch		4												12					16
Small blades with retouch		1												12					13
Microblades with retouch														1					1
Hammerstones																		2	2
Small round smooth stones											2								2
Pumice with grooves																	2		2
Sum	6	220	1	1	1	20	10	1	20	12	142	1	1	1093	11	3	2	2	1547
Percentage	0,4	14,2	0,1	0,1	0,1	1,3	0,6	0,1	1,3	0,8	9,2	0,1	0,1	70,7	0,7	0,2	0,1	0,1	100,0

Nilsvik 4 structure 30																			
Types	Greenstone	Diabase	Flint	Fine black quartzite 1	Medium green quartzite	Medium grey quartzite 10	Medium grey quartzite 11	Quartz crystal	Fine white quartz	Coarse quartz	Blue mylonite	Other mylonites	Rhyolite	Green slate	Anorthosite	Sandstone	Pumice	Undetermined	Pottery	Sum
Flakes > 4cm						5	1			6			1							13
Flakes 1 >< 4cm	4		108	1	4	79	7		24	212	1		236	12		1				689
Flakes < 1cm	1		58			8		1	19	96		1	49	1		3				237
Ordinary blades > 12mm													2							2
Small blades 8 >< 12mm			11							1			18							30
Microblades <8mm			5			1				1			9							16
Crested blades			1										6							7
Other core fragments			6			3			1	1			18					1		30
Cylindrical cores													3							3
Other cores with two platforms													1							1
Bipolar cores			7						8	10			2							27
Other cores			2							6										8
Flakes struck from ground artefacts	4	2												1						7
Ground adzes - rectangular cross-section (Vestland/Vespestad adzes)																		1		1
Tanged points (A-type)			6										10							16
Slate points with bi-convex cross section														1						1
Slate points - chipped blanks														3	1					4
Retouched flake	1		22			2			1	6			35							67
Ordinary blades with retouch			2										5							7
Small blades with retouch			1										1							2
Microblades with retouch			1										3							4
Retouched crested blades													2							2
Grinding slabs																		4		4
Hammerstones																		3		3
Grinding stones																		1		1
Pumice without grooves																	8			8
Pumice with grooves																	6			6
Pottery with no/undetermined decoration (locally produced)																			53	53
Other artefacts	2																			2
Sum	12	2	230	1	4	98	8	1	53	339	1	1	401	18	1	4	14	10	53	1251
Percentage	1,0	0,2	18,4	0,1	0,3	7,8	0,6	0,1	4,2	27,1	0,1	0,1	32,1	1,4	0,1	0,3	1,1	0,8	4,2	100,0

Austvik III																
Types	Basaltic rock	Greenstone	Fine flint	Medium flint	Coarse flint	Fine grey quartzite 11	Fine blue quartzite 8	Fine white quartzite 16	Medium grey quartzite 11	Coarse quartzite	Quartz crystal	Fine white quartz	Coarse quartz	Rhyolite	Pumice	Sum
Flakes > 4cm		5		4					1	1				26		37
Flakes 1 >< 4cm	1	7	27	195	14	1	8		98		4	30	14	867		1266
Flakes < 1cm			16	100	4		1	1	60		6	27	7	524		746
Ordinary blades > 12mm				2							1			32		35
Small blades 8 >< 12mm			1	45			1		1		1			168		217
Microblades <8mm				8							2	3		19		32
Crested blades				2										3		5
Other core fragments				2										10		12
One sided cores with one platform				1										1		2
Cylindrical cores												1	1	15		17
Bipolar cores			3	16									3	2		24
Other cores				2	1	1										4
Struck chisels/blanks (not ground)		1														1
Vespestad adzes		2														2
Tanged points (A-type)				2										22		24
Retouched flake			1	64		1						2		21		89
Small blades with retouch				4										1		5
Pumice with grooves															4	4
Flint nodules				1	1											2
Sum	1	15	48	448	20	3	10	1	160	1	14	63	25	1711	4	2524
Percentage	0,0	0,6	1,9	17,8	0,8	0,1	0,4	0,0	6,3	0,0	0,6	2,5	1,0	67,8	0,2	100,0

88 Føyno

Types	Basaltic rock	Greenstone	Fine flint	Medium flint	Coarse flint	Other fine grey quartzites	Fine black quartzite 1	Medium grey quartzite 10	Coarse quartzite	Fine white quartz	Rhyolite	Sandstone	Pumice	Undetermined	Sum
Flakes > 4cm		3	2		1						14				20
Flakes 1 >< 4cm		5	10	153				1	1	1	340				511
Flakes < 1cm			7	28		1		1	1	6	95				139
Ordinary blades > 12mm											2				2
Small blades 8 >< 12mm			4	13							59				76
Microblades <8mm			1	2					1	1	34				39
Crested blades				2							2				4
Other core fragments											5				5
Cylindrical cores				1							5				6
Other cores with two platforms											1				1
Bipolar cores				7							2				9
Other cores							1				7				8
Flakes struck from ground stone artefacts		2													2
Vespestad adzes		1													1
Struck adze blanks - short and broad		3													3
Tanged points (A-type)				2							13				15
Retouched flake			13	39	1						30				83
Ordinary blades with retouch				1							1				2
Small blades with retouch			7	10							18				35
Microblades with retouch											2				2
Grinding slabs												2			2
Hammerstones														1	1
Small round smooth stones	3														3
Pumice without grooves													12		12
Pumice with grooves													12		12
Flint nodules				2	1										3
Sum	3	14	44	260	3	1	1	2	3	8	630	2	24	1	996
Percentage	0,3	1,4	4,4	26,1	0,3	0,1	0,1	0,2	0,3	0,8	63,3	0,2	2,4	0,1	100,0

115 Føyno													
Types	Basaltic rock	Greenstone	Fine flint	Medium flint	Coarse flint	Coarse quartzite	Quartz crystal	Fine white quartz	Coarse quartz	Rhyolite	Pumice	Undetermined	Sum
Flakes > 4cm	4	4	1	3	7	3		1		5			28
Flakes 1 >< 4cm		6	17	269	41	18		43	6	368			768
Flakes < 1cm			22	107	9		2	18	1	129			288
Ordinary blades > 12mm				3						15			18
Small blades 8 >< 12mm			1	15	1					55			72
Microblades <8mm				5						18			23
Crested blades										1			1
Other core fragments				1									1
Conical cores				1									1
Cylindrical cores										3			3
Bipolar cores			4	7	1			3					15
Other cores				3		1				1			5
Flakes struck from ground stone artefacts		1											1
Struck adze blanks - short and broad		4											4
Tanged points (A-type)				1						9			10
Retouched flake			8	42		5		3		8			66
Ordinary blades with retouch			1	1						2			4
Small blades with retouch			1							2			3
Hammerstones												2	2
Pumice without grooves											15		15
Pumice with grooves											11		11
Flint nodules				2	1								3
Sum	4	15	55	460	60	27	2	68	7	616	26	2	1342
Percentage	0,3	1,1	4,1	34,3	4,5	2,0	0,2	5,1	0,5	45,9	1,9	0,2	100,0

Sokkamyro 1901-1902																				
Types	Greenstone production	Basaltic rock	Greenstone	Flint	Fine grey quartzite 11	Fine blue quartzite 7	Other fine grey quartzites	Coarse quartz	Quartz crystal	Fine white quartz	Coarse quartz	Jasper	Rhyolite	Blue slate	Green slate	Grey slate	Sandstone	Undetermined	Pottery	Sum
Flakes > 4cm	710			11						1	3		34							49
Flakes 1 >< 4cm	52			597	1	3		6	8	2	16	1	373							1007
Ordinary blades > 12mm													9							9
Small blades 8 >< 12mm				28	2			2					98							130
Microblades <8mm				4									9							13
Microblades <8mm (microblade technique)				1																1
Crested blades				3																3
Conical cores				2																2
Cylindrical cores					32			1					27							60
Bipolar cores				15					1	4	2									22
Other cores	1			12			4	2	1	1	1		25							46
Flakes struck from ground stone artefacts	3													1						1
Four-sided ground adzes			1																	1
Other ground adzes			3																	3
Ground adzes - round cross-section			2																	2
Pecked adzes - round cross-section		1	1																	2
Vespestad adzes			9																	9
Struck adze blanks - very large	17																			0
Struck adze blanks - long and slim	62																			0
Struck adze blanks - short and broad	360																			0
Ground spearheads															1					1
Tanged points (A-type)													3							3
Slate points with diamond cross section - undetermined shoulders															2					2
Slate points - ground blanks																1				1
Slate points - other/undeterminded cross sections															2					2
Retouched flakes			3	110			2	1		1			46							163
Small blades with retouch				12									7							19
Grinding slabs																	1	81		82
Hammerstones			1															124		125
Grinding stones																		22		22
Flint nodules				13																13
Pottery with no/undetermined decoration (locally produced)																			1	1
Other artefacts			1																	1
Sum	1205	1	21	808	35	3	6	12	10	9	22	1	631	1	5	1	1	227	1	1795
Percentage		0,1	1,2	45,0	2,0	0,2	0,3	0,7	0,6	0,5	1,2	0,1	35,2	0,1	0,3	0,1	0,1	12,6	0,1	100,0

3 Tjernagel

Types	Basaltic rock	Fine flint	Medium flint	Coarse flint	Fine blue quartzite 8	Quartz crystal	Fine white quartz	Coarse quartz	Rhyolite	Sandstone	Undetermined	Sum
Flakes > 4cm	3		1		6				3	2		15
Flakes 1 >< 4cm	56	14	318	33	303	1	3	126	57	1		912
Flakes < 1cm	12	17	502	11	202		6	259	49			1058
Ordinary blades > 12mm		1	2		10							13
Small blades 8 >< 12mm		5	26		40				8			79
Microblades <8mm		5	9		15							29
Crested blades			1									1
Other core fragments					1				1			2
One sided cores with one platform					2							2
Other cores with one platform									1			1
Cylindrical cores					2							2
One-sided cores with two platforms									1			1
Other cores with two platforms					1							1
Bipolar cores		2	10									12
Other cores			2		1			2				5
Vespestad adzes	1											1
Tanged points (A-type)			7						2			9
Retouched flake		2	50						1			53
Small blades with retouch		2	4									6
Grinding slabs											1	1
Flint nodules			1									1
Sum	72	48	933	44	583	1	9	387	123	3	1	2204
Percentage	3,3	2,2	42,3	2,0	26,5	0,1	0,4	17,6	5,6	0,1	0,1	100,0

Styggvasshelleren																		
Types	Fine flint	Medium flint	Fine blue quartzite 1	Fine green quartzite 1	Fine grey quartzite 2	Fine white quartzite 16	Fine black quartzite 1	Other fine black quartzites	Medium grey quartzite 9	Quartz crystal	Fine white quartz	Coarse quartz	Rhyolite	Blue slate	Green slate	Grey slate	Anorthosite	Sum
Flakes > 4cm	2		2								3					1		8
Flakes 1 >< 4cm	44	3	11	2		2	8	3	36	52	522	172				2		857
Flakes < 1cm	27	1	4	1			1		10	21	440	48						553
Small blades 8 >< 12mm	6	3			1				1		6		3					20
Microblades <8mm	4		1								1							6
Other cores with two platforms											1							1
Bipolar cores	2	2	1		2				2	15	68	34						126
Other cores			1							1	5	1						8
Blade tanged points (A-type)		1	1						1				2					5
Slate points with diamond cross section - hanging shoulders															2			2
Slate points with diamond cross section - square shoulders															3		1	4
Slate points with diamond cross section - contracting shoulders														1	1			2
Slate points with diamond cross section - undetermined shoulders														1	18		3	22
Slate points - ground blanks															2			2
Slate points - chipped blanks																	1	1
Slate points - other/undeterminded cross sections															19			19
Retouched flake	6	1	2			1				2	9							21
Small blades with retouch	1						1											2
Quartz crystal, entire crystal										2								2
Sum	92	11	23	3	3	3	10	3	50	93	1055	255	5	2	45	3	5	1661
Percentage	5,5	0,7	1,4	0,2	0,2	0,2	0,6	0,2	3,0	5,6	63,5	15,4	0,3	0,1	2,7	0,2	0,3	100,0

Mørkedøla I										
Types	Fine flint	Medium flint	Fine green quartzite 1	Fine white quartzite 16	Fine brown quartzite 1	Fine black quartzite 1	Quarz crystal	Green slate	Red slate	Sum
Flakes > 4cm			205		8				1	214
Flakes 1 >< 4cm	10	8	1016	2	50		1			1087
Flakes < 1cm			373		4					377
Ordinary blades > 12mm	1		34		1					36
Small blades 8 >< 12mm			15		3	1				19
Microblades <8mm	8	1	11				1			21
One sided cores with one platform			4							4
Other cores with one platform			6		1					7
Other cores			22	1						23
Blade tanged points (A-points)	1				1					2
Flake tanged points (A-points)	1		1							2
Transverse points	1									1
Slate points with diamond cross section - contracting shoulders									1	1
Slate points - chipped blanks								1		1
Retouched flakes	3	1	129		1					134
Sum	25	10	1816	3	69	1	2	1	2	1929
Percentage	1,3	0,5	94,1	0,2	3,6	0,1	0,1	0,1	0,1	100,0

Gyrinos IV

Types	Fine flint	Medium flint	Fine green quartzite 1	Fine brown quartzite 1	Other fine grey quartzites	Fine white quartz	Blue slate	Grey slate	Undetermined	Sum
Flakes > 4cm	1	1	36	5	1					44
Flakes 1 >< 4cm	209	254	505	900	10	29	1	6	12	1926
Flakes < 1cm	9	9	30	50		2				100
Ordinary blades > 12mm	7		15	6						28
Small blades 8 >< 12mm	15	7	39	8						69
Microblades <8mm	21	3	1	3						28
Microblades <8mm (microblade technique)	23	2								25
Other cores with one plattform	1	1	2							4
Other cores with two plattforms				6						6
Bipolar cores	5				1					6
Other cores	4	1	9	3						17
Single-edged points		3	8							11
Blade tanged points (A-points)	3		3		1					7
Flake tanged points (A-points)			4							4
Transverse points		3	8							11
Triangular bifacial points	1									1
Flake drills				1						1
Slate points with diamond cross section - undetermined shoulders							1			1
Retouched flakes	26		73	28						127
Ordinary blades with retouch	4									4
Sum	329	284	733	1010	13	31	2	6	12	2420
Percentage	13,6	11,7	30,3	41,7	0,5	1,3	0,1	0,2	0,5	100,0

Blånut IV

Types	Fine flint	Medium flint	Fine green quartzite 1	Medium grey quartzite 8	Coarse quartzite	Quartz crystal	Fine white quartz	Blue slate	Green slate	Grey slate	Undetermined	Sum
Flakes > 4cm			15	2	1			7				25
Flakes 1 >< 4cm	27	20	376	20	6		12	9	20		6	496
Flakes < 1cm	2	1	13				1					17
Ordinary blades > 12mm	1		1	2								4
Small blades 8 >< 12mm			1									1
Microblades <8mm	1		11				1					13
Other cores			27		4							31
Flakes struck from ground flint artefacts	1	3										4
Thin-butted axes (fragment)		1										1
Slate points with diamond cross section - square shoulders									2			2
Slate points with diamond cross section - undetermined shoulders									19			19
Slate points - ground blanks										8		8
Slate points - chipped blanks								2				2
Slate points - other/undeterminded cross sections									1			1
Retouched flakes	4	6	9			1						20
Ordinary blades with retouch	9											9
Small blades with retouch	2		1									3
Sum	47	31	454	24	11	1	14	18	42	8	6	656
Percentage	7,2	4,7	69,2	3,7	1,7	0,2	2,1	2,7	6,4	1,2	0,9	100,0

526 Nordmannslågen

Types	Fine flint	Medium flint	Fine grey quartzite 3	Medium green quartzite	Other medium quartzites	Coarse quartzite	Quartz crystal	Fine white quartz	Coarse quartz	Grey slate	Pottery	Sum
Flakes 1 >< 4cm	173	258	12				113	188	2			746
Flakes < 1cm	89	79	5		1		146	177				497
Ordinary blades > 12mm	3						1					4
Small blades 8 >< 12mm	12	4										16
Microblades <8mm	20	7					5					32
Microblades <8mm (microblade technique)	10	1										11
Crested blades	3			1								4
One sided cores with one platform									2			2
Cylindrical cores		1					1					2
One-sided cores with two platforms		1					1					2
Bipolar cores	1	2			1		12	2	4			22
Flakes struck from ground flint artefacts	2											2
Single edged points	5	5				1	2	1				14
Transverse points		1										1
Blade tanged points (A-type)	7											7
Flake tanged points (A-type)	8	5					2	1				16
Slate points - ground blanks										1		1
Retouched flake	36	23	1				19	2				81
Small blades with retouch	5	1										6
Microblades with retouch	2											2
Quartz crystal, entire crystal							3					3
Pottery with impressions from real cord											10	10
Pottery with no/undetermined decoration											34	34
Sum	376	388	18	1	2	1	305	371	8	1	44	1515
Percentage	24,8	25,6	1,2	0,1	0,1	0,1	20,1	24,5	0,5	0,1	2,9	100,0

760 Finnsbergvatn

Types	Fine flint	Medium flint	Fine green quartzite 1	Fine grey quartzite 3	Fine white quartz	Coarse quartz	Rhyolite	Blue slate 2	Grey slate	Sum
Flakes > 4cm				24	11			2		37
Flakes 1 >< 4cm	100	163	4	887	1187	8		102		2451
Flakes < 1cm	173	199		1229	1972			28		3601
Ordinary blades > 12mm	4	1		2	1					8
Small blades 8 >< 12mm	4	2		6	7		1			20
Microblades <8mm	3	4		5	9		1			22
Crested blades					1					1
One sided cores with one platform					2					2
Other cores with one platform					1					1
One-sided cores with two platforms				4						4
Other cores with two platforms					1					1
Bipolar cores		1		14	51					66
Other cores				3	4					7
Single edged points	2	2		3	6					13
Transverse points	1									1
Tanged blade points (A-type)	2			2	2		1			7
Tanged flake points (A-type)				1	3		1			5
Slate points with diamond cross section - undetermined shoulders									1	1
Slate points with diamond cross section - pointed base and lashing notches									1	1
Slate points with planoconvex cross section - contracting shoulders									1	1
Slate points - other/undeterminded cross sections								2		2
Retouched flake	7	2		16	11					36
Flake drills					1					1
Ordinary blades with retouch	1				2					3
Small blades with retouch	4	1		1	1					7
Microblades with retouch	1				1					2
Sum	302	375	4	2197	3274	8	4	134	3	6301
Percentage	4,8	6,0	0,1	34,9	52,0	0,1	0,1	2,1	0,1	100,0

1020 Bjornesfjorden

Types	Basaltic rock	Fine flint	Medium flint	Fine white quartzite 16	Coarse quartzite	Quartz crystal	Fine white quartz	Rhyolite	Sandstone	Sum
Flakes 1 >< 4cm		224	328	2		8	6			568
Flakes < 1cm		127	150	1		4	1	2		285
Ordinary blades > 12mm		6								6
Small blades 8 >< 12mm		5	4							9
Microblades <8mm		1								1
Bipolar cores		4	2			1	1			8
Other cores		3	10			1				14
Flakes struck from ground flint artefacts		3								3
Four-sided ground adzes	1									1
Single edged points					1					1
Transverse points		9	3							12
Flake tanged points (A-points)		4								4
Leaf-shaped bifacial points							1			1
Retouched flakes		39	60	2			2			103
Ordinary blades with retouch		2	1							3
Grinding slabs									3	3
Quartz crystal, entire crystal						1				1
Sum	1	427	558	5	1	15	11	2	3	1023
Percentage	0,1	41,7	54,5	0,5	0,1	1,5	1,1	0,2	0,3	100,0

Vivik

Types	Fine flint	Medium flint	Quartz crystal	Fine white quartz	Coarse quartz	Blue slate 1	Grey slate	Sum
Flakes > 4cm	2	1	1		1	5	18	28
Flakes 1 >< 4cm	117	185	600		154	39	133	1228
Flakes < 1cm	9	42	617		8		2	678
Ordinary blades > 12mm	16							16
Small blades 8 >< 12mm	22		1					23
Microblades <8mm	2	1	6					9
Bipolar cores	1		30	1	11			43
Other cores		1	42	2	30			75
Single-edged points	7		3					10
Flakes struck from ground flint artefacts	3							3
Thin-butted flint axe (fragment)	1							1
Blade tanged points (A-points)	3							3
Flake tanged points (A-points)	1							1
Transverse points			1		2			3
Blade tanged points (B-points)	3							3
Slate points with diamond cross section - square shoulders							1	1
Slate points with diamond cross section - undetermined shoulders						1	1	2
Slate points - ground blanks						5		5
Slate points - chipped blanks						2	5	7
Retouched flakes	22	3	4	1	1			31
Ordinary blades with retouch	6							6
Quartz crystal, entire crystal			1					1
Sum	215	233	1306	4	207	52	160	2177
Percentage	9,9	10,7	60,0	0,2	9,5	2,4	7,4	100,0

APPENDIX 4

CALIBRATED RADIOCARBON DATES

Excavated sites

Quarries

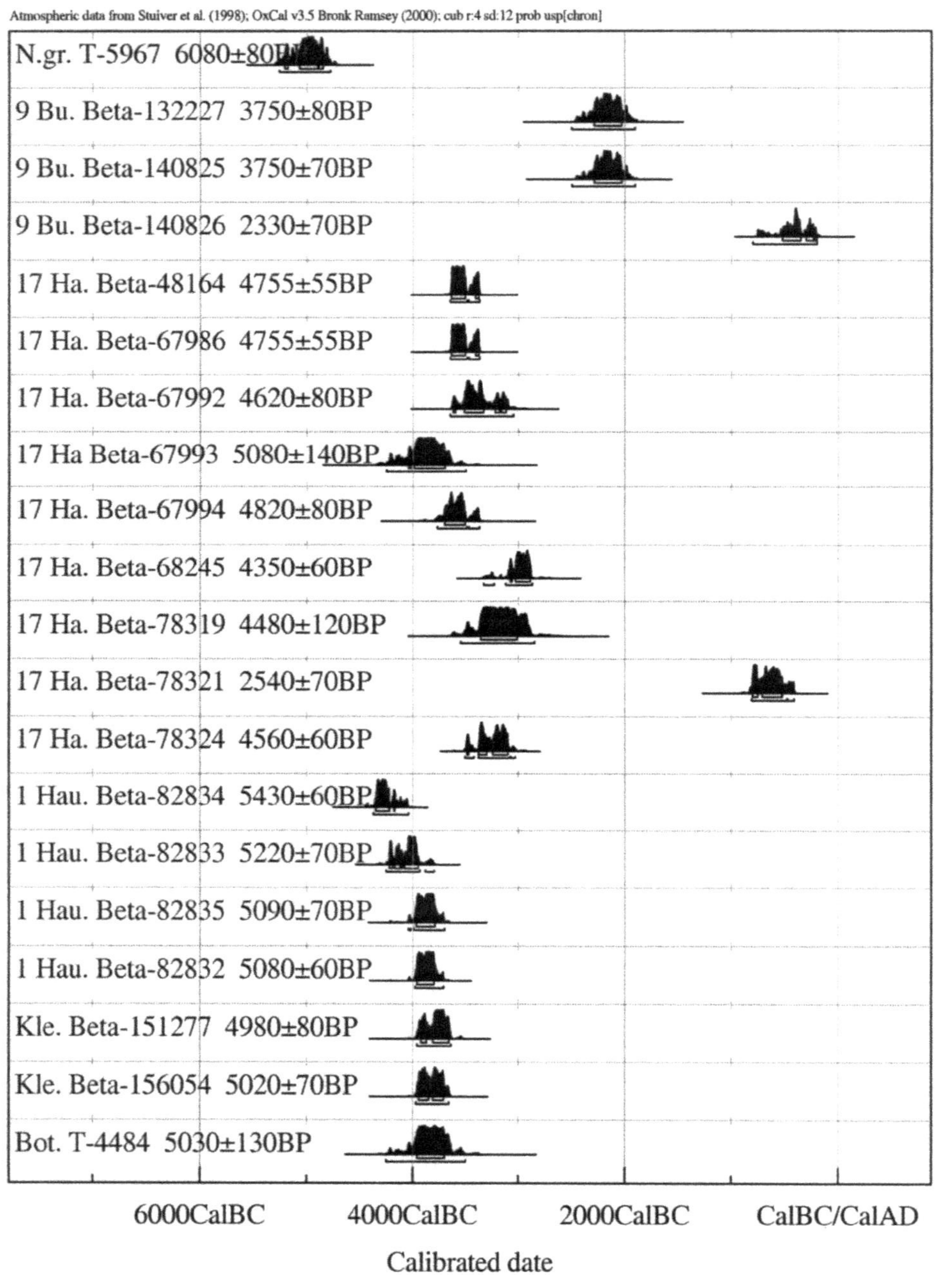

Calibrated radiocarbon dates from excavated sites Nordøy Grendahus to Botnaneset (Chapter 6).

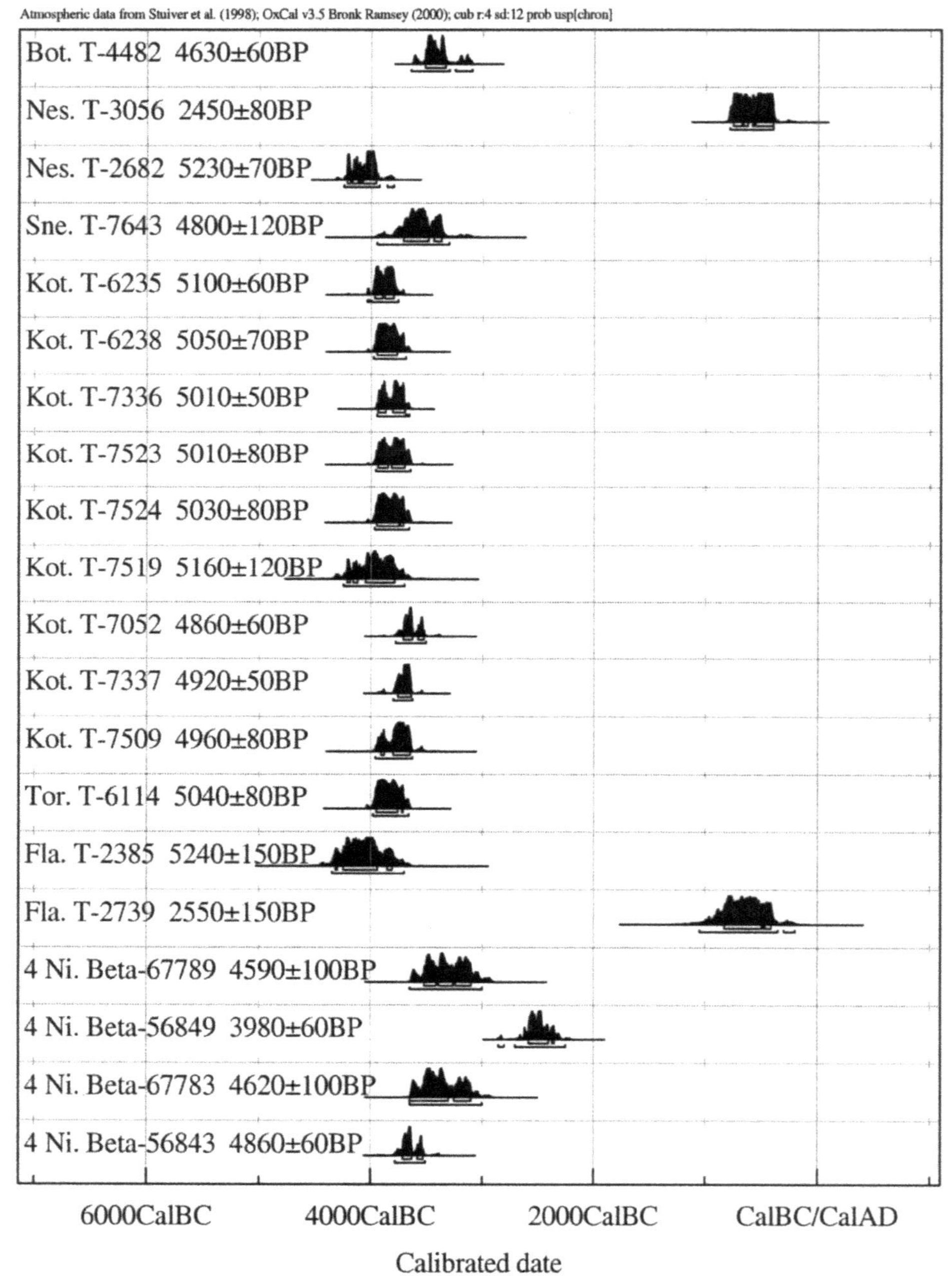

Calibrated radiocarbon dates from excavated sites Botnaneset to 4 Nilsvik (Chapter 6).

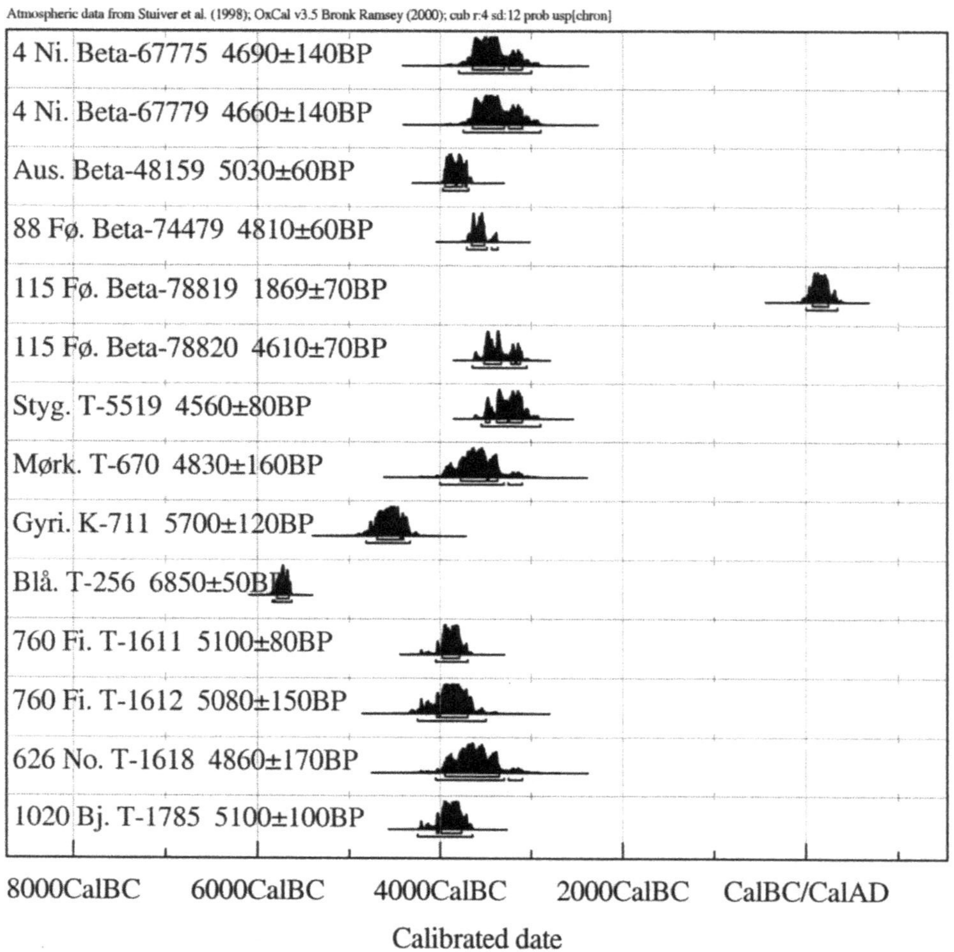

Calibrated radiocarbon dates from excavated sites 4 Nilsvik to 1020 Bjornesfjorden Nordøy Grendahus to Botnaneset (Chapter 6).

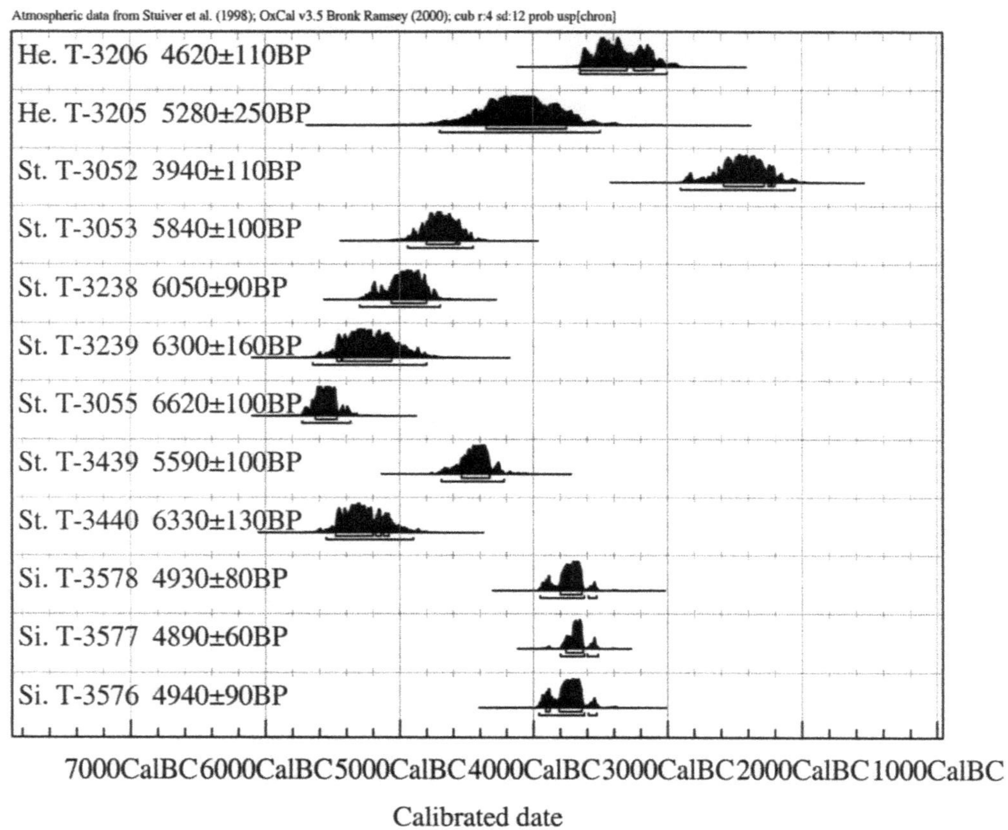

Calibrated radiocarbon dates from the quarries Hespriholmen and Siggjo (Chapter 8).

APPENDIX 5

CATALOGUES
OF VESPESTAD ADZES
IN THE MUSEUM COLLECTIONS

A. Bergen University Museum and Sunnmøre Museum

B. Museum of Natural History and Archaeology, Trondheim

C. Archaeological museum in Stavanger

D. Museum of Cultural History, Oslo

A. Vespestad adzes at Bergen University Museum

B no.	Unit/no.	Excavated archaeological site	Farm	Municipality	District	Raw material	Geological sample/ Museum collection
13582	5	Bjørnøy	Bjørnøy	Haram	Northern Sunnmøre	Diabase	
7659			Fylingen	Sula	Northern Sunnmøre	Diabase	
10107			Fylingen	Sula	Northern Sunnmøre	Diabase	Sunnmøre museum A.B.Olsen's database 1981
3261			Giske	Giske	Northern Sunnmøre	Diabase	
13473		N. Grendahus	Longva	Haram	Northern Sunnmøre	Diabase	
13824	100x/109y lag 5	N. Grendahus	Longva	Haram	Northern Sunnmøre	Diabase	
10008			Nedregarden	Ålesund	Northern Sunnmøre	Diabase	Sunnmøre museum A.B.Olsen's database 1981
10107			Fylingen	Sula	Northern Sunnmøre	Greenstone	Sunnmøre museum A.B.Olsen's database
8529			Kvassnes	Sula	Northern Sunnmøre	Greenstone	Sunnmøre museum A.B.Olsen's database 1981
13540			Mauseidvåg	Sula	Northern Sunnmøre	Greenstone	
9800			Skarbøvik	Ålesund	Northern Sunnmøre	Greenstone	Sunnmøre museum A.B.Olsen's database
10005			Valderhaug	Giske	Northern Sunnmøre	Greenstone	Sunnmøre museum A.B.Olsen's database
10005			Valderhaug	Giske	Northern Sunnmøre	Undetermined rock	Sunnmøre museum A.B.Olsen's database 1981
13939		Volsvikja	Fjørtoft	Haram	Northern Sunnmøre	Weathered, undetermined rock	
9719			Hareid	Hareid	Southern Sunnmøre	Dark blue/grey coarse rock	
9719	d		Hareid	Hareid	Southern Sunnmøre	Dark blue/ grey coarse, sedimented rock	
12827		Molvikshaugen	Berkneset	Volda	Southern Sunnmøre	Diabase	
12773	1		Haddal	Ulstein	Southern Sunnmøre	Diabase	
12997	1	Gråtane	Hareid	Hareid	Southern Sunnmøre	Diabase	
14155			Hustad	Ørsta	Southern Sunnmøre	Diabase	
8836			Igesund	Herøy	Southern Sunnmøre	Diabase	
9144			Liavåg	Hareid	Southern Sunnmøre	Diabase	
11340			Pilskog	Hareid	Southern Sunnmøre	Diabase	Sunnmøre museum A.B.Olsen's database 1981
12998			Sundet	Ulstein	Southern Sunnmøre	Diabase	
9451			Tjørvåg	Herøy	Southern Sunnmøre	Diabase	Sunnmøre museum A.B.Olsen's database 1981

A. Vespestad adzes at Bergen University Museum (continued)

B no.	Unit/no.	Excavated archaeological site	Farm	Municipality	District	Raw material	Geological sample/ Museum collection
8508			Hareid	Ulstein	Southern Sunnmøre	Green basalt with white, heterogenous intrusions	
8241	a		Furneset	Volda	Southern Sunnmøre	Greenstone	
8241	c		Furneset	Volda	Southern Sunnmøre	Greenstone	
10581			Gjerde	Sande	Southern Sunnmøre	Greenstone	
12831			Gnr.	Herøy	Southern Sunnmøre	Greenstone	
8508			Hareid	Ulstein	Southern Sunnmøre	Greenstone	Isotope sample nr. 46
6814			Hjertnes	Herøy	Southern Sunnmøre	Greenstone	
7627			Igesund	Herøy	Southern Sunnmøre	Greenstone	Sunnmøre museum A.B.Olsen's database 1981
9270			Igesund	Herøy	Southern Sunnmøre	Greenstone	Sunnmøre museum A.B.Olsen's database
10781			Jøsåk	Herøy	Southern Sunnmøre	Greenstone	Sunnmøre museum A.B.Olsen's database
7172			Kvalsund	Herøy	Southern Sunnmøre	Greenstone	Isotope sample nr. 45
6872			Kyrkjesande	Sande	Southern Sunnmøre	Greenstone	
8480			Kyrkjesande	Sande	Southern Sunnmøre	Greenstone	
9138			Liavåg	Hareid	Southern Sunnmøre	Greenstone	
9142			Liavåg	Hareid	Southern Sunnmøre	Greenstone	Sunnmøre museum A.B.Olsen's database 1981
9143			Liavåg	Hareid	Southern Sunnmøre	Greenstone	Isotope sample nr. 47. Thin-sectioned by A.B. Olsen 1981
9425			Nordre Sande	Sande	Southern Sunnmøre	Greenstone	Sunnmøre museum A.B.Olsen's database
9547			Røyset	Sande	Southern Sunnmøre	Greenstone	Sunnmøre museum A.B.Olsen's database
5823			Steinnes	Ørsta	Southern Sunnmøre	Greenstone	Isotope sample nr. 44
8486			Voksa	Sande	Southern Sunnmøre	Greenstone	Sunnmøre museum A.B.Olsen's database 1981
11462			Ytre Sande	Sande	Southern Sunnmøre	Greenstone	
10783			Åram	Sande	Southern Sunnmøre	Greenstone	Sunnmøre museum A.B.Olsen's database
10653			Årset	Volda	Southern Sunnmøre	Greenstone	Sunnmøre museum A.B.Olsen's database

***A.** Vespestad adzes at Bergen University Museum (continued)*

B no.	Unit/no.	Excavated archaeological site	Farm	Municipality	District	Raw material	Geological sample/ Museum collection
9928			Furneset	Volda	Southern Sunnmøre	Light green basalt with many yellow intrusions	
9927			Furnes	Volda	Southern Sunnmøre	Undetermined rock	Sunnmøre museum A.B.Olsen's database 1981
10644			Indre Steinnes	Ørsta	Southern Sunnmøre	Undetermined rock	Sunnmøre museum A.B.Olsen's database 1981
8726			Nordre Sande	Sande	Southern Sunnmøre	Undetermined rock	Sunnmøre museum A.B.Olsen's database 1981
8528			Ose	Ørsta	Southern Sunnmøre	Undetermined rock	Sunnmøre museum A.B.Olsen's database 1981
8528			Ose	Ørsta	Southern Sunnmøre	Undetermined rock	Sunnmøre museum A.B.Olsen's database 1981
8505			Bringsvoren	Sande	Southern Sunnmøre	Weathered, undetermined rock	
9161	1916?			Hareid	Southern Sunnmøre	Weathered, undetermined rock	
5523			Hove	Selje	Outer Nordfjord	Diabase	Thin-section/ trace element analysis S.Alsaker 1982
13769			Selje krk.	Selje	Outer Nordfjord	Diabase	
9924			Barmen	Selje	Outer Nordfjord	Greenstone	
3029			Grotle	Bremanger	Outer Nordfjord	Greenstone	
14588		1 Haukedal	Haukedal	Bremanger	Outer Nordfjord	Greenstone	Isotope sample nr. 51
14604	35000	17 Havnen	Havnen	Bremanger	Outer Nordfjord	Greenstone	Isotope sample nr. 49
14604	897	17 Havnen	Havnen	Bremanger	Outer Nordfjord	Greenstone	Isotope sample nr. 50
B no.	Unit/no.	Excavated archaeological site	Farm	Municipality	District	Raw material	Comment
14604	35001, 102x52y Ae4 SV	17 Havnen	Havnen	Bremanger	Outer Nordfjord	Greenstone	
8919			Sandvikja	Selje	Outer Nordfjord	Greenstone	
11124	b		Åsmundvåg	Vågsøy	Outer Nordfjord	Greenstone	
11124	c		Åsmundvåg	Vågsøy	Outer Nordfjord	Greenstone	Isotope sample nr. 48

A. *Vespestad adzes at Bergen University Museum (continued)*

B no.	Unit/no.	Excavated archaeological site	Farm	Municipality	District	Raw material	Geological sample/ Museum collection
204			Hove	Selje	Outer Nordfjord	Weathered, undetermined rock	
13818			Brandsøy	Flora	Sunnfjord	Dark blue/green basalt (2)	
13787		Nakkaneset II	Brandsøy	Flora	Sunnfjord	Dark blue/green basalt (3)	
13313			Grøneng	Flora	Sunnfjord	Dark blue/green basalt (3)	
13787		Nakkaneset II	Brandsøy	Flora	Sunnfjord	Greenstone	Atypical adze
230			Florø	Flora	Sunnfjord	Greenstone	
14839			Unneset	Askvoll	Sunnfjord	Greenstone	
13975		Keilevågen III	Florø	Flora	Sunnfjord	Weathered, undetermined rock	
8227			Eide	Gulen	Outer Sogn	Greenstone	Trace element analysis 1982 S.Alsaker. Atypical adze.
7811		Masnesberget	Masnesberget	Høyanger	Outer Sogn	Light basalt, somewhat coarse	
11671		Grønehelleren	Ytrøyna	Solund	Outer Sogn	Light green basalt. Many light intrusions	
7811		Masnesberget	Masnesberget	Høyanger	Outer Sogn	Weathered, undetermined rock	
5799			Dyrdal	Aurland	Inner Sogn	Dark blue/green basalt (1)	
10513			Bale	Balestrand	Inner Sogn	Diabase	
10513			Bale	Balestrand	Inner Sogn	Greenstone	
10794			Sandviki	Luster	Inner Sogn	Greenstone	
8237			Tjugum	Balestrand	Inner Sogn	Greenstone	
12609			Ulvastad	Balestrand	Inner Sogn	Greenstone	
12860			Ulvastad	Balestrand	Inner Sogn	Greenstone	
13332			Vangsnes	Vik	Inner Sogn	Light green basalt. Spots	
13063		Haglesundet 3	Isdal	Lindås	Nordhordland	Dark blue/green basalt (3)	
13048			Flatøy	Meland	Nordhordland	Diabase	
8034			Hjelmås	Lindås	Nordhordland	Diabase	
11511	J22III	Ramsvikneset	Straume	Radøy	Nordhordland	Diabase	
11511	H11IV	Ramsvikneset	Straume	Radøy	Nordhordland	Diabase	
11511	H15V	Ramsvikneset	Straume	Radøy	Nordhordland	Diabase	
14500		Kotedalen	Straume	Radøy	Nordhordland	Diabase	
10660		Grøndalen	Straume	Radøy	Nordhordland	Green basalt with small blisters	
10677			Bergsvik	Austrheim	Nordhordland	Greenstone	
14700			Hopland	Austrheim	Nordhordland	Greenstone	
7796			Krossøyni	Austrheim	Nordhordland	Greenstone	
11257			Soltveit	Radøy	Nordhordland	Greenstone	Trace element analysis 1982 S. Alsaker.

__A.__ Vespestad adzes at Bergen University Museum (continued)

B no.	Unit/no.	Excavated archaeological site	Farm	Municipality	District	Raw material	Geological sample/ Museum collection
11258			Soltveit	Radøy	Nordhordland	Greenstone	
7803		Grøndalen	Straume	Radøy	Nordhordland	Greenstone	
10660		Grøndalen	Straume	Radøy	Nordhordland	Greenstone	
11255		Grøndalen	Straume	Radøy	Nordhordland	Greenstone	
11256	a	Grøndalen	Straume	Radøy	Nordhordland	Greenstone	
11369		Ramsvikneset	Straume	Radøy	Nordhordland	Greenstone	
11450	tt	Ramsvikneset	Straume	Radøy	Nordhordland	Greenstone	
11511	O30I	Ramsvikneset	Straume	Radøy	Nordhordland	Greenstone	
11511	F30I	Ramsvikneset	Straume	Radøy	Nordhordland	Greenstone	
11511	J19II	Ramsvikneset	Straume	Radøy	Nordhordland	Greenstone	
11511	I23II	Ramsvikneset	Straume	Radøy	Nordhordland	Greenstone	
11511	F19I	Ramsvikneset	Straume	Radøy	Nordhordland	Greenstone	
11511	L29III	Ramsvikneset	Straume	Radøy	Nordhordland	Greenstone	
11511	F28IV	Ramsvikneset	Straume	Radøy	Nordhordland	Greenstone	
11511	H3IV	Ramsvikneset	Straume	Radøy	Nordhordland	Greenstone	
11511	M26II	Ramsvikneset	Straume	Radøy	Nordhordland	Greenstone	
11511	P13VI	Ramsvikneset	Straume	Radøy	Nordhordland	Greenstone	
11511	L8VII	Ramsvikneset	Straume	Radøy	Nordhordland	Greenstone	
11511	F31 II	Ramsvikneset	Straume	Radøy	Nordhordland	Greenstone	
12029		Ramsvikneset	Straume	Radøy	Nordhordland	Greenstone	
14500	22724	Kotedalen	Straume	Radøy	Nordhordland	Greenstone	
14500	29519	Kotedalen	Straume	Radøy	Nordhordland	Greenstone	
14500	17510	Kotedalen	Straume	Radøy	Nordhordland	Greenstone	
11449	c		Ytstebø	Radøy	Nordhordland	Greenstone	
13062	95x99y 2/3	Haglesundet 2	Isdal	Lindås	Nordhordland	Light brown basalt, lines and spots.	
14522	2	Rappeneset 23	Straume	Radøy	Nordhordland	Light green basalt with undular structure.	
12604			Helland	Lindås	Nordhordland	Light green basalt. Small spots with mica.	
12888		Flatøy XIV	Flatøy	Meland	Nordhordland	Weathered, undetermined rock	
11660			Krossøyni	Austrheim	Nordhordland	Weathered, undetermined rock	
11511	K5IV	Ramsvikneset	Straume	Radøy	Nordhordland	Weathered, undetermined rock	
11511	L29XI	Ramsvikneset	Straume	Radøy	Nordhordland	Weathered, undetermined rock	
14500	91550	Kotedalen	Straume	Radøy	Nordhordland	Weathered, undetermined rock	
11993			Vikebø	Meland	Nordhordland	Weathered, undetermined rock	
14571	14864	4 Nilsvik	Bjorøy	Fjell	Midthordland	Dark blue/green basalt (3)	
13430			Eide	Fjell	Midthordland	Dark blue/green basalt with white lines/spots	
11295			Hundhamar	Osterøy	Midthordland	Diabase	Thin-section by A.B.Olsen 1981

A. Vespestad adzes at Bergen University Museum (continued)

B no.	Unit/no.	Excavated archaeological site	Farm	Municipality	District	Raw material	Geological sample/ Museum collection
195			Åstveit	Bergen	Midthordland	Diabase	
6600		Garnes	Garnes	Bergen	Midthordland	Green basalt with large blisters	
5923			Ytre Arna	Bergen	Midthordland	Green basalt with small blisters	
13500		10 Karidalen	Toftøy	Øygarden	Midthordland	Green basalt with small blisters	
13452			?	Sund	Midthordland	Greenstone	
13404	a		Ask	Askøy	Midthordland	Greenstone	
10714			Bergen	Bergen	Midthordland	Greenstone	
10716			Bergen	Bergen	Midthordland	Greenstone	
5156			Bjorøy	Fjell	Midthordland	Greenstone	
14571	37	4 Nilsvik	Bjorøy	Fjell	Midthordland	Greenstone	
12541	1	Garnes	Garnes	Bergen	Midthordland	Greenstone	
8843			Hamre	Bergen	Midthordland	Greenstone	
12805			Hundhamar	Osterøy	Midthordland	Greenstone	Trace element analysis 1982 S. Alsaker.
11234			Lægreid	Ulvik	Midthordland	Greenstone	
8080			Mildo	Bergen	Midthordland	Greenstone	
8896	c		Nese	Fjell	Midthordland	Greenstone	
10718	a		Skansen	Askøy	Midthordland	Greenstone	
8600	I8VI		Straume	Vaksdal	Midthordland	Greenstone	
12816			Toft	Sund	Midthordland	Greenstone	
5772			Ytre Arna	Bergen	Midthordland	Greenstone	
1786			Åstveit	Bergen	Midthordland	Greenstone	
14617	2520	5 Revarvika	Breivik	Øygarden	Midthordland	Greenstone	Possibly diabase.
8945			Eide	Øygarden	Midthordland	Greenstone	
9101			Rong	Øygarden	Midthordland	Greenstone	
8369			Sele	Øygarden	Midthordland	Greenstone	
14300	64x51y 4	2 Eidet	Sture	Øygarden	Midthordland	Greenstone	
14300	62x52y 5	2 Eidet	Sture	Øygarden	Midthordland	Greenstone	
13501	49x49yE	1 Svelgen	Toftøy	Øygarden	Midthordland	Greenstone	
13501	50x50y	1 Svelgen	Toftøy	Øygarden	Midthordland	Greenstone	
13501	49x49y A	1 Svelgen	Toftøy	Øygarden	Midthordland	Greenstone	
13505		5 Ingerdalen	Toftøy	Øygarden	Midthordland	Greenstone	
14200	45x51y 2	11 Torsteinsvik	Toftøy	Øygarden	Midthordland	Greenstone	
12070			Myro	Bergen	Midthordland		
14572	33	5 Nilsvik	Bjorøy	Fjell	Midthordland	Light green basalt with undular structure	
13038			Gnr. 22	Bergen	Midthordland	Light green basalt with undular structure	
7130			Hanevik	Askøy	Midthordland	Light green basalt with undular structure	
11875	a	Risøyna II	Toft	Sund	Midthordland	Light green basalt with undular structure	

A. Vespestad adzes at Bergen University Museum (continued)

B no.	Unit/no.	Excavated archaeological site	Farm	Municipality	District	Raw material	Geological sample/ Museum collection
14571	42	4 Nilsvik	Bjorøy	Fjell	Midthordland	Weathered, undetermined rock	
14571	11882	4 Nilsvik	Bjorøy	Fjell	Midthordland	Weathered, undetermined rock	
14571	103x97y 7	4 Nilsvik	Bjorøy	Fjell	Midthordland	Weathered, undetermined rock	
11190	F - 2	Nesvikja II	Hummlesund	Sund	Midthordland	Weathered, undetermined rock	
13438			N. Bildøy	Fjell	Midthordland	Weathered, undetermined rock	
196			Åstveit	Bergen	Midthordland	Weathered, undetermined rock	
14617	3927	5 Revarvika	Breivik	Øygarden	Midthordland	Weathered, undetermined rock	
14671	35	5 Revarvika	Breivik	Øygarden	Midthordland	Weathered, undetermined rock	
11668	a		Tjeldstø	Øygarden	Midthordland	Weathered, undetermined rock	
13501		1 Svelgen	Toftøy	Øygarden	Midthordland	Weathered, undetermined rock	
13501		1 Svelgen	Toftøy	Øygarden	Midthordland	Weathered, undetermined rock	
14200	18x48y 4	11 Torsteinsvik	Toftøy	Øygarden	Midthordland	Weathered, undetermined rock	
8051			Breivikjo	Fitjar	Sunnhordland	Dark blue/green basalt (3)	
13310			Stangeland	Austevoll	Sunnhordland	Dark blue/green basalt (3)	
8978			Røyksund	Bømlo	Sunnhordland	Green basalt with large blisters	
6180			Vespestad	Bømlo	Sunnhordland	Green basalt with red ligaments	
7671			Eidsvåg	Sveio	Sunnhordland	Green basalt with small blisters	
7253	b	Nappen	Aukland	Sveio	Sunnhordland	Greenstone	
10840			Byrkjeland	Stord	Sunnhordland	Greenstone	
5683			Eide	Bømlo	Sunnhordland	Greenstone	
8459			Eide	Bømlo	Sunnhordland	Greenstone	
9252			Eide	Kvinnherad	Sunnhordland	Greenstone	
12129			Hufthamar	Austevoll	Sunnhordland	Greenstone	Trace element analysis 1982 S. Alsaker.
12313			Kuleseid	Bømlo	Sunnhordland	Greenstone	
9228			Kverno	Fitjar	Sunnhordland	Greenstone	
9172			Naustheller	Austevoll	Sunnhordland	Greenstone	
12950	3	1 Sydnes	Rubbestadneset	Bømlo	Sunnhordland	Greenstone	

A. Vespestad adzes at Bergen University Museum (continued)

B no.	Unit/no.	Excavated archaeological site	Farm	Municipality	District	Raw material	Geological sample/ Museum collection
8659		Røykjenes	Røykjenes	Sveio	Sunnhordland	Greenstone	
9667	i	Røykjenes	Røykjenes	Sveio	Sunnhordland	Greenstone	
9668	d	Røykjenes	Røykjenes	Sveio	Sunnhordland	Greenstone	
9674		Røykjenes	Røykjenes	Sveio	Sunnhordland	Greenstone	
12332			Røykjenes	Sveio	Sunnhordland	Greenstone	
4212			Sakseid	Bømlo	Sunnhordland	Greenstone	
7715			Skogseid	Fusa	Sunnhordland	Greenstone	
12167			Storavik	Fitjar	Sunnhordland	Greenstone	
14585			Storebø	Austevoll	Sunnhordland	Greenstone	
5683			Vespestad	Bømlo	Sunnhordland	Greenstone	
5751	2a	Sokkamyro	Vespestad	Bømlo	Sunnhordland	Greenstone	
5751	19	Sokkamyro	Vespestad	Bømlo	Sunnhordland	Greenstone	
5751	22	Sokkamyro	Vespestad	Bømlo	Sunnhordland	Greenstone	
5751	23	Sokkamyro	Vespestad	Bømlo	Sunnhordland	Greenstone	
5751	21	Sokkamyro	Vespestad	Bømlo	Sunnhordland	Greenstone	
5751	23	Sokkamyro	Vespestad	Bømlo	Sunnhordland	Greenstone	
5751	20	Sokkamyro	Vespestad	Bømlo	Sunnhordland	Greenstone	
5850	c	Sokkamyro	Vespestad	Bømlo	Sunnhordland	Greenstone	
7021	GI3	Sokkamyro	Vespestad	Bømlo	Sunnhordland	Greenstone	
8658	a	Røykjenes	Røykjenes	Sveio	Sunnhordland	Grey/brown basalt	
9674		Røykjenes	Røykjenes	Sveio	Sunnhordland	Light basalt, somewhat coarse	
11254			Valo	Sveio	Sunnhordland	Light basalt, somewhat coarse	
7224	a	Nappen	Aukland	Sveio	Sunnhordland	Light green basalt with undular structure	
12719			Hovland	Bømlo	Sunnhordland	Light green basalt with undular structure	
7416	b	Nappen	Aukland	Sveio	Sunnhordland	Weathered, undetermined rock	
7441	a	Nappen	Aukland	Sveio	Sunnhordland	Weathered, undetermined rock	
8422	c	Nappen	Aukland	Sveio	Sunnhordland	Weathered, undetermined rock	
11253	h	Nappen	Aukland	Sveio	Sunnhordland	Weathered, undetermined rock	
12968	1	Vetlehagen	Djupvik	Fusa	Sunnhordland	Weathered, undetermined rock	
12968	2	Vetlehagen	Djupvik	Fusa	Sunnhordland	Weathered, undetermined rock	
9067	b		Eide	Bømlo	Sunnhordland	Weathered, undetermined rock	
11456			Gripne	Tysnes	Sunnhordland	Weathered, undetermined rock	

***A.** Vespestad adzes at Bergen University Museum (continued)*

B no.	Unit/no.	Excavated archaeological site	Farm	Municipality	District	Raw material	Geological sample/ Museum collection
7786			Hillestveit	Bømlo	Sunnhordland	Weathered, undetermined rock	
14388			Kverno	Fitjar	Sunnhordland	Weathered, undetermined rock	
11500	a		Mehus	Kvinnherad	Sunnhordland	Weathered, undetermined rock	
7116	n		Nes	Kvinnherad	Sunnhordland	Weathered, undetermined rock	
12119			Osnes	Etne	Sunnhordland	Weathered, undetermined rock	
8329			Rubbestadneset	Bømlo	Sunnhordland	Weathered, undetermined rock	
9667	q	Røykjenes	Røykjenes	Sveio	Sunnhordland	Weathered, undetermined rock	
12335			Solheim	Sveio	Sunnhordland	Weathered, undetermined rock	
12267	a		Storavik	Fitjar	Sunnhordland	Weathered, undetermined rock	
13518		Torgjo	Torgjo	Fitjar	Sunnhordland	Weathered, undetermined rock	
8421	b		Valo	Sveio	Sunnhordland	Weathered, undetermined rock	
7117	F - 2	Sokkamyro	Vespestad	Bømlo	Sunnhordland	Weathered, undetermined rock	
1401			Opedal	Ullensvang	Hardanger-Voss	Greenstone	
4890			Tu	Klepp		Greenstone	
220			Unknown	Unknown		Greenstone	
4363			Unknown	Unknown		Greenstone	
5231			Unknown	Unknown		Greenstone	
5577			Unknown	Unknown		Greenstone	
6172			Unknown	Unknown		Grey/black stone, White small spots.	
5230			Unknown	Unknown		Grey/brown basalt	
3909			Unknown	Unknown		Weathered, undetermined rock	
5198			Unknown	Unknown		Weathered, undetermined rock	

B. Asle Bruen Olsen's catalogue (1981) of Vespestad adzes from Nordmøre and Romsdal (kept at Museum of Natural history and Archaeology, Trondheim)

T no.	Site unit/ Sub.no.	Farm	Municipality	District	Raw material
15337	c	Strøm	Averøy	Nordmøre	Greenstone
16500	a	Strøm	Averøy	Nordmøre	Greenstone
16683		Mork	Averøy	Nordmøre	Undetermined rock
18685		Breivik	Averøy	Nordmøre	Undetermined rock
11096		Tingvoll	Tingvoll	Nordmøre	Undetermined rock
16145	b	Aulset	Tingvoll	Nordmøre	Undetermined rock
11266	i	Allanhenget IV	Kristiansund	Nordmøre	Undetermined rock
11266	i	Allanhenget IV	Kristiansund	Nordmøre	Undetermined rock
11266	i	Allanhenget IV	Kristiansund	Nordmøre	Undetermined rock
11266	i	Allanhenget IV	Kristiansund	Nordmøre	Undetermined rock
11266	i	Allanhenget IV	Kristiansund	Nordmøre	Undetermined rock
11266	i	Allanhenget IV	Kristiansund	Nordmøre	Undetermined rock
11266	i	Allanhenget IV	Kristiansund	Nordmøre	Undetermined rock
11266	i	Allanhenget IV	Kristiansund	Nordmøre	Undetermined rock
11266	i	Allanhenget IV	Kristiansund	Nordmøre	Undetermined rock
11436		Ødegården	Kristiansund	Nordmøre	Undetermined rock
9469		Golma	Tustna	Nordmøre	Undetermined rock
13762		Lesund	Tustna	Nordmøre	Undetermined rock
16921		Espeset	Aure	Nordmøre	Undetermined rock
14515	e	Vågstranda	Rauma	Romsdal	Diabase
13266	c	Hogsnes	Aukra	Romsdal	Diabase
13809		Hogsnes	Aukra	Romsdal	Diabase
11903		Øvre Tornes	Fræna	Romsdal	Diabase
14515	a	Vågstranda	Rauma	Romsdal	Greenstone
13897	a	Øvre Tornes	Fræna	Romsdal	Greenstone
13897	b	Øvre Tornes	Fræna	Romsdal	Greenstone
13897	c	Øvre Tornes	Fræna	Romsdal	Greenstone
10720		Sør-Hegdal	Midsund	Romsdal	Undetermined rock
12348	a	Sør-Hegdal	Midsund	Romsdal	Undetermined rock
13799	a	Bjørerem	Midsund	Romsdal	Undetermined rock
13038		Jenset	Midsund	Romsdal	Undetermined rock
12909	a	Gjermundsnes	Vestnes	Romsdal	Undetermined rock
14458		Nerås	Vestnes	Romsdal	Undetermined rock
3986		Straume	Rauma	Romsdal	Undetermined rock
14515	f	Vågstranda	Rauma	Romsdal	Undetermined rock
12173	a	Eikrem	Aukra	Romsdal	Undetermined rock
12383	b	Eikrem	Aukra	Romsdal	Undetermined rock
12536	b	Eikrem	Aukra	Romsdal	Undetermined rock
14997	a	Eikrem	Aukra	Romsdal	Undetermined rock
13387		Haukebø	Aukra	Romsdal	Undetermined rock
13266	a	Hogsnes	Aukra	Romsdal	Undetermined rock
13266	b	Hogsnes	Aukra	Romsdal	Undetermined rock
11419		Øvre Tornes	Fræna	Romsdal	Undetermined rock
12335		Øvre Tornes	Fræna	Romsdal	Undetermined rock
13722		Vestad	Fræna	Romsdal	Undetermined rock

***C.** Sigmund Alsaker's catalogue (1982) of Vespestad adzes from Northern Rogaland and Southern Rogaland (kept at Arkeological museum in Stavanger)*

S no.	Site unit/ Sub.no.	Farm	Municipality	District	Raw material
5779		Tveit	Nedstrand	Northern Rogaland	Greenstone
6956	a	Valvik	Skjold	Northern Rogaland	Greenstone
9009		Våge	Bokn	Northern Rogaland	Greenstone
8126		Risøy	Haugesund	Northern Rogaland	Greenstone
8211		Torvestad	Torvestad	Northern Rogaland	Greenstone
7342		Mæland	Avaldsnes	Northern Rogaland	Greenstone
5661		Bergsøy	Hjelmeland	Northern Rogaland	Greenstone
8808		Meling	Finnøy	Northern Rogaland	Greenstone
8808	a	Meling	Finnøy	Northern Rogaland	Greenstone
8808	b	Meling	Finnøy	Northern Rogaland	Greenstone
6311		Utstein kloster	Rennesøy	Northern Rogaland	Greenstone
9472		Vatsendvik	Tysvær	Northern Rogaland	Undetermined rock
2209		Gard	Haugesund	Northern Rogaland	Undetermined rock
7480		Fjøløy	Rennesøy	Northern Rogaland	Undetermined rock
4038		Tjøtta	Klepp	Southern Rogaland	Diabase
B 4890		Tu	Klepp	Southern Rogaland	Greenstone
8630		Skogbakka	Høgsfjord	Southern Rogaland	Greenstone
5887		Sømme	Håland	Southern Rogaland	Greenstone
5684		Svanevågen	Egersund	Southern Rogaland	Greenstone
5747	b	Sandbakken	Egersund	Southern Rogaland	Greenstone
6114		Re	Time	Southern Rogaland	Greenstone
4571		Obrestad	Hå	Southern Rogaland	Greenstone
4281	a	Randaberg	Stavanger	Southern Rogaland	Greenstone
3592		Hellesøy	Stavanger	Southern Rogaland	Greenstone
9240		Meling	Stavanger	Southern Rogaland	Greenstone
5162		Hole, Bore	Klepp	Southern Rogaland	Greenstone
8028		Årstad	Sokndal	Southern Rogaland	Undetermined rock
4019		Møgedal	Høyland	Southern Rogaland	Undetermined rock
8445	b	Ystebø	Egersund	Southern Rogaland	Undetermined rock
1956		Mauland	Time	Southern Rogaland	Undetermined rock
9115		Ø. Haugland	Time	Southern Rogaland	Undetermined rock
1585		Sola	Håland	Southern Rogaland	Undetermined rock
8007		Nedre Horpestad	Klepp	Southern Rogaland	Undetermined rock
4312		Grude	Klepp	Southern Rogaland	Undetermined rock
5214		Grude	Klepp	Southern Rogaland	Undetermined rock

***D.** Catalogue of Vespestad adzes kept at the Museum of Cultural History, Oslo (Oldsaksamlingen)*

C no	Site unit/ Sub.no.	Municipality	District	Raw material
10220		Hole	Buskerud	Undetermined rock
24941		Mandal	Vest-Agder	Undetermined rock
10843		Mandal	Vest-Agder	Greenstone
20631	a	Sande	Vestfold	Undetermined rock

APPENDIX 6

CATALOGUE
OF THIN-BUTTED AXES
AT BERGEN UNIVERSITY MUSEUM

Catalogue of thin-butted axes from the district of Bergen University Museum

B No.	Farm	Municipality	District	Sub-type	Raw material	Lengdh mm	Width mm	Thickness mm	Edge angle	Condition	Comment
3221	Giske-Øygarden	Giske	Northern Sunnmøre	1A	Stone		60	>36			Has had convex sides
9281	Berstad	Selje	Outer Nordfjord	1A	Stone	221	35	78			Coarse form, not very marked sides
9911	Hove	Selje	Outer Nordfjord	1A	Stone	233	75	33		Many traces after the blank-form	
10031	Hovland	Fjaler	Inner Sogn	1A	Stone	115	69	53		Neck and part of the body missing	Unsecure type. Pecked body.
8304	Litle Lindås	Austrheim	Nordhordland	1A	Stone		65	33		Neck and part of body missing, a large piece has flaked off on one of the sides	
11026	Flesland	Bergen	Midthordland	3	Flint	176	78	29	10°		
11643	Brattholmen	Fjell	Midthordland	1B	Stone	60	68	26		Neck and part of the body missing	Unsecure type. The sides are only slightly marked
9227	Folkedal	Granvin	Hardanger	1A	Stone	162	68	33		Many traces after the blank-form	
12214a	Fylkesnes	Bømlo	Sunnhordland	1B	Stone	123	55	29			Unsecure type
12037	Eidet	Fitjar	Sunnhordland		Stone	121	66	36		Probably resharpened	Unsecure type
4867	Håvik	Kvinnherad	Sunnhordland	"Old type"	Flint		>50		7°	Neck and edge missing	Burned middle fragment. All four sides ground. Possible MNb because of coarse grinding.
11261	Bjelland	Stord	Sunnhordland	3	Flint	142	74	32	8°	Neck and part of the body missing	Unsecure sub-type
10694	Aukland	Sveio	Sunnhordland	1A	Stone	76	69	31		Neck and part of the body missing	
9977	Øvre Dalland	Tysnes	Sunnhordland	1A	Stone	181	62	46			Unsecure type (almost thick-butted). Convex sides, pecked body, ground neck and edge.
13201	Klungeland	Ølen	Sunnhordland	1B	Stone	11	94	24		Edge destroyed, body slightly reshaped	Unsecure type
219	Unknown			1	Flint		62		7°		Sharp neck, convex sides, resharped, transitional form
5155	Unknown			1A	Stone	160	75	34			Somewhat irregular form.
5778	Unknown			1A	Stone	209	64	38			Quite convex sides. Possibly Malmers type b - but too long

APPENDIX 7

CORRESPONDENCE ANALYSES

Matrix types/techniques (presence/absence)

Coordinates types/techniques (presence/absence)

Matrix types/techniques (values)

Coordinates types/techniques (values)

Matrix raw materials all sites (values)

Coordinates raw materials all sites (values)

Matrix raw materials coastal sites (values)

Coordinates raw materials coastal sites (values)

Matrix correspondence analysis of types and techniques (presence/absence)

	A-points	Transverse points	Single-edged points	Slate points	Early Neolithic stone adzes	Ground stone flakes	Early Neolithic flint axes/ground flint flakes	Struck blanks - short and broad	Retouched flakes and blades	Pottery (TRB)	Grindstone slabs	Small round smooth stones	Bipolar cores	Cylindrical cores	Other cores	Platform cores	Small blades
Nordøy grendahus, ph.2	1	1	0	0	0	1	0	1	1	0	0	0	1	0	1	0	1
Synnaland	0	0	0	1	1	0	0	1	1	0	0	0	1	1	1	0	1
Valderøya vest 13, ph.2	1	0	0	1	1	1	0	0	1	0	1	0	1	0	1	1	1
Valderøya vest 26, ph 1	1	0	0	1	1	1	0	0	1	0	1	0	1	0	1	1	1
9 Bustadvika	1	0	0	1	1	1	0	0	1	0	1	1	1	0	1	1	1
Holvikshaugen	1	0	0	1	1	1	0	1	1	0	1	1	1	0	1	1	1
17 Havnen, phase 3	1	0	0	1	1	1	0	0	1	0	1	1	1	2	1	1	1
1 Haukedal, phase 2	1	1	0	1	0	1	0	1	1	0	1	0	1	0	1	1	1
1 Haukedal, phase 3	0	0	0	1	1	1	0	0	1	0	1	0	1	1	1	1	1
Kleiva	1	0	0	0	1	1	0	1	1	0	1	0	1	0	1	0	1
Botnaneset VIII, ph. 3	1	0	0	0	0	0	0	1	1	0	1	0	1	1	1	1	1
Vikja I, phase 2	1	0	0	0	0	1	0	1	1	0	1	0	1	1	1	1	1
Vikja I, phase 3	1	1	0	1	1	1	0	0	1	0	1	1	1	1	1	1	1
Neset II	1	0	0	0	0	0	0	0	1	0	1	1	1	1	1	1	1
Masnesberget	0	0	0	1	1	0	0	0	1	0	1	1	1	1	1	1	1
Snekkevik 1, phase 3	1	0	0	0	0	0	0	0	1	0	0	0	1	1	1	1	1
Kotedalen phase 12	1	0	0	1	0	1	0	0	1	0	1	0	1	1	1	1	1
Kotedalen phase 13	1	0	0	1	1	1	0	0	1	1	1	1	1	1	1	1	1
Torsteinsvik 11, ph. 1	1	0	0	1	1	1	0	0	1	0	0	0	1	1	1	0	1
Flatøy VIII, phase 2	1	0	0	0	1	1	0	0	1	0	1	0	1	1	1	0	1
17 Nilsvik	1	0	0	0	1	1	0	0	1	0	1	1	1	1	1	0	1
4 Nilsvik str. 10	1	0	0	1	0	1	0	0	1	0	1	0	0	1	1	0	1
4 Nilsvik str. 29	1	0	0	1	0	1	0	0	1	0	1	1	1	1	1	0	1
4 Nilsvik str. 30	1	0	0	1	1	1	0	0	1	0	0	0	1	1	1	1	1
Austvik III	1	0	0	0	1	0	0	1	1	0	0	0	1	1	1	1	1
88 Føyno	1	0	0	0	1	1	0	1	1	0	1	1	1	1	1	1	1
115 Føyno	1	0	0	0	0	1	0	1	1	0	0	0	1	1	1	0	1
Sokkamyro 1901-2	1	0	0	1	1	0	0	0	1	0	1	0	1	1	1	0	1
3 Tjernagel	1	0	0	0	1	0	0	0	1	0	1	0	1	1	1	1	1
Styggvasshelleren, ph. 2	1	0	0	1	0	0	0	0	1	0	0	0	1	0	1	1	1
Mørkedøla I	1	1	0	1	0	0	0	0	1	0	0	0	0	0	1	1	1
Gyrinos IV	1	1	1	1	0	0	0	0	1	0	0	0	1	0	1	1	1
Blånut IV	0	0	0	1	0	0	1	0	1	0	0	0	0	0	1	0	1
760 Finnsbergvatn	1	1	1	1	0	0	0	0	1	0	0	0	1	0	1	1	1
526 Nordmannslågen	1	1	1	1	0	0	1	0	1	1	0	0	1	1	0	1	1
1020 Bjornesfjorden	1	1	1	0	1	0	1	0	1	0	1	0	1	0	1	0	1
Vivik	1	1	0	1	0	0	1	0	1	0	0	0	1	0	1	0	1

Coordinates for the first three axis of the correspondence analysis of types/techniques (presence/absence) (Appendix 10, Colour Figure 4).

Types/techniques	Axis 1	Axis 2	Axis 3
A-points	0,035	-0,125	-0,087
Transverse points	1,388	-0,565	-0,469
Single-edged points	2,210	-0,030	-1,164
Slate points	0,242	0,216	0,085
Early Neolithic stone adzes	-0,353	0,227	0,120
Ground stone flakes	-0,433	-0,034	0,048
Early Neolithic stone axes/ground stone flakes	2,415	0,498	1,946
Struck blanks - short and broad	-0,437	-1,218	0,139
Retouched flakes and blades	0,053	-0,069	0,093
Pottery (TRB)	1,368	1,925	-0,856
Grindstone slabs	-0,358	0,199	0,020
Small round smooth stones	-0,541	0,788	-0,212
Bipolar cores	-0,029	-0,081	-0,046
Cylindrical cores	-0,350	0,224	0,020
Other cores	-0,024	-0,117	0,118
Platform cores	-0,008	0,043	-0,399
Small blades	0,053	-0,069	0,093
Sites	**Axis 1**	**Axis 2**	**Axis 3**
Nordøy grendahus, ph.2	0,156	-0,865	-0,047
Synnaland	-0,217	-0,337	0,264
Valderøya vest 13, ph.2	-0,169	0,056	0,015
Valderøya vest 26, ph 1	-0,169	0,056	0,015
9 Bustadvika	-0,255	0,268	-0,051
Holvikshaugen	-0,309	-0,061	-0,007
17 Havnen, phase 3	-0,327	0,331	-0,032
1 Haukedal, phase 2	0,090	-0,503	-0,124
1 Haukedal, phase 3	-0,248	0,162	0,052
Kleiva	-0,342	-0,435	0,188
Botnaneset VIII, ph. 3	-0,244	-0,409	-0,018
Vikja I, phase 2	-0,308	-0,379	-0,000
Vikja I, phase 3	-0,051	0,147	-0,160
Neset II	-0,267	0,266	-0,150
Masnesberget	-0,271	0,412	-0,036
Snekkevik 1, phase 3	-0,079	-0,085	-0,101
Kotedalen phase 12	-0,168	0,055	-0,018
Kotedalen phase 13	-0,054	0,728	-0,261
Torsteinsvik 11, ph. 1	-0,184	0,056	0,168
Flatøy VIII, phase 2	-0,322	0,050	0,143
17 Nilsvik	-0,401	0,284	0,057
4 Nilsvik str. 10	-0,201	0,084	0,166
4 Nilsvik str. 29	-0,278	0,281	0,045
4 Nilsvik str. 30	-0,167	0,064	0,015
Austvik III	-0,242	-0,400	0,019
88 Føyno	-0,411	-0,059	-0,026
115 Føyno	-0,291	-0,565	0,160
Sokkamyro 1901-2	-0,167	0,135	0,157
3 Tjernagel	-0,224	0,077	-0,025
Styggvasshelleren, ph. 2	0,095	-0,088	-0,069
Mørkedøla I	0,512	-0,298	-0,274
Gyrinos IV	0,897	-0,270	-0,670
Blånut IV	1,129	0,277	1,587
760 Finnsbergvatn	0,897	-0,270	-0,670
526 Nordmannslågen	1,382	0,541	-0,242
1020 Bjornesfjorden	1,111	-0,041	0,212
Vivik	1,065	-0,119	0,735

Matrix correspondence analysis of types/techniques (values)

	A-points	Transverse points	Single-edged points	Slate points	Early Neolithic stone adzes	Ground stone flakes	Early Neolithic flint axes/ground flint flakes	Struck blanks - short and broad	Retouched flakes and blades	Pottery (TRB)	Grindstone slabs	Small round smooth stones	Bipolar cores	Cylindrical cores	Other cores	Platform cores	Small blades
Nordøy grendahus, ph.2	1	1	0	0	0	1	0	1	99	0	0	0	287	0	31	0	20
Synnaland	0	0	0	1	1	0	0	1	45	0	0	0	179	1	3	0	4
Valderøya vest 13, ph.2	1	0	0	2	3	1	0	0	17	0	2	0	83	0	2	3	9
Valderøya vest 26, ph 1	2	0	0	5	1	2	0	0	41	0	3	0	245	0	2	1	14
9 Bustadvika	2	0	0	1	3	8	0	0	14	0	3	1	42	0	1	4	10
Holvikshaugen	1	0	0	2	2	1	0	1	22	0	3	1	67	0	8	1	12
17 Havnen, phase 3	5	0	0	5	1	10	0	0	36	0	14	1	13	2	7	7	70
1 Haukedal, phase 2	2	1	0	4	0	5	0	1	7	0	11	0	1	0	5	2	6
1 Haukedal, phase 3	0	0	0	1	1	4	0	0	2	0	5	0	3	1	2	3	7
Kleiva	3	0	0	0	1	1	0	9	17	0	3	0	5	0	13	0	6
Botnaneset VIII, ph. 3	2	0	0	0	0	0	0	1	16	0	1	0	4	2	9	31	40
Vikja I, phase 2	3	0	0	0	0	2	0	2	45	0	14	0	21	11	30	19	92
Vikja I, phase 3	3	1	0	18	11	2	0	0	50	0	8	1	24	1	33	14	49
Neset II	8	0	0	0	0	0	0	0	38	0	7	1	8	1	54	1	119
Masnesberget	0	0	0	3	5	0	0	0	25	0	21	1	4	4	14	9	28
Snekkevik 1, phase 3	3	0	0	0	0	0	0	0	21	0	0	0	4	3	3	8	20
Kotedalen phase 12	43	0	0	5	0	24	0	0	119	0	6	0	11	14	8	24	297
Kotedalen phase 13	56	0	0	22	3	54	0	0	284	22	10	1	26	5	3	20	319
Torsteinsvik 11, ph. 1	5	0	0	1	1	2	0	0	48	0	0	0	19	3	2	0	65
Flatøy VIII, phase 2	8	0	0	0	1	1	0	0	28	0	1	0	6	5	3	0	96
17 Nilsvik	29	0	0	0	3	6	0	0	118	0	7	2	21	2	6	0	75
4 Nilsvik str. 10	6	0	0	3	0	1	0	0	43	0	10	0	0	2	1	0	27
4 Nilsvik str. 29	14	0	0	5	0	5	0	0	105	0	3	2	2	5	2	0	39
4 Nilsvik str. 30	16	0	0	1	1	7	0	0	80	0	0	0	27	4	8	1	30
Austvik III	24	0	0	0	2	0	0	1	94	0	0	0	24	17	4	2	217
88 Føyno	15	0	0	0	1	2	0	3	122	0	2	3	9	6	8	1	76
115 Føyno	10	0	0	0	0	1	0	4	73	0	0	0	15	3	5	0	72
Sokkamyro 1901-2	3	0	0	5	9	0	0	0	182	0	82	0	22	60	46	0	130
3 Tjernagel	9	0	0	0	1	0	0	0	59	0	1	0	12	2	5	5	79
Styggvasshelleren, ph. 2	5	0	0	42	0	0	0	0	23	0	0	0	126	0	8	1	20
Mørkedøla I	2	1	0	1	0	0	0	0	134	0	0	0	0	0	23	11	19
Gyrinos IV	11	11	11	1	0	0	0	0	124	0	0	0	6	0	17	10	69
Blånut IV	0	0	0	22	0	0	5	0	32	0	0	0	0	0	31	0	1
760 Finnsbergvatn	12	1	13	5	0	0	0	0	48	0	0	0	66	0	7	8	20
526 Nordmannslågen	23	1	14	1	0	0	2	0	89	44	0	0	22	2	0	4	16
1020 Bjornesfjorden	4	12	1	0	1	0	3	0	106	0	3	0	8	0	14	0	9
Vivik	4	3	0	3	0	0	4	0	37	0	0	0	43	0	75	0	23

Coordinates for the first three axis of the correspondence analysis of types/techniques (values) (Appendix 10, Colour Figure 5)			
Types/techniques	**Axis 1**	**Axis 2**	**Axis 3**
A-points	-0,403	0,421	-0,220
Transverse points	-0,213	0,457	1,369
Single-edged points	0,020	2,344	0,853
Slate points	0,414	-0,318	0,860
Early Neolithic stone adzes	-0,004	-0,510	0,186
Ground stone flakes	-0,398	0,067	-0,490
Early Neolithic stone axes/ground stone flakes	0,037	0,142	3,062
Struck blanks - short and broad	-0,077	-0,534	0,532
Retouched flakes and blades	-0,207	0,154	0,123
Pottery (TRB)	-0,391	3,523	0,752
Grindstone slabs	-0,43	-0,738	0,297
Small round smooth stones	-0,252	-0,147	-0,063
Bipolar cores	1,351	0,004	-0,178
Cylindrical cores	-0,562	-0,455	-0,108
Other cores	-0,024	-0,567	0,992
Platform cores	-0,382	-0,150	-0,068
Small blades	-0,467	-0,11	-0,341
Sites	Axis 1	Axis 2	Axis 3
Nordøy grendahus, ph.2	1,227	-0,013	-0,078
Synnaland	1,490	0,037	-0,247
Valderøya vest 13, ph.2	1,265	-0,050	-0,240
Valderøya vest 26, ph 1	1,514	0,01	-0,296
9 Bustadvika	0,725	-0,070	-0,297
Holvikshaugen	0,982	-0,119	-0,055
17 Havnen, phase 3	-0,327	-0,238	-0,189
1 Haukedal, phase 2	-0,336	-0,559	0,510
1 Haukedal, phase 3	-0,247	-0,531	-0,037
Kleiva	-0,091	-0,456	0,724
Botnaneset VIII, ph. 3	-0,446	-0,280	-0,127
Vikja I, phase 2	-0,294	-0,363	0,013
Vikja I, phase 3	-0,043	-0,365	0,424
Neset II	-0,394	-0,389	0,172
Masnesberget	-0,359	-0,591	0,337
Snekkevik 1, phase 3	-0,351	-0,071	-0,141
Kotedalen phase 12	-0,533	-0,062	-0,450
Kotedalen phase 13	-0,426	0,251	-0,234
Torsteinsvik 11, ph. 1	-0,195	-0,011	-0,326
Flatøy VIII, phase 2	-0,505	-0,122	-0,499
17 Nilsvik	-0,282	0,097	-0,147
4 Nilsvik str. 10	-0,466	-0,086	0,016
4 Nilsvik str. 29	-0,396	0,134	0,003
4 Nilsvik str. 30	-0,050	0,118	-0,052
Austvik III	-0,429	-0,064	-0,449
88 Føyno	-0,375	0,051	-0,074
115 Føyno	-0,291	0,017	-0,193
Sokkamyro 1901-2	-0,389	-0,444	0,199
3 Tjernagel	-0,352	-0,019	-0,273
Styggvasshelleren, ph. 2	1,154	-0,141	0,183
Mørkedøla I	-0,334	0,058	0,444
Gyrinos IV	-0,352	0,313	0,286
Blånut IV	0,023	-0,478	1,888
760 Finnsbergvatn	0,539	0,441	0,109
526 Nordmannslågen	-0,172	2,186	0,568
1020 Bjornesfjorden	-0,198	0,209	0,764
Vivik	0,292	-0,434	1,063

Matrix correspondence analysis raw material values at all sites

	Basaltic rock	Greenstone	Diabase	Flint	Fine green quartzite 1	Fine blue quartzite 1	Fine blue quartzite 7	Fine blue quartzite 8	Fine grey quartzite 2	Fine grey quartzite 3	Fine grey quartzite 11	Fine white quartzite 16	Fine brown quartzite 1	Fine black quartzite 1	Medium grey quartzite 1	Medium grey quartzite 9	Medium grey quartzite 10 and 11	Coarse quartzite	Quartz crystal	Fine white quartz	Medium and coarse quartz	Blue mylonite	Medium and coarse grey mylonite	Chert	Rhyolite	Brown slate	Green slate	Grey slate	Red slate	Blue slate 1	Blue slate 2	Anorthosite	Sandstone	Pumice with grooves
Nordøy grendahus, ph.2	4	0	0	2046	1	0	1	0	0	0	0	0	0	0	0	0	0	1	31	44	184	2	0	0	0	0	2	0	0	0	0	0	3	37
Synnaland	2	0	0	1063	0	0	0	0	0	0	0	0	0	0	0	0	0	0	8	3	95	5	0	0	0	0	2	1	0	1	0	0	0	0
Valderøya vest 13, ph.2	7	0	0	624	0	0	0	0	0	0	0	2	0	0	0	0	0	0	5	1	3	1	0	0	0	1	1	0	1	0	0	0	3	0
Valderøya vest 26, ph 1	7	0	0	2253	0	0	0	0	0	0	0	0	0	0	0	2	0	2	12	16	63	1	0	1	0	0	30	1	0	0	0	0	3	1
9 Bustadvika	15	4	0	609	0	0	0	0	0	0	0	0	0	0	0	0	0	0	442	343	25	31	0	1	2	0	0	120	0	0	0	0	17	210
Holvikshaugen	3	1	1	237	0	0	1	0	0	0	0	0	0	0	0	1	0	0	30	38	170	0	0	1	7	0	6	0	0	0	0	0	0	0
17 Havnen, phase 3	2	5	0	253	2	0	1	0	370	0	0	2	0	1	0	1	0	14	17	166	217	90	27	39	66	29	42	49	83	0	0	6	18	6
1 Haukedal, phase 2	9	1	7	37	0	0	0	0	2	0	0	0	0	1	0	5	0	2	4	29	175	19	0	27	7	6	64	23	1	0	0	3	18	0
1 Haukedal, phase 3	17	3	4	33	0	0	0	0	2	0	0	0	0	0	0	0	0	7	2	14	115	39	1	271	2	1	37	8	0	0	0	0	5	7
Kleiva	2	2	10	24	0	0	1	0	0	0	0	0	0	0	0	11	0	18	4	8	52	386	188	0	18	0	0	10	0	0	0	0	5	1
Botnaneset VIII, ph. 3	2	0	6	246	0	0	0	0	0	0	0	0	0	0	231	0	0	12	0	0	2	1	287	1280	0	0	0	0	0	0	0	0	3	0
Vikja I, phase 2	10	0	0	158	1	0	1	0	0	0	0	1	0	0	273	0	0	4	33	132	165	453	164	7	7	0	13	1	31	0	0	0	25	1
Vikja I, phase 3	0	0	21	217	1	0	0	0	0	0	0	11	0	0	679	0	0	5	3	141	82	330	23	4	5	0	145	0	126	0	0	0	20	11
Neset II	0	0	0	503	0	0	9	0	0	0	0	0	0	0	1376	0	0	0	72	240	212	1766	107	79	224	0	0	9	6	0	0	0	41	5
Masnesberget	18	7	13	146	0	0	0	0	0	0	0	0	0	155	19	0	0	1	2	0	0	0	3	0	32	0	8	0	0	0	0	0	24	0
Snekkevik 1, phase 3	0	0	0	174	0	0	0	0	1	0	0	0	0	0	0	0	0	0	2	12	7	1	0	0	161	0	23	0	0	0	0	2	0	0
Kotedalen phase 12	19	6	1	317	0	3	0	0	4	0	0	1	0	3	0	4	0	5	17	33	134	702	0	2	2649	3	49	0	6	0	0	30	3	10
Kotedalen phase 13	52	10	6	1008	3	17	0	0	0	0	0	18	0	6	0	14	0	9	11	113	417	839	17	17	2923	180	1292	0	17	0	0	178	10	43
Torsteinsvik 11, ph. 1	3	0	0	103	0	0	0	0	0	0	0	0	0	0	0	2	0	2	0	1	5	5	0	0	548	0	5	2	0	0	0	1	1	1
Flatøy VIII, phase 2	6	2	0	229	0	3	0	0	0	0	0	10	0	0	0	3	0	1	3	33	14	12	0	7	279	0	0	0	0	0	0	0	0	1
17 Nilsvik	14	4	0	400	0	0	0	211	0	0	0	0	0	0	0	1	54	0	3	73	302	0	0	0	1291	0	11	0	0	0	0	0	7	35
4 Nilsvik str. 10	0	3	0	157	0	0	0	0	0	0	0	0	0	0	0	0	26	0	2	38	114	0	0	0	460	0	8	0	0	0	0	0	1	1
4 Nilsvik str. 29	0	6	0	220	0	1	0	0	0	0	0	0	0	1	0	0	30	0	1	20	150	1	0	0	1093	0	11	0	0	0	0	0	3	2
4 Nilsvik str. 30	0	12	2	230	0	0	0	0	0	0	0	0	0	1	0	0	106	0	1	53	339	1	0	0	401	0	18	0	0	0	0	1	4	14
Austvik III	1	15	0	516	0	0	0	10	0	0	3	10	0	0	0	0	160	1	14	63	25	0	0	0	1711	0	0	0	0	0	0	0	0	4
88 Føyno	3	14	0	307	0	0	0	0	0	0	0	0	0	1	0	0	2	1	0	8	0	0	0	0	630	0	0	0	0	0	0	0	2	24
115 Føyno	4	15	0	575	0	0	0	0	0	0	0	0	0	0	0	0	0	27	2	68	7	0	0	0	616	0	0	0	0	0	0	0	0	26
Sokkamyro 1901-2	1	21	0	808	0	0	3	0	0	0	35	0	0	0	0	0	0	12	10	9	22	0	0	0	631	0	5	1	0	1	0	0	1	0
3 Tjernagel	72	0	0	1025	0	0	0	583	0	0	0	0	0	0	0	0	0	0	1	9	387	0	0	0	123	0	0	0	0	0	0	0	3	0
Styggvasshelleren, ph. 2	0	0	0	103	3	23	0	0	3	0	0	3	0	10	0	50	0	0	93	1055	255	0	0	0	5	0	45	3	0	2	0	5	0	0
Mørkedøla I	0	0	0	35	1816	0	0	0	0	0	0	0	69	1	0	0	0	0	2	0	0	0	0	0	0	0	1	0	2	0	0	0	0	0
Gyrinos IV	0	0	0	613	733	0	0	0	0	0	0	0	1010	0	0	0	0	0	0	31	0	0	0	0	0	0	0	6	0	2	0	0	0	0
Blånut IV	0	0	0	78	454	0	0	0	0	0	0	0	0	0	0	0	0	0	1	14	0	0	0	0	0	0	42	8	0	18	0	0	0	0
760 Finnsbergvatn	0	0	0	677	4	0	0	0	0	2197	0	0	0	0	0	0	0	0	0	3274	8	0	0	0	4	0	0	3	0	0	134	0	0	0
526 Nordmannslågen	0	0	0	764	0	0	0	0	0	18	0	0	0	0	0	0	0	1	305	371	8	0	0	0	0	0	0	1	0	0	0	0	0	0
1020 Bjornesfjorden	1	0	0	985	0	0	0	0	0	0	0	5	0	0	0	0	0	1	15	10	0	0	0	0	2	0	0	0	0	0	0	0	3	0
Vivik	0	0	0	448	0	0	0	0	0	0	0	0	0	0	0	0	0	0	1306	4	207	0	0	0	0	0	0	160	0	52	0	0	0	0

Coordinates on the three first axis of the correspondence analysis raw material values all sites (Appendix 10, Colour Figure 6).			
Raw materials	**Axis 1**	**Axis 2**	**Axis 3**
Basaltic rock	-0,307	0,107	-0,155
Greenstone	-0,421	0,162	-0,361
Diabase	-0,508	-0,152	0,598
Flint	0,230	0,035	-0,139
Fine, green quartzite 1	1,312	-0,528	0,310
Fine, blue quartzite 1	-0,189	-0,234	0,079
Fine, blue quartzite 7	-0,140	-0,057	0,235
Fine, blue quartzite 8	-0,506	0,972	-2,172
Fine, grey quartzite 2	-0,097	-0,347	0,452
Fine, grey quartzite 3	3,046	3,737	1,460
Fine, grey quartzite 11	-0,115	0,648	-2,030
Fine, white quartzite 16	-0,229	-0,128	0,147
Fine, brown quartzite 1	3,392	-3,418	-0,803
Fine, black quartzite 1	-0,048	-0,520	0,065
Medium, grey quartzite 1	-0,450	-0,214	1,091
Medium, grey quartzite 9	-0,261	-0,071	0,11
Medium, grey quartzite 10 and 11	-0,544	0,533	-1,438
Coarse quartzite	-0,237	0,174	0,122
Quartz crystal	0,056	-0,017	-0,194
Fine, white quartz	0,228	0,146	-0,218
Medium and coarse quartz	0,040	0,274	-0,057
Blue mylonite	-0,265	-0,115	0,332
Medium and coarse grey mylonite	-0,414	-0,199	0,868
Chert	-0,378	-0,141	0,624
Rhyolite	-0,147	0,281	-0,128
Brown slate	-0,380	-0,347	0,633
Green slate	0,035	-0,279	-0,049
Grey slate	0,871	-0,007	0,049
Red slate	-0,009	-0,688	0,687
Blue slate 1	1,798	-0,955	-1,002
Blue slate 2	4,056	4,758	2,184
Anorthosite	-0,156	-0,280	0,286
Sandstone	-0,350	0,051	-0,012
Pumice	-0,369	0,133	-0,170
Sites	Axis 1	Axis 2	Axis 3
Nordøy grendahus, ph.2	0,035	-0,012	0,000
Synnaland	0,548	-0,188	-0,358
Valderøya vest 13, ph.2	-0,158	-0,182	0,198
Valderøya vest 26, ph 1	-0,099	0,047	0,041
9 Bustadvika	-0,124	0,159	-0,081
Holvikshaugen	-0,240	0,050	0,049
17 Havnen, phase 3	-0,11	-0,206	0,321
1 Haukedal, phase 2	-0,193	-0,181	0,297
1 Haukedal, phase 3	-0,246	-0,027	0,273
Kleiva	-0,228	0,097	0,123
Botnaneset VIII, ph. 3	-0,484	-0,037	0,735
Vikja I, phase 2	-0,050	-0,129	0,395
Vikja I, phase 3	-0,157	-0,168	0,503
Neset II	-0,144	-0,077	0,476
Masnesberget	-0,392	-0,099	0,319
Sites	Axis 1	Axis 2	Axis 3
Snekkevik 1, phase 3	-0,015	-0,070	0,070
Kotedalen phase 12	-0,314	-0,186	0,270
Kotedalen phase 13	-0,221	-0,212	0,329
Torsteinsvik 11, ph. 1	-0,097	0,072	-0,005
Flatøy VIII, phase 2	-0,295	0,090	-0,001
17 Nilsvik	-0,327	0,375	-0,862
4 Nilsvik str. 10	-0,227	0,276	-0,622

Coordinates on the three first axis of the correspondence analysis raw material values all sites (Appendix 10, Colour Figure 6) (continued).

4 Nilsvik str. 29	-0,246	0,072	-0,396
4 Nilsvik str. 30	-0,290	0,038	-0,238
Austvik III	-0,327	0,535	-1,174
88 Føyno	-0,361	0,231	-0,547
115 Føyno	-0,189	0,302	-0,325
Sokkamyro 1901-2	0,201	0,084	-0,632
3 Tjernagel	-0,173	0,484	-0,864
Styggvasshelleren, ph. 2	0,559	-0,325	-0,025
Mørkedøla I	1,302	-1,619	-0,04
Gyrinos IV	2,395	-1,648	-0,675
Blånut IV	1,188	-0,480	-0,399
760 Finnsbergvatn	2,210	2,274	0,972
526 Nordmannslågen	1,110	1,298	0,328
1020 Bjornesfjorden	-0,173	0,170	-0,162
Vivik	0,986	-0,182	-0,585

Matrix correspondence analysis raw material values at coastal sites	Basaltic rock	Greenstone	Diabase	Flint	Fine green quartzite 1	Fine blue quartzite 1	Fine blue quartzite 8	Fine grey quartzite 2	Fine grey quartzite 11	Fine white quartzite 16	Fine black quartzite 1	Medium grey quartzite 1	Medium grey quartzite 9	Medium grey quartzite 10 and 11	Coarse quartzite	Quartz crystal	Fine white quartz	Medium and coarse quartz	Blue mylonite	Medium and coarse grey mylonite	Chert	Rhyolite	Brown slate	Green slate	Grey slate	Red slate	Blue slate 1	Anorthosite	Sandstone	Pumice with grooves
Nordøy grendahus, ph.2	4	0	0	2046	1	0	0	0	0	0	0	0	0	0	1	31	44	184	2	0	0	0	0	2	0	0	0	0	3	37
Synnaland	2	0	0	1063	0	0	0	0	0	0	0	0	0	0	0	8	3	95	5	0	0	0	0	2	1	0	1	0	0	0
Valderøya vest 13, ph.2	7	0	0	624	0	0	0	0	0	2	0	0	0	0	0	5	1	3	1	0	0	0	1	1	0	1	0	0	3	0
Valderøya vest 26, ph 1	7	0	0	2253	0	0	0	0	0	0	0	0	2	0	2	12	16	63	1	0	1	0	0	30	1	0	0	0	3	1
9 Bustadvika	15	4	0	609	0	0	0	0	0	0	0	0	0	0	0	442	343	25	31	0	1	2	0	0	120	0	0	0	17	210
Holvikshaugen	3	1	1	237	0	0	0	0	0	0	0	0	1	0	0	30	38	170	0	0	1	7	0	6	0	0	0	0	0	0
17 Havnen, phase 3	2	5	0	253	2	0	0	370	0	2	1	0	1	0	14	17	166	217	90	27	39	66	29	42	49	83	0	6	18	6
1 Haukedal, phase 2	9	1	7	37	0	0	0	2	0	0	1	0	5	0	2	4	29	175	19	0	27	7	6	64	23	1	0	3	18	0
1 Haukedal, phase 3	17	3	4	33	0	0	0	2	0	0	0	0	0	0	7	2	14	115	39	1	271	2	1	37	8	0	0	0	5	7
Kleiva	2	2	10	24	0	0	0	0	0	0	0	0	11	0	18	4	8	52	386	188	0	18	0	0	10	0	0	0	5	1
Botnaneset VIII, ph. 3	2	0	6	246	0	0	0	0	0	0	0	231	0	0	12	0	0	2	1	287	1280	0	0	0	0	0	0	0	3	0
Vikja I, phase 2	10	0	0	158	1	0	0	0	0	1	0	273	0	0	4	33	132	165	453	164	7	7	0	13	1	31	0	0	25	1
Vikja I, phase 3	0	0	21	217	1	0	0	0	0	11	0	679	0	0	5	3	141	82	330	23	4	5	0	145	0	126	0	0	20	11
Neset II	0	0	0	503	0	0	0	0	0	0	0	1376	0	0	0	72	240	212	1766	107	79	224	0	0	9	6	0	0	41	5
Masnesberget	18	7	13	146	0	0	0	0	0	0	155	19	0	0	1	2	0	0	0	3	0	32	0	8	0	0	0	0	24	0
Snekkevik 1, phase 3	0	0	0	174	0	0	0	1	0	0	0	0	0	0	0	2	12	7	1	0	0	161	0	23	0	0	0	2	0	0
Kotedalen phase 12	19	6	1	317	0	3	0	4	0	1	3	0	4	0	5	17	33	134	702	0	2	2649	3	49	0	6	0	30	3	10
Kotedalen phase 13	52	10	6	1008	3	17	0	0	0	18	6	0	14	0	9	11	113	417	839	17	17	2923	180	1292	0	17	0	178	10	43
Torsteinsvik 11, ph. 1	3	0	0	103	0	0	0	0	0	0	0	0	2	0	2	0	1	5	5	0	0	548	0	5	2	0	0	1	1	1
Flatøy VIII, phase 2	6	2	0	229	0	3	0	0	0	10	0	0	3	0	1	3	33	14	12	0	7	279	0	0	0	0	0	0	0	1
17 Nilsvik	14	4	0	400	0	0	211	0	0	0	0	0	1	54	0	3	73	302	0	0	0	1291	0	11	0	0	0	0	7	35
4 Nilsvik str. 10	0	3	0	157	0	0	0	0	0	0	0	0	0	26	0	2	38	114	0	0	0	460	0	8	0	0	0	0	1	1
4 Nilsvik str. 29	0	6	0	220	0	1	0	0	0	0	1	0	0	30	0	1	20	150	1	0	0	1093	0	11	0	0	0	0	3	2
4 Nilsvik str. 30	0	12	2	230	0	0	0	0	0	0	1	0	0	106	0	1	53	339	1	0	0	401	0	18	0	0	0	1	4	14
Austvik III	1	15	0	516	0	0	10	0	3	10	0	0	0	160	1	14	63	25	0	0	0	1711	0	0	0	0	0	0	0	4
88 Føyno	3	14	0	307	0	0	0	0	0	0	1	0	0	2	1	0	8	0	0	0	0	630	0	0	0	0	0	0	2	24
115 Føyno	4	15	0	575	0	0	0	0	0	0	0	0	0	0	27	2	68	7	0	0	0	616	0	0	0	0	0	0	0	26
Sokkamyro 1901-2	1	21	0	808	0	0	0	0	35	0	0	0	0	0	12	10	9	22	0	0	0	631	0	5	1	0	1	0	1	0
3 Tjernagel	72	0	0	1025	0	0	583	0	0	0	0	0	0	0	0	1	9	387	0	0	0	123	0	0	0	0	0	0	3	0

Coordinates on the three first axis of the correspondence analysis raw material values coastal sites (Appendix 10, Colour Figure 7).			
Raw materials	Axis 1	Axis 2	Axis 3
Basaltic rock	-0,135	0,413	0,314
Greenstone	-0,415	0,259	-0,237
Diabase	0,828	-0,373	0,464
Flint	-0,286	0,489	0,571
Fine green quartzite 1	0,108	-0,507	0,289
Fine blue quartzite 1	-0,413	-0,235	-0,821
Fine blue quartzite 8	-0,562	0,863	0,527
Fine grey quartzite 2	0,290	-0,343	1,079
Fine grey quartzite 11	-0,612	0,567	-0,109
Fine white quartzite 16	-0,161	-0,379	-0,300
Fine black quartzite 1	-0,182	0,546	2,196
Medium grey quartzite 1	1,164	-1,401	0,419
Medium grey quartzite 9	0,066	-0,417	-0,323
Medium grey quartzite 10 and 11	-0,644	0,246	-0,793
Coarse quartzite	0,510	0,090	-0,100
Quartz crystal	-0,012	-0,093	1,428
Fine white quartz	0,115	-0,430	0,563
Medium and coarse quartz	-0,093	0,051	0,110
Blue mylonite	0,487	-1,176	-0,007
Medium and coarse grey mylonite	1,929	-0,155	-0,149
Chert	3,200	1,787	-0,860
Rhyolite	-0,521	-0,001	-0,778
Brown slate	-0,258	-0,312	-0,546
Green slate	-0,175	-0,348	-0,547
Grey slate	0,19	-0,087	1,362
Red slate	0,598	-1,197	0,671
Blue slate 1	-0,512	0,786	0,639
Anorthosite	-0,363	-0,325	-0,806
Sandstone	0,408	-0,524	0,679
Pumice with grooves	-0,215	0,125	0,888
Sites	Axis 1	Axis 2	Axis 3
Nordøy grendahus ph. 2	-0,325	0,627	0,857
Synnaland	-0,334	0,654	0,837
Valderøya vest 13, ph. 2	-0,348	0,694	0,884
Valderøya vest 26, ph. 1	-0,343	0,678	0,854
9 Bustadvika	-0,100	0,073	1,320
Holsvikhaugen	-0,207	0,319	0,661
17 Havnen, phase 3	0,229	-0,240	0,718
1 Haukedal ph. 2	0,209	0,005	0,162
1 Haukedal ph. 3	1,931	1,168	-0,508
Kleiva	0,943	-0,972	0,004
Botnaneset VII, ph. 3	2,968	1,468	-0,68
Vikja I, ph. 2	0,730	-0,962	0,359
Vikja I, ph. 3	0,711	-1,235	0,447
Neset II	0,732	-1,210	0,303
Masnesberget	-0,131	0,411	1,572
Snekkevik 1, ph. 3	-0,452	0,271	-0,116
Kotedalen ph. 12	-0,364	-0,264	-0,724
Kotedalen ph. 13	-0,296	-0,223	-0,538
Torsteinsvik 11, ph. 1	-0,583	0,091	-0,841
Flatøy 8, ph. 2	-0,386	0,231	-0,187
17 Nilsvik	-0,510	0,233	-0,386
4 Nilsvik str. 10	-0,484	0,127	-0,492
4 Nilsvik str. 29	-0,548	0,104	-0,732
4 Nilsvik str. 30	-0,402	0,160	-0,252
Austvik III	-0,576	0,161	-0,680
88 Føyno	-0,544	0,231	-0,456
115 Føyno	-0,450	0,292	-0,106
Sokkamyro 1902-3	-0,476	0,398	-0,017
3 Tjernagel	-0,418	0,710	0,617

APPENDIX 8

SAMPLES FOR GEOLOGICAL ANALYSES

Test-samples for geological analysis							
Sample	**B-no.**	**Int. no.**	**Artefact type**	**Raw material**	**Site**	**District**	**County**
1	12782		Flake	Chert	Flatøy VIII	Midthordland	Hordaland
2	12782		Flake	Chert	Flatøy VIII	Midthordland	Hordaland
3	14200		Flake	Rhyolite	Torsteinsvik 11	Midthordland	Hordaland
4	14200		Flake	Rhyolite	Torsteinsvik 11	Midthordland	Hordaland
5	14571	140	Projectile point	Brown slate	4 Nilsvik, str 10	Midthordland	Hordaland
6	14571	10308	Flake	Grey slate	4 Nilsvik, str 29	Midthordland	Hordaland
7	14571	650	Grinding slab	Sandstone	4 Nilsvik, str 29	Midthordland	Hordaland
8	14571	616	Grinding slab	Gneiss	4 Nilsvik, str 10	Midthordland	Hordaland
9	14571	8988	Grinding slab	Gneiss	4 Nilsvik, str 30	Midthordland	Hordaland
10	14571	5105	Flake	Rhyolite	4 Nilsvik, str 10	Midthordland	Hordaland
11	14403		Grinding slab	Sandstone	Valderøya Vest 13	N. Sunnmøre	Møre og Romsd.
12	14415		Flake	Chert	Valderøya Vest 26	N. Sunnmøre	Møre og Romsd.
13	14415		Grinding slab	Sandstone	Valderøya Vest 26	N. Sunnmøre	Møre og Romsd.
14	14415		Grinding slab	Green slate	Valderøya Vest 26	N. Sunnmøre	Møre og Romsd.
15	14415		Projectile point	Green slate	Valderøya Vest 26	N. Sunnmøre	Møre og Romsd.
16	12827		Flake	Rhyolite	Holvikshaugen	S. Sunnmøre	Møre og Romsd.
17	12827		Flake	Chert	Holvikshaugen	S. Sunnmøre	Møre og Romsd.
18	12827		Grinding slab	Sandstone	Holvikshaugen	S. Sunnmøre	Møre og Romsd.
19	12827		Slate point	Green slate	Holvikshaugen	S. Sunnmøre	Møre og Romsd.
20	12827		Flake	Green slate	Holvikshaugen	S. Sunnmøre	Møre og Romsd.
21	15241	277	Grinding slab	Sandstone	88 Føyno	Sunnhordland	Hordaland
22	14849		Blade	Rhyolite	Styggvasshelleren	Breheimen	Sogn og Fj.
23	14849		Projectile point	Green slate	Styggvasshelleren	Breheimen	Sogn og Fj.
24	14849		Projectile point	Green slate	Styggvasshelleren	Breheimen	Sogn og Fj.
25	15747	914	Blade	Chert	9 Bustadvika	S. Sunnmøre	Møre og Romsd.
26	15747	314	Core	Rhyolite	9 Bustadvika	S. Sunnmøre	Møre og Romsd.
27	14157		Flake	Chert	Botnaneset VIII	Sunnfjord	Sogn og Fj.
28	14157		Ret. flake	Chert	Botnaneset VIII	Sunnfjord	Sogn og Fj.
29	14500	91450	Grinding slab	Sandstone	Kotedalen	Nordhordland	Hordaland
30	14500	12707	Grinding slab	Sandstone	Kotedalen	Nordhordland	Hordaland
31	14500	16513	Grinding slab	Sandstone	Kotedalen	Nordhordland	Hordaland
32	14500	16510	Point blank	Green slate	Kotedalen	Nordhordland	Hordaland
33	14500	20414	Point blank	Green slate	Kotedalen	Nordhordland	Hordaland
34	14500	21404	Core	Blue mylonite	Kotedalen	Nordhordland	Hordaland
35	14500	24117	Ret. flake	Blue mylonite	Kotedalen	Nordhordland	Hordaland
36	14500	23835	Ret. flake	Blue mylonite	Kotedalen	Nordhordland	Hordaland
37	14500	1534	Core	Rhyolite	Kotedalen	Nordhordland	Hordaland
38	14500	16012	Ret. flake	Rhyolite	Kotedalen	Nordhordland	Hordaland
39	14604	27131	Flake	Chert	17 Havnen	Nordfjord	Sogn og Fj.
40	14604	56121	Flake	Chert	17 Havnen	Nordfjord	Sogn og Fj.
41	14604	27137	Flake	Rhyolite	17 Havnen	Nordfjord	Sogn og Fj.
42	14604	31321	Flake	Rhyolite	17 Havnen	Nordfjord	Sogn og Fj.
43	14604	3983	Flake	Chert	1 Haukedal	Nordfjord	Sogn og Fj.
44	5823		Vespestad adze	Greenstone	Stray find	S. Sunnmøre	Møre og Romsd.
45	7172		Vespestad adze	Greenstone	Stray find	S. Sunnmøre	Møre og Romsd.
46	8508		Vespestad adze	Greenstone	Stray find	S. Sunnmøre	Møre og Romsd.
47	9143		Vespestad adze	Greenstone	Stray find	S. Sunnmøre	Møre og Romsd.
48	11124		Vespestad adze	Greenstone	Stray find	Nordfjord	Sogn og Fj.
49	14604	35000	Vespestad adze	Greenstone	17 Havnen	Nordfjord	Sogn og Fj.
50	14604	897	Vespestad adze	Greenstone	17 Havnen	Nordfjord	Sogn og Fj.
51	14588		Vespestad adze	Greenstone	1 Haukedal	Nordfjord	Sogn og Fj.
52	11716		Point blank	Green slate	Mørkedøla 1	Inner Sogn	Sogn og Fj.
53	11257		Vespestad adze	Greenstone	Stray find	Nordhordland	Hordaland
54	8227		Vespestad adze	Greenstone	Stray find	Outer Sogn	Sogn og Fj.
55	12129		Vespestad adze	Greenstone	Stray find	Sunnhordland	Hordaland
56	11805		Vespestad adze	Greenstone	Stray find	Sunnhordland	Hordaland
57	Aks.		Flake	Greenstone	Hespriholmen	Sunnhordland	Hordaland
58	Aks.		Adze blank	Greenstone	Hespriholmen	Sunnhordland	Hordaland

Test-samples for geological analysis (continued)

59	Aks.		Adze blank	Greenstone	Hespriholmen	Sunnhordland	Hordaland
60	9311		Flake	Greenstone	Hespriholmen	Sunnhordland	Hordaland
Sample	B-no.	Int. no.	Artefact type	Raw material	Site	District	County
61	14428		Adze blank	Greenstone	Hespriholmen	Sunnhordland	Hordaland
62	14588	3434	Projectile point	Anorthosite	1 Haukedal	Nordfjord	Sogn og Fj.
63	14604	27224	Point blank	Anorthosite	17 Havnen	Nordfjord	Sogn og Fj.
64	14849		Point blank	Anorthosite	Styggvasshelleren	Breheimen	Sogn og Fj.
65	14849		Projectile point	Anorthosite	Styggvasshelleren	Breheimen	Sogn og Fj.
66	14189	70	Point blank	Anorthosite	Snekkevik 1	Nordhordland	Hordaland
67	14500	17513	Projectile point	Anorthosite	Kotedalen	Nordhordland	Hordaland
68	14500	23604	Point blank	Anorthosite	Kotedalen	Nordhordland	Hordaland
69	14500	2957	Point blank	Green slate	Kotedalen	Nordhordland	Hordaland
70	14500	2526	Point blank	Green slate	Kotedalen	Nordhordland	Hordaland
71	13075		Point blank	Green slate	Vikja I	Sunnfjord	Sogn og Fj.
72	13075		Point blank	Red slate	Vikja I	Sunnfjord	Sogn og Fj.
73	14588	3289	Point blank	Green slate	1 Haukedal	Nordfjord	Sogn og Fj.
74	14588	4165	Point blank	Green slate	1 Haukedal	Nordfjord	Sogn og Fj.
75	14604	28251	Point blank	Green slate	17 Havnen	Nordfjord	Sogn og Fj.
76	14604	56530	Point blank	Green slate	17 Havnen	Nordfjord	Sogn og Fj.
77	14849		Point blank	Green slate	Styggvasshelleren	Breheimen	Sogn og Fj.
78	12263	5	Point blank	Blue slate 2	760 Finnsbergvatn	Hardangervidda	Hordaland
79	12263	1	Point blank	Blue slate 2	760 Finnsbergvatn	Hardangervidda	Hordaland
80	12263	3	Piercer	Blue slate 2	760 Finnsbergvatn	Hardangervidda	Hordaland
81	13075		Grinding slab	Sandstone	Vikja I	Sunnfjord	Sogn og Fj.
82	14604	27502	Grinding slab	Sandstone	17 Havnen	Nordfjord	Sogn og Fj.
83	14604	35010	Grinding slab	Sandstone	17 Havnen	Nordfjord	Sogn og Fj.
84	5751	1901:71	Projectile point	Green slate	Sokkamyro	Sunnhordland	Hordaland
85	5811		Projectile point	Green slate	Sokkamyro	Sunnhordland	Hordaland
86	C29998	m	Flake	Blue slate 1	Blånut IV	Hemsedal	Buskerud
87	C29998	m	Flake	Blue slate 1	Blånut IV	Hemsedal	Buskerud
A	HO1		Block	Red slate	Hovden	Sunnfjord	Sogn og Fj.
B	KA1		Block	Red slate	Kalvåg, Frøya	Sunnfjord	Sogn og Fj.
C	MU1		Block	Green slate	Mulevika	Sunnfjord	Sogn og Fj.
D	MU2		Block	Green slate	Mulevika	Sunnfjord	Sogn og Fj.
88 (E)	Aks. 03		Flake	Diabase	Stakaneset	Sunnfjord	Sogn og Fj.
89 (F)	Aks. 03		Adze blank	Diabase	Stakaneset	Sunnfjord	Sogn og Fj.
90	Aks. 03		Flake	Greenstone	Stegahaugen A	Sunnhordland	Hordaland
91	Aks. 03		Flake	Greenstone	Stegahaugen A	Sunnhordland	Hordaland
92	Aks. 03		Flake	Diabase	Stegahaugen B	Sunnhordland	Hordaland
93	Aks. 03		Flake	Diabase	Stegahaugen B	Sunnhordland	Hordaland
94	Aks. 03		Block	Rhyolite	Siggjo, cons. A	Sunnhordland	Hordaland
95	Aks. 03		Block	Rhyolite	Siggjo, cons. A	Sunnhordland	Hordaland

APPENDIX 9

CATALOGUE OF SLATE SPEARS AT BERGEN UNIVERSITY MUSEUM

Catalogue of slate spears from the district of Bergen University Museum						
B No.	**Unit/No.**	**Archaeological site**	**Farm**	**Municipality**	**District**	**Comment**
12838	1		Godøy	Giske	Northern Sunnmøre	
12838	2		Godøy	Giske	Northern Sunnmøre	
6335			Haram	Haram	Northern Sunnmøre	
6532			Flem	Haram	Northern Sunnmøre	
8863			Eik	Haram	Northern Sunnmøre	dagger?
9276			Austnes	Haram	Northern Sunnmøre	
10516			Ulla	Haram	Northern Sunnmøre	
12402	1		Longva	Haram	Northern Sunnmøre	
13583	6		Bjørnøy	Haram	Northern Sunnmøre	
13583	3		Bjørnøy	Haram	Northern Sunnmøre	
Å1335			Austnes	Haram	Northern Sunnmøre	
Å1395			Syvikja	Haram	Northern Sunnmøre	
10099			Kvassnes	Sula	Northern Sunnmøre	Zig-zag pattern
13543			Mauseid	Sula	Northern Sunnmøre	
13427			Abelset	Ørskog	Northern Sunnmøre	
4252			Ålesund	Ålesund	Northern Sunnmøre	
6566			Hov	Ålesund	Northern Sunnmøre	
9832			Ålesund	Ålesund	Northern Sunnmøre	
9147			Pilskog	Hareid	Southern Sunnmøre	
7083			Vågso	Sande	Southern Sunnmøre	
8773			Kyrkjesande	Sande	Southern Sunnmøre	
9412			Vågso	Sande	Southern Sunnmøre	
9569			Kvamme	Sande	Southern Sunnmøre	
10705			Kvamme	Sande	Southern Sunnmøre	
12840			Våge	Sande	Southern Sunnmøre	
9794			Tjønes	Sykkylven	Southern Sunnmøre	
10140			Jarnes	Sykkylven	Southern Sunnmøre	
9713			Dimnasund	Ulstein	Southern Sunnmøre	
9713			Ertesvåg	Ulstein	Southern Sunnmøre	
9936	a		Gardshol	Ulstein	Southern Sunnmøre	
9936	b		Gardshol	Ulstein	Southern Sunnmøre	
11334			Osnes	Ulstein	Southern Sunnmøre	
9335			Tunheim	Vanylven	Southern Sunnmøre	
10886			Rotevatn	Volda	Southern Sunnmøre	
7966			Steinnes	Ørsta	Southern Sunnmøre	
8471			Setre	Ørsta	Southern Sunnmøre	
10755			Lianeset	Ørsta	Southern Sunnmøre	
11148			Leikneset	Ørsta	Southern Sunnmøre	
8471			Setre	Østra	Southern Sunnmøre	Notches along back
11670		26 Gloføyk	Gloføyk	Bremanger	Outer Nordfjord	
11670		26 Gloføyk	Gloføyk	Bremanger	Outer Nordfjord	
14588		1 Haukedal	Haukedal	Bremanger	Outer Nordfjord	
14588		1 Haukedal	Haukedal	Bremanger	Outer Nordfjord	
14601		14 Havnen	Havnen	Bremanger	Outer Nordfjord	
14601		14 Havnen	Havnen	Bremanger	Outer Nordfjord	
14601		14 Havnen	Havnen	Bremanger	Outer Nordfjord	
14601		14 Havnen	Havnen	Bremanger	Outer Nordfjord	
14601		14 Havnen	Havnen	Bremanger	Outer Nordfjord	
14601		14 Havnen	Havnen	Bremanger	Outer Nordfjord	
14610		23 Gloføyk	Gloføyk	Bremanger	Outer Nordfjord	
14610		23 Gloføyk	Gloføyk	Bremanger	Outer Nordfjord	
14741		30 Nygård	Nygård	Bremanger	Outer Nordfjord	
8680			Barmen	Selje	Outer Nordfjord	Drilled holes at base
9177			Skjåstad	Selje	Outer Nordfjord	dagger?
9465			Berstad	Selje	Outer Nordfjord	
10334			Nordpollen	Selje	Outer Nordfjord	
4323			Kvalheim	Vågsøy	Outer Nordfjord	
4323			Røysa	Vågsøy	Outer Nordfjord	

Catalogue of slate spears from the district of Bergen University Museum (continued)

B number	Unit/Nr.	Archaeologi cal site	Farm	Municipality	District	Comment
11072	a	Åsmundvåg	Åsmundvåg	Vågsøy	Outer Nordfjord	
11991	e	Åsmundvåg	Åsmundvåg	Vågsøy	Outer Nordfjord	
12299		Åsmundvåg	Åsmundvåg	Vågsøy	Outer Nordfjord	
5718			Hilda	Stryn	Inner Nordfjord	
8451			Rake	Stryn	Inner Nordfjord	
10949			Sæta	Stryn	Inner Nordfjord	
10973			Kinn	Flora	Sunnfjord	
8101			Kvamsøy	Balestrand	Inner Sogn	
8189			Mundal	Balestrand	Inner Sogn	
12098			Osen	Gaular	Inner Sogn	
12098			Osen	Gaular	Inner Sogn	
8665			Djupvik	Vik	Inner Sogn	
8448			Høgfjellet	Årdal	Inner Sogn	
14150			Tyin	Årdal	Inner Sogn	
8948			Mykletun	Lindås	Nordhordland	
11496			Spurkland	Lindås	Nordhordland	
11449	f		Ytstebø	Radøy	Nordhordland	
11511	M28VI	Ramsvikneset	Straume	Radøy	Nordhordland	
11511		Ramsvikneset	Straume	Radøy	Nordhordland	
11511		Ramsvikneset	Straume	Radøy	Nordhordland	
15001	14101	Kotedalen	Straume	Radøy	Nordhordland	
6367	d		Sekkingstad	Fjell	Midthordland	
9010			Vik	Øygarden	Midthordland	
8159			Lemme	Voss	Voss	
13700			Kalve	Austevoll	Sunnhordland	
5683		Sokkamyro	Vespestad	Bømlo	Sunnhordland	
11252			Aukland	Sveio	Sunnhordland	

APPENDIX 10

COLOUR FIGURES

Colour Figure 1. *Raw materials in reference system (1): 1. Bluish greenstone, 2. Greenstone (from Hespriholmen, Bømlo. The drilled hole is a test-sample from a geological analysis by Sigmund Alsaker (1987), 3. Diabase, 4. Coarse flint, 5. Medium flint, 6. Fine flint, 7. Fine green quartzite 1, 8. Fine brown quartzite 1, 9. Fine black quartzite 1, 10. Fine blue quartzite 1, 11. Fine blue quartzite 7, 12. Fine blue quartzite 8, 13. Fine grey quartzite 2, 14. Fine grey quartzite 3, 15. Fine white quartzite 16.*

Colour Figure 2. *Raw materials in reference system (2): 1. Medium grey quartzite 1, 2. Medium grey quartzite 8, 3. Medium grey quartzite 9, 4. Medium grey quartzite 10, 5. Medium grey quartzite 11, 6. Coarse quartzite, 7. Quartz crystal, 8. Fine white quartz, 9. Fine yellow quartz (example of other fine quartz), 10. Medium white quartz, 11. Coarse white quartz, 12. Medium grey mylonite 1, Medium grey mylonite 3, Coarse grey mylonite 2,*

Colour Figure 3. *Raw materials in reference system (3): 1. Blue mylonite, 2. Green mylonite, 3. Rhyolite 1, 4. Fine grey quartzite 11, 5. Chert, 6. Jasper, 7. Red slate, 8. Green slate, 9. Grey slate, 10. Blue slate 1, 11. Blue slate 2, 12. Anorthosite, 13. Sandstone, 14. Pumice.*

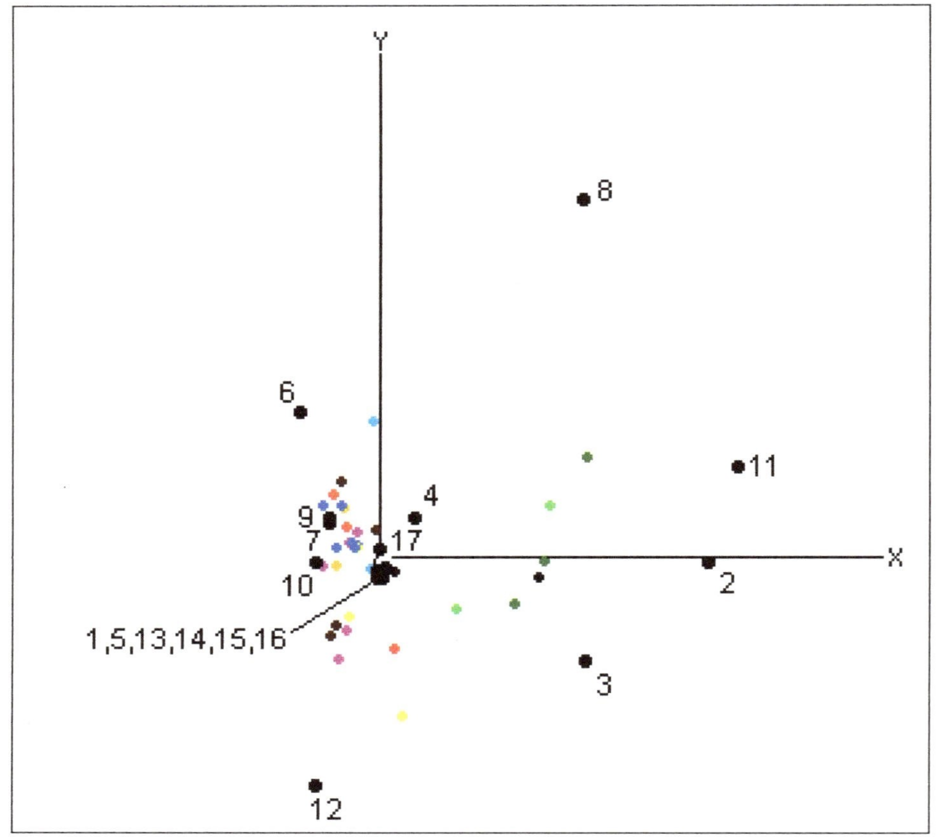

Colour Figure 4. *Correspondence analysis. Plot of presence/absence of types/techniques at all sites (see Table 3 for a list of types/techniques to which the numbers refer). Yellow symbols: sites at N. Sunnmøre, orange: S. Sunnmøre, red: Nordfjord, brown: Sunnfjord/Sogn, light blue: Nordhordland, dark blue: Midthordland, pink: Sunnhordland, light green: Lærdal/Hemsedal, dark green: Hardangervidda, black: Vivik.*

1	Blade points and flake points (A-points)	10	Ground stone adze flakes
2	Single-edged points	11	Ground flint axe flakes
3	Transverse points	12	Struck adze blanks
4	Slate points	13	Cylindrical cores
5	Retouched flakes and blades	14	Small blades
6	Small round smooth stones	15	Bipolar cores
7	Grindstone slabs	16	Other cores
8	TRB sherds	17	Platform cores
9	Neolithic adzes		

Table 3. *Types and techniques referred to in correspondence analyses Colour Figures 4 and 5a.*

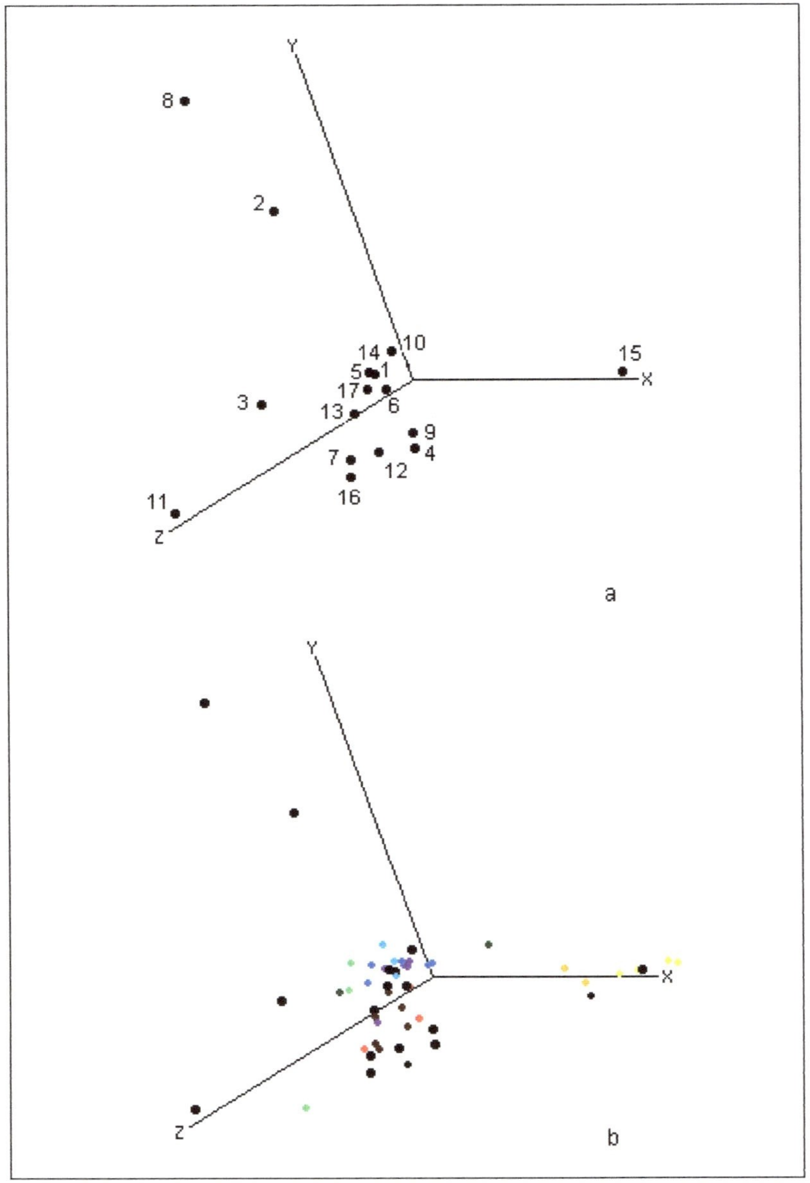

Colour Figure 5. *Correspondence analysis. Plot of values of types/technologies at all sites, (a) only types/techniques (see Table 3 for a list of types/techniques to which the numbers refer), and (b) plot of types/techniques and sites. Yellow symbols: sites at N. Sunnmøre, orange: S. Sunnmøre, red: Nordfjord, brown: Sunnfjord/Sogn, light blue: Nordhordland, dark blue: Midthordland, purple: Sunnhordland, green: Lærdal/Hemsedal, dark green: Hardangervidda, black: Styggvasshelleren/Vivik.*

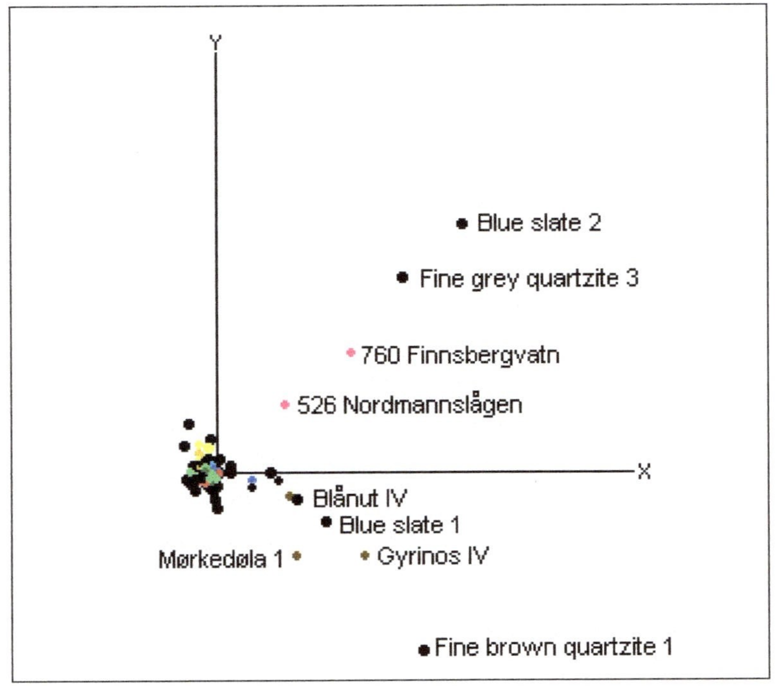

Colour Figure 6. *Correspondence analysis. Plot of values of raw materials at all sites.*

1	Diabase	18	Coarse quartzites
2	Greenstone	19	Quartz crystal
3	Basaltic rock	20	Fine, white quartz
4	Flint	21	Coarse and medium quartz
5	Fine, grey quartzite 2	22	Blue mylonite
6	Fine, grey quartzite 3	23	Coarse and medium grey mylonite
7	Fine grey quartzite 11	24	Chert
8	Fine, blue quartzite 1	25	Rhyolite
9	Fine, blue quartzite 7	26	Red slate
10	Fine, blue quartzite 8	27	Green slate
11	Fine, black quartzite 1	28	Grey slate
12	Fine, brown quartzite 1	29	Brown slate
13	Fine, green quartzite 1	30	Blue slate 1
14	Fine, white quartzite 16	31	Blue slate 2
15	Medium, grey quartzite 1	32	Anorthosite
16	Medium, grey quartzite 9	33	Sandstone
17	Medium, grey quartzite 10 and 11	34	Pumice

Table 4. *Raw materials referred to in correspondence analysis Colour Figure 7a.*

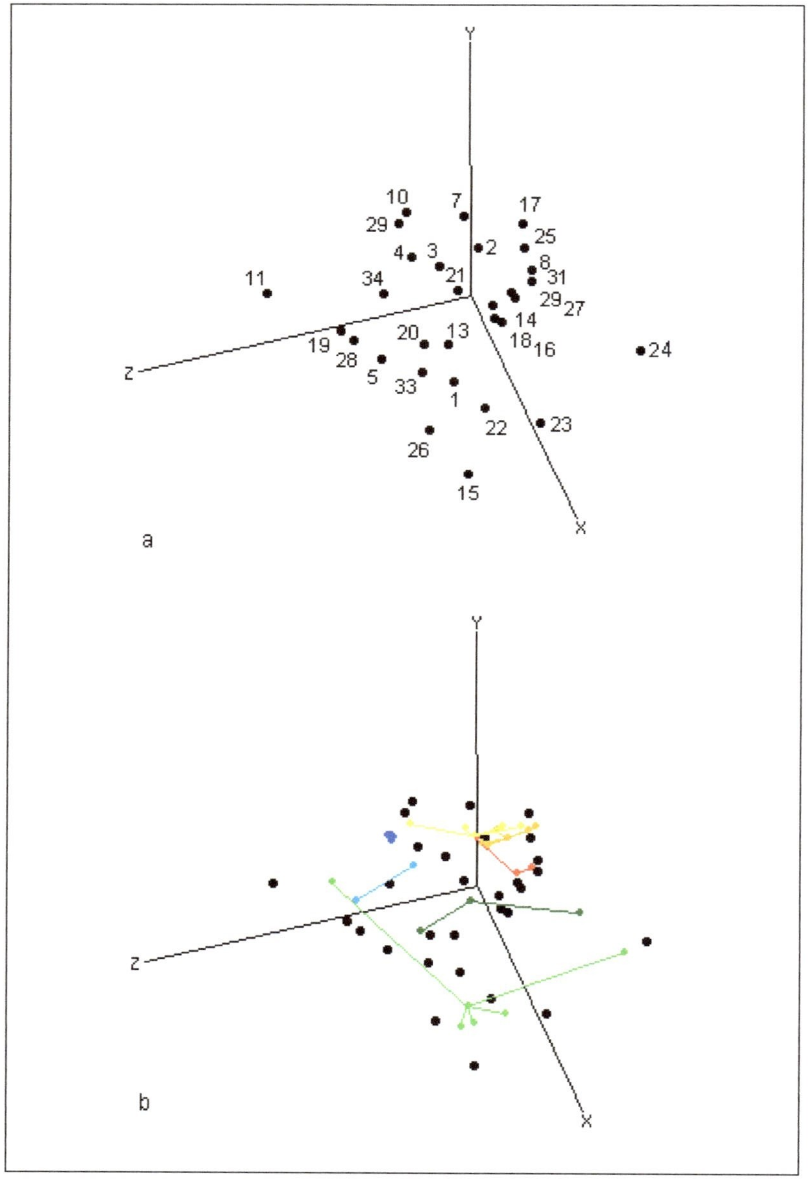

Colour Figure 7. *Correspondence analysis. Plot of values of raw materials at coastal sites, (a) only raw materials (see table 4 for a list of the raw materials to which the numbers refer), and (b) plot of raw materials and sites. Dark blue symbols: sites at N. Sunnmøre, light blue: S. Sunnmøre, dark green: Nordfjord, light green: Sunnfjord/Sogn, red: Nordhordland, orange: Midthordland, yellow: Sunnhordland.*

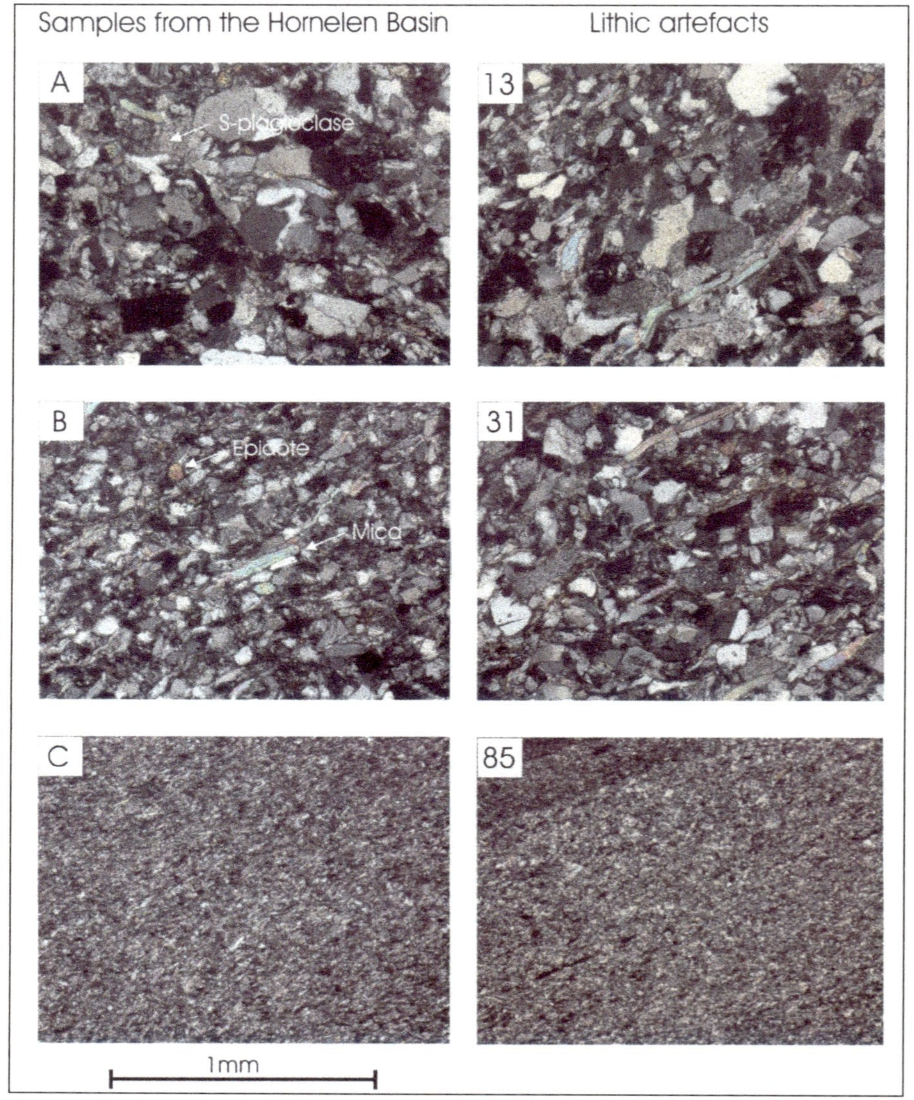

Colour Figure 8. *Microscope pictures of Devonian sand (A,B) and siltstones (C) from the Hornelen Basin, and typical examples of sand (13,31) and siltstones (85) artefacts (Table 4). Slices of the samples were prepared as thin sections and photographed in a standard petrographic microscope equipped with a digital camera. The most characteristic features of the Devonian sandstones are: 1) abundant clastic plagioclase, which in general are altered to a fine-grained aggregate of secondary minerals (saussurite), 2) clastic epidote, 3) secondary chlorite, and 4) white mica that often show deformation features. The siltstones exhibit the same characteristic features, but these cannot be well illustrated by photomicrographs (C) due to the fine grain size. Most of the sand and siltstone artefacts that have been investigated under microscope in this study, exhibit the characteristic features as the Devonian sand and siltstones (13, 31, 85).*

www.ingramcontent.com/pod-product-compliance
Lightning Source LLC
LaVergne TN
LVHW071205210926
848482LV00015B/727
* 9 7 8 1 8 4 1 7 1 9 8 3 2 *